Theme Issue Honoring Professor Robert Verpoorte's 75th Birthday: Past, Current and Future of Natural Products Research

Theme Issue Honoring Professor Robert Verpoorte's 75th Birthday: Past, Current and Future of Natural Products Research

Editors

Young Hae Choi
Young Pyo Jang
Yuntao Dai
Luis Francisco Salomé-Abarca

MDPI • Basel • Beijing • Wuhan • Barcelona • Belgrade • Manchester • Tokyo • Cluj • Tianjin

Editors
Young Hae Choi
Natural Products Laboratory,
Institute of Biology Leiden,
Leiden University
The Netherlands

Young Pyo Jang
College of Pharmacy,
Kyung Hee University
Korea

Yuntao Dai
Institute of Chinese Materia
Medica, China Academy of
Chinese Medical Sciences
China

Luis Francisco Salomé-Abarca
Laboratorío de Ecología Química
de Insectos, Campus Montecillo,
Colegio de Postgraduados
Mexico

Editorial Office
MDPI
St. Alban-Anlage 66
4052 Basel, Switzerland

This is a reprint of articles from the Special Issue published online in the open access journal *Molecules* (ISSN 1420-3049) (available at: https://www.mdpi.com/journal/molecules/special_issues/Molecules_Professor_Robert_Verpoorte).

For citation purposes, cite each article independently as indicated on the article page online and as indicated below:

LastName, A.A.; LastName, B.B.; LastName, C.C. Article Title. *Journal Name* **Year**, *Volume Number*, Page Range.

ISBN 978-3-0365-2742-0 (Hbk)
ISBN 978-3-0365-2743-7 (PDF)

Contents

molecules

MDPI

Editorial

A Theme Issue to Celebrate Professor Robert Verpoorte's 75th Birthday: "The Past, Current, and Future of Natural Products"

Yuntao Dai [1], Luis Francisco Salomé Abarca [2], Young Pyo Jang [3] and Young Hae Choi [3,4,*]

1   Institute of Chinese Materia Medica, China Academy of Chinese Medical Sciences, Beijing 100700, China; ytdai@icmm.ac.cn
2   Colegio de Postgraduados Campus Montecillo, Km 36.5 Carretera, Texcoco 56230, Mexico; luis.salome@colpos.mx
3   College of Pharmacy, Kyung Hee University, Seoul 02447, Korea; ypjang@khu.ac.kr
4   Natural Products Laboratory, Institute of Biology, Leiden University, Sylviusweg 72, 2333 BE Leiden, The Netherlands
*   Correspondence: y.choi@chem.leidenuniv.nl; Tel.: +31-71-514-2276

Citation: Dai, Y.; Salomé Abarca, L.F.; Jang, Y.P.; Choi, Y.H. A Theme Issue to Celebrate Professor Robert Verpoorte's 75th Birthday: "The Past, Current, and Future of Natural Products". *Molecules* **2021**, *26*, 7226. https://doi.org/10.3390/molecules26237226

Received: 19 November 2021
Accepted: 24 November 2021
Published: 29 November 2021

Publisher's Note: MDPI stays neutral with regard to jurisdictional claims in published maps and institutional affiliations.

Dear Colleagues,

Dr. Robert Verpoorte, Emeritus Professor at the Institute of Biology of Leiden University is among the most recognized world-wide experts in natural product chemistry. In recognition of his outstanding lifetime scientific contribution, some of his many former students and colleagues decided to honor this occasion by organizing a Special Issue of this renowned journal.

Prof. Dr. Verpoorte was born in 1946 in Eindhoven, The Netherlands. He graduated as a pharmacist in 1967 from Leiden University, and received a master's degree in Pharmaceutical Sciences in 1970 from the same university. He continued his studies under the supervision of Prof. Dr. A. Baerheim Svendsen (Faculty of Pharmacy, Leiden University, The Netherlands) and Prof. Dr. F. Sandberg (Faculty of Pharmacy, Stockholm University, Sweden), graduating with a PhD with his thesis on "Pharmacognostical studies of some African *Strychnos* species". He held a position as a senior faculty member of the Faculty of Pharmaceutical Sciences, Leiden University, from 1976 until 1987, when he was appointed as Professor and Head of the Department of Pharmacognosy/Plant Cell Biotechnology at Leiden University. In 2003, his group moved to the Institute of Biology of the same university, and, until his retirement in 2011, he served as Professor and Head of the Department of Pharmacognosy/Plant Cell Biotechnology, Metabolomics Section, Institute of Biology, Leiden University. He is currently an Emeritus Professor of the Natural Products Laboratory, Institute of Biology, Leiden University.

Prof. Verpoorte was a highly prolific scientist throughout his academic career, producing over 800 scientific publications, including research papers, books, and book chapters. His research interests have been highly diverse, covering all topics related to natural products, including plant cell biotechnology, biosynthesis, metabolomics, genetic engineering, and green technology, as well as the isolation of previously unknown biologically active compounds from natural products.

Moreover, his contribution to science greatly exceeds research. He has additionally been actively involved in the organization of many courses on natural products research as a way of transmitting and sharing knowledge and experience of participants and lecturers alike. These courses were taught in many countries all over the world, and dealt with topics such as biotechnology, biosynthesis, extraction and separation, and metabolomics. Throughout these years, he has regularly delivered hundreds of lectures in conferences and scientific meetings (15–20 times per year). In addition, he has served as an editorial board member for numerous scientific journals, having been the Editor-in-Chief of the *Journal of Ethnopharmacology* (2003–2016) and currently of *Phytochemistry Reviews* (since 2001) and *Biotechnology Letters* (since 2006).

For his excellent achievements, Prof. Verpoorte has received numerous scientific distinctions, honorary doctorates, and professorships. Since his retirement, he has continued to lead an active life devoted to science, having been invited to attend many scientific meetings across the world and delivering lectures on the topics of his expertise: natural products chemistry, biotechnology, metabolomics, and green technology.

As close friends and colleagues who have been in nearly daily contact with him over the last 20 years viewing all of these remarkable scientific contributions, we felt compelled to recognize this by the publication of a Special Issue of this journal dedicated to him.

Thus, this Special Issue has now finally been released with the help of many of his colleagues and former students as a token of our gratitude to his impressive work.

The Special Issue covers five main natural products topics: (1) chemical profiling and metabolomics, (2) separation/isolation and identification of plant specialized metabolites, (3) pharmacognosy of natural products to identify bioactive molecules from natural products, (4) novel formulation of natural products, and (5) overview of natural products as a source of bioactive molecules.

One of the new trends in natural products chemistry is the application of a holistic approach to natural products research, consisting of an untargeted analysis that can reveal whole hidden systems in organisms behind essential physiological functions (e.g., interactions between chemicals and both biotic and abiotic factors, and compound–compound interactions such as synergisms, which is the uncovering complex mechanisms of physiological roles of natural products). This approach has resulted in several scientific papers in the last decades responding to keywords, such as chemical profiling, or (the more recently coined term) metabolomics since the end of the 1990s, when this discipline was introduced in the field of life sciences.

In this Special Issue, Velasco-Azorsa et al. [1] report the application of a chemical profiling tool, an NMR-based method, to discover nematocidal metabolites in plants. Graczyk and colleagues further report an interesting approach based on the chemical profiling of the fruits of a traditional tonic plant, *Eleutherococcus senticosus,* a species that is endangered by overexploitation [2]. Their results provided valuable data for the quality control of the plant and the basis for alternatives to this plant in the future. The special issue additionally provided interesting examples of the application of diverse analytical platforms for successful chemical profiling or metabolomics studies. The potential of high-performance thin layer chromatography (HPTLC) was evaluated by Morloch and co-workers using fortified plant extracts [3]. Hyphenated techniques, such as GC-FID-MS, were additionally investigated as a tool for a metabolomics-based method to control the quality of a plum liquor by Ivanović et al., which allowed the selection and identification of 12 flavor-related volatiles as chemomarkers [4]. Another interesting NMR- and LC-MS based metabolomics approach was applied to reveal the metabolites responsible for the osmotic stress in lignan-deficient flax [5]. An interesting development was reported by Yun and colleagues, who analyzed the feasibility of using direct mass analysis; in this case direct analysis in real-time-time combined with flight-mass spectrometry (DART-TOF-MS) as a chemical screening tool for the detection of *Ephedra* alkaloids [6].

The second topic of the Special Issue is related to the basics of pharmacognosy, i.e., linking metabolites to certain bioactivities. Graziani et al. employed NMR-based chemical profiling of *Ononius diffusa* to identify cytotoxic compounds with activity against colorectal cancer cell lines. This approach was found to be much more successful, in terms of time and costs in the detection and identification of bioactive compounds, as compared to classical bioactivity-guided fractionation [7]. A similar design, integrating chemical profiling with bioactivity tests, allowed Nipun et al. [8] to identify α-glucosidase inhibitors from *Psychotria malayana* leaves using an LC-MS platform [8]. Using a more classical natural products approach, the therapeutic potential of well-known-plant-sourced compounds berberine [9], a prenyl chalcone [10], and phloridzin [11] was evaluated for their bioactivity against colorectal cancer, inflammation, and diabetes, respectively.

The isolation and identification of metabolites in plant material is undoubtedly the key step in current natural products-related research. Almost daily advances in analytical technologies are being applied to isolate natural chemicals and to elucidate their structures, reinforcing the use of natural products as a plentiful source of useful chemicals. Van der Klift and co-workers reported a technical improvement that allowed the extraction of reduced amounts of plant material (in the 1×10^{-3} g range) with extremely low volumes of solvents (mL range), exemplifying its successful applicability with the identification of flavones in luteola-dyed wool [12]. Lianza and colleagues provided an interesting report on the rapid identification of natural products, based on a database with an example of alkaloids from *Urceolina peruviana* [13]. The same group (Canton et al.) additionally reported the results of the application of a less common hyphenated method, centrifugal partition chromatography (CPC) with presaturation of solvent signals in ^{13}C NMR, for the identification of metabolites in cosmetic samples, in which glycerin, as a residual solvent, posed a challenging analytical problem [14].

Alongside these main streams of natural products research, two multidisciplinary investigations were reported: a study on natural products formulation with ionic liquids that are known as promising green media, Kiyonga et al. [15] and a chemical ecology study related to pheromones by Pineda-Rios [16].

Finally, this Special Issue includes two interesting review papers on the current progress of the research of *Agastache* species, one of the most studied Mexican medicinal plants [17] and the chemical diversity of Datura tropane alkaloids [18].

Thus, with its 18 articles, this Special Issue covers a broad range of topics in natural products, from classical studies to new trends, with the hope of providing an overview of the past and present of research in this field, and most importantly, the promising future perspectives now possible thanks to the exciting new developments in analytical technologies.

Conflicts of Interest: The authors declare no conflict of interest.

References

1. Velasco-Azorsa, R.; Cruz-Santiago, H.; Cid del Prado-Vera, I.; Ramirez-Mares, M.V.; Gutiérrez-Ortiz, M.d.R.; Santos-Sánchez, N.F.; Salas-Coronado, R.; Villanueva-Cañongo, C.; Lira-de León, K.I.; Hernández-Carlos, B. Chemical Characterization of Plant Extracts and Evaluation of their Nematicidal and Phytotoxic Potential. *Molecules* **2021**, *26*, 2216. [CrossRef]
2. Graczyk, F.; Strzemski, M.; Balcerek, M.; Kozłowska, W.; Mazurek, B.; Karakuła, M.; Sowa, I.; Ptaszyńska, A.A.; Załuski, D. Pharmacognostic Evaluation and HPLC–PDA and HS–SPME/GC–MS Metabolomic Profiling of *Eleutherococcus senticosus* Fruits. *Molecules* **2021**, *26*, 1969. [CrossRef] [PubMed]
3. Morlock, G.E.; Heil, J.; Bardot, V.; Lenoir, L.; Cotte, C.; Dubourdeaux, M. Effect-Directed Profiling of 17 Different Fortified Plant Extracts by High-Performance Thin-Layer Chromatography Combined with Six Planar Assays and High-Resolution Mass Spectrometry. *Molecules* **2021**, *26*, 1468. [CrossRef]
4. Ivanović, S.; Simić, K.; Tešević, V.; Vujisić, L.; Ljekočević, M.; Gođevac, D. GC-FID-MS Based Metabolomics to Access Plum Brandy Quality. *Molecules* **2021**, *26*, 1391. [CrossRef]
5. Hamade, K.; Fliniaux, O.; Fontaine, J.-X.; Molinié, R.; Otogo Nnang, E.; Bassard, S.; Guénin, S.; Gutierrez, L.; Lainé, E.; Hano, C.; et al. NMR and LC-MS-Based Metabolomics to Study Osmotic Stress in Lignan-Deficient Flax. *Molecules* **2021**, *26*, 767. [CrossRef]
6. Yun, N.; Kim, H.J.; Park, S.C.; Park, G.; Kim, M.K.; Choi, Y.H.; Jang, Y.P. Localization of Major Ephedra Alkaloids in Whole Aerial Parts of Ephedrae Herba Using Direct Analysis in Real Time-Time of Flight-Mass Spectrometry. *Molecules* **2021**, *26*, 580. [CrossRef]
7. Graziani, V.; Potenza, N.; D'Abrosca, B.; Troiani, T.; Napolitano, S.; Fiorentino, A.; Scognamiglio, M. NMR Profiling of *Ononis diffusa* Identifies Cytotoxic Compounds against Cetuximab-Resistant Colon Cancer Cell Lines. *Molecules* **2021**, *26*, 3266. [CrossRef]
8. Nipun, T.S.; Khatib, A.; Ibrahim, Z.; Ahmed, Q.U.; Redzwan, I.E.; Saiman, M.Z.; Supandi, F.; Primaharinastiti, R.; El-Seedi, H.R. Characterization of α-Glucosidase Inhibitors from *Psychotria malayana* Jack Leaves Extract Using LC-MS-Based Multivariate Data Analysis and In-Silico Molecular Docking. *Molecules* **2020**, *25*, 5885. [CrossRef] [PubMed]
9. Samad, M.A.; Saiman, M.Z.; Abdul Majid, N.; Karsani, S.A.; Yaacob, J.S. Berberine Inhibits Telomerase Activity and Induces Cell Cycle Arrest and Telomere Erosion in Colorectal Cancer Cell Line, HCT 116. *Molecules* **2021**, *26*, 376. [CrossRef] [PubMed]
10. Zeraik, M.L.; Pauli, I.; Dutra, L.A.; Cruz, R.S.; Valli, M.; Paracatu, L.C.; de Faria, C.M.Q.G.; Ximenes, V.F.; Regasini, L.O.; Andricopulo, A.D.; et al. Identification of a Prenyl Chalcone as a Competitive Lipoxygenase Inhibitor: Screening, Biochemical Evaluation and Molecular Modeling Studies. *Molecules* **2021**, *26*, 2205. [CrossRef]
11. Yoon, S.-Y.; Yu, J.S.; Hwang, J.Y.; So, H.M.; Seo, S.O.; Kim, J.K.; Jang, T.S.; Chung, S.J.; Kim, K.H. Phloridzin Acts as an Inhibitor of Protein-Tyrosine Phosphatase MEG2 Relevant to Insulin Resistance. *Molecules* **2021**, *26*, 1612. [CrossRef] [PubMed]

12. van der Klift, E.; Villela, A.; Derksen, G.C.H.; Lankhorst, P.P.; van Beek, T.A. Microextraction of *Reseda luteola*-Dyed Wool and Qualitative Analysis of Its Flavones by UHPLC-UV, NMR and MS. *Molecules* **2021**, *26*, 3787. [CrossRef] [PubMed]

13. Lianza, M.; Leroy, R.; Machado Rodrigues, C.; Borie, N.; Sayagh, C.; Remy, S.; Kuhn, S.; Renault, J.-H.; Nuzillard, J.-M. The Three Pillars of Natural Product Dereplication. Alkaloids from the Bulbs of Urceolina peruviana (C. Presl) J.F. Macbr. as a Preliminary Test Case. *Molecules* **2021**, *26*, 637. [CrossRef] [PubMed]

14. Canton, M.; Hubert, J.; Poigny, S.; Roe, R.; Brunel, Y.; Nuzillard, J.-M.; Renault, J.-H. Dereplication of Natural Extracts Diluted in Glycerin: Physical Suppression of Glycerin by Centrifugal Partition Chromatography Combined with Presaturation of Solvent Signals in ^{13}C-Nuclear Magnetic Resonance Spectroscopy. *Molecules* **2020**, *25*, 5061. [CrossRef] [PubMed]

15. Kiyonga, A.N.; Hong, G.; Kim, H.S.; Suh, Y.-G.; Jung, K. Facile and Rapid Isolation of Oxypeucedanin Hydrate and Byakangelicin from *Angelica dahurica* by Using [Bmim]Tf$_2$N Ionic Liquid. *Molecules* **2021**, *26*, 830. [CrossRef] [PubMed]

16. Pineda-Ríos, J.M.; Cibrián-Tovar, J.; Hernández-Fuentes, L.M.; López-Romero, R.M.; Soto-Rojas, L.; Romero-Nápoles, J.; Llanderal-Cázares, C.; Salomé-Abarca, L.F. α-Terpineol: An Aggregation Pheromone in *Optatus palmaris* (Coleoptera: Curculionidae) (Pascoe, 1889) Enhanced by Its Host-Plant Volatiles. *Molecules* **2021**, *26*, 2861. [CrossRef] [PubMed]

17. Palma-Tenango, M.; Sánchez-Fernández, R.E.; Soto-Hernández, M. A Systematic Approach to *Agastache mexicana* Research: Biology, Agronomy, Phytochemistry, and Bioactivity. *Molecules* **2021**, *26*, 3751. [CrossRef] [PubMed]

18. Cinelli, M.A.; Jones, A.D. Alkaloids of the Genus *Datura*: Review of a Rich Resource for Natural Product Discovery. *Molecules* **2021**, *26*, 2629. [CrossRef] [PubMed]

molecules

Article

Chemical Characterization of Plant Extracts and Evaluation of their Nematicidal and Phytotoxic Potential

Raúl Velasco-Azorsa [1], Héctor Cruz-Santiago [2], Ignacio Cid del Prado-Vera [3], Marco Vinicio Ramirez-Mares [4], María del Rocío Gutiérrez-Ortiz [5], Norma Francenia Santos-Sánchez [2], Raúl Salas-Coronado [2], Claudia Villanueva-Cañongo [2], Karla Isabel Lira-de León [6] and Beatriz Hernández-Carlos [2,*]

[1]   Instituto de Recursos, Universidad del Mar, Puerto Ángel, San Pedro Pochutla, Oaxaca 70902, Mexico; rvazorsa@gmail.com

[2]   Instituto de Agroindustrias, Universidad Tecnológica de la Mixteca, Acatlima, Huajuapan de León, Oaxaca 69000, Mexico; sacrumix@hotmail.com (H.C.-S.); nsantos@mixteco.utm.mx (N.F.S.-S.); rsalas@mixteco.utm.mx (R.S.-C.); villanueva.vc@gmail.com (C.V.-C.)

[3]   Colegio de Postgraduados, km 36.5 Carretera México-Texcoco, Montecillos, Estado de Mexico, Texcoco 56230, Mexico; icid@colpos.mx

[4]   Departamento de Ingeniería Química y Bioquímica, Tecnológico Nacional de México/I.T. Morelia, Av. Tecnológico 1500, Lomas de Santiaguito, Morelia 58120, Mexico; jvinicio2000@yahoo.com.mx

[5]   Instituto de Ecologia, Universidad del Mar, Puerto Ángel, San Pedro Pochutla, Oaxaca 70902, Mexico; rocio@angel.umar.mx

[6]   Facultad de Química, Universidad Autónoma de Querétaro, Las Campanas, Querétaro 76010, Mexico; liraleonki@gmail.com

*   Correspondence: bhcarlos@mixteco.utm.mx; Tel.: +52-9535320399

Citation: Velasco-Azorsa, R.; Cruz-Santiago, H.; Cid del Prado-Vera, I.; Ramirez-Mares, M.V.; Gutiérrez-Ortiz, M.d.R.; Santos-Sánchez, N.F.; Salas-Coronado, R.; Villanueva-Cañongo, C.; Lira-de León, K.I.; Hernández-Carlos, B. Chemical Characterization of Plant Extracts and Evaluation of their Nematicidal and Phytotoxic Potential. *Molecules* **2021**, *26*, 2216. https://doi.org/10.3390/molecules26082216

Academic Editors: Young Pyo Jang, Yuntao Dai and Luis Francisco Salomé-Abarca

Received: 1 March 2021
Accepted: 9 April 2021
Published: 12 April 2021

Publisher's Note: MDPI stays neutral with regard to jurisdictional claims in published maps and institutional affiliations.

Abstract: *Nacobbus aberrans* ranks among the "top ten" plant-parasitic nematodes of phytosanitary importance. It causes significant losses in commercial interest crops in America and is a potential risk in the European Union. The nematicidal and phytotoxic activities of seven plant extracts against *N. aberrans* and *Solanum lycopersicum* were evaluated in vitro, respectively. The chemical nature of three nematicidal extracts ($EC_{50,48h} \leq 113$ µg mL^{-1}) was studied through NMR analysis. Plant extracts showed nematicidal activity on second-stage juveniles (J2): ($\geq 87\%$) at 1000 µg mL^{-1} after 72 h, and their EC_{50} values were 71.4–468.1 and 31.5–299.8 µg mL^{-1} after 24 and 48 h, respectively. Extracts with the best nematicidal potential ($EC_{50,48h} < 113$ µg mL^{-1}) were those from *Adenophyllum aurantium*, *Alloispermum integrifolium*, and *Tournefortia densiflora*, which inhibited *L. esculentum* seed growth by 100% at 20 µg mL^{-1}. Stigmasterol (**1**), β-sitosterol (**2**), and α-terthienyl (**3**) were identified from *A. aurantium*, while **1**, **2**, lutein (**4**), centaurin (**5**), patuletin-7-β-O-glucoside (**6**), pendulin (**7**), and penduletin (**8**) were identified from *A. integrifolium*. From *T. densiflora* extract, allantoin (**9**), 9-O-angeloyl-retronecine (**10**), and its N-oxide (**11**) were identified. The present research is the first to report the effect of *T. densiflora*, *A. integrifolium*, and *A. aurantium* against *N. aberrans* and chemically characterized nematicidal extracts that may provide alternative sources of botanical nematicides.

Keywords: *Nacobbus aberrans*; nematicidal activity; α-terthienyl; stigmasterol; quercetagetin derivatives; 9-O-angeloyl-retronecine N-oxide

1. Introduction

The plant parasite nematode, *Nacobbus aberrans* (Thorne, 1935; Thorne & Allen, 1994), also known as false root-knot, is distributed in Argentina, Bolivia, Chile, Ecuador, Mexico, Peru, and the USA. It reduces crop yield in tomatoes (*Solanum lycopersicum* L.), potatoes (*Solanum tuberosum*), beans (*Phaseolus vulgaris* L.), and sugar beets (*Beta vulgaris*). The nematode ranks among the "top ten" plant-parasitic nematodes of phytosanitary importance [1]. It is estimated it meets the criteria to be a potential risk in the EU [2]. Management strategies of plant-parasitic nematodes are cultural practices (crop rotation), mixed-cropping, organic amendments, resistant crop cultivars, biological control [3–6], chemical nematicides, and bioactive products of plant origin. Among these strategies, natural product usage represents a vital option for controlling phytopathogenic nematodes because of their low impact on the environment and non-target organisms. In the search for botanic nematicides, some of the most recent proposals are using *Stevia rebaudiana* and *Origanum vulgare* to control *Meloidogyne*; in vivo experiments showed this effect [7,8]. In the case of *N. aberrans*, crude herbal extracts from *T. lunulate* [9], *Cosmos sulphureus* [6], *Senecio salignus* [6], *Witheringia stramoniifolia* [6], and *Lantana cámara* [6] showed in vitro nematicidal activity at 500 µg mL^{-1} (>70%) to second-stage juvenile (J2) individuals. Simultaneously, in vivo protection from infection of *Lycopersicum esculentum* Mill and *Capsicum annumm* plants occurred with extracts of *Tagetes erecta* [9] and *Trichilia galuca* [10], respectively. There are only two reports of natural compounds with toxic potential for the control of *N. aberrans*: capsidiol acts as a nematostatic on *N. aberrans* J2 (90% immobility) at 1.5 µg mL^{-1} after 72 h [11]; and various cadinenes affect immobility-mortality (LC$_{50}$ 25.4–111.4 µg mL^{-1}) and inhibit eggs hatching (IC$_{50}$ 31.23–56.71 µg mL^{-1}) [12]. Identification of substances from botanical origins capable of controlling *N. aberrans* lies partially on differences with other common plant-parasitic nematodes (e.g., *Meloidogyne* spp.), such as its endoparasitic stage (migratory and sedentary), several infective stages, and 15–30 days dehydration tolerance of J3 and J4 larvae [13]. This research identifies botanical sources of compounds with nematicidal or nematostatic characteristics by screening seven plant species for nematicidal activities against *N. aberrans* J2 individuals and phytotoxicity to tomato seeds. NMR analysis established the chemical nature of nematicidal extracts (EC$_{50,48h}$ ≤ 112.3 µg mL^{-1}) as sterol, thiophene, flavonoid, and alkaloid-like compounds (Figure 1). Some of these compounds are known for their nematicidal or nematostatic activity against other nematode species.

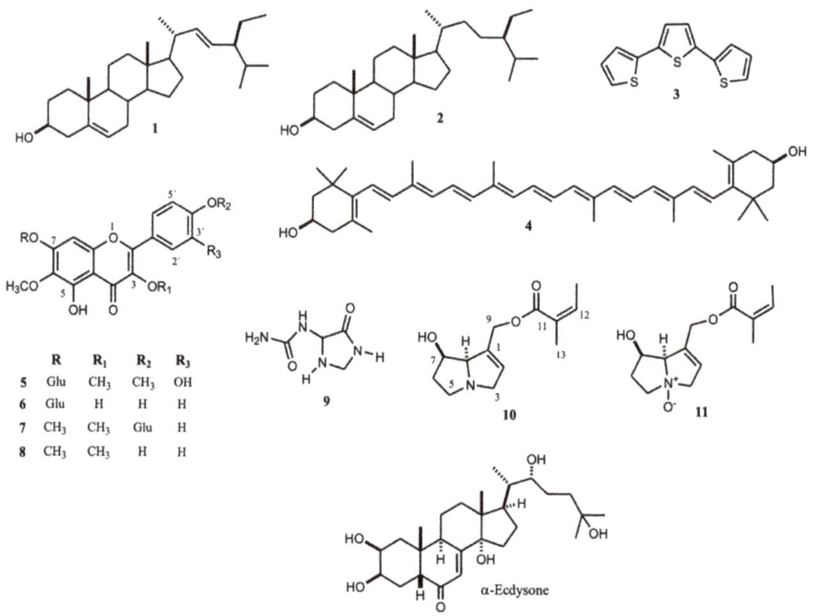

	R	R$_1$	R$_2$	R$_3$
5	Glu	CH$_3$	CH$_3$	OH
6	Glu	H	H	H
7	CH$_3$	CH$_3$	Glu	H
8	CH$_3$	CH$_3$	H	H

Figure 1. Structure of compounds identified from nematicidal extracts against J$_2$ individuals of *N. aberrans* (**1–11**) and α-ecdysone structure.

2. Results and Discussion

2.1. Nematostatic and Nematicidal Effect and EC$_{50}$ of Extracts

From the nine extracts evaluated in this study (Table 1), *G. mexicanum* and *A. integrifolium* achieved the maximum paralysis effect (immobility = 85.3 ± 10% and 94.1 ± 3%, respectively) at 1000 µg mL^{-1} after 24 h. The remaining extracts showed their highest nematostatic effects after 36–60 h (immobility = 91.3%–100%). The immobility percentage after 72 h reached >95% for all extracts. Nematicidal (72 h *) and nematostatic effects (72 h) at 1000 µg mL^{-1} occurred at similar immobility percentages for A. aurantium (roots), A. integrifolium, A. subviscida, G. mexicanum, and H. terebinthinaceus, while the rest of the extracts showed their nematicidal effects at lower immobility percentages due to recovery of mobility (6.3–11%) (Table 1).

Table 1. Effect of plant extracts at 1000 µg mL^{-1} on the immobility of *N. aberrans* J2s individual (100–150) after different exposure times.

Extract	% Immobility J2s						
	12 h	24 h	36 h	48 h	60 h	72 h	72 h *
A. aurantium A	84.6 ± 6 a	89.32 ± 1 b	97.0 ± 2 bc	98.1 ± 3 bc	99.3 ± 1.1 c	99.2 ± 1.5 c	96.4 ± 5.4 bc
A. aurantium R	61.0 ± 16 a	65.97 ± 6 a	95.3 ± 4 b	99.0 ± 0 b	100 ± 0 b	100 ± 0.0 b	99.8 ± 0.5 b
A. cuspidata	82.3 ± 5 a	86.6 ± 4 a	94.6 ± 3 ab	96.0 ± 1 b	98.3 ± 2 b	98.9 ± 2 b	90.6 ± 5 a
A. integrifolium	61.8 ± 13 a	94.1 ± 3 b	93.3 ± 4 b	96.5 ± 2 b	99.6 ± 1 b	99.8 ± 0.4 b	99.4 ± 1 b
A. subviscida	78.9 ± 10 a	79.3 ± 10 a	91.3 ± 6 b	93.5 ± 6 b	98.1 ± 2 b	99.3 ± 1 b	95.4 ± 2 b
G. mexicanum	62.9 ± 23 a	85.3 ± 10 b	93.7 ± 3 b	96.2 ± 3 b	98.5 ± 2 b	98.6 ± 2 b	94.2 ± 3 b
H. terebinthinaceous	86.5 ± 7 a	91.0 ± 16 ab	99.2 ± 1.2 b	97.5 ± 4 ab	99.6 ± 0.8 b	96.9 ± 11 ab	87 ± 11 ab
T. densiflora R	80.3 ± 6 a	87.7 ± 13 ab	92.9 ± 2 bc	96.3 ± 2 c	98.1 ± 3 c	98.8 ± 1 c	92.5 ± 3 bc
T. densiflora A	79.0 ± 9 a	85.9 ± 9 a	93.2 ± 3 ab	98.9 ± 2 b	98.1 ± 2 b	95.6 ± 3.2 b	88.0 ± 3 ab

* Immobility after 24 h. A: Stem part, R: roots. Data shown correspond to the average of all values ± sd. According to Tukey's test, the same letter indicates data in each row is not significantly different ($p < 0.05$).

Extracts at 100 µg mL^{-1} produced the highest immobility rates after 24 h (*A. aurantium* A), 36 h (*A. integrifolium*), 48 h (*A. aurantium* R), 60 h (*A. cuspidata* and *G. mexicanum*), and 72 h (*A. subviscida*, *H. terebinthinaceus*, and *T. densiflora* R and A) with immobility percentages ranging from 54.0 to 73.6 (Table 2). Lower concentration extracts (10 µg mL^{-1}) produced immobility percentages <26% (Supplementary Table S1). Table 3 shows calculated EC$_{50}$ values that cause 50% immobility in *N. aberrans* J2 individuals at 24 and 48 h. EC$_{50}$ values for 24 h ranged from 53.5 to 468.1 µg mL^{-1} (ci, α = 0.05) and for 48 h varied from 31.5 to 299.8 µg mL^{-1}. The most effective extracts (EC$_{50,48h}$ ≤ 112.3 µg mL^{-1}) were *A. aurantium* R (62.3–88.3 µg mL^{-1}), *A. aurantium* A (31.5–110.4 µg mL^{-1}), *A. integrifolium* (47.4–107.1 µg mL^{-1}), and *T. densiflora* R (59.0–112.3 µg mL^{-1}). Chemical, non-fumigant nematicides acted as positive controls (Fluopyram and Abamectin) with EC$_{50}$ values of 27.8–31.3 and 25.0–27.4 µg mL^{-1} at 24 and 48 h, respectively. Fluopyram is a succinate dehydrogenase inhibitor that blocks cellular respiration, while abamectin acts on glutamate-gated chloride channels, causing nematode death [14].

Table 2. Effect of plant fractions and extracts at 100 µg mL^{-1} on the immobility of *N. aberrans* J2s individuals (100–150) after different exposure times.

Extract	% Immobility J2s					
	12 h	24 h	36 h	48 h	60 h	72 h
A. aurantium A	44.1 ± 6 a	54.02 ± 7 b	68.8 ± 5 b	68.3± 6 b	73.8 ± 6.1 ab	70.8 ± 10 b
A. aurantium R	40.1 ± 14 a	34.1 ± 6 ab	52.1 ± 6 b	63.1 ± 4 c	70.2 ± 6 c	71.3 ± 7 c
A. cuspidata	44.3 ± 15 b	29.3 ± 7 a	19.6 ± 9 a	25.2 ± 10 a	64.2 ± 5 c	73.7 ± 4 c
A. integrifolium	17.6 ± 8 a	42.0 ± 14 b	62.31 ± 11 c	75.3 ± 4 c	68.0 ± 10 c	72.7 ± 10 c
A. subviscida	31.4 ± 13 ab	46.0 ± 10 abc	29.6 ± 19 ab	24.50 ± 4 a	44.4 ± 26 bc	60.2 ± 17 c
G. mexicanum	20.1 ± 8 a	28.8 ± 5 a	26.4 ± 8 a	40.4 ± 6 b	58.4 ± 11 c	66.9 ± 9 c
H. terebinthinaceous	31.1 ± 10 a	62.9 ± 8 bc	62.9 ± 8 bc	61.0 ± 5 b	69.4 ± 6 bc	73.6 ± 3 c
T. densiflora R	46.5 ± 20 bc	23.2 ± 12 a	33.0 ± 8 ab	49.1 ± 6 cd	63.2 ± 8 de	66.7 ± 10 e
T. densiflora A	45.9 ± 9 a	35.7 ± 16 a	36.5 ± 15 a	37.0 ± 11 a	52.7 ± 17 ab	64.6 ± 13 b

Data shown correspond to the average of all values ± sd. According to Tukey's test, the same letter indicates data in each row is not significantly different ($p < 0.05$).

Table 3. Effective concentrations (50% immobility) at 24 and 48 h on *N. aberrans*.

Extract	EC$_{50,24h}$ µg mL^{-1}	EC$_{50,48h}$ µg mL^{-1}
A. aurantium A	53.5–187.4	31.5–110.4
A. aurantium R	289.4–468.1	63.2–88.3
A. cuspidata	132.1–282.8	54.1–199.3
A. integrifolium	71.4–214.1	47.4–107.1
A. subviscida	124.7–342.2	60.0–299.8
G. mexicanum	93.0–319.6	74.4–183.4
H. terebinthinaceous	84.5–183.4	56.0–164.4
T. densiflora R	170.3–301.8	59.0–112.3
T. densiflora A	91.3–184.7	81.9–166.1
Fluopyram	27.8–28.4	25.0–26.4
Abamectin	30.6–31.3	26.9–27.4

Confidence limit ($\alpha = 0.05$).

2.2. Compounds Identified in A. aurantium and their Nematostatic Effects

Comparison of ^{1}H and ^{13}C NMR data allowed identification of stigmasterol (**1**) and β-sitosterol (**2**) [15,16]. Previously, we described the isolation and identification of α-terthienyl (**3**) and 5-(4″-hydroxy-1′-butynyl)-2–2′-bithiophene from *A. aurantium* roots [17]. In the present research, nematicidal extracts against *N. aberrans* contained 3.

Several treatments were tested at 50 and 100 µg mL^{-1}: pure compounds **1** and **3** (Figure 1); a mixture of 1 and 2 (Mixt); α-terthienyl (aT); stigmasterol (St); and β-sitosterol (bS) (Figure 2). At 100 µg mL^{-1}, **1**, **3**, and Mixt reached maximum inhibition at 72 h (93.3 ± 3%), 36 h (99.4 ± 0.6%), and 24 h (88.3 ± 8%), respectively. Commercial compounds, β-sitosterol, α-terthienyl, and stigmasterol showed lower nematostatic effects (50.6–80.7%), mainly during the first few hours of observation (12–36 h) (Figure 2). After 72 h, isolated and commercial β-stigmasterol showed 100 ± 0.0 and 94.5 ± 5.3% nematostatic effects, respectively. Treatments of isolated and commercial α-terthienyl showed the same immobility percentages ($\alpha = 0.01$) after 72 h with values of 93.3 ± 3.1 and 90.6 ± 5%, respectively. Commercial α-terthienyl, stigmasterol, and β-sitosterol showed nematodes recovery values of 0.83, 0.61, and 0.61%, respectively (Table 4). Therefore, these compounds are nematostatic and nematicides at the same concentrations ($\alpha = 0.05$).

Thiophenes are common in the *Tagetes* genera [18,19]; this is a cover crop against plant-parasitic nematodes *Meloidogyne* and *Pratilenchus* (*T. erecta, T. patula, T. tenuifolia,* and *T. minuta*) [20]. Some thiophenes described as nematicide are 1-phenylhepta-1,3,5-triyne, and 5-phenyl-2-(1′-propynyl)-thiophene. These thiophenes, isolated from *Coreopsis lanceolata* L., effectively treated the pinewood nematode *Bursaphelenchus xylophillus* at 2 mM [21]. A commercial sample of α-terthienyl caused 100% mortality after 24 h at 0.125% against the infective larval stage of the cyst nematode *Heterodera zeae* [22]. In our experiments, α-terthienyl showed 83.2 ± 5.2% immobility after 24 h at 100 µg mL^{-1}, or 0.01% against

N. aberrans, while the commercial compound showed less effectiveness (71.74 ± 8.8%) (Figure 2). Sterols functioned on nematicidal activity as has been documented: β-sitosterol showed 60% mortality in *M. incognita* at 1% (after 12 h) [23] and together with stigmasterol at 5 µg mL^{-1} caused 74.4% and 55.3% mortality in *M. incognita* and *Heterodera glycines*, respectively [24]. In this work, isolated stigmasterol showed maximum activity after 36 h (99.4 ± 0.56%), while the commercial compound achieved it after 72 h (94.5 ± 5.3%). Synergy from impurities in the natural stigmasterol could account for the observed stronger activity. The impurities were β-sitosterol and a very similar compound but with 2 oxygen atoms, which was not identified. Similar results caused by β-sitosterol were observed with increased immobility from 68.7 ± 8.5% to 75.8 ± 12.5% when stigmasterol is present (Figure 2). A recent report about synergistic effects discusses nematicidal activity against *Meloidogyne incognita* of a mixture of 23a-homostigmast-5-en-3β-ol and nonacosan-10-ol. The mixture, at 50 µg mL^{-1}, showed 93.7% mortality after 24 h, while the compounds individually, at 100 µg mL^{-1}, exerted 50% mortality (24 h) [25,26]. The sterol kill mechanism on nematodes may disrupt steroid metabolism as stigmasterol (**1**) possesses a chemical similarity to α-ecdysone (Figure 1). α-ecdysone is involved in the biosynthesis and metabolism of molting and sex hormones of nematodes [27]. Such a function took part in accelerating the development of *M. incognita* by applying an ecdysone derivate (0.5 mM) on tomato seeds [28].

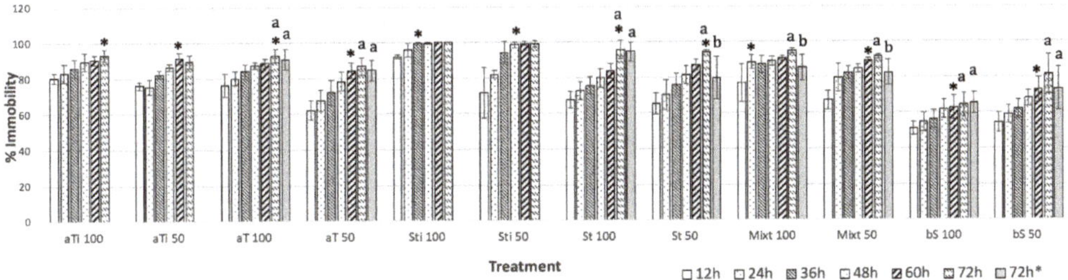

Figure 2. Effect of compounds isolated α-terthienyl (aTi), stigmasterol (Sti), a mixture of 1 and 2 (Mixt), and commercial compounds: stigmasterol (St), α-terthienyl (aT) and β-sitosterol (bS) on the immobility of *N. aberrans* J$_2$ individuals. * Maximum percentage of immobility. According to Tukey's test, columns followed by the same letter are not significantly different (*p* < 0.05). Concentration 100 and 50 µg mL^{-1}.

Table 4. Effect of mixture stigmasterol/β-sitosterol, commercial and isolated compounds from *A. aurantium* roots on the immobility of *N. aberrans* J$_2$ individuals after 72 h.

Treatment	Concentration µg mL^{-1}	% Immobility	% Immobility after Washing
β-sitosterol	100 †	68.7 ± 8.5 a	68.12 ± 8.0
stigmasterol/β-sitosterol	100	88.3 ± 8.1 b	—
α-terthienyl	100	93.3 ± 3.1 bc	–
	100 †	90.6 ± 5.60 bc	89.72 ± 5.5
stigmasterol	100	100.0 ± 0.0 c	–
	100 †	94.5 ± 5.3 bc	93.85 ± 5.6

Data shown correspond to the average of all values ± sd. According to Tukey's test, the same letter indicates data in each row is not significantly different (*p* < 0.01). † Commercial compounds.

2.3. Compounds Identified in A. integrifolium

Structural identification of 1, 2, 4–8 was made by comparing their ^{1}H and ^{13}C NMR data with those described [15,16,29–34]. Identification of 5–8 required HSQC and HMBC experiments to confirm the structures and assign ^{1}H and ^{13}C signals (Supplementary Table S2).

Previous studies of *A. integrifolium* include the presence of an acetylene compound from stems [35]. Additionally, methanol extracts from stems showed inhibition of a topoisomerase enzyme (JN394, $-81.19 \pm 2.12\%$ and JN362a, $126.06 \pm 12.02\%$), and no significant antioxidant (AE = 4.18%) nor antimicrobial activities (MIC $\geq$ 6250 μg mL^{-1}) [36]. In this research, the identification of quercetagetin derivatives **5–8** in *A. integrifolium* matches the genus's phytochemistry (also named *Calea* genus) [37]. Previous studies have shown the nematicidal potential of flavonoids. For example, rutin exhibited the same mortality percentage as the positive control (carbofuran): 100% mortality at a 0.5% concentration (24 h) against the cyst nematode *Heterodera zeae*. Moreover, quercetin showed 50% mortality, and patuletin 7-β-O-glycoside showed no significant activity (20%, 24 h, light) [22]. However, this compound exerts moderate activity against *Meloidogyne incognita* J2 individuals (LC$_{50,48h}$ = 0.506%), while kaempferol, isorhamnetin, rutin, myricetin, and fisetin, among others, exhibited an LC$_{50,48h}$ value similar to carbofuran (LC$_{50,48h}$ = 0.0506%) [38]. Nematicidal activity of flavonoids may be due to acetylcholinesterase inhibition (AChE) because nematodes possess some neurotransmitters common in mammals like acetylcholine, serotonin, or glutamate [39]. Flavonoids as quercetin, genistein, and luteolin 7-O-glycoside inhibited AChE by 76.2, 65.7, and 54.9%, respectively [40], while a methoxylated quercetagetin showed a minor effect (inhibition $23.73 \pm 1.94\%$) [41]. However, a glycosylated derivative, quercetagetin-7-O-(6-O-caffeoyl-β-D-glucopyranoside, possessed significant activity (IC$_{50}$ 12.54 ± 0.50 μg mL^{-1}) against AChE isolated from *Caenorhabditis elegans* and *Spodoptera litura*. The IC$_{50}$ value found came close to the value for chloropyrifos (2.32 ± 0.06 μg mL^{-1}) [42]. We hypothesize flavonoid (**5–8**) isolated from *A. integrifolium* exerted their effects on the cholinergic nervous system of *N. aberrans* J2 individuals, while sterols (triterpenes) (**1, 2**) acted on the hormonal system. Therefore, the nematostatic (EC$_{50,48h}$ = 47.4–107.1 μg mL^{-1}) and nematicidal (at 1000 μg mL^{-1}, $99.4 \pm 1\%$) effects observed could be the synergetic interaction between sterols and flavonoids. Synergic effects have been described on triterpenes (as saponins) and polyphenols combinations, which improved nematicidal activity in vivo against various nematode species (*Meloidogyne, Xiphinema, Tylenchorhynchus, Criconemoides,* and *Pratylenchus*) [43].

2.4. Compounds Identified in T. densiflora

Studies of the chemical composition of extracts from the *Tournefortia* genera revealed phenolic compounds [44] and alkaloids [45]. We identified allantoin (**9**) and pyrrolizidine alkaloids **10** and **11** (PAs), initially identified in the methanolic extract from its NMR data. The ^{13}C NMR spectrum showed signals at δ 96.70, 96.28, and 96.21 (Figure 3) possibly related to C-8 of N-oxide PAs [46,47] and signals at δ 76.74 and 76.66 probably due to C-8 of PAs [48]. The detection of PAs in *T. densiflora* roots required alkaloid extraction and chromatographic separation to identify **10** and **11**. ^{1}H and ^{13}C NMR data comparison with similar data from retronecine and PAs N-oxides [48–51], as well as HMQC and HMBC experiments, identified 9-O-angeloyl-retronecine N-oxide (**11**). The DEPTQ spectrum revealed the following functional groups: two CH$_3$, four CH$_2$, four CH, and three quaternary carbon atoms. Two double bonds were observed from signals at δ 132.9 (C), 121.4 (CH), 127.6 (C), and 139.0 (CH), while an angeloyl ester moiety was deduced from the ^{13}C signal at δ 167.1 and the ^{1}H NMR signal at δ 6.16 as qq (J = 1.6, 8.4Hz, H-12), which showed HMBC correlation with signals at δ 20.1 (C-12) and 16.0 (C-13). The signal at δ 95.5, caused by CH (C-8) next to quaternary nitrogen (N-oxide), and the signals at δ 34.8 (C-6), 68.5 (C-5), 77.8 (C-3), 60.9 (C-9), and 69.8 (C-7) were consistent with a necine base (as N-oxide) carrying a double-bond between C1 (132.9) and C2 (121.4). HMBC correlations support this proposal, and the presumed position of the angeloyl group at C-9, as the correlation of H-9 (δ 4.78 and 4.72) with the ester carbon C-10 (δ 167.1) showed (Table 5). Compound **10** displayed a ^{13}C NMR spectrum very similar to **11**. The main differences lay in the chemical shifts of C-3, C-5, and C-8 (atoms near the quaternary nitrogen), which appear in **10** at 60.4, 53.4, and 77.8 ppm, respectively. In **11**, these signals shifted to higher frequencies at δ 77.8 (C-3), 68.5 (C-5), and 95.5 (C-8) ppm (Table 5).

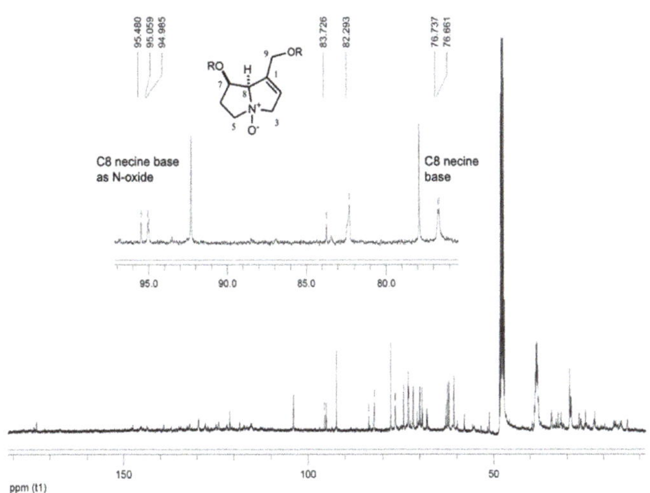

Figure 3. ^{13}C NMR spectrum of methanol extract of *T. densiflora* roots. 100 MHz, CD$_3$OD.

Table 5. ^{1}H (400 MHz) and ^{13}C NMR (100 MHz) data for **9** (MeOD), **10** and **11** (DMSO-d_6).

Atom	9		10		11	
	δ ^{1}H	δ ^{13}C	δ ^{1}H	δ ^{13}C	δ ^{1}H	δ ^{13}C
1	10.54 s	-		133.5	-	132.9
2	-	157.8	5.8 d (1.3 Hz)	122.9	5.80 bs	121.4
3	8.05 s	-	4.15 d (14.7 Hz) 3.77 d (14.7 Hz)	60.4	4.28 d (16.0 Hz) 4.56 so	77.8
4	-	174.1	-	-	-	-
5	5.24 d (8.0)	62.9	3.61 so 3.09 ddd (6.4 Hz)	53.4	3.71 so 3.60 so	68.5
6	6.89 d (8.0)	-	1.98 so 1.98 so	35.8	2.45 bs 1.92 so	34.8
7	-	157.2	4.43 bs	68.9	4.57 so	69.8
8	5.79 s	-	4.61 bs	77.8	4.57 so	95.5
9	-	-	4.78 bs 4.78 bs	60.3	4.78 d (14 Hz) 4.72 d (14 Hz)	60.9
10	-	-	-	166.9	-	167.1
11	-	-	-	127.2	-	127.6
12	-	-	6.18 qq (1.3, 7.3)	138.3	6.16 qq (1.6, 8.4 Hz)	139.0
13	-	-	1.87 q (1.3)	20.5	1.87 q (1.6 Hz)	20.1
14	-	-	1.93 dq (1.3, 7.3)	15.9	1.95 dq (1.6, 8.4 Hz)	16.0

Previous research demonstrated toxicity to *Saccharomyces cerevisae* from methanolic extracts from roots (70 mgmL^{-1}) [52] and mycelial growth inhibition of *Alternaria alternata* (69.07 ± 2.0%) and *Fusarium solani* (52.42 ± 2.0%) [46]. Pyrrolizidine alkaloids (PAs) play a defensive role in the plant as antifeedants against herbivores [53]. Research shows PAs and PAs N-oxides are nematicides against *Meloidogyne incognita*, *Heterodera schachtii*, *Pratylenchus penetrans*, *Plasmarhabditis hermaphrodita*, and *Rhabditis* sp. [54]. Additionally, allantoin (**9**) possesses nematicidal activity (51.3% mortality) in *Meloidogyne incognita* and *Heterodera glycines* at 5 µg mL^{-1} [24]. These toxic activities and common occurrence of PAs in the *Boraginaceae* family [45] confirmed that PAs and allantoin are responsible for the toxic effects of the *T. densiflora* root extract observed on *N. aberrans* J2 individuals. PAs toxicity relies on their oxidation to pyrrolizinium ions by the cytochrome-P450 enzyme in the nematode. These ions are highly reactive as electrophiles and react with DNA, proteins, and other important macromolecules [55].

2.5. Phytotoxicity Test

Most nematicidal extracts against *N. aberrans* extracts showed 100% inhibition (-100%) of *L. esculentum* radicle growth at 20 µg mL^{-1}, except extracts from *H. terenbinthinaceus* (-37%) (Figure 4). *T. densiflora* R extracts were the most phytotoxic with 40 and 38% inhibition at 0.02 µg mL^{-1} (Figure 4). Potentially, these extracts could act as soil disinfection agents. Also, *A. cuspidata*, *A. subviscida*, and *T. densiflora* A extracts showed hormetic effects: a 20 µg mL^{-1} solution inhibited radicle growth inhibition while a 0.02 µg mL^{-1} solution promoted it.

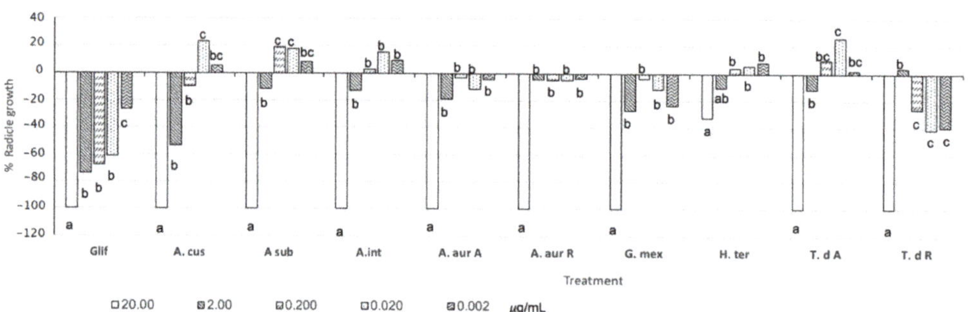

Figure 4. Phytotoxic activity on *L. esculentum* of plant extracts.

Nematostatic and nematicidal effects observed by treatments with EC$_{50,48h}$ < 113 µg mL^{-1} relate to secondary metabolites like sterols, flavonoids, thiophenes, or alkaloids (PAs and allantoin), possibly biosynthesized in plants as a stress response. In general, sterols are involved in plants' growth and fertility as hormonal precursors and cell membranes' functional components. Nematodes also need sterols for their survival, but they cannot biosynthesize them de novo, so the nematodes readily absorb sterols. For example, *Meloidogyne arenaria*, *M. incognita*, and *Pratilenchus agilis* incorporate and transform sterols into necessary derivatives in their growth and reproduction [56]. Thus, nematodes should elicit a biological response to some sterols. Also, plant–nematode interactions require flavonoids and might be required for nematode reproduction. However, some flavonoids with specific structural arrangements have shown toxic effects on specific targets such as enzymes. Finally, thiophenes could inhibit enzymes like superoxide dismutase [57] and damage DNA [58]. The transformation of secondary metabolites to more toxic compounds also happened with PAs, as mentioned before.

3. Materials and Methods

3.1. General Experimental Procedures

NMR measurements were carried out on Bruker ASCEND^{TM} 400 (400 MHz proton frequency) spectrometer (Bruker, Germany) at 298 K using 5 mm probes at 22 °C from CD$_3$OD or DMSOd$_6$ solutions. Chemical shifts (δ = ppm) were referenced to 2.50 (^{1}H) and 39.43 (^{13}C) ppm (DMSOd$_6$) or to 3.30 (^{1}H) and 36.067 (^{13}C) ppm (CD$_3$OD). Coupling constants are given in Hz. Signals are described as s (singlet), d (double), t (triple), and q (quartet).

3.2. Chemicals

All reagents and solvents (ACS grade), LiChroprep RP-18, and SiO$_2$ supports for column and plate chromatography were obtained from Merck (MA, USA). Amberlite

XAD16, α-terthienyl, β-sitosterol, stigmasterol, deuterated solvents, and dimethyl sulfoxide (DMSO-Hybri-Max) were obtained from Sigma Chemical (St. Louis, MO, USA).

3.3. Plant Species

The plant species were collected in Oaxaca, Mexico (See Table 6), and voucher specimens were deposited in the Herbarium of Forest Sciences, Universidad Autonoma de Chapingo, Texcoco (Estado de México, México). The scientific name, collection site, voucher number, plant part used, and extraction solvent are listed in Table 6.

Table 6. Plants used in experiments.

Specie (Family)	Collection Site	Voucher Number	Part Plant Used	Extraction Solvent
Acalypha cuspidata Jacq. (Euphorbiaceae)	B	25068	Stem	MeOH
Acalypha subviscida S. Watson var. Lovelanddii McVaugh (Euphorbiaceae)	A	24007	Stem	MeOH
Alloispermum integrifolium (DC.) H. Rob. (Asteraceae)	A	24024	Stem	MeOH
Adenophyllum aurantium (L.) Strother (Asteraceae)	C	25173	Stem Root	MeOH MeOH
Galium mexicanum Kunth (Rubiaceae)	A	23994	Stem	MeOH
Heliocarpus terebinthinaceus (DC.) Hochr. (Tiliaceae)	D	25225	Seeds	H_2O
Tournefortia densiflora M. Martens & Galeotti (Boraginaceae)	C	25221	Stem Root	MeOH MeOH

MeOH: methanol. A: San Miguel Suchixtepec, Miahuatlán; B: Candelaria Loxicha, San Pedro Pochutla. C: Chepilme Garden (Universidad del Mar), SanPedro Pochutla, D: Huajuapan.

3.4. Preparation of Extracts

Extracts were prepared according to procedures previously described [52]. All extracts were kept at 4 °C and protected from light and moisture until further use.

3.5. Isolation of 1–3 from A. aurantium Extract

The methanol extract (0.582 g) from aerial parts of *A. aurantium* was subjected to column chromatography (CC) using *n*-hexane-ethyl acetate mixtures. The fractions eluted with an 8:2 mixture, were re-chromatographed on CC with *n*-hexane-ethyl acetate (95:5) to yield 63.2 mg (10.8%), 30 mg (5.15%), and 3.02 mg (0.52%) of stigmasterol (**1**) and a mixture of stigmasterol (**1**)/β-sitosterol (**2**), and α-terthienyl (**3**), respectively (Figure 1). The purity of stigmasterol and α-terthienyl was approximately 95 and 98%, respectively. Purity was approximate since ^{1}H NMR spectra by comparison of integration areas of **1** and **3** with those corresponding to impurities. The methanol extract from roots (7.215 g) was dissolved in acetone, and the solution yielded a solid residue (0.347 g), which was subjected to CC using *n*-hexane-ethyl acetate mixtures to obtain 40 mg (0.55%) and 23.7 mg (0.32%) of **1** and **3** respectively.

3.6. Identification of Compounds from A. integrifolium

Methanol extract of *A. integrifolium* (23.1 g) was partitioned with ethyl acetate (3 times) to obtain 10.1 g of ethyl acetate soluble fraction (ESF) and 13 g of methanol soluble fraction (MSF). ESF was subjected to column chromatography (SiO_2) and eluted with mixtures of AcOEt: *n*-hexanes to obtain a mixture of chlorophylls "a" and "b" (80 mg), and a dark solid (155.8 mg). The solid was re-chromatographed (SiO_2) with the same eluents to obtain lutein (**4**, 9.3 mg) (Figure 1) and a mixture (27.1 mg) of stigmasterol (**1**) and β-sitosterol (**2**). MFS (13 g) was solubilized in water and supported on a column of Amberlite XAD16; after two washes with water, the compounds retained were eluted with methanol to obtain a residue

(1.5 g) free from simple carbohydrates. The residue (1.0 g) was eluted in a chromatography column (C_{18}) using methanol:water mixtures as eluent. Chromatographic separation yielded 72 mg of four quercetagetin derivatives in binaries mixtures; its approximate composition was calculated by integrating ^{1}H NMR areas of their characteristic signals. These compounds were identified as centaurin (**5**, 28.1 mg), patuletin-7-β-*O*-glucoside (**6**, 1.7 mg), pendulin (**7**, 6.4 mg), and penduletin (**8**, 1.0 mg) (Figure 1) from the analysis of their NMR data (Supplementary Table S2).

3.7. Identification of Compounds from T. densiflora Roots

The methanol extract (1.14 g), previously defatted with *n*-hexane and AcOEt, was subjected to C_{18} column chromatography and eluted with water. The eluent was identified by its NMR data [59] as allantoin (9, 33 mg, Figure 1, Table 5). PAs extraction required 8 g of the methanolic extract to be stirred with 1 M HCl (44 mL, 20 min); the mixture was filtered, the filtrate adjusted to pH 10 (KOH 1 M), and extracted with ethyl acetate. Organic fraction (180 mg) was subjected to chromatographic column (CC-SiO$_2$) and eluted with CHCl$_3$: methanol mixtures to obtain 9-*O*-angeloyl-retronecine (**10**, 11.5 mg, approx. 80% purity) and their N-oxide (**11**, 12.1 mg, approx. 90% purity).

3.8. Screening of Nematicidal and Nematostatic Activities

3.8.1. Nematodes

Mature egg masses of *N. aberrans* were extracted from infected roots of tomato plants (*Lycopersicum esculentum* Mill., 1768 or *Solanum lycopersicum*), propagated at Colegio de Postgraduados, Montecillo, Texcoco, Mexico. Egg masses were gently washed with water to remove adhered soil and a NaOCl 0.53% solution until the gelatinous matrix dissolved. Then they were washed with distilled water on a mesh sieve (#400) and incubated in distilled water at 25 °C for 5 days. Emerging J2 individuals were used in all experiments.

3.8.2. Assay

Test solutions were prepared in DMSO with 0.5% Tween 20 at 10, 100, 1000 µg mL^{-1} for extracts, while concentrations at 100 µg mL^{-1} were used for compounds and fractions. Fluopyram 50% (Verango, Bayer) and abamectin 5.41% (Oregon 60C-FMC) at 5, 10, 15, 25, 30 and 50 µg mL^{-1} (dissolved in distilled water) were tested as positive control. Treatments (5 µL) and between 100 and 150 J$_2$ individuals in 95 µL of water were added to 96-well plates (Falcon, USA) and incubated at 25 °C. DMSO with 0.5% Tween 20 (5 µL) in 95 µL of water was used as blank. Previously, non-effect on J2 mobility was shown at 24, 36, 48, 60, and 72 h with the solvents used (Supplementary Table S3). Percentages of J2 immobility were recorded after 12, 24, 36, 48, 60, and 72 h by counting mobile and immobile J$_2$ individuals under a stereomicroscope at 240X. A nematode was considered immobile if the nematode failed to respond to stimulation with a needle. After that, the J2 individuals at 1000 µg mL^{-1} (extracts) and 100 µg mL^{-1} (commercial stigmasterol, α-terthienyl, and β-sitosterol) were washed on a 400-mesh filter with distilled water to remove the excess test substance (extracts and commercial compounds). The treatments were replaced with distilled water to allow a possible recovery of the J2 individuals after 24 h. If they remained immobile, they were assumed to be dead, and the effect was considered nematicide. If any J$_2$ individual regained mobility, the effect was considered nematostatic (paralysis). All treatments (extracts, isolated and commercial compounds) and control were replicated five times, and the experiments were performed two times. The immobility percentage was calculated using the equation: i = 100 × (1 − n_t/n_c); where i = immobility percentage, n_t = active J$_2$ in the treatment, and n_c = active J$_2$ in the blank [60].

3.9. Phytotoxicity Assay

Experiments were conducted with *L. esculentum* F1 seeds var. Sheva according to the methodology described [61]. Prior to evaluation, all extracts were dissolved in a 0.5% DMSO/H$_2$O solution at 20, 2.0, 0.2, 0.02, and 0.002 µg mL^{-1} concentrations to obtain solids-free solutions. Commercial herbicide (Glyphosate) was used as a positive control at the same concentrations, while 0.5% DMSO/H_2O was used as blank (100% growth).

3.10. Statistical Analysis

All experimental data were subjected to an analysis of variance (ANOVA) using Statistica Pro (Stat Soft, Japan). Treatment means were tested with Tukey's HSD multiple comparison test at 0.05% or 0.01% probability levels.

4. Conclusions

To our knowledge, our results show for the first time the nematicidal activity against *N. aberrans* from *T. densiflora*, *A. integrifolium*, and *A. aurantium* extracts. In this research, we identified several compounds present in the nematicidal extracts against J2 individuals of *N. aberrans* containing: (a) flavonoids (*A. integrifolium*); (b) triterpene-type compounds (*A. aurantium*, *A. integrifolium*), (c) thiophene-type compounds (*A. aurantium*) and (d) alkaloids (*T. densiflora*). We identify **5–8** and **9–10** from *A. integrifolium* and *T. densiflora*, respectively. Moreover, we described the phytotoxic effect of all extracts on tomato radicle growth. Further research of these plant extracts will allow us to identify more compounds responsible for the nematicidal activity and provide alternative nontoxic crop protection chemicals.

Supplementary Materials: The following are available online. Table S1: Effect of plant extracts at 10 µg mL^{-1} on the immobility of *N. aberrans* J2s individuals after different exposure times; Table S2: ^1H and ^{13}C data for compounds **5–8**. 400 MHz, 100 MHz CD_3OD; Table S3: Effect of DMSO with 0.5% Tween on immobility of *N. aberrans* J2s after different exposure times.

Author Contributions: Conceptualization, R.V.-A., I.C.d.P.-V., and B.H.-C.; Formal analysis, B.H.-C.; Funding acquisition, B.H.-C.; Investigation, R.V.-A., H.C.-S., M.d.R.G.-O., R.S.-C., C.V.-C., and K.I.L.-d.L.; Methodology, R.V.-A., H.C.-S., M.d.R.G.-O., N.F.S.-S., and R.S.-C.; Project administration, B.H.-C.; Resources, I.C.d.P.-V., N.F.S.-S., and R.S.-C.; Supervision, M.V.R.-M.; Validation, M.V.R.-M.; Visualization, I.C.d.P.-V.; Writing–original draft, R.V.-A., H.C.-S., and B.H.-C.; Writing–review & editing, M.V.R.-M., I.C.d.P.-V., and B.H.-C. All authors have read and agreed to the published version of the manuscript.

Funding: This research was funded by CONACyT grant number 225188, SAGARPA-CONACyT grant number 2016-1-277609, and SEP-PROMEP/103.5/12/6525.

Institutional Review Board Statement: Not applicable.

Informed Consent Statement: Not applicable.

Data Availability Statement: Not applicable.

Acknowledgments: Raul Velasco Azorsa thanks CONACyT for Masters in Science Fellowship (No. 318148). C.V.-C. thanks Prodep-SEP for postdoctoral Fellowship. We thank M.Sc. Ernestina Cedillo-Portugal from Universidad Autónoma Chapingo, Gerardo A. Salazar-Chávez, and Jaime Jiménez-Ramirez from the Universidad Nacional Autónoma de México for the identification of the plant species studied.

Conflicts of Interest: The authors declare no conflict of interest.

Sample Availability: Samples of the compounds are available from the corresponding author.

References

1. Singh, S.K.; Hodda, M.; Ash, G.J. Plant-parasitic nematodes of potential phytosanitary importance, their main hosts and reported yield losses. *EPPO Bull.* **2013**, *43*, 334–374. [CrossRef]
2. Flores-Camacho, R.; Manzanilla-López, R.H.; Cid del Prado-Vera, I.; Martínez-Garza, A. Control de *Nacobbus aberrans* (Thorne) Thorne y Allen con *Pochonia chlamydosporia* (Goddard) Gams y Zare. *Rev. Mex. Fitopatol.* **2007**, *25*, 26–34.
3. Pérez-Rodríguez, I.; Franco-Navarro, F.; Cid del Prado-Vera, I.; Zavaleta-Mejía, E. Control de *Nacobbus aberrans* en chile ancho (*Capsicum annuum* L.) mediante el uso combinado de enmiendas orgánicas, hongos nematófagos y nematicidas. *Nematropica* **2011**, *41*, 122–129.

4. Lax, P.; Marro, N.; Agaras, B.; Valverde, C.; Doucet, M.E.; Becerra, A. Biological control of the false root-knot nematode *Nacobbus aberrans* by *Pseudomonas protegens* under controlled conditions. *Crop Prot.* **2013**, *52*, 97–102. [CrossRef]

5. Caccia, M.; Lax, P.; Doucet, M.E. Effect of entomopathogenic nematodes on the plants-parasitic nematode *Nacobbus aberrans*. *Biol. Fertil. Soils* **2013**, *49*, 105–109. [CrossRef]

6. Vázquez-Sánchez, M.; Medina-Medrano, J.R.; Cortez-Madrigal, H.; Angoa-Pérez, M.V.; Muñoz-Ruíz, C.V.; Villar-Luna, E. Nematicidal activity of wild plant extracts against second-stage juveniles of *Nacobbus aberrans*. *Nematropica* **2018**, *48*, 136–144.

7. Ntalli, N.; Kasiotis, K.M.; Baira, E.; Stamatis, C.L.; Machera, K. Nematicidal activity of *Stevia rebaudiana* (Bertoni) assisted by phytochemical analysis. *Toxins* **2020**, *12*, 319. [CrossRef] [PubMed]

8. Ntalli, N.G.; Ozalexandridou, E.X.; Kasiotis, K.M.; Samara, M.; Golfinopoulos, S.K. Nematicidal activity and phytochemistry of Greek Lamiaceae species. *Agronomy* **2020**, *10*, 1119. [CrossRef]

9. Zavaleta-Mejía, E.; Gómez, R.O. Effect of *Tagetes erecta* L.-Tomato (*Lycopersicon esculentum* Mill.) intercropping on some tomato pests. *Fitopatología* **1995**, *30*, 33–45.

10. Mareggiani, G.; Zamuner, N.; Michetti, M.; Franzetti, D.; Collavino, M. Impact of natural extracts on target and non-target soil organisms. *Bol. San. Veg. Plagas* **2005**, *31*, 443–448.

11. Godínez-Vidal, D.; Soto-Hernández, M.; Rocha-Sosa, M.; Lozoya-Gloria, E.; Rojas-Martínez, R.I.; Guevara-Olvera, L.; Zavaleta-Mejía, E. Contenido de capsidiol en raíces de chile CM-334 infectadas por *Nacobbus aberrans* y su efecto en juveniles del segundo estadio. *Nematropica* **2010**, *40*, 227–237.

12. Rodríguez-Chávez, J.L.; Franco-Navarro, F.; Delgado, G. In vitro nematicidal activity of natural and semisynthetic cadinenes from *Heterotheca inuloides* against the plant-parasitic nematode *Nacobbus aberrans* (Tylenchida: Pratylenchidae). *Pest Manag. Sci.* **2019**, *75*, 1734–1742. [CrossRef]

13. Cristóbal-Alejo, J.; Mora-Aguilera, G.; Manzanilla-Lopez, R.H.; Marbán-Méndoza, N.; Sánchez-Garcia, P.; Del Prado-Vera, I.C.; Evans, K. Epidemiology and integrated control of *Nacobbus aberrans* on tomato in Mexico. *Nematology* **2006**, *8*, 727–737. [CrossRef]

14. Feist, E.; Kearn, J.; Gaihre, Y.; O'Connor, V.; Holden-Dye, L. The distinct profiles of the inhibitory effects of fluensulfone, abamectin, aldicarb and fluopyram on *Globodera pallida* hatching. *Pestic. Biochem. Physiol.* **2020**, *165*, 104541. [CrossRef]

15. Forgo, P.; Kövér, K.E. Gradient enhanced selective experiments in the ^{1}H NMR chemical shift assignment of the skeleton and side-chain resonances of stigmasterol, a phytosterol derivative. *Steroids* **2004**, *69*, 43–50. [CrossRef] [PubMed]

16. Ododo, M.M.; Choudhury, M.K.; Dekebo, A.H. Structure elucidation of β-sitosterol with antibacterial activity from the root bark of *Malva parviflora*. *SpringerPlus* **2016**, *5*, 1–11. [CrossRef] [PubMed]

17. Herrera-Martínez, M.; Hernández-Ramírez, V.I.; Hernández-Carlos, B.; Chávez-Munguía, B.; Calderón-Oropeza, M.A.; Talamás-Rohana, P. Antiamoebic activity of *Adenophyllum aurantium* (L.) Strother and its effect on the actin cytoskeleton of *Entamoeba histolytica*. *Front. Pharmacol.* **2016**, *7*, 169. [CrossRef] [PubMed]

18. Downum, K.R.; Keil, D.J.; Rodríguez, E. Distribution of acetylenic thiophenes in the pectidinae. *Biochem. Syst. Ecol.* **1985**, *13*, 109–113. [CrossRef]

19. Marotti, I.; Marotti, M.; Piccaglia, R.; Nastri, A.; Grandi, S.; Dinelli, G. Thiophene occurrence in different *Tagetes* species: Agricultural biomasses as sources of biocidal substances. *J. Sci. Food Agric.* **2010**, *90*, 1210–1217. [CrossRef]

20. Hooks, C.R.R.; Wang, K.H.; Ploeg, A.; McSorley, R. Using marigold (*Tagetes* spp.) as a cover crop to protect crops from plant-parasitic nematodes. *Appl. Soil Ecol.* **2010**, *46*, 307–320. [CrossRef]

21. Kimura, Y.; Hiraoka, K.; Kawano, T.; Fujioka, S.; Shimada, A. Nematicidal activities of acetylene compounds from *Coriopsis lanceolata* L. *Zeitschrift für Naturforschung C* **2008**, *63*, 843–847. [CrossRef]

22. Faizi, S.; Fayyaz, S.; Bano, S.; Yawar Iqbal, E.; Lubna, L.; Siddiqi, H.; Naz, A. Isolation of nematicidal compounds from *Tagetes patula* L. yellow flowers: Structure-activity relationship studies against cyst nematode *Heterodera zeae* infective stage larvae. *J. Agric. Food Chem.* **2011**, *59*, 9080–9093. [CrossRef] [PubMed]

23. Ferheen, S.; Akhtar, M.; Ahmed, A.G.; Anwar, M.A.; Kalhoro, M.A.; Afza, N.; Malik, A. Nematicidal potential of the *Galinsoga parviflora*. *Pak. J. Sci. Ind. Res. Ser. B Biol. Sci.* **2011**, *54*, 83–87.

24. Barbosa, L.C.A.; Barcelos, F.F.; Demuner, A.J.; Santos, M.A. Chemical constituents from *Mucuma aterrima* with activity against *Meloidogyne incognita* and *Heterodera glycines*. *Nematropica* **1999**, *29*, 81–88.

25. Naz, I.; Khan, M.R. Nematicidal activity of nonacosane-10-ol and 23a-homostigmast-5-en-3β-ol isolated from the roots of *Fumaria parviflora* (Fumariaceae). *J. Agric. Food Chem.* **2013**, *61*, 5689–5695. [CrossRef]

26. Naz, I.; Saifullah, S.; Palomares-Rius, J.E.; Ahmad, M.; Ali, A.; Rashid, M.U.; Bibi, F. Combined nematocidal effect of nonacosan-10-ol and 23a-homostigmast-5-en-3β-ol on *Meloidogyne incognita* (kofoid and white) chitwood. *J. Anim. Plant Sci.* **2016**, *26*, 1633–1640.

27. Chitwood, D.J.; McClure, M.A.; Feldlaufer, M.F.; Lusby, W.R.; Oliver, T.E. Sterol composition and ecdysteroid content of eggs of the root-knot nematodes *Meloidogyne incognita* and *M. arenaria*. *J. Nematol.* **1987**, *19*, 352.

28. Udalova, Z.V.; Zinov'eva, S.V.; Valis'eva, I.S.; Paseshnichenko, V.A. Correlation between the structure of plant steroids and their effects on phytoparasitic nematodes. *Appl. Biochem. Microbiol.* **2004**, *40*, 93–97. [CrossRef]

29. Moss, G.P. Carbon-13 NMR spectra of carotenoids. *Pure Appl. Chem.* **1976**, *47*, 97–102. [CrossRef]

30. El-Raey, M.A.; Ibrahim, G.E.; Eldahshan, O.A. Lycopene and lutein; A review for their chemistry and medicinal uses. *J. Pharmacogn. Phytochem.* **2013**, *2*, 245–254.

31. Chiang, Y.M.; Chuang, D.Y.; Wang, S.Y.; Kuo, Y.H.; Tsai, P.W.; Shyur, L.F. Metabolite profiling and chemopreventive bioactivity of plant extracts from *Bidens pilosa*. *J. Ethnopharmacol.* **2004**, *95*, 409–419. [CrossRef] [PubMed]

32. Schmeda-Hirschmann, G.; Tapia, A.; Theoduloz, C.; Rodríguez, J.; López, S.; Feresin, G.E. Free radical scavengers and antioxidants from *Tagetes mendocina*. *Zeitschrift für Naturforschung C* **2004**, *59*, 345–353. [CrossRef]

33. Arisawa, M.; Hatashita, T.; Numata, Y.; Tanaka, M.; Sasaki, T. Cytotoxic principles from *Chrysosplenium flagelliferum*. *Int. J. Pharmacogn.* **1997**, *35*, 141–143. [CrossRef]

34. Bai, N.; He, K.; Zhou, Z.; Lai, C.S.; Zhang, L.; Quan, Z.; Shao, X.; Pan, M.-H.; Ho, C.T. Flavonoids from *Rabdosia rubescens* exert anti-inflammatory and growth inhibitory effect against human leukemia HL-60 cells. *Food Chem.* **2010**, *122*, 831–835. [CrossRef]

35. Bohlmann, F.; Zdero, C. C17-Acetylenverbindungen aus *Calea integrifolia*. *Phytochemistry* **1976**, *15*, 1177. [CrossRef]

36. Lira-De León, K.I.; Herrera-Martínez, M.; Ramírez-Mares, M.V.; Hernández-Carlos, B. Evaluation of anticancer potential of eight vegetal species from the state of Oaxaca. *Afr. J. Tradit. Complement. Altern. Med.* **2017**, *14*, 61–73. [CrossRef] [PubMed]

37. Lima, T.C.; de Jesus Souza, R.; da Silva, F.A.; Biavatti, M.W. The genus *Calea* L.: A review on traditional uses, phytochemistry and biology activities. *Phytother. Res.* **2018**, *32*, 769–795. [CrossRef]

38. Bano, S.; Iqbal, E.Y.; Lubna; Zil-ur-Rehman, S.; Fayyaz, S.; Faizi, S. Nematicidal activity of flavonoids with structure activity relationship (SAR) studies against root knot nematode *Meloidogyne incognita*. *Eur. J. Plant Pathol.* **2020**, *157*, 299–309. [CrossRef]

39. Isaac, R.E.; MacGregor, D.; Coates, D. Metabolism and inactivation of neurotransmitters in nematodes. *Parasitology* **1996**, *113*, S157–S173. [CrossRef] [PubMed]

40. Orhan, I.; Kartal, M.; Tosun, F.; Şener, B. Screening of various phenolic acids and flavonoid derivatives for their anticholinesterase potential. *Zeitschrift für Naturforschung C* **2007**, *62*, 829–832. [CrossRef]

41. Promchai, T.; Saesong, T.; Ingkaninan, K.; Laphookhieo, S.; Pyne, S.G.; Limtharakul, T. Acetylcholinesterase inhibitory activity of chemical constituents isolated from *Miliusa thorelii*. *Phytochem. Lett.* **2018**, *23*, 33–37. [CrossRef]

42. Li, M.; Gao, X.; Lan, M.; Liao, X.; Su, F.; Fan, L.; Zhao, Y.; Hao, X.; Wu, G.; Ding, X. Inhibitory activities of flavonoids from *Eupatorium adenophorum* against acetylcholinesterase. *Pestic. Biochem. Physiol.* **2020**, *170*, 104701. [CrossRef] [PubMed]

43. San Martín, R.; Magunacelaya, J.C. Control of plant-parasitic nematodes with extracts of *Quillaja saponaria*. *Nematology* **2005**, *7*, 577–585.

44. Correia Da Silva, T.B.; Souza, V.K.T.; Da Silva, A.P.F.; Lyra-Lemos, R.P.; Conserva, L.M. Determination of the phenolic content and antioxidant potential of crude extracts and isolated compounds from leaves of *Cordia multispicata* and *Tournefortia bicolor*. *Pharm. Biol.* **2010**, *48*, 63–69. [CrossRef]

45. El-Shazly, A.; Wink, M. Diversity of pyrrolizidine alkaloids in the Boraginaceae structures, distribution, and biological properties. *Diversity* **2014**, *6*, 188–282. [CrossRef]

46. Ogihara, K.; Miyagi, Y.; Higa, M.; Yogi, S. Pyrrolizidine alkaloids from *Messerschmidia argentea*. *Phytochemistry* **1997**, *44*, 545–547. [CrossRef]

47. Constantinidis, T.; Harvala, C.; Skaltsounis, A.L. Pyrrolizidine *N*-oxide alkaloids of *Heliotropium hirsutissimum*. *Phytochemistry* **1993**, *32*, 1335–1337. [CrossRef]

48. Molyneux, R.J.; Roitman, J.N.; Benson, M.; Lundin, R.E. ^{13}C NMR spectroscopy of pyrrolizidine alkaloids. *Phytochemistry* **1982**, *21*, 439–443. [CrossRef]

49. Hammouda, F.M.; Ismail, S.I.; Hassan, N.M.; Tawfiq, W.A.; Kamel, A. Pyrrolizidine alkaloids from *Alkanna orientalis* (L.) Boiss. *Qatar Univ. Sci. J.* **1992**, *12*, 80–82.

50. Logie, C.G.; Grue, M.R.; Liddell, J.R. Proton NMR spectroscopy of pyrrolizidine alkaloids. *Phytochemistry* **1994**, *37*, 43–109. [CrossRef]

51. Roeder, E. Carbon-13 NMR spectroscopy of pyrrolizidine alkaloids. *Phytochemistry* **1990**, *29*, 11–29. [CrossRef]

52. Lira-De León, K.I.; Ramírez-Mares, M.V.; Sánchez-López, V.; Ramírez Lepe, M.; Salas-Coronado, R.; Santos-Sánchez, N.F.; Valadez-Blanco, R.; Hernández-Carlos, B. Effect of crude plant extracts from some Oaxacan flora on two deleterious fungal phytopathogens and extracts compatibility with a biofertilizer strain. *Front. Microbiol.* **2014**, *5*, 383. [CrossRef]

53. van Dam, N.M.; Vuister, L.W.; Bergshoeff, C.; de Vos, H.; Van der Meijden, E.D. The "Raison D'être" of pyrrolizidine alkaloids in *Cynoglossum officinale*: Deterrent effects against generalist herbivores. *J. Chem. Ecol.* **1995**, *21*, 507–523. [CrossRef] [PubMed]

54. Thoden, T.C.; Boppré, M.; Hallmann, J. Effects of pyrrolizidine alkaloids on the performance of plant-parasitic and free-living nematodes. *Pest Manag. Sci.* **2009**, *65*, 823–830. [CrossRef]

55. Schramm, S.; Köhler, N.; Rozhon, W. Pyrrolizidine alkaloids: Biosynthesis, biological activities and occurrence in crop plants. *Molecules* **2019**, *24*, 498. [CrossRef]

56. Chitwood, D.J.; Lusby, W.R. Metabolism of plants sterol by Nematodes. *Lipids* **1991**, *26*, 619–626. [CrossRef]

57. Nivsarkar, M.; Kumar, G.P.; Laloraya, M.; Laloraya, M.M. Superoxide dismutase in the anal gills of the mosquito larvae of *Aedes aegypti*: Its inhibition by α-terthienyl. *Arch. Insect Biochem. Physiol.* **1991**, *16*, 249–255. [CrossRef]

58. MacRae, W.D.; Chan, G.F.Q.; Wat, C.K.; Towers, G.H.N.; Lam, J. Examination of naturally occurring polyacetylenes and α-terthienyl for their ability to induce cytogenetic damage. *Experientia* **1980**, *36*, 1096–1097. [CrossRef]

59. Sripathi, S.K.; Gopal, P.; Lalitha, P. Allantoin from the leaves of *Pisonia grandis* R. Br. *Int. J. Pharm. Life Sci.* **2011**, *2*, 815–817.

60. Argentieri, M.P.; D'Addabbo, T.; Tava, A.; Agostinelli, A.; Jurzysta, M.; Avato, P. Evaluation of nematicidal properties of saponins from *Medicago* spp. *Eur. J. Plant Pathol.* **2008**, *120*, 189–197. [CrossRef]

61. Hernández-Carlos, B.; González-Coloma, A.; Orozco-Valencia, A.U.; Ramírez-Mares, M.V.; Andrés-Yeves, M.F.; Joseph-Nathan, P. Bioactive saponins from *Microsechium helleri* and *Sycios bulbosus*. *Phytochemistry* **2011**, *72*, 743–751. [CrossRef] [PubMed]

molecules

Article

Pharmacognostic Evaluation and HPLC–PDA and HS–SPME/GC–MS Metabolomic Profiling of *Eleutherococcus senticosus* Fruits

Filip Graczyk [1,*], Maciej Strzemski [2], Maciej Balcerek [1], Weronika Kozłowska [3], Barbara Mazurek [4], Michał Karakuła [2], Ireneusz Sowa [2], Aneta A. Ptaszyńska [5] and Daniel Załuski [1]

1   Department of Pharmaceutical Botany and Pharmacognosy, Ludwik Rydygier Collegium Medicum, Nicolaus Copernicus University, Marie Curie-Skłodowska 9, 85-094 Bydgoszcz, Poland; balcerek@cm.umk.pl (M.B.); daniel_zaluski@onet.eu (D.Z.)
2   Department of Analytical Chemistry, Medical University of Lublin, Chodźki 4a, 20-093 Lublin, Poland; maciej.strzemski@poczta.onet.pl (M.S.); michal.karakula@umlub.pl (M.K.); i.sowa@umlub.pl (I.S.)
3   Department of Pharmaceutical Biology, Wroclaw Medical University, Borowska 211, 50-556 Wroclaw, Poland; weronika.kozlowska@umed.wroc.pl
4   Analytical Department, New Chemical Syntheses Institute, Aleja Tysiąclecia Państwa Polskiego 13a, 24-110 Puławy, Poland; barbara.mazurek@ins.lukasiewicz.gov.pl
5   Department of Immunobiology, Institute of Biological Sciences, Faculty of Biology and Biotechnology, Maria Curie-Skłodowska University, Akademicka 19 Str., 20-033 Lublin, Poland; anetaptas@wp.pl
*   Correspondence: filip.graczk@gmail.com; Tel.: +48-795672587

Citation: Graczyk, F.; Strzemski, M.; Balcerek, M.; Kozłowska, W.; Mazurek, B.; Karakuła, M.; Sowa, I.; Ptaszyńska, A.A.; Załuski, D. Pharmacognostic Evaluation and HPLC–PDA and HS–SPME/GC–MS Metabolomic Profiling of *Eleutherococcus senticosus* Fruits. *Molecules* **2021**, *26*, 1969. https://doi.org/10.3390/molecules26071969

Academic Editors: Young Hae Choi, Young Pyo Jang, Yuntao Dai and Luis Francisco Salomé-Abarca

Received: 5 March 2021
Accepted: 27 March 2021
Published: 31 March 2021

Publisher's Note: MDPI stays neutral with regard to jurisdictional claims in published maps and institutional affiliations.

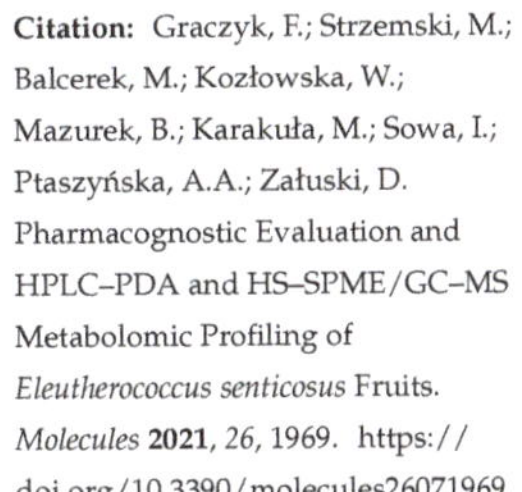

Abstract: *Eleutherococcus senticosus* (Rupr. et Maxim.) Maxim. is a medicinal plant used in Traditional Chinese Medicine (TCM) for thousands of years. However, due to the overexploitation, this species is considered to be endangered and is included in the Red List, e.g., in the Republic of Korea. Therefore, a new source of this important plant in Europe is needed. The aim of this study was to develop pharmacognostic and phytochemical parameters of the fruits. The content of polyphenols (eleuthero-sides B, E, E1) and phenolic acids in the different parts of the fruits, as well as tocopherols, fatty acids in the oil, and volatile constituents were studied by the means of chromatographic techniques [HPLC with Photodiode-Array Detection (PDA), headspace solid-phase microextraction coupled to gas chromatography-mass spectrometry (HS–SPME/GC–MS)]. To the best of our knowledge, no information is available on the content of eleutherosides and phenolic acids in the pericarp and seeds. The highest sum of eleutheroside B and E was detected in the whole fruits (1.4 mg/g), next in the pericarp (1.23 mg/g) and the seeds (0.85 mg/g). Amongst chlorogenic acid derivatives (3-CQA, 4-CQA, 5-CQA), 3-CQA was predominant in the whole fruits (1.08 mg/g), next in the pericarp (0.66 mg/g), and the seeds (0.076 mg/g). The oil was rich in linoleic acid (C18:3 (n-3), 18.24%), ursolic acid (35.72 mg/g), and α-tocopherol (8.36 mg/g). The presence of druses and yellow oil droplets in the inner zone of the mesocarp and chromoplasts in the outer zone can be used as anatomical markers. These studies provide a phytochemical proof for accumulation of polyphenols mainly in the pericarp, and these structures may be taken into consideration as their source subjected to extraction to obtain polyphenol-rich extracts.

Keywords: *Eleutherococcus senticosus*; fruits; eleutherosides; nutri-pharmacological; metabolomics; herbs

1. Introduction

Plant-based metabolites have served as lead compounds for many important drugs, such as morphine, digoxin, quinine, hyoscyamine, salicylic acid, and artemisinin [1]. According to the WHO (World Health Organization), about 80% of the world's population use plants in the primary health system, both in the developing and developed countries. For instance, 14% of the Russian population use them regularly and 44% from time-to-time [2].

One of the best-known plants, used as a source of pharmacologically active and nutritional molecules, is *Eleutherococcus senticosus* (Siberian ginseng). The plant is native to Russia, the Far East, China, Korea, and Japan. The fruits of *E. senticosus* have been used in Russia for many years as a tonic on the central nervous system and as an adaptogen. Modern research is focused on elucidation of the pharmacological activities of wild fruits, including their antioxidant, antimicrobial, and anticancer effects. An acidified 80% methanol extract showed high xanthine oxidase and AChE inhibitory activities. The anticancer activity of the extract was also proven by screening various human cell lines, including LNCaP (prostate cells), MOLT-4F (leukemia cells), A549 (lung cells), ACHN (renal cells) HCT-15 (colon cells), and SW-620 (colon cells). The latest reports indicate their immunostimulatory and anti-inflammatory activities and an increase in the number of leukocytes. The intractum from the fruits was found to stimulate human leukocyte resistance to the VSV (*Vesicular Stomatitis Virus*) infection via reduction of viral replication, which might be associated with increased secretion of interleukin 10 (IL-10). Besides medical applications, the fruits are industrially used in the production of phyto-jam (0.8–1.6 kg/100 kg of the product). Considering the data mentioned above, the consumption of wild edible fruits of *E. senticosus* by local communities in many developing countries is gaining increasing interest [3–7].

According to the European Medicines Agency (EMA), the pharmacological activity of *E. senticosus* in traditional applications is attributed to secondary metabolites called eleutherosides. They are very varied with saponins, lignans, coumarins, and phenylpropanoids as their aglycons. Eleutheroside B (syringin 4-β-D-glucoside) and E ((−)-syringaresinol 4,4″-O-β-D-diglucoside) are the major compounds and, according to the European Pharmacopoeia, their sum should not be less than 0.08%. Eleutherosides are thought to be the most active compounds present mainly in the roots. Although the fruits are used as ingredients in the production of galenic formulations, the roots are a major subject of research while the fruits remain largely unknown. The fruits are rich in flavonoids (13.4–14.4 mg/g ext.), polyphenols (38.5–41.1 mg/g ext.), minerals (Ca 3730–4495, Mg 1430–1540, Fe 35.4–53, Mn 75.2–88.3, Zn 18.9–41.0, Cu 3.34–13, Se 0.19–061; mg/kg respectively), and essential oil (0.3%, v/d.w.). Interestingly, a large amount of *myo*-inositol and D-mannitol was found as well (267.5 and 492.5 mg/g dry extract, respectively) [8–10].

As demonstrated by previous studies conducted by Załuski, *E. senticosus* is successfully cultivated in the botanical garden in Rogów (Central Polish Lowlands), and some methods required for determination of phenolic metabolites in the plant material and the biological activity of extracts have been developed [11,12]. The observations have also been confirmed by Bączek, who studied the impact of growth conditions on accumulation of biologically active compounds in organs of two-, three-, and four-year-old plants [13]. Nevertheless, the species cultivated in Bydgoszcz (North Polish Lowlands) has not yet been studied in detail (Figure 1). Preliminary information on chemical compounds in the *E. senticosus* fruits intractum have been previously reported by Graczyk et al. [8]. To the best of our knowledge, no data are available on the localization of eleutherosides and phenolic acids in anatomical structures or on pharmacognostic features that can be used as markers in quality assessment. It is well known that phytochemicals can be accumulated in different fruit parts, e.g., *Citrus limon* Burm. contains the highest amount of chlorogenic or caffeic acids or hesperidin in the peel [14–17]. In addition, there is no information about the oil in the fruits, which may be of interest e.g., for the food industry [10,18–20].

Figure 1. Morphological characteristics of the investigated species; (**A**) 1-year-old plant, general morphology; (**B1,B2**) 4-year-old plants growing in the crop field, general morphology; (**C**) flowers; (**D**) mature fruits.

To confirm our hypothesis, this study was focused on the quantitative analysis of phytochemicals with their localization in the anatomical fruit parts (seeds, pericarp) and on anatomical features that can be possibly used in the quality control of this pharmaceutically important plant.

2. Results and Discussion

2.1. Microscopic Pharmacognostic Features of Fruits

The quality control of plant-based material is important to ensure the best quality of products. For many years, plant materials imported to Europe have been of bad quality (adulterated or substituted) [21]; therefore, to ensure good quality, new protocols for a new herbal material are needed. The fruits of *Eleutherococcus senticosus* species growing in Asia and cultivated in Poland or other European countries have not yet been evaluated in terms of pharmacognostic features. The poor-quality plant material now offered on the market necessitates development of proper well-controlled production procedures. The development of a pharmacognostic protocol including microscopic and phytochemical studies will help in identification of these fruits and protect from adulteration.

The characteristic anatomical features of the fruits are shown in Figures 2 and 3. The anatomical structure of the *E. senticosus* fruits is characteristic for most species from the Araliaceae family. The microscopic studies of the transverse section showed the presence of secretory canals filled with yellow oil droplets in the inner zone of the mesocarp. The presence of druses and yellow oil droplets in the inner zone of the mesocarp and red chromoplasts in the outer zone of the mesocarp visible in the transverse fruit section are distinguishing features that can be used as anatomical markers. This study is a complement of the research conducted by Solomonowa et al. [20], in which the morphological-anatomical and morphometric study of fruits were carried out. The morphometric indicators of the fruits are as follows: fruit length 9.95 mm, fruit diameter 4.65, stone length 5.25 mm, and stone width 1.98 mm.

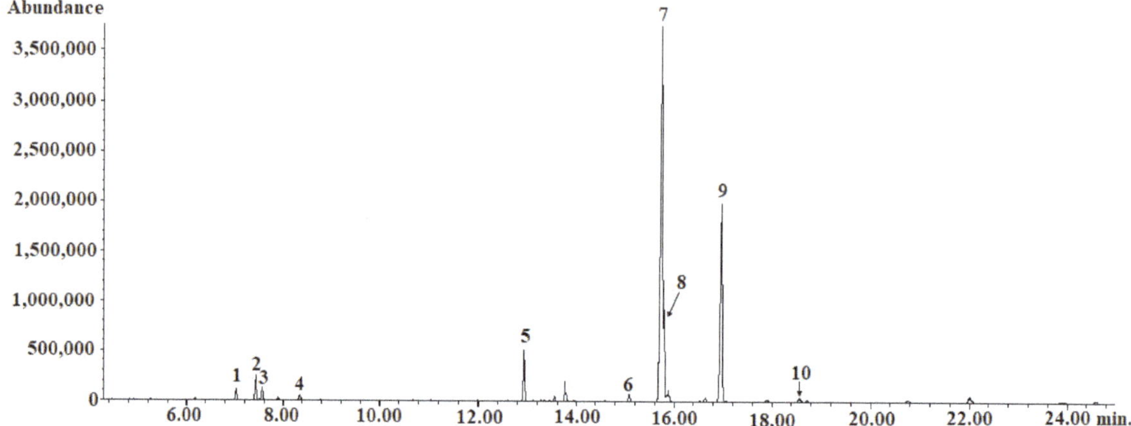

Figure 2. (**A**). Morphological structure and anatomical features of *E. senticosus* fruits. (**B**). **1**. Transverse section of the fruit, **2–3**. Seed, **4–5**. Seed coat and endosperm, **6**. Pericarp, **7–8**. Outer zone of mesocarp, **9–10**. Mesocarp with druses, **11–13**. Inner zone of mesocarp with secretory canals and oil droplets (transverse section), **14**. Inner zone of mesocarp with secretory canals and oil droplets (longitudinal section), **15**. Endocarp (transverse section), **16**. Endocarp (tangential section). end.—endosperm, s.c.—seed coat, en.—endocarp, i.mes.—inner zone of mesocarp, o.mes.—outer zone of mesocarp, ex.—exocarp.

The powder microscopy of the fruits (Figure 3) shows epidermis cells tightly adjacent to each other. The endocarp contains fibers oriented askew to the stone axis. The secretory canals in the mesocarp are filled with oil, probably essential oil, rounded parenchymatous cells. In turn, the druses and chromoplasts in the mesocarp and fatty oil drops in the endosperm can be considered as diagnostic features. The parameters analyzed here are useful for identification and authentication of this medicinally important plant and will provide the latest knowledge for the development of herbal monographs as recommended by the European Medicines Agency.

2.2. HPLC–PDA and GC–MS Metabolite Profiling of the Fruits

2.2.1. HPLC–PDA Analysis of Eleutherosides B, E, and E1 in the Anatomical Structures of the Fruits

For the analysis of the metabolome, such approaches as HPLC, TLC–UV, GC–MS, LC-MS, MSⁿ, and NMR-spectrometry are currently used. TLC coupled with densitometry and HPLC or HPLC–MS have usually been applied to analyze eleutherosides. High-performance liquid chromatography (HPLC) is a leading technique for detection of plant-based metabolites and is very often used as a pilot technique for separation and identification of phytochemicals [10–12].

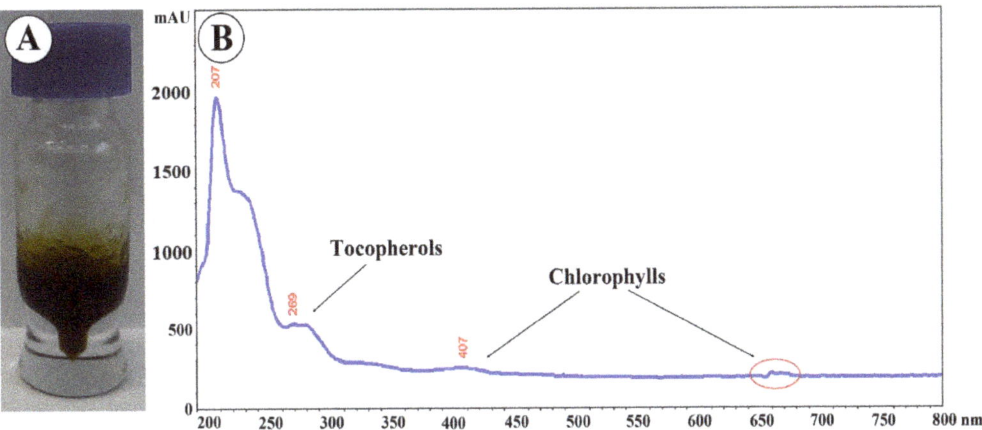

Figure 3. Diagnostic features of powdered fruits. **1.** Polygonal cells of fruit epidermis (exocarp). **2.** Inner layers of endocarp cells are represented by fibers, oriented askew to stone axis (2A—tangential section, 2B—transverse section). **3.** Secretory canal (cross-section) in mesocarp, parenchymal cells (3A—visible oil content). **4.** Outer part of the mesocarp composed of parenchymal cells containing red plastids (chromoplasts). **5.** Druses in mesocarp cells located closer to the endocarp. **6.** Drops of fatty oil from the endosperm. **7.** Seed coat built of large cells with brown contents. Diagnostic features of the fruits. s.c.—seed coat, end.—endosperm, i.mes.—inner zone of mesocarp, s.ch.—secretory canals, s.—secretion (essential oil), dr.—druses, ch.—chromoplasts.

The extraction of the fruits, seeds, and pericarp obtained from the 4-year-old plant resulted in 20, 26, and 33% dry extract yield, respectively. No data on the characterization of the anatomical parts of *E. senticosus* fruits and the distribution of their active constituents have been published before. Plant metabolites exhibit a very broad range of polarities. This means that only part of the plant's chemodiversity is present in any plant extract. Among the three studied eleutherosides (eleutheroside B, E, and E1), only eleutheroside B and E have been found in the largest amount in the whole fruits i.e., 0.66 and 0.74 mg/g d. ext., respectively (Figures 4 and 5). Considering the distribution of these compounds, a higher quantity was noticed in the pericarp than in the seeds; however, there was no significant difference between the pericarp and seeds, especially in the case of eleutheroside E. The latest studies conducted by Graczyk et al. [8] have revealed the absence of eleutherosides in the intractum made of the fresh fruits. This may result from their medium polarity. Eleutherosides are usually well extracted with 75% (*v/v*) methanol or ethanol; therefore, we suppose that the ethanol concentration (40% *v/v*) used to prepare the intractum may have been too low to extract eleutherosides. Moreover, the extraction of the whole fruits also means that some compounds will not be extracted, and the intractum was prepared through maceration, while ultrasounds were applied in this study.

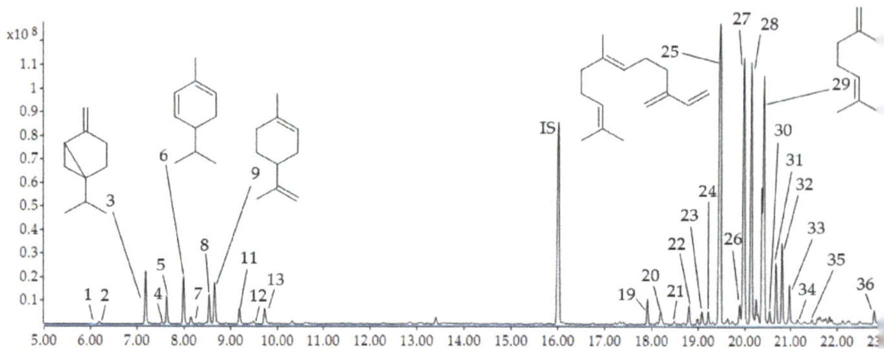

Figure 4. Contents of eleutherosides (mg/g extract): B—eleutheroside B, E—eleutheroside E, B + E—sum of eleutheroside B and eleutheroside E.

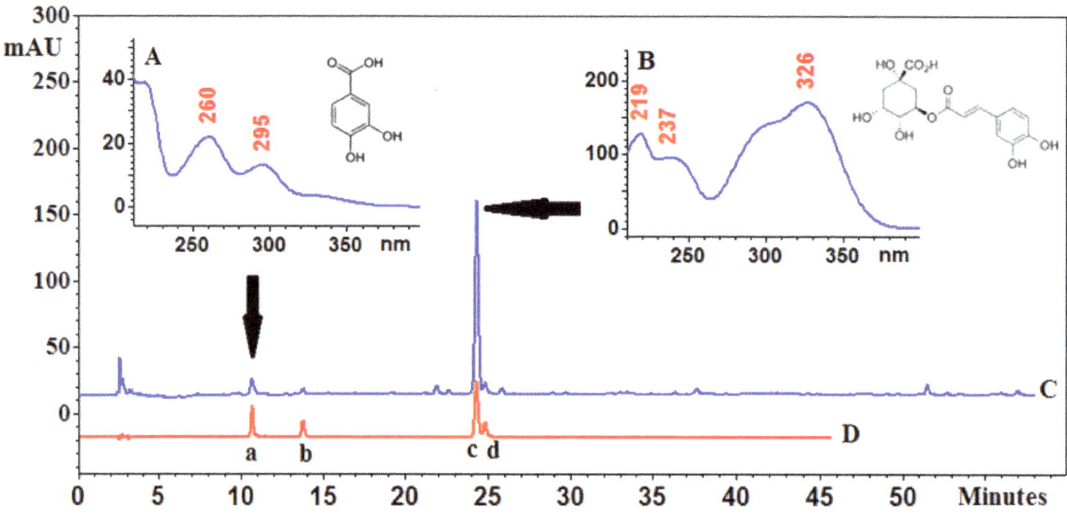

Figure 5. Chromatogram of *Eleutherococcus senticosus* fruit and reference standards with PDA spectra and structures of determined compounds: (**A**)—PDA spectrum and chemical structure of eleutheroside B; (**B**)—PDA spectrum and chemical structure of eleutheroside E; (**C**)—chromatogram of the extract of *Eleutherococcus senticosus* fruit; (**D**)—chromatogram of the mixture of standards (**a**—eleutheroside B, **b**—eleutheroside E, **c**—eleutheroside E1). RP18e LiChrospher 100 column (Merck, Darmstadt, Germany) (25 cm × 4.0 mm i.d., 5 μm particle size) at 25 °C; mixtures of water (solvent A) acetonitrile (solvent B) and both acidified with 0.025% of trifluoroacetic acid were used as the mobile phase. The compounds were separated by gradient elution with the following program: 0.0–8.0 min A 90%, B 10%; 8.1–18.0 min A 90–80%; B 10–20%, 18.1–30.0 min A 80%, B 20%. The flow rate was 1.0 mL/min.

Bączek [13] showed that the content of eleutherosides B and E in methanol fruit extract obtained from a 4-year-old plant was 35.6 and 29.7 (mg/100g), respectively. It should be mentioned that eleutherosides were not present in the fruits of 2- and 3-year-old plants, respectively. Unfortunately, the authors did not provide information about the calculation, i.e., it was not specified whether 100 g referred to the plant material or the extract. This makes the comparison difficult and, in many cases, unreliable. Taking into consideration the sum of eleutheroside B and E, the extract from the whole fruits contains a higher quantity of eleutherosides. However, it is very important that the pericarp contains more eleutherosides than the seeds, which means that the seeds may be used in the micropropagation of the species. *Eleutherococcus senticosus* is included in the Red List in some countries [8]; in this case, the pericarp may serve as a source of eleutherosides, while the seed (embryo) may be used for in vitro germination to develop seedlings and increase the number of botanical specimens.

2.2.2. HPLC–PDA Analysis of Phenolic Acids in the Anatomical Structures of the Fruits

Phenolic acids play a vital function as non-nutritional constituents of human diet and pharmacologically active compounds. In some fruits, they are accumulated in different parts. In this study, it has been found that the whole fruits contain a high amount of chlorogenic acid; 3-CQA (1.08 mg/g d. ext.), in comparison to the pericarp (0.66 mg/g d. ext.) and seeds (00.76 mg/g d. ext.) (Table 1, Figure 6).

Table 1. Contents of phenolic acids (mg/g extract ±SD), n = 3 (3-CQA—chlorogenic acid, 4-CQA—cryptochlorogenic acid, 5-CQA—neochlorogenic acid, PA—protocatechuic acid. Additional information on HPLC–DAD data is given in the supporting information part).

Sample	3-CQA	4-CQA	5-CQA	PA	Total Content
Fruit	1.08 ± 0.96	0.07 ± 0.01	0.030 ± 0.01	0.08 ± 0.07	1.26
Pericarp	0.66 ± 0.2	0.03 ± 0.02	0.01 ± 0.001	0.04 ± 0.03	0.74
Seed	0.076 ± 0.03	0.008 ± 0.001	0.004 ± 0.002	0.008 ± 0.001	0.096

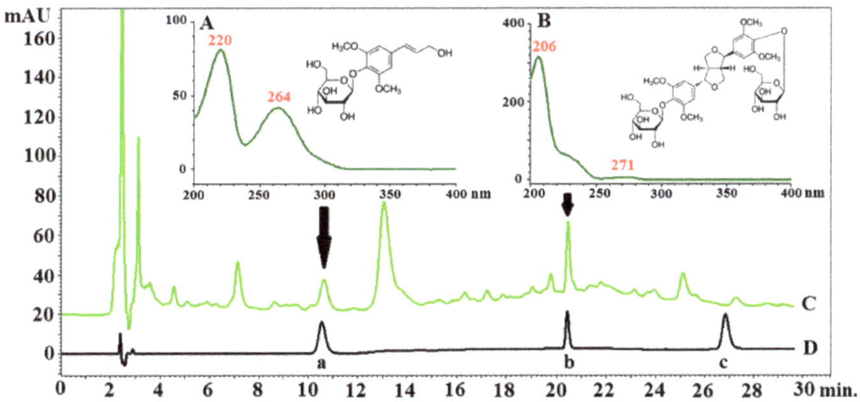

Figure 6. Chromatogram of *Eleutherococcus senticosus* fruits (**C**) and a mixture of reference standards (**D**): a—protocatechuic acid, b—neochlorogenic acid, c—chlorogenic acid and d—cryptochlorogenic acid with PDA spectra and structures of the main compounds: (**A**)—protocatechuic acid and (**B**)—chlorogenic acid. C18 core-shell column (Kinetex, Phenomenex, Aschaffenburg, Germany) (25 cm × 4.6mm i.d., 5 μm particle size) at 25 °C, a mixture of water (solvent A) and acetonitrile (solvent B) with 0.025% of trifluoroacetic acid; gradient elution program: 0.0–5.0 min A 95%, B 5%; 5–60 min A from 5 to 20%, and B from 95–80%. The flow rate was 1.0 mL/min.

Chlorogenic acid is the main phenolic component in coffee and beverages prepared from herbs, fruits, and vegetables. The results presented in literature revealed that the content of chlorogenic acid in methanol extracts of fruits from a 4-year-old plant was 409.2 mg/100g [13]. No chlorogenic acid has been detected in the fruits of 2- and 3-year-old plants, respectively. Other studies have shown that the water extract from the fruits of *E. senticosus* contained 0.57 mg/g dry weight of chlorogenic acid [9].

2.3. HS–SPME/GC–MS Analysis of Volatile Metabolites in the Fruits

HS–SPME coupled with the GC–MS analysis identified 38 volatile constituents (VOCs) in the *E. senticosus* fruits, representing 93.9% of the total number. The detected compounds belong to several classes, such as monoterpene hydrocarbons and oxygenated monoterpenes (13.2%) and sesquiterpene hydrocarbons and oxygenated sesquiterpenes (76.6%). The percentage composition of the identified compounds is listed in the elution order in Table 2 and Figure 7 and Figure S2. (E)-β-farnesene (19.46 ± 0.78%), germacrene D (12.88 ± 1.10%), (E,Z)-α-farnesene (12.84 ± 0.85%), and β-bisabolene (12.64 ± 0.70%) were the main compounds identified in the *E. senticosus* fruits.

Table 2. Percentage of volatile compounds identified by HS–SPME/GC–MS in *E. senticosus* fruits (% ± SD) (RT—retention time, RI calc—retention index, calculated, RI lit—retention index, literature. Internal standard—2-undecanone. All analyses were performed in triplicate).

No.	RI Lit.	RI Calc.	RT (Min)	Compound Name	Formula	Compound m/z	Match Factor (%)	Content
1	930	926	6.01	α-Thujene	$C_{10}H_{16}$	93; 91; 77	94.55	0.05 ± 0.02
2	939	933	6.18	α-Pinene	$C_{10}H_{16}$	93; 91; 77	95.00	0.12 ± 0.09
3	975	973	7.18	(Z)-Sabinene	$C_{10}H_{16}$	93; 91; 77	97.84	2.46 ± 1.19
4	985	987	7.52	6-methyl-5-hepten-2-one	$C_8H_{14}O$	43; 108; 69; 55	97.50	0.15 ± 0.03
5	990	991	7.63	β-Myrcene	$C_{10}H_{16}$	93; 69; 41	96.20	1.43 ± 0.44
6	1002	1005	7.99	α-Phellandrene	$C_{10}H_{16}$	93; 77; 136	96.90	2.22 ± 0.6
7	1011	1010	8.15	3-Carene	$C_{10}H_{16}$	93; 79; 121	97.90	0.37 ± 0.012
8	1026	1025	8.54	O-Cymene	$C_{10}H_{14}$	119; 134; 91	97.90	1.40 ± 0.26
9	1029	1029	8.65	Limonene	$C_{10}H_{16}$	93; 68; 79	98.50	2.81 ± 0.54
10	1037	1038	8.91	(Z)-β-Ocimene	$C_{10}H_{16}$	93; 91; 79	93.24	0.05 ± 0.01
11	1050	1048	9.19	(E)-β-Ocimene	$C_{10}H_{16}$	93; 91; 79	97.10	0.86 ± 0.15
12	1059	1059	9.49	γ-Terpinene	$C_{10}H_{16}$	93; 136; 77	98.60	0.11 ± 0.01
13	1070	1067	9.73	(Z)-Sabinene hydrate	$C_{10}H_{18}O$	71; 93; 43	98.50	0.90 ± 0.09
14	1088	1089	10.33	Terpinolene	$C_{10}H_{16}$	121; 93; 136	95.04	0.16 ± 0.01
15	1159	1159	12.30	Sabina ketone	$C_9H_{14}O$	81; 96; 67; 55	83.10	0.08 ± 0.01
16	1177	1179	12.85	Terpinen-4-ol	$C_{10}H_{18}O$	71; 111; 93	96.52	0.15 ± 0.02
17	1294	1295	15.99	2-Undecanone (IS)	$C_{11}H_{22}O$	58; 43; 71	93.20	3.79 ± 6.31
18	1294	1299	16.11	Methyl myrtenate	$C_{11}H_{16}O_2$	105; 137; 91;77	94.50	0.07 ± 0.02
19	1376	1381	17.91	α-Copaene	$C_{15}H_{24}$	161;119; 105; 93	98.20	1.05 ± 0.16
20	1388	1390	18.11	β-Bourbonene	$C_{15}H_{24}$	81; 123; 161	94.90	0.12 ± 0.02
21	1408	1409	18.49	7-epi-Sesquithujene	$C_{15}H_{24}$	119; 93; 91; 69	93.00	0.14 ± 0.02
22	1419	1426	18.81	β-Caryophyllene	$C_{15}H_{24}$	93; 91; 133; 79	97.90	0.91 ± 0.14
23	1432	1436	18.99	(Z)-β-Copaene	$C_{15}H_{24}$	161; 105; 91; 119	99.00	0.25 ± 0.04
24	1434	1441	19.08	(E)-α-Bergamotene	$C_{15}H_{24}$	119; 93; 91; 69	97.60	0.54 ± 0.07
25	1456	1463	19.49	(E)-β-Farnesene	$C_{15}H_{24}$	69; 93; 133; 161	90.00	19.46 ± 0.78
26	1466	1471	19.64	(Z)-Muurola-4,5-diene	$C_{15}H_{24}$	161; 105; 91; 204	94.20	0.26 ± 0.04
27	1481	1490	20.00	Germacrene D	$C_{15}H_{24}$	161; 105; 91; 119	94.60	12.88 ± 1.10
28	1497	1498	20.16	(E,Z)-α-Farnesene	$C_{15}H_{24}$	93; 119; 69; 107	92.10	12.84 ± 0.85
29	1511	1515	20.43	(E)-β-Bisabolene	$C_{15}H_{24}$	93; 69; 94; 109	95.70	12.64 ± 0.70
30	1513	1522	20.55	γ-Cadinene	$C_{15}H_{24}$	161; 93; 105; 119	97.70	0.55 ± 0.06
31	1522	1530	20.68	β-Sesquiphellandrene	$C_{15}H_{24}$	69; 161; 93; 91	95.70	2.6 ± 0.22
32	1532	1538	20.81	(E)-γ-Bisabolene	$C_{15}H_{24}$	93; 107; 135; 119	94.60	3.26 ± 0.43
33	1540	1548	20.97	(E)-α-Bisabolene	$C_{15}H_{24}$	93; 119; 121; 80	95.70	1.63 ± 0.20
34	1563	1559	21.15	(E)-Nerolidol	$C_{15}H_{26}O$	69; 93; 81; 83	85.30	0.23 ± 0.06
35	1607	1619	22.12	β-Oplopenone	$C_{15}H_{24}O$	177; 43; 93; 79	93.40	0.15 ± 0.02
36	1658	1665	22.80	α-Bisabolol oxide B	$C_{15}H_{26}O_2$	143; 105; 85; 81	97.66	0.76 ± 0.12
37	1685	1692	23.20	α-Bisabolol	$C_{15}H_{26}O$	109; 119; 69; 43	98.90	6.35 ± 0.9
38	2104	2107	27.78	6-Octadecenoic acid, methyl ester, (Z)-	$C_{19}H_{36}O_2$	55; 74; 84; 96	94.52	0.13 ± 0.10
Monoterpene hydrocarbons and oxygenated monoterpenes								13.17
Sesquiterpene hydrocarbons and oxygenated sesquiterpenes								76.62
Other compounds								4.07
Total								93.86

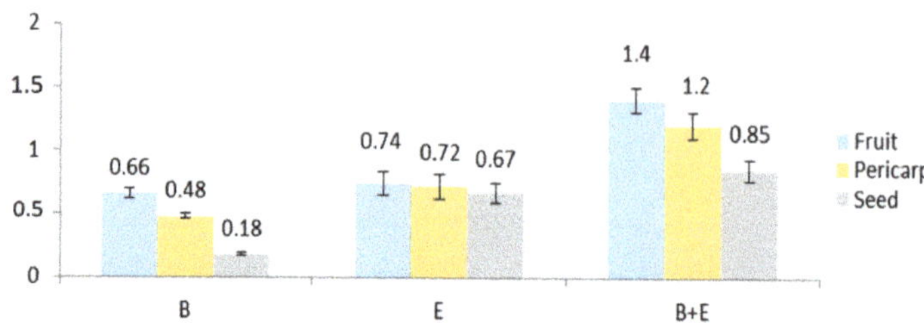

Figure 7. HS–SPME/GC–MS chromatogram of volatile compounds in *E. senticosus* fruits. The numbers correspond to the compounds in Table 2; IS—internal standard.

Previous reports on the essential oil from *E. senticosus* fruits indicated β-caryophyllene and humulene [7] or spathulenol [10] as the main compounds. However, the results of the present study revealed farnesene and bisabolene-type derivatives as dominant compounds showing more similarity to the essential oil from leaves described by Zhai et al. [22]. Studies conducted on *E. senticosus* root material obtained from different regions (Russia, China— two different sources) showed significant fluctuations in the quantity of the compounds, which may be connected with differentiation of the chemotypes [23]. Moreover, Załuski and Janeczko reported differences in the composition of fruit essential oil connected with duration of storage [10].

2.4. GC–MS and HPLC–PDA Profiling of the Fruit Fatty Oil

The oil obtained by hexane extraction of ground fruits constituted $5.4 \pm 0.015\%$ of their weight. The oil is a semi-solid, yellow-green substance with a characteristic aroma (Figure 8A) absorbing UV–VIS radiation in the range of 190–430 nm and at a wavelength of 660–680 nm (Figure 8B). Absorbance at a wavelength of 269 nm may indicate the presence of tocopherols, especially α-tocopherol absorbing radiation in the range of 265–310 nm [24]. Absorbance at 407 and 660–680 nm may indicate the presence of chlorophylls [25]. It was shown that the content of chlorophyll A and chlorophyll B was 12.43 mg/g and 3.67 mg/g, respectively, and the content of carotenoids in the tested oil was 10.13 mg/g.

2.4.1. GC–MS Analysis of Fatty Acids

The analysis of the oil by gas chromatography with mass spectrometry showed the presence of five fatty acids, including three unsaturated acids (linoleic, oleic, and α-linolenic acids), the total content of which was 24% (Table 3). The dominant fatty acid was linoleic acid (over 18%), while the content of α-linolenic acid was only 0.4%. The content of saturated acids was only 2.5%. Thus, the *E. senticosus* fruit oil is a rich source of unsaturated acids, but the proportion of n-3 acids to n-6 acids is not beneficial for human health [26]. In addition, the analysis showed the presence of four components of the essential oil and trimethylsulfonium ursolate—a derivative of ursolic acid and TMSH. An example of a chromatogram is presented in Figure 9, and the mass spectrum for fatty acids is shown in Figure S1.

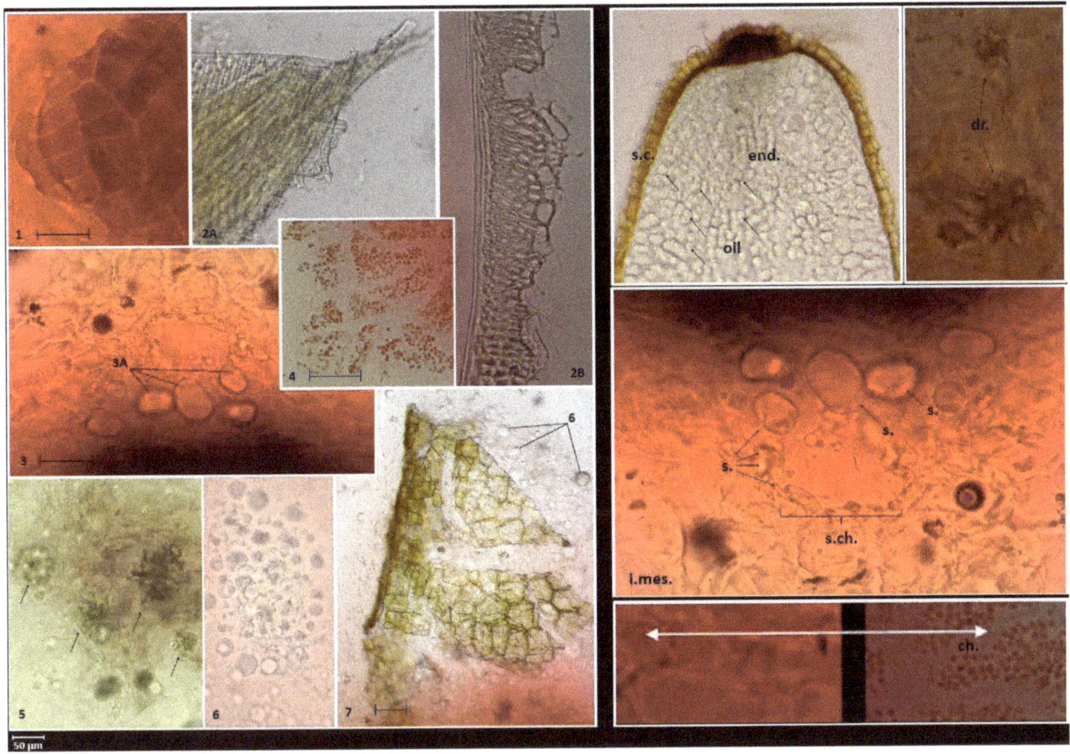

Figure 8. (**A**) Photography and (**B**) UV–VIS spectroscopic fingerprint of *Eleutherococcus senticosus* fruit oil.

Table 3. Content of fatty acids (% *w/w* ±SD) in *Eleutherococcus senticosus* seed oil (*n* = 3) (RT—retention time, FAME—fatty acid methyl esters, FA—fatty acid, M—molecular ion), LOD—limit of detection, LOQ—limit of quantification.

Compound Name	FA Abbreviation	RT FAME (Min.)	Mass Data FAME	LOD (% *w/w*)	LOQ (% *w/w*)	Content of FA
Palmitic acid	C16:0	12.94 ± 0.1	270 (M), 239,227,213,185,143, 129,11197,87,74,69,55	0.044	0.132	2.18 ± 0.04
Stearic acid	C18:0	15.08 ± 0.1	298 (M), 267,255,241,227, 213,199,185,171,157, 143,129,115,97, 87,83,74,69,55	0.026	0.078	0.29 ± 0.00
Oleic acid	C18:1 (*n-9*)	15.77 ± 0.1	296 (M), 278,264,235, 222,180,166, 137,123,110, 97,83,69,55	0.037	0.111	5.36 ± 0.05
Linoleic acid	C18:2 (*n-6*)	16.95 ± 0.2	294 (M), 263,233,220,205, 191,178,164,150, 135,123,109,95, 81,67,59,55	0.013	0.039	18.24 ± 0.26
α-Linolenic acid	C18:3 (*n-3*)	18.56 ± 0.2	292 (M), 261,236,193,163, 149,121,108, 95,79,67,55	0.015	0.045	0.40 ± 0.04
Total						26.47

Figure 9. GC–MS chromatogram of *Eleutherococcus senticosus* fruit oil. 1-(Z)-β-farnesene; 2-α-bergamotene; 3-α-farnesene; 4-β-cubebene; 5-methyl palmitate; 6-methyl stearate; 7-trimethylsulfonium ursolate; 8-methyl oleate; 9-methyl linoleate; 10-methyl linolenate. HP-88 Agilent 45 capillary column (60 m × 0.25 mm; 0.20 μm film thickness). The oven temperature was programmed from 110 °C to 190 °C with 8 °C/min held for 2 min at 110 °C and 13 min at 190 °C. The temperature of the injector was 250 °C. Helium was used as a carrier gas at a flow rate of 1.2 mL/min A quadrupole mass spectrometer with electron ionization (EI) at 70 eV and with a full scan type acquisition mode (50 *m/z* to 500 *m/z*) was used as a detector connected with GC. The temperature of the MS source and the MS quadrupole was set to 230 °C and 150 °C, respectively.

The chromatographic analysis (HPLC–PDA) revealed a dominant peak with a retention time of 7.25 min. This peak was identified as a mixture of β- and γ-tocopherols, and their content was determined to be 1.36 mg/g oil. The α- and δ-tocopherol content was 0.29 and 0.33 mg/g of oil, respectively (Table 4). Grilo et al. assessed the content of tocopherols in commonly used oils (rapeseed, sunflower, corn, and soybean). The content of α-tocopherol ranged from 0.071 to 0.43 mg/g and γ-tocopherol ranged from 0.092 to 0.27 mg/g for soybean and sunflower oils, respectively [27]. Thus, it can be concluded that the *E. senticosus* seed oil contains a moderate amount of α-tocopherol. On the other hand, the tested oil may be a valuable source of β- and γ-tocopherols; however, the assessment of the mutual quantitative relationships of these two compounds requires further research and other methods. Ergönül and Köseoğlu have shown that the content of δ-tocopherol in unrefined soybean oil is 0.12 mg/g, while in rapeseed oil it is 0.012 mg/g [28]. Thus, it can be concluded that the *E. senticosus* fruit oil is quite a rich source of δ-tocopherol, as it contains nearly three-fold higher amounts of this compound than soybean oil.

Table 4. Content of selected biologically active compounds (mg/g ± SD) in *Eleutherococcus senticosus* seed oil ($n = 3$) (RT—retention time, LOD—limit of detection, LOQ—limit of quantification. Tocopherol content).

Compound Name	RT (Min.)	Theoretical Plates	Linear Regression Equation	Concentration Range (μg/mL)	Correlation Coefficient (r)	LOD	LOQ	Content
Fluorescence Detection						(ng/mL)	(ng/mL)	
α-Tocopherol	8.36 ± 0.1	6158	y = 1851639x + 17320 31	0.48–7.20	0.9979	12.0	38.0	0.29 ± 0.00
β- + γ-Tocopherol	7.25 ± 0.1	5779	y = 4739064x + 115013	0.59–8.88	0.9997	6.0	20.0	1.36 ± 0.07
δ-Tocopherol	6.30 ± 0.1	5527	y = 4041480x + 112005	0.55–8.34	0.9999	5.5	18.5	0.33 ± 0.00
Spectrophotometric Detection						(μg/mL)	(μg/mL)	

2.4.2. HPLC–PDA Analysis of Ursolic Acid

Pentacyclic triterpenes are lipophilic compounds present in the oil fraction [29]. As shown in literature, *E. senticosus* fruits contain ursolic acid [7]. Therefore, the content of this triterpene in *E. senticosus* seed oil was evaluated in the present study. Its content was shown to be 35.72 mg/g of oil (Table 4). Given the oil content in the fruits, it can be concluded that the ursolic acid content in the fruits is over 1.9 mg/g. Our research shows that *E. senticosus* fruits are a much richer source of ursolic acid than previously thought. Jang et al. found that 100 g of fruits contain about 3.5 mg of ursolic acid [7]. The results obtained by Yang's team were on lower possibly due to the use of methanol in fruit extraction.

3. Materials and Methods

3.1. Chemicals and Reagents

The standards of eleutheroside B ($\geq$98.0%), eleutheroside E ($\geq$98.0%), eleutheroside E1 ($\geq$98.0%), protocatechuic acid ($\geq$97%), ursolic acid ($\geq$90%), δ-tocopherol, (+)-γ-tocopherol, ($\pm$)-α-tocopherol (analytical standards), methyl palmitate ($\geq$99.0%), methyl stearate ($\geq$99.5%), methyl oleate ($\geq$99.0%), methyl linoleate ($\geq$98.5%), methyl linolenate ($\geq$99.0%), 2-propanol (99.9%), hexane ($\geq$95%), phosphoric acid ($\geq$85%), trimethylsulfonium hydroxide (TMSH) (0.25 M methanolic solution), gradient grade acetonitrile, and trifluoroacetic acid ($\geq$99%) were obtained from Sigma-Aldrich (St. Louis, MO, USA). LC grade methanol (MeOH) was purchased from J.T. Baker (Phillipsburg, NJ, USA). Water for HPLC was purified by ULTRAPURE Millipore Direct-Q® 3UV-R (Merck Millipore, Billerica, MA, USA). Tert-butyl methyl ether (TBME) (99.8%) was purchased from Avantor Performance Materials Poland S.A. (Gliwice, Poland). All other reagents were of analytical grade. Eleutheroside standards were dissolved in 75% methanol (final concentration of 0.16, 0.16, and 0.36 mg/mL for eleutheroside B, E, and E1, respectively). Phenolic acid standards and ursolic acid were dissolved in methanol (final concentration of 0.24, 1.04, 0.23, 0.30, and 0.50 mg/mL for neochlorogenic, chlorogenic, cryptochlorogenic, protocatechuic, and ursolic acid, respectively). Standard solutions were prepared by dilution of stock solutions to appropriate concentrations.

3.2. Plant Material and Preparation of the Extract

Mature fruits of 4-year-old *Eleutherococcus senticosus* were collected in the Garden of Medicinal and Cosmetic Plants in Bydgoszcz (Poland) in September 2017 (N: 53°07′36.55″ E: 18°01′51.64″). The plant sample was deposited at the Department of Pharmaceutical Botany and Pharmacognosy, Collegium Medicum, Bydgoszcz, Poland, Cat. Nr. ES 10/2018. Air-dried and powdered fruits (10 g each) were soaked in 100 mL of 75% ethanol for 24 h. Next, the samples were subjected to triple UAE-type extraction (ultrasonic bath—Polsonic, Warsaw, Poland) using 100 and 2 × 50 mL of 75% ethanol. The extraction was performed at room temperature for 15 min for each cycle. Finally, 200 mL of each extract was obtained. After that, the extract was filtered through Whatman no. 4 filter paper. The solvent was dried with an evaporator in vacuum conditions at 45 °C, frozen at −20 °C, and subjected to lyophilization. The dried residue was stored in an exsiccator at 4 °C. The extraction yield was calculated based on the dry weight of the extract [%]. The same steps were made for extraction of the pericarp and seeds.

3.3. Microscopic Analysis

The anatomical structure of the fruits was examined microscopically using a compound microscope coupled with a camera, evaluated, and photographed (40X). Chloral hydrate was used as a reagent. Computer images were captured using software ProgRes CapturePro 2.8—Jenoptik optical system.

3.4. HPLC–PDA Analysis of Eleutherosides B, E, and E1

The analyses were performed on an EliteLaChrom chromatograph with a PDA detector and EZChrom Elite software (Merck, Darmstadt, Germany). The following chromatographic system was used: an RP18e LiChrospher 100 column (Merck, Darmstadt, Germany) (25 cm × 4.0 mm i.d., 5 μm particle size) at 25 °C; mixtures of water (solvent A) and acetonitrile (solvent B) both acidified with 0.025% of trifluoroacetic acid were used as the mobile phase. The compounds were separated by gradient elution with the following program: 0.0–8.0 min A 90%, B 10%; 8.1–18.0 min A 90–80%, B 10–20%; 18.1–30.0 min A 80%; B 20%; 30.1–45.0 min A 0%, B 100%; 45.1–60.0 min A 90%, B 10%. The flow rate was 1.0 mL/min. Data were collected between 190 and 400 nm. The identity of compounds was established by comparison of retention times and UV spectra with the corresponding standards. Quantitative analysis was performed at λ = 264 nm for eleutheroside B and λ = 206 nm for eleutheroside E and E1. The chromatographic parameters and calibration data for quantification of the investigated eleutherosides are provided in Table S1.

3.5. HPLC–PDA Analysis of Phenolic Acids

The analyses were performed on an EliteLaChrom chromatograph with a PDA detector and EZChrom Elite software (Merck, Darmstadt, Germany). The chromatographic system was as follows: a C18 reversed phase core-shell column (Kinetex, Phenomenex, Aschaffenburg, Germany) (25 cm × 4.6 mm i.d., 5 μm particle size), a mixture of water (solvent A) and acetonitrile (solvent B) with 0.025% of trifluoroacetic acid, and the following gradient elution program: 0.0–5.0 min A 95%, B 5%; 5–60 min A from 5 to 20% and B from 95–80%. The flow rate of the mobile phase was 1.0 mL/min and the temperature of thermostat was set at 25 °C. Data were collected between 210 and 400 nm. The identity of compounds was established by comparison of retention times and UV spectra with the corresponding standards. Quantitative analysis was performed at λ = 326 nm for chlorogenic acids and λ = 260 nm for protocatechuic acid. The chromatographic parameters and calibration data for quantification of the investigated phenolic acids are provided in Table S1.

3.6. HS–SPME/GC–MS Analysis of Volatile Compounds

Head space-solid phase microextraction (HS–SPME) was conducted according to Zielińska et al. [30] with slight modifications. Briefly, 100 mg of dry *E. senticosus* fruits were placed in a 15 mL sealed vial and extracted using a fiber coated with 50/30 μm divinylbenzene–carboxen–polydimethylsiloxane (DVB/CAR/PDMS; Supelco, Bellefonte, PE, USA). The 2-undecanone–2 mg/mL in water (Merck, Poland) was used as an internal standard. Equilibration was performed at 60 °C for 15 min, the fiber exposition time was 15 min, and the thermal desorption was 3 min at 250 °C directly in the gas chromatography (GC) injection port. All analyses were performed in triplicate. The gas chromatography (GC) analysis was performed using Agilent 7890B GC coupled with a 7000GC/TQ system mass spectrometer (Agilent Technologies, Paolo Alto, CA, USA). Separation was carried out on an HP-5 MS column; 30m × 0.25 mm × 0.25 μm (J&W, Agilent Technologies, Palo Alto, CA, USA) at a constant helium flow of 1 mL/min. The injector temperature was set at 250 °C and the sample was applied in a split mode (70:1). The temperature program was 50 °C for 1 min, followed by 5 °C/min to 120 °C, 8 °C/min to 200 °C, then to 250 °C in 16 °C/min and held isothermal for 2 min. The MS source, transfer line, and quadrupole temperature were set at 230 °C, 320 °C, and 150 °C, respectively. The mass spectra were collected in a scan mode from *m/z* 30–400 and the ionization voltage was 70 eV. Data acquisition was performed using Agilent MassHunter Workstation software (version B.08.00). The identification of the compounds was based on a comparison of fragmentation patterns with the NIST17 mass spectra library and they were matched with retention index (RI) obtained by calculation relative to the n-alkane standard (C8–C20; Merc, Poland). The quantitative analysis (expressed as percentages of each compound) was carried out by peak area normalization measurements without a correction factor.

3.7. Isolation of Oil and Preparation of Samples

Air-dried and pulverized fruits (5 g) were extracted four times with hexane (4 × 30 mL) using an ultra-sonic bath (4 × 15 min.) The extracts were combined and evaporated in a rotary evaporator. The oil (50 mg) was placed in a volumetric flask (5 mL) and dissolved in 2-propanol. The solution was filtered through a 0.25 μm polyamide membrane filter before HPLC analysis. The oil (10 mg) was dissolved in 500 μL of TBME and derivatized by the addition of 250 μL of TMSH. The whole sample was shaken vigorously, and the GC–MS analysis was performed.

3.7.1. HPLC–PDA Analysis of Ursolic Acid and Tocopherols in the Oil

The analysis was performed on a VWR Hitachi Chromaster 600 chromatograph with a 5430 Diode Array Detector, a 5440 FL Detector, and EZChrom Elite software (Merck, Darmstadt, Germany). A RP18e LiChrospher 100 column (Merck, Darmstadt, Germany) (25 cm × 4.0 mm i.d., 5 μm particle size) was used for the analyses. The identity of compounds was established by comparison of retention times and PDA spectra with the corresponding standards. Ursolic acid was determined using a previously published methodology [31]. An isocratic system was used with the basic chromatographic conditions: the mobile phase consisted of acetonitrile, water, and a 1% aqueous phosphoric acid solution (75:25:0.5 *v/v/v*); eluent flow rate 1.0 mL/min; column temperature 10 °C. The injection volume was 10 μL. The analysis was based on chromatograms recorded with the PDA detector. Data were collected between 200 and 400 nm. The quantitative analysis was performed at $\lambda = 200$ nm.

Tocopherols were determined using an isocratic system. The mobile phase consisted of acetonitrile and methanol (5:95 *v/v*). The eluent flow rate was 1.2 mL/min. The column temperature was set at 30 °C. The injection volume was 5 μL. The quantitative analysis was performed using a fluorescence detector with an excitation wavelength at $\lambda = 296$ nm and an emission wavelength at $\lambda = 330$ nm.

3.7.2. GC–MS Analysis of Fatty Acids in the Oil

The analysis was performed using an Agilent GC–MSD system (GC/MSD 6890N/5975) equipped with a HP-88 Agilent capillary column (60 m × 0.25 mm; 0.20 μm film thickness), MSD ChemStation ver. E.02.02.1431 software (Agilent Technologies, Santa Clara, CA, USA), and a split–splitless injector. The oven temperature was programmed from 110 °C to 190 °C with 8 °C/min, held for 2 min at 110 °C and 13 min at 190 °C. The temperature of the injector was 250 °C. The injection volume was 1 μL (split ratio 150:1; split flow 180 mL/min). Helium was used as a carrier gas at a flow rate of 1.2 mL/min. A quadrupole mass spectrometer with electron ionization (EI) at 70 eV and with a full scan type acquisition mode (50 *m/z* to 500 *m/z*) was used as a detector connected with the GC. The temperature of the MS source and the MS quadrupole was set at 230 °C and 150 °C, respectively. Identification of the constituents was based on a comparison of their mass spectra with the mass spectra library NIST resources and retention times with standards.

3.8. Statistical Analysis

Determinations were performed in triplicate. The data were subjected to statistical analysis using Statistica 7.0. (StatSoft, Cracow, Poland). The evaluations were analyzed for one-way analysis of variance. Statistical differences between the treatment groups were estimated by Spearman's (R) and Person's (r) test. All statistical tests were carried out at a significance level of $\alpha = 0.05$.

4. Conclusions

The practical aspect of the present results may be the application of the fruits as an ingredient of plant-based products used to treat immune-related diseases, as confirmed by Graczyk et al. [8]. With their chlorogenic acid content, the fruit extracts can be examined as a skin-whitening agent acting as tyrosinase inhibitor and possibly used in the cosmetic industry. In addition to this, we did not find information on doses used in ethnomedicine, showing the need of re-confirmation of the fruits' activity with regard to therapeutically active doses. Based on these results, the fruits may be a substitution for the roots to prevent exploitation of this endangered plant in some countries. This study has clearly shown that the species can be cultivated in Europe, producing biologically active metabolites.

Supplementary Materials: The following are available online, Figure S1: Mass spectra for standards (based on the NIST database) and investigated fatty acids. Figure S2: Mass spectra of HS-SPME/GC-MS investigated compounds. Table S1: Chromatographic parameters and calibration data for quantification of investigated eleutherosides and phenolic acids.

Author Contributions: Conceptualization, F.G. and D.Z.; methodology, F.G.; software, M.S.; validation, M.S.; formal analysis, F.G.; investigation, F.G., M.B., W.K., M.S., B.M., M.K.; data curation, M.S.; writing—original draft preparation, F.G., M.K., I.S., W.K., A.A.P.; visualization, F.G.; supervision, D.Z.; project administration, D.Z.; funding acquisition, D.Z. All authors have read and agreed to the published version of the manuscript.

Funding: Publication of the article was financed by the Polish National Agency for Academic Exchange under the Foreign Promotion Programme (NAWA), for A.A.P. (bee-research.umcs.pl; Api Lab UMCS PPI/PZA/2019/1/00039). This work was partially supported by the Nicolaus Copernicus University under the programme The Polish–Dutch Platform of plant-based medicines. M.S. is a Scholarship holder of the Polish Minister of Science and Higher Education for Outstanding Young Scientists 2020.

Institutional Review Board Statement: Not applicable.

Informed Consent Statement: Not applicable.

Data Availability Statement: Not applicable.

Conflicts of Interest: The authors declare no conflict of interest.

Sample Availability: Samples of the compounds are not available from the authors.

References

1. Verpoorte, R. Secondary metabolism. In *Metabolic Engineering of Plant Secondary Metabolism*; Kluwer Academic Publishers: Dordrecht, The Netherlands, 2000; pp. 1–29.
2. Shikov, A.N.; Pozharitskaya, O.N.; Makarov, V.G.; Wagner, H.; Verpoorte, R.; Heinrich, M. Medicinal plants of the Russian Pharmacopoeia; Their history and applications. *J. Ethnopharmacol.* **2014**, *154*, 481–536. [CrossRef] [PubMed]
3. Li, T.S.C. Comprehensive crop reports: Siberian ginseng. *Horttechnology* **2001**, *11*, 79–85. [CrossRef]
4. Schmidt, M.; Thomsen, M.; Kelber, O.; Kraft, K. Myths and facts in herbal medicines: *Eleutherococcus senticosus* (Siberian ginseng) and its contraindication in hypertensive patients. *Bot. Targets Ther.* **2014**, *4*, 27–32. [CrossRef]
5. Załuski, D.; Olech, M.; Galanty, A.; Verpoorte, R.; Kuźniewski, R.; Nowak, R.; Bogucka-Kocka, A. Phytochemical content and pharma-nutrition study on *Eleutherococcus senticosus* fruits intractum. *Oxidative Med. Cell. Longev.* **2016**, *2016*. [CrossRef]
6. Lee, J.H.; Lim, J.D.; Choung, M.G. Studies on the anthocyanin profile and biological properties from the fruits of *Acanthopanax senticosus* (Siberian Ginseng). *J. Funct. Foods* **2013**, *5*, 380–388. [CrossRef]
7. Jang, D.; Lee, J.; Eom, S.H.; Lee, S.M.; Gil, J.; Lim, H.B.; Hyun, T.K. Composition, antioxidant and antimicrobial activities of *Eleutherococcus senticosus* fruit extracts. *J. Appl. Pharm. Sci.* **2016**, *6*, 125–130. [CrossRef]
8. Graczyk, F.; Orzechowska, B.; Franz, D.; Strzemski, M.; Verpoorte, R.; Załuski, D. The intractum from the *Eleutherococcus senticosus* fruits affects the innate immunity in human leukocytes: From the ethnomedicinal use to contemporary evidence-based research. *J. Ethnopharmacol.* **2021**, *268*. [CrossRef]
9. Kim, Y.H.; Cho, M.L.; Kim, D.B.; Shin, G.H.; Lee, J.H.; Lee, J.S.; Park, S.O.; Lee, S.J.; Shin, H.M.; Lee, O.H. The antioxidant activity and their major antioxidant compounds from *Acanthopanax senticosus* and *A. koreanum*. *Molecules* **2015**, *20*, 13281–13295. [CrossRef]
10. Załuski, D.; Janeczko, Z. Variation in phytochemicals and bioactivity of the fruits of *Eleutherococcus* species cultivated in Poland. *Nat. Prod. Res.* **2015**, *29*, 2207–2211. [CrossRef]
11. Załuski, D.; Kuźniewski, R.; Janeczko, Z. HPTLC-profiling of eleutherosides, mechanism of antioxidative action of eleutheroside E1, the PAMPA test with LC/MS detection and the structure–activity relationship. *Saudi J. Biol. Sci.* **2018**, *25*, 520–528. [CrossRef]
12. Załuski, D.; Olech, M.; Kuźniewski, R.; Verpoorte, R.; Nowak, R.; Smolarz, H.D. LC-ESI-MS/MS profiling of phenolics from *Eleutherococcus* spp. *inflorescences*, structure-activity relationship as antioxidants, inhibitors of hyaluronidase and acetyl-cholinesterase. *Saudi Pharm. J.* **2017**, *25*, 734–743. [CrossRef]

13. Bączek, K. Accumulation of biologically active compounds in Eleuthero (*Eleutherococcus senticosus*/Rupr. et Maxim./Maxim.) grown in Poland. *Herba Pol.* **2009**, *55*, 7–13.
14. Gorshkova, T.A.; Salnikov, V.V.; Pogodina, N.M.; Chemikosova, S.B.; Yablokova, E.V.; Ulanov, A.V.; Ageeva, M.V.; Van Dam, J.E.G.; Lozovaya, V.V. Composition and distribution of cell wall phenolic compounds flax (*Linum usitatissimum* L.) stem tissues. *Ann. Bot.* **2000**, *85*, 477–486. [CrossRef]
15. Hutzler, P.; Fischbach, R.; Heller, W.; Jungblut, T.P.; Reuber, S.; Schmitz, R.; Veit, M.; Weissenböck, G.; Schnitzler, J.P. Tissue localization of phenolic compounds in plants by confocal laser scanning microscopy. *J. Exp. Bot.* **1998**, *49*, 953–965. [CrossRef]
16. Iglesias-Carres, L.; Mas-Capdevila, A.; Bravo, F.I.; Aragonès, G.; Muguerza, B.; Arola-Arnal, A. Optimization of a polyphenol extraction method for sweet orange pulp (*Citrus sinensis* L.) to identify phenolic compounds consumed from sweet oranges. *PLoS ONE* **2019**, *14*. [CrossRef]
17. Xi, W.; Lu, J.; Qun, J.; Jiao, B. Characterization of phenolic profile and antioxidant capacity of different fruit part from lemon (*Citrus limon* Burm.) cultivars. *J. Food Sci. Technol.* **2017**, *54*, 1108–1118. [CrossRef] [PubMed]
18. Ahn, J.; Um, M.Y.; Lee, H.; Jung, C.H.; Heo, S.H.; Ha, T.Y. Eleutheroside E, an active component of *Eleutherococcus senticosus*, ameliorates insulin resistance in type 2 diabetic db/db mice. *Evid. Based Complement. Altern. Med.* **2013**, *2013*. [CrossRef]
19. Bączek, K. Diversity of *Eleutherococcus* genus in respect of biologically active compounds accumulation. *Herba Pol.* **2015**, *60*, 34–43. [CrossRef]
20. Solomonova, E.; Trusov, N.; Nozdrina, T. Opportunities for using of eleutherococcuses fruits as a new food raw material. In Proceedings of the 1st International Symposium Innovations in Life Sciences (ISILS 2019), Belgorod, Russia, 10–11 October 2019.
21. Wang, Y.H.; Meng, Y.; Zhai, C.; Wang, M.; Avula, B.; Yuk, J.; Smith, K.M.; Isaac, G.; Khan, I.A. The chemical characterization of *Eleutherococcus senticosus* and Ci-wu-jia tea using UHPLC-UV-QTOF/MS. *Int. J. Mol. Sci.* **2019**, *20*, 475. [CrossRef]
22. Zhai, C.; Wang, M.; Raman, V.; Rehman, J.U.; Meng, Y.; Zhao, J.; Avula, B.; Wang, Y.H.; Tian, Z.; Khan, I.A.; et al. *Eleutherococcus senticosus* (Araliaceae) leaf morpho-anatomy, essential oil composition, and its biological activity against *Aedes aegypti* (Diptera: Culicidae). *J. Med. Entomol.* **2017**, *54*, 658–669. [CrossRef] [PubMed]
23. Richter, R.; Hanssen, H.P.; Koenig, W.A.; Koch, A. Essential oil composition of *Eleutherococcus senticosus* (Rupr. et Maxim.) Maxim roots. *J. Essent. Oil Res.* **2007**, *19*, 209–210. [CrossRef]
24. Strzemski, M.; Płachno, B.J.; Mazurek, B.; Kozłowska, W.; Sowa, I.; Lustofin, K.; Załuski, D.; Rydzik, Ł.; Szczepanek, D.; Sawicki, J.; et al. Morphological, anatomical, and phytochemical studies of *Carlina acaulis* L. cypsela. *Int. J. Mol. Sci.* **2020**, *21*, 9230. [CrossRef]
25. Blicharska, E.; Tatarczak-Michalewska, M.; Plazińska, A.; Plaziński, W.; Kowalska, A.; Madejska, A.; Szymańska-Chargot, M.; Sroka-Bartnicka, A.; Flieger, J. Solid-phase extraction using octadecyl-bonded silica modified with photosynthetic pigments from *Spinacia oleracea* L. for the preconcentration of lead(II) ions from aqueous samples. *J. Sep. Sci.* **2018**, *41*, 3129–3142. [CrossRef]
26. Sonestedt, E.; Ericson, U.; Gullberg, B.; Skog, K.; Olsson, H.; Wirfält, E. Do both heterocyclic amines and omega-6 polyunsaturated fatty acids contribute to the incidence of breast cancer in postmenopausal women of the Malmö diet and cancer cohort? *Int. J. Cancer* **2008**, *123*, 1637–1643. [CrossRef] [PubMed]
27. Grilo, E.C.; Costa, P.N.; Gurgel, C.S.S.; de Lima Beserra, A.F.; de Souza Almeida, F.N.; Dimenstein, R. α-tocopherol and gamma-tocopherol concentration in vegetable oils. *Food Sci. Technol.* **2014**, *34*, 379–385. [CrossRef]
28. Ergönül, P.G.; Köseoğlu, O. Changes in α-, β-, γ- And δ-tocopherol contents of mostly consumed vegetable oils during refining process. *CYTA J. Food* **2014**, *12*, 199–202. [CrossRef]
29. Madrigal, R.V.; Plattner, R.D.; Smith, C.R. *Carduus nigrescens* seed oil-A rich source of pentacyclic triterpenoids. *Lipids* **1975**, *10*, 208–213. [CrossRef]
30. Zielińska, S.; Dąbrowska, M.; Kozłowska, W.; Kalemba, D.; Abel, R.; Dryś, A.; Szumny, A.; Matkowski, A. Ontogenetic and trans-generational variation of essential oil composition in *Agastache rugosa*. *Ind. Crop. Prod.* **2017**, *97*, 612–619. [CrossRef]
31. Strzemski, M.; Wójciak-Kosior, M.; Sowa, I.; Rutkowska, E.; Szwerc, W.; Kocjan, R.; Latalski, M. *Carlina* species as a new source of bioactive pentacyclic triterpenes. *Ind. Crop. Prod.* **2016**, *94*. [CrossRef]

*molecules*

Article

Effect-Directed Profiling of 17 Different Fortified Plant Extracts by High-Performance Thin-Layer Chromatography Combined with Six Planar Assays and High-Resolution Mass Spectrometry

Gertrud E. Morlock [1,*], Julia Heil [1], Valérie Bardot [2], Loïc Lenoir [2], César Cotte [2] and Michel Dubourdeaux [2]

[1] TransMIT Center for Effect-Directed Analysis, and Chair of Food Science, Institute of Nutritional Science, Justus Liebig University Giessen, Heinrich-Buff-Ring 26–32, 35392 Giessen, Germany; Julia.Heil@ernaehrung.uni-giessen.de

[2] PiLeJe Industrie, Naturopôle Nutrition Santé, Les Tiolans, 03800 Saint-Bonnet-de-Rochefort, France; v.bardot@pileje.com (V.B.); l.lenoir@pileje.com (L.L.); c.cotte@pileje-industrie.com (C.C.); m.dubourdeaux@pileje.com (M.D.)

* Correspondence: gertrud.morlock@uni-giessen.de; Tel.: +49-641-9939141

Citation: Morlock, G.E.; Heil, J.; Bardot, V.; Lenoir, L.; Cotte, C.; Dubourdeaux, M. Effect-Directed Profiling of 17 Different Fortified Plant Extracts by High-Performance Thin-Layer Chromatography Combined with Six Planar Assays and High-Resolution Mass Spectrometry. *Molecules* **2021**, *26*, 1468. https://doi.org/10.3390/molecules26051468

Academic Editors: Young Hae Choi, Young Pyo Jang, Yuntao Dai and Luis Francisco Salomé-Abarca

Received: 23 January 2021
Accepted: 22 February 2021
Published: 8 March 2021

Publisher's Note: MDPI stays neutral with regard to jurisdictional claims in published maps and institutional affiliations.

Abstract: An effect-directed profiling method was developed to investigate 17 different fortified plant extracts for potential benefits. Six planar effect-directed assays were piezoelectrically sprayed on the samples separated side-by-side by high-performance thin-layer chromatography. Multipotent compounds with antibacterial, α-glucosidase, β-glucosidase, AChE, tyrosinase and/or β-glucuronidase-inhibiting effects were detected in most fortified plant extracts. A comparatively high level of antimicrobial activity was observed for *Eleutherococcus*, hops, grape pomace, passiflora, rosemary and *Eschscholzia*. Except in red vine, black radish and horse tail, strong enzyme inhibiting compounds were also detected. Most plants with anti-α-glucosidase activity also inhibited β-glucosidase. Green tea, lemon balm and rosemary were identified as multipotent plants. Their multipotent compound zones were characterized by high-resolution mass spectrometry to be catechins, rosmarinic acid, chlorogenic acid and gallic acid. The results pointed to antibacterial and enzymatic effects that were not yet known for plants such as *Eleutherococcus* and for compounds such as cynaratriol and caffeine. The nontarget effect-directed profiling with multi-imaging is of high benefit for routine inspections, as it provides comprehensive information on the quality and safety of the plant extracts with respect to the global production chain. In this study, it not only confirmed what was expected, but also identified multipotent plants and compounds, and revealed new bioactivity effects.

Keywords: enzyme inhibition assay; antibacterial assay; effect-directed analysis; bioassay; botanicals; health food

1. Introduction

The complexity of the composition of plants and plant extracts gives them properties that are different from and even superior to those of isolated compounds [1,2]. However, in terms of extracts, each extraction process is selective. Distinct secondary metabolites are more or less extracted depending on the conditions used [3,4]. This means that the phytochemical composition of the extracts differs from the plant totum, defined as the entire set of compounds contained in the plant part (used for extraction). Especially, the bioactive metabolites on which the effects in humans and animals depend can highly vary in quantity or be lost or degraded during the production of plant extracts [5,6]. It depends on the quality of the raw material, the applied preprocessing steps, the extraction method used and further treatments during production. Therefore, it becomes clear that extracts of the same plant currently on the market, have not the same quality and composition [6,7]. Information generally given on the production of the so-called botanicals on the market is sparse; almost all specifications lack crucial and more detailed information on the production and final composition of such commercially available plant extracts. This applies in particular to the most important bioactive ingredient profile, which has not yet been harmonized and matched to all relevant active ingredients.

The ipowder® technology was developed to extract as many compounds as possible from the original plant and to increase their content in the final product, solely made of the plant (no further additions such as carrier or filling agents, etc.) [8]. The essential steps of this patented process include contact between the plant material and a solvent in at least one extraction step, and the spray-drying of the extract rich in active compounds on a new batch of the same plant material. The resulting fortified plant material is crushed to form the ipowder®, used in dietary supplements and foods with added value. Benefits have been shown in terms of the enrichment of active compounds [5]. While the complex phytochemical composition of such plant extracts can be interesting with regard to biological and therapeutic effects, it makes their analysis and quality assessment extremely challenging. Currently, plant extracts are characterized by identifying and quantifying the well-known characteristic markers (usually one or two) of the selected plant species, whereas in such complex extracts other compounds may also contribute to the effects and overall product quality. Hence, high-performance thin-layer chromatography (HPTLC) hyphenated with planar effect-directed assays (EDA) was found to be a powerful tool for the rapid screening and assessment of such plant extracts in relation to the total bioactivity profile. Two of its many advantages are that it allows the analysis of many plant extracts in parallel in the same chromatographic run and imaging of the different types of the activity responses of the individual compounds without requiring their tedious individual isolation from the plant totum [6,9,10].

In this study, a generic effect-directed profiling was developed for the evaluation and comparison of potential beneficial effects of 17 different ipowder® extracts. HPTLC was combined with five different enzyme inhibition assays and an antibacterial bioassay. The instant bioluminescence of the Gram–negative *Aliivibrio fischeri* bacteria is widely used for the detection of bioactive or antibacterial compounds in environmental samples. The inhibition assays against α-glucosidase, β-glucosidase, acetylcholinesterase, β-glucuronidase and tyrosinase have proven their worth in the search for active plant extracts and secondary metabolites against widespread diseases, such as diabetes, Alzheimer's and Parkinson's diseases [6,11,12]. It was hypothesized that these fortified extracts would show pronounced effects in the six selected assays, new information would be obtained by the effect-directed profiles and such bioanalytical profiling would complement the actual analytical tools for product quality control. The important multipotent zones detected were characterized further by the straightforward elution to high-resolution mass spectrometry (HRMS).

2. Results and Discussion

2.1. HPTLC-UV/Vis/FLD Method Development

A generic nontarget effect-directed profiling was developed for the analysis of 18 plant extracts (IDs **1–18**, Table S1). Analytical methanolic extracts were prepared, as methanol was found to be a good compromise in polarity, based on our experience in other effect-directed studies [6]. No particular analytes were selected or targeted at this stage. The nontarget HPTLC method was developed with the aim of spreading the inherent compounds along the developing distance. Among the different mobile phases tested, two complementary mixtures were selected, i.e., ethyl acetate–toluene–formic acid–water, 16:4:3:2 (MP 1, more polar) versus cyclohexane–ethyl acetate–formic acid, 6:3.8:0.2 (MP 2, more apolar). The combination of the two widened the polarity range and enabled more comprehensive profiling. The physicochemical fingerprints of the inherent compounds in each methanolic extract were detected via multi-imaging at white light illumination (Vis), 254 nm (UV) and 366 nm (FLD). The vertical UV/Vis/FLD profiles (band pattern along each sample track) showed clear differences between the methanolic extracts (Figure 1, IDs **1–18**). This outcome was considered a good start to continue with the six effect-directed assays for the detection of the antibacterials and the inhibitors of AChE, α-glucosidase, β-glucosidase, β-glucuronidase, and tyrosinase.

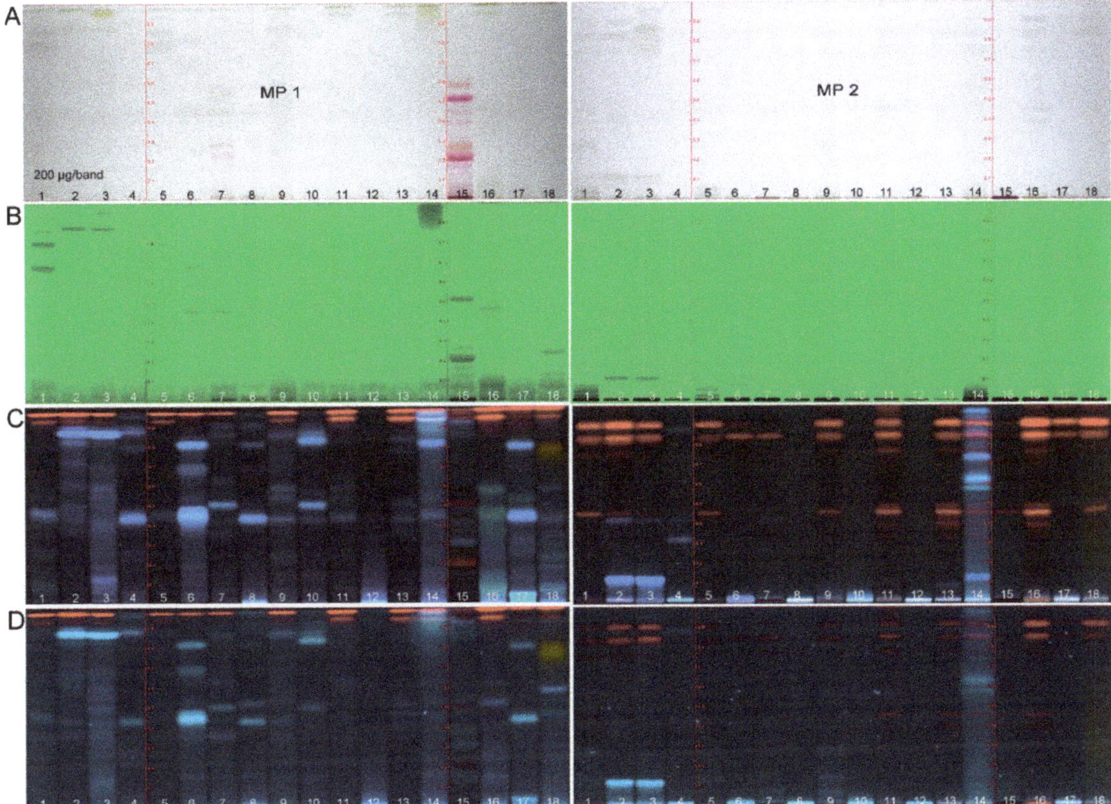

Figure 1. High-performance thin-layer chromatography (HPTLC) chromatograms of the 17 fortified (ipowder®) plant extracts (ID **1–4** and **6–18**, Table S1) and the nonfortified extract (ID **5**, Table S1), all 200 µg each on the HPTLC plate silica gel 60 F_{254} MS grade with the mobile phase mixtures MP 1 (ethyl acetate–toluene–formic acid–water 16:4:3:2, $v/v/v/v$) and MP 2 (cyclohexane–ethyl acetate–formic acid 6:3.8:0.2, $v/v/v$) at white light illumination (**A**), UV 254 nm (**B**), FLD 366 nm before (**C**) and after neutralization (**D**).

2.2. Development of the Effect-Directed Profiling

A neutralization of residual acidic traces on the chromatogram by applying a buffer solution was relevant for both mobile phase mixtures containing formic acid. The comparison of the FLD images before versus after the neutralization step (Figure 1C versus Figure 1D) showed that the additional buffer salt load on the plate reduced the fluorescence intensity of most zones. In addition, it altered the fluorescence nuance and intensity of some zones (e.g., IDs **16–18** in MP 1). Such fluorescence signal shift can be explained by the pH change (from acidic to neutral) or inorganic traces in the buffer solution. MS grade HPTLC plates have been available on the market for some years. These plates are known to have a lower layer thickness and were therefore investigated for their applicability in the effect-directed profiling. The rule of thumb was that half the volume of (expensive) assay solution was needed as with normal plates. However, the correct volume also depended on the humidity of the day, and thus layer. On humid days, the application of even less assay solution volume was recommended to avoid an excessive liquid layer. The more water was preadsorbed from the ambient air, the less assay volume could be adsorbed and penetrate to the deeper layer. On days with higher humidity (>50%), the plate was therefore dried (e.g., 40 °C, 1 min). The subsequent assay application started with the *A. fischeri* bioassay.

This bioassay is frequently used [13], and gives a first overview on sample compounds with influence on the energy metabolism of the nonpathogenic Gram–negative bacteria. Based on our experience, this bioassay often indicates most bioactive compounds present, if compared to other assays. Thus, this instant bioassay is helpful for a cost-efficient selection of the first effect-relevant sample amount to be applied. For applied 200 µg extract per band, antibacterial compound zones were detected. This was a good start for this and all the other assays, although later adjustment or fine-tuning may be required.

2.2.1. Antibacterial Activity

In the Gram–negative *A. fischeri* bioautogram, active compounds can have a positive or negative influence on the bioluminescence of the bacteria, resulting in bright or dark zones. Of each plant extract, 200 µg were applied which allowed a direct comparison of the antibacterial activity between the different samples (Table 1). Many antibacterial bioluminescence-reducing compound zones were observed in 15 out of the 17 fortified plant extracts (Figure 2A, depicted as greyscale image, IDs **1**–**18** in Table S1, respective UV/Vis images in Figure S1). In particular, the fortified extracts of *Eleutherococcus* (**4**), grape pomace (**15**), passiflora (**16**), artichoke (**17**) and *Eschscholzia* (**18**) had a strong Gram–negative antibacterial activity. Strong antibacterial zones were also detected for the fortified green tea extract (**1**), which were 2–3-fold stronger, when visually compared to the nonfortified green tea extract (**5**). This was in good agreement with the specified drug-to-extract ratio of 2:1 for the production of the fortified green tea extract (Table S1). An increased content of chemical markers was recently reported for ipowder® extracts [5], which was also confirmed by this comparison (ID **5** versus **1**). In both green tea samples, two dominant antibacterial zones were evident at hR_F 80 and 89 (the latter marked as zone **V**) in the stronger eluting MP 1 system, which were located at hR_F 5 and 11 (zone **V**) in the weaker eluting MP 2 system. This illustrates clearly that the polarity range was complementary to each other, and thus selected well for a comprehensive profiling. Comparatively fewer and also less intense antibacterial zones were observed in the fortified extracts of lemon balm (**2**), rosemary (**3**), red vine (**7**), meadowsweet (**9**) and *Echinacea* (**10**). However, additional apolar antibacterial compounds of especially hops (**14**) and also rosemary (**3**), and to some extent for valerian (**8**) and blackcurrant (**11**) were first revealed with the apolar MP 2 system. Dark zones, which eluted almost with the solvent front of the stronger eluting MP 1, were now well separated and spread along the developing distance. In contrast to these positive findings and considering the same sample amount applied, black radish (**12**) and horse tail (**13**) were the only two fortified extracts with no obvious antibacterial activity in both investigated polarity ranges. It is worth noting that an enhancement of the bacterial bioluminescence was observed for more apolar compound zones in the MP 2 bioautogram for yerba mate (**6**) at hR_F 83 and meadowsweet (**9**) at hR_F 77 as well as for a few further plants such as lemon balm (**2**) and rosemary (**3**). The bright zones may highlight a prebiotic activity. Recently, some fatty acids have been proven to enhance the bioluminescence of the Gram–negative *A. fischeri* bacteria [7].

While the antibacterial effect is well known for some of these plants, it is much less described for others. Among the plants with the highest activity against *A. fischeri* bacteria, the hop plant is recognized as an excellent source of antibacterial compounds [14,15]. This is confirmed by our study in view of the number and intensity of antibacterial zones observed for hops (**14**) in the MP 2 bioautogram. The high antibacterial activity of numerous compounds in *Eleutherococcus* root (**4**) is shown for the first time. Data on the antibacterial effect of this plant are rare [16,17]. The strong antibacterial effect observed for *Passiflora incarnata* (**16**), is only supported by a reported effect against *Helicobacter pylori* [18]. As for the observed antibacterial activity of *Eschscholzia* (**18**), its alkaloids have demonstrated antifungal [19] and some even antibacterial potential [20,21]. In accordance with our findings is also the well documented antibacterial activity of green tea [22,23] and grape pomace [24]. Extracts and essential oils of lemon balm [25] and rosemary [26] have also proven antibacterial activity, which confirms our results. A few more studies confirm the outcome of our profiling, which have shown antibacterial effects for artichoke [27], meadowsweet [28,29], red vine leaf [30], *Echinacea* species [31], blackcurrant [32], yerba mate [33], and valerian species, whereby the only study found for *Valeriana officinalis* was on the essential oil of its root [34].

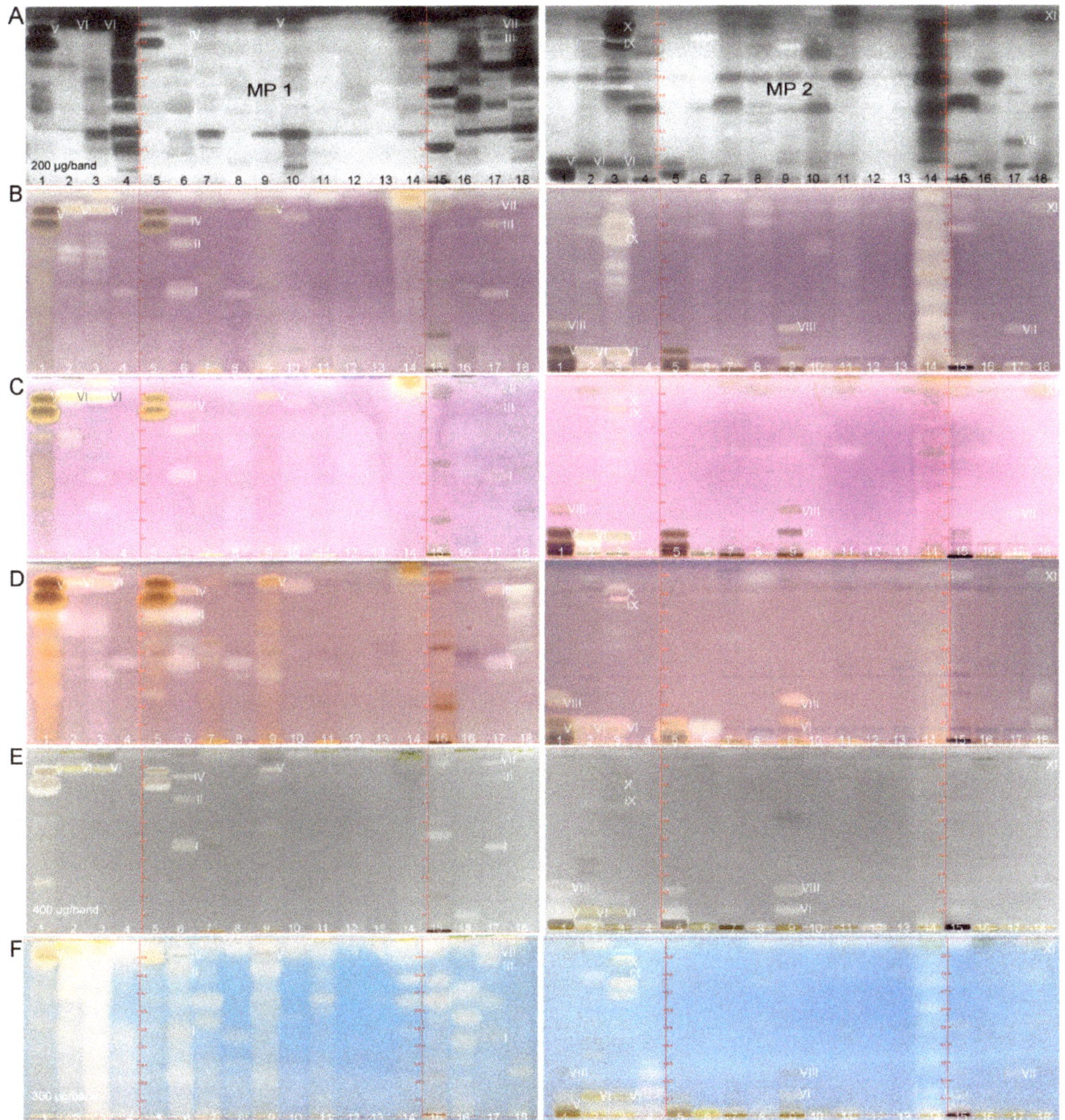

Figure 2. HPTLC autograms of the 17 fortified (ipowder®) plant extracts (ID **1–4** and **6–18**, Table S1) and the nonfortified extract (ID **5**, Table S1) separated as in Figure 1 and detected after the *A. fischeri* bioassay ((**A**), bioluminescence as grey-scale image) and α-glucosidase (**B**), β-glucosidase (**C**), AChE ((**D**); **A–D** 200 µg each), tyrosinase ((**E**), 400 µg each) and β-glucuronidase inhibition assays ((**F**), 300 µg each; **B–F** at white light illumination); multipotent zones **I–XI** characterized by HPTLC–HRMS in Figures 3 and 4.

Table 1. Summary on the effect-directed profiling of the 17 fortified (ipowder®) plant extracts (ID **1–4** and **6–18**, Table S1) and the nonfortified extract (ID **5**, Table S1), showing weak (+), moderate (++), high (+++), very high (++++) and no activity (-) in the respective assay.

	Assay	*A. fischeri*	α-Glucosidase	β-Glucosidase	AChE	Tyrosinase	β-Glucuronidase
ID	**Plant**	Amount applied (µg/band): 200				400	300
1	Green tea	+++	++++	+++	+++	++	+++
2	Lemon balm	++	++	+++	++	++	+++
3	Rosemary	++++	+++	+++	++	++	++++
4	*Eleutherococcus*	++++	+	-	+	-	++
5	Green tea	++	++	++	++	++	++
6	Yerba mate	+	++	+++	++	++	+++
7	Red vine	++	+	-	-	+	++
8	Valerian	++	+	-	+	+	+
9	Meadowsweet	++	++	++	++	+	+++
10	Echinacea	+++	+	+	+	-	-
11	Blackcurrant	++	+	+	-	-	+
12	Black radish	-	-	-	-	+	-
13	Horse tail	-	-	-	-	-	+
14	Hops	++++	++++	++	-	-	++
15	Grape pomace	++++	+	+	+	+	++
16	Passiflora	++++	+	+	-	+	++
17	Artichoke	+++	++	++	+	+	++
18	*Eschscholzia*	++++	+	+	++	+	+

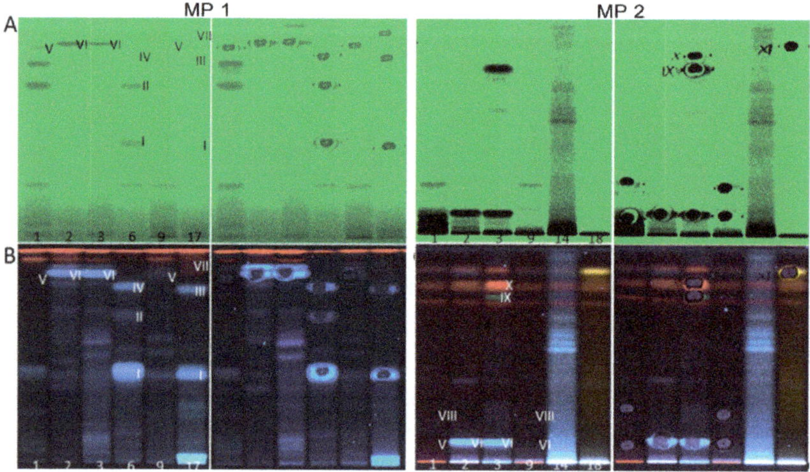

Figure 3. HPTLC chromatograms of eight selected fortified plant extracts (same chromatographic conditions as in Figure 1) with the multipotent zones **I–XI** marked at UV 254 nm (**A**) and FLD 366 nm (**B**) before and after the HPTLC–HRMS recording (with elution head imprint).

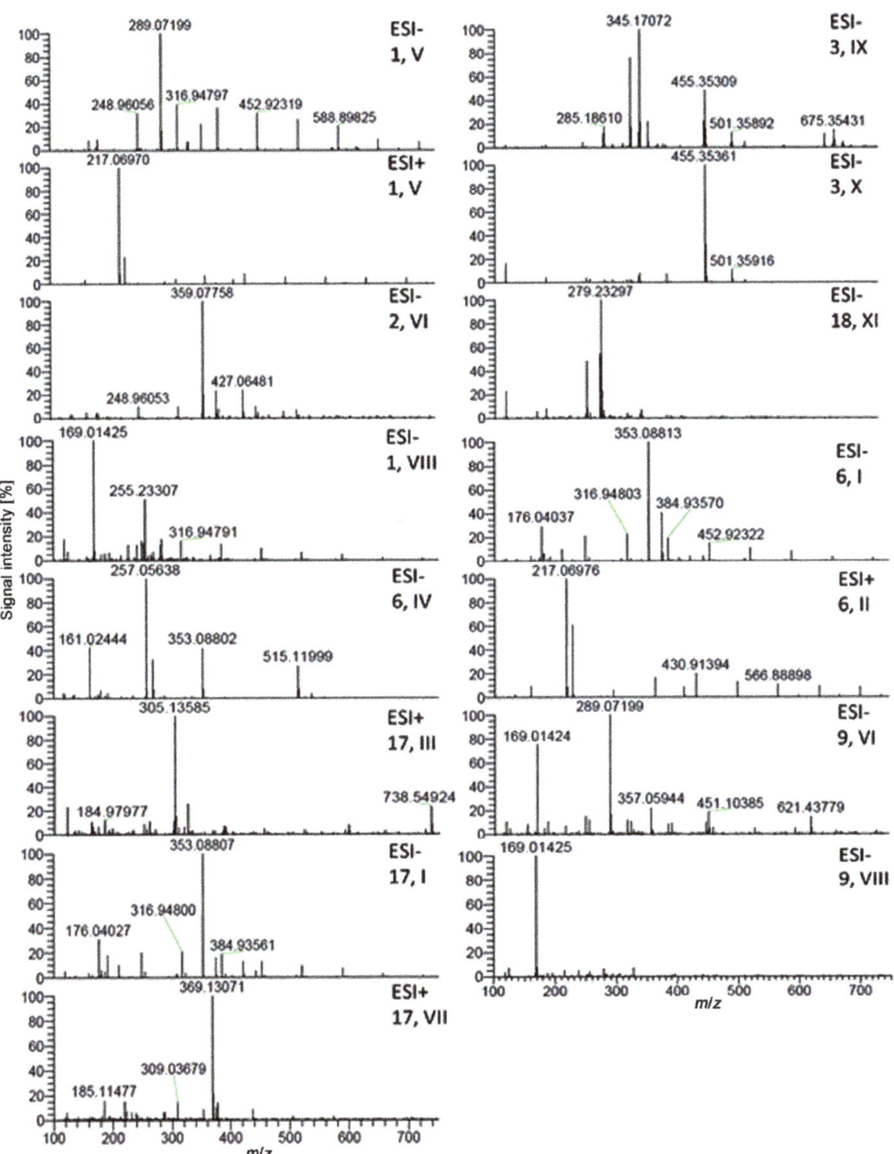

Figure 4. HPTLC–HESI–HRMS full scan spectra of the eleven multipotent zones **I–XI** in the fortified plant extracts (same chromatographic conditions as in Figure 1).

2.2.2. Enzyme Inhibiting Activity

In the five different enzyme inhibition assay autograms (Figure 2B–F), enzyme inhibitors appeared as colorless (white or yellowish) zones on a violet, grey or blue background. A pronounced enzyme inhibition was observed for the majority of the 18 tested plant extracts. For the same plant extract, the brightest inhibitory zones were often detected in several enzyme assays, and thus found to be multipotent. In contrast, red vine (**7**), black radish (**12**) and horse tail (**13**) were three plant examples with comparatively very weak enzyme inhibition profiles across all five enzyme assays performed. The comparison of the fortified green tea extract (**1**) versus nonfortified extract (**5**) confirmed that the same enzyme inhibiting zones were present. However, in each case, a stronger effect was evident for the fortified extract across all five inhibition assays. This shows the potential of this bioanalytical profiling for quantitative studies [7]. This stronger effect of the fortified extract observed in this profiling was also consistent with target analysis in that a higher concentration of chemical markers was found in the fortified extracts [5].

In the α-glucosidase inhibition autograms, the bright zones indicated α-glucosidase inhibitors, and thus potential antihyperglycemic components in the 18 tested plant extracts (200 µg each; Figure 2B; Table 1, respective UV/FLD images in Figure S2). Inhibiting compound zones were detected for all plants but black radish (**12**) and horse tail (**13**). Interestingly, most of the plant ingredients with anti-α-glucosidase activity were also able to inhibit β-glucosidase (200 µg/band; Figure 2C). The horizontal pattern recognition (along the so-called substance windows in HPTLC) showed that several intense inhibition zones observed in the α-glucosidase assay were also detected in the β-glucosidase assay, indicating the same main inhibitors. However, a different pattern or less intense β-glucosidase inhibitory zones were detected for example for rosemary (**3**) and hops (**14**) in the MP 2 autograms (Figure 2C versus Figure 2B). Depending on the plant, the selective α- or β-glucosidase inhibitions by compounds were similar, but also different. Many studies have been conducted on the anti-α-glucosidase activity of plant extracts and their compounds, while comparatively less literature is available on the also important anti-β-glucosidase effects [35,36]. In the two glucosidase autograms, the green tea profiles (ID **1** and **5**) showed two very intense (more brownish) active zones. These were identical to the two previously discussed antibacterial zones at hR_F 80 and 89 (zone **V**, MP 1) and hR_F 5 and 11 (zone **V**, MP 2). Studies have shown that green tea contains potent α-glucosidase inhibitors [37–39]. To the best of our knowledge, there is no data showing an effect of green tea on β-glucosidase in the literature but the present study. Meanwhile, this effect was studied and assigned to the catechins in green tea [6]. Zones at a similar hR_F value were identified in both glucosidase assays for meadowsweet (**9**) which has known effects on α-glucosidase [40], but none on β-glucosidase. Hence, the anti-β-glucosidase activity of meadowsweet is evident for the first time. An intense inhibiting compound zone at hR_F 90 (zone **VI**) was observed for lemon balm (**2**) and rosemary (**3**) in both MP 1 autograms. The same intense zone was evident at hR_F 11 (zone **VI**) in both MP 2 autograms. Two further less intense zones were observed at hR_F 62 and 67 (MP 1) for lemon balm, and much weaker in the response for rosemary. However, in both MP 2 autograms of rosemary, two further intense α-glucosidase and comparatively weaker β-glucosidase inhibitors were revealed at hR_F 73 (zone **IX**) and 83 (zone **X**); both eluted with the solvent front in the MP 1 system. The inhibitory effects of lemon balm and rosemary on α-glucosidase have been previously observed [12,41], whereas to the best of our knowledge, their anti-β-glucosidase activity is reported here for the first time. The profiles of yerba mate (**6**) and artichoke (**17**) each show an inhibitory zone at hR_F 45 (zone **I**, MP 1) in both glucosidase assays. Two further inhibiting zones were revealed at hR_F 71 (zone **II**) and 85 (zone **IV**) for yerba mate (**6**), and at hR_F 82 (zone **III**) and 92 (zone **VII**) for artichoke (**17**) in both MP 1 autograms. While the α-glucosidase inhibitory activity of yerba mate and artichoke has been previously shown [42–44], their β-glucosidase inhibition has not been reported so far. Hops (**14**) showed the strongest α-glucosidase inhibition profile across the tested 17 fortified extracts in the MP 2 autogram, and thus, hops were found to be a comparatively strong α-glucosidase inhibitor. A previous study has demonstrated the α-glucosidase inhibitory activity of hops [45]. The comparatively weaker β-glucosidase inhibition has not been reported so far.

AChE inhibitory zones in the applied extracts (200 µg/band; Figure 2D; Table 1) were especially observed for green tea (**1** and **5**), lemon balm (**2**), rosemary (**3**), yerba mate (**6**), meadowsweet (**9**), and *Eschscholzia* (**18**). Meanwhile, the AChE inhibitory effect was studied and assigned to the catechins in green tea [6]. For *Eschscholzia*, three active zones, at hR_F 52, 71 and 84, were evident in the MP 1 autogram and three further, at hR_F 10, 26 and 90, in the

MP 2 autogram. In *Eschscholzia*, the AChE inhibiting activity was strongest among the five enzyme assays examined. Further weaker AChE inhibiting zones were detected for valerian (**8**), *Echinacea* (**10**), grape pomace (**15**) and artichoke (**17**). Green tea [46], lemon balm [47], rosemary [48], yerba mate [49], meadowsweet [50], artichoke [44] and grape pomace [51] were reported to inhibit AChE, which confirms our profiling results. For *Eschscholzia*, no prominent AChE inhibition was observed for 14 isolated alkaloids [52]; however, using a simple methanol extraction of the plant-fortified extract in our study, an intense AChE inhibiting zone at hR_F 84 (MP 1) and five weaker zones (MP 1/2) were observed.

For twice the sample amount applied (400 µg/band; Figure 2E; Table 1), the tyrosinase inhibitory activity was considered to be comparatively lower, if compared to the previous assays. Distinct tyrosinase inhibition zones were detected for green tea (**1** and **5**), lemon balm (**2**), rosemary (**3**), yerba mate (**6**), meadowsweet (**9**), grape pomace (**15**) and artichoke (**17**). Antityrosinase activity was reported for green tea [53], lemon balm [54], rosemary essential oil [55], meadowsweet [50] and grape pomace [56]. No proof of the tyrosinase inhibition was found in the literature for yerba mate, artichoke and *Eschscholzia*. In this profiling study on fortified extracts, tyrosinase inhibitors were detected for the first time. For comparatively weak responses (e.g., for *Eschscholzia*), higher sample amounts should be applied in future studies. Minor compounds could be better detected when higher sample amounts were applied on the plate, but the autogram was then overloaded for the most active samples.

For the 1.5-fold extract amount applied (300 µg/band; Figure 2F; Table 1), the β-glucuronidase was inhibited by almost all fortified plant extracts, as evident especially in the stronger eluting MP 1 autogram. The β-glucuronidase inhibition was comparatively more pronounced (the profile was already overloaded by the strong signals) for rosemary (**3**), a bit less intense for lemon balm (**2**) and then green tea (**1**), yerba mate (**6**) and meadowsweet (**9**). The inhibitory activity was more moderate for the other plants, but for *Echinacea* (**10**), black radish (**12**) and horse tail (**13**) an effect was not or hardly observed. In rats, the administration of green tea extract reduced the activity of bacterial β-glucuronidase [57], which confirms our profiling result. Blackcurrant (**11**) was also shown to inhibit this enzyme [58], which also showed a moderate β-glucuronidase inhibition in our study. For the other plants tested, a β-glucuronidase inhibition has not been reported so far. It is first recognized and illustrated by this effect-directed profiling.

2.3. Characterization of Active Zones I–XI by HPTLC–HESI–HRMS

Eleven prominent multipotent compound zones detected in the different assays (**I–XI**, numbered from polar to apolar) were further characterized by HPTLC–HESI–HRMS. Therefore, six plant extract samples were applied twice (two sets) on the same plate. After plate cut, one plate part was subjected to the bioassay. On the other plate part (Figure 3), the eleven zones were marked according to their bioautogram responses. The high-resolution mass spectra of these zones were recorded and evaluated (Figure 4).

The exact mass signals found were assigned to molecular formulas and tentative candidates in agreement with the literature data (Table 2). In green tea (**1**) and meadowsweet (**9**), the zone **V** at hR_F 89 (MP 1) and 11 (MP 2) was a coelution of catechins, whereby the most pronounced mass signal (base peak) was obtained at m/z 289.0720 $[M - H]^-$, tentatively assigned as (epi)catechin ($C_{15}H_{14}O_6$). In the MP 1 system, the different catechins in green tea also coeluted in two zones—the zone **V** and the active zone below (hR_F 80). Catechins in green tea and meadowsweet were therefore in part responsible for the activities of both plants. The antibacterial activity of green tea is ascribed in particular to galloylated catechins such as (−)-epicatechin gallate and (−)-epigallocatechin gallate [6,22,23].

Table 2. HPTLC–HESI–HRMS analysis of the multipotent zones **I–XI** and their tentative assignment, confirmed by literature data and effect-directed profiling via the (**a**) *A. fischeri*, (**b**) α-glucosidase, (**c**) β-glucosidase, (**d**) AChE, (**e**) tyrosinase and (**f**) β-glucuronidase assays (active +; not active −).

ID	Plant Extract	Band	Mass Signal, *m/z*		Molecular Formula	Mass Error Δ ppm	Tentative Assignment	Compound Activity Found in Assay Literature						
								a	b	c	d	e	f	
1 + 5	Green tea	V	289.07199	[M − H]⁻	$C_{15}H_{14}O_6$	2.65	(epi)catechin	+	+	+	+	+	+	[22,23,53,59–62]
		VIII	169.01419	[M − H]⁻	$C_7H_6O_5$	0.54	gallic acid	−	+	+	+	+	+	[63–66]
2	Lemon balm	VI	359.07754	[M − H]⁻	$C_{18}H_{16}O_8$	2.36	rosmarinic acid	+	+	+	+	+	+	[12,40,41,67–73]
			383.07388	[M + Na]⁺										
			405.05585	[M + 2Na − H]⁺										
			381.05942	[M + Na − 2H]⁻										
3	Rosemary	VI	359.07754	[M − H]⁻	$C_{18}H_{16}O_8$		rosmarinic acid	+	+	+	+	+	+	[12,40,41,67–73]
			381.05942	[M − 2H + Na]⁻										
		IX	329.17612	[M − H]⁻	$C_{20}H_{26}O_4$	0.81	carnosol	+	+	+	+	+	+	[71,74–77]
			345.17087	[M + O − H]⁻										
			359.18661	[M + O + CH₂ − H]⁻										
			455.35339	coeluted										
		X	455.35339	[M − H]⁻	$C_{30}H_{48}O_3$	0.70	oleanolic/ursolic acid	+	+	+	+	+	+	[75,78–84]
6	Yerba mate	I	353.08807	[M − H]⁻	$C_{16}H_{18}O_9$	0.81	chlorogenic acid	+	+	+	+	+	+	[85–91]
			375.06992	[M − 2H + Na]⁻										
		II	217.06967	[M + Na]⁺	$C_8H_{10}O_2N_4$	0.48	caffeine	+	+	+	+	+	+	[92,93]
			411.15007	[2M + Na]⁺										
		IV	257.05630	[M − 2H]²⁻	$C_{25}H_{24}O_{12}$									
			353.08801	[M − C₉H₆O₃ − H]⁻	$C_{16}H_{18}O_9$		fragment: chlorogenic acid							
			515.11999	[M − H]⁻	$C_{25}H_{24}O_{12}$	1.11	dicaffeoylquinic acid	+	+	+	+	+	+	[94–98]
			537.10193	[M + Na − 2H]⁻										
9	Meadowsweet	V	289.07199	[M − H]⁻	$C_{16}H_9O_2N_4$	1.97								
					$C_{15}H_{14}O_6$	2.65	(epi)catechin	+	+	+	+	+	+	[22,23,53,59–62]
		VIII	169.01425	[M − H]⁻	$C_7H_6O_5$		gallic acid	−	+	+	+	+	+	[63–66]
17	Artichoke	I	353.08807	[M − H]⁻	$C_{16}H_{18}O_9$	0.81	chlorogenic acid	+	+	+	+	+	+	[85–91]
		III	307.06430	[M − 2H]²⁻										
			615.13568	[M − H]⁻										
			305.13589	[M + Na]⁺	$C_{15}H_{22}O_5$	2.11	cynaratriol	+	+	+	+	+	+	
		VII	369.13070	[M + H]⁺	$C_{17}H_{20}O_9$	34 *	3-*O*-feruloyl quinic acid	+	+	+	−	+	+	
18	*Eschscholzia*	XI	279.23296 \\ 277.21747	[M − H]⁻ \\ [M − H]⁻	$C_{18}H_{32}O_2$ \\ $C_{18}H_{30}O_2$	1.99	linoleic acid and linolenic acid **	+	+	+	+	+	+	[99,100]

* Unsure assignment due to high mass error; ** or fragment of *m/z* 342.17 or 356.18, both [M + H]⁺.

The reported antibacterial activity of meadowsweet has not been attributed to a structural class or particular molecule [28,101,102]. Of note is the fact that meadowsweet flowers, the plant part used to make the fortified extract, were shown to contain catechins [40], which confirms our assignment.

For zone **VIII** at hR_F 100 (MP 1) and 24 (MP 2) in green tea (**1**) and meadowsweet (**9**), the base peak at *m/z* 169.0142 [M − H]⁻ was assigned to gallic acid ($C_7H_6O_5$). Opposite to zone **V** (catechins), the gallic acid zone **VIII** showed no antibacterial effect against Gram-negative bacteria. However, both zones (**V**, catechins and **VIII**, gallic acid) were identified as potent inhibitors of α-glucosidase, β-glucosidase, AChE and tyrosinase. The respective β-glucuronidase inhibition was comparatively low for zone **V**. A more brownish instead of bright inhibition zone was revealed for the catechins (zone **V** and zone below). This was caused by the substrate for high catechin amounts, which was proven in detail in another study and then substituted by another substrate [6]. Our results were confirmed by reports on catechins and derivatives to be responsible for the inhibition of α-glucosidase [59] and β-glucosidase [60]. Gallic acid was reported as inhibitor of α-glucosidase [63,64]. The β-glucosidase inhibition was, to the best of our knowledge, first reported here. Catechins and gallic acid were reported to be AChE and tyrosinase inhibitors [53,61,65,66]. The β-glucuronidase inhibition was reported for catechins [62]. Gallic acid has no reported effect on this enzyme, which complies with our result. For zone **VI** at hR_F 90 (MP 1) and 11 (MP 2) in lemon balm (**2**) and rosemary (**3**), the base peak at m/z 359.0775 $[M - H]^-$ was assigned to rosmarinic acid ($C_{18}H_{16}O_8$), which was reported as a marker compound in both plants [103]. The rosmarinic acid zone **VI** was clearly detected in all six assays.

Zone **IX** at hR_F 100 (MP 1) and 73 (MP 2) in rosemary (**3**) showed a mass signal at m/z 329.1761 $[M - H]^-$ and as base peak its oxidized form at m/z 345.1709 $[M + O - H]^-$. These mass signals were assigned to carnosol ($C_{20}H_{26}O_4$). The zone **X** at hR_F 100 (MP 1) and 83 (MP 2) showed a base peak at m/z 455.3534 $[M - H]^-$ and was assigned to oleanolic/ursolic acid ($C_{30}H_{48}O_3$). The carnosol zone **IX** and oleanolic/ursolic acid zone **X** were very intense in the antibacterial bioassay as well as α-glucosidase and β-glucuronidase inhibition assays, whereas both were much less intense inhibitors in the other assays. Rosmarinic, oleanolic and ursolic acids as well as carnosol have been identified as antibacterial agents [67–71,74,75,78,104], which confirms our results. The inhibitory effects of rosmarinic acid (pure and from rosemary and lemon balm) have been observed on α-glucosidase [12,40,41], which is also in agreement with the intense inhibition observed in our study. Oleanolic/ursolic acid are known α-glucosidase inhibitors [79,80], but in our study, the effect of rosemary is directly related to the presence of these compounds for the first time. In line with our results, carnosol from rosemary was identified as an α-glucosidase inhibitor; its oral administration significantly reduced the postprandial blood glucose levels of normal mice [76]. Rosmarinic, oleanolic and ursolic acids are also known inhibitors of AChE [72,81,82,104] and tyrosinase [73,83,84,104], which was confirmed by our results. Their ability to inhibit β-glucosidase and β-glucuronidase was not known so far. The direct effect of carnosol on AChE had not been shown so far, whereas its effect on tyrosinase has been reported [77].

Zones **I**, **II** and **IV** (hR_F 45, 71 and 85 in MP 1, respectively; all hR_F 0 in MP 2) observed in yerba mate (**6**) were assigned to chlorogenic acid ($C_{16}H_{18}O_9$, base peak at m/z 353.0881 $[M - H]^-$), caffeine ($C_8H_{10}N_4O_2$, base peak at m/z 217.0697 $[M + Na]^+$) and dicaffeoylquinic acid ($C_{25}H_{24}O_{12}$, base peak at m/z 257.0563 $[M - 2H]^{2-}$), respectively. The chlorogenic acid zone **I** was also observed in artichoke (**17**), in addition to zone **III** (hR_F 83 in MP 1 and hR_F 0 in MP 2) assigned to cynaratriol ($C_{15}H_{22}O_5$, base peak at m/z 305.1359 $[M + Na]^+$) and zone **VII** (hR_F 92 in MP 1 and hR_F 24 in MP 2) assigned as 3-*O*-feruloyl quinic acid ($C_{17}H_{20}O_9$, base peak at m/z 369.1307 $[M + H]^+$). These zones were detected in all five enzyme assays, except for the lack of AChE inhibition of 3-*O*-feruloyl quinic acid (zone **VII**). In previous studies, chlorogenic acid showed antibacterial activity against Gram–negative bacteria [85] and caffeoylquinic acids were in the top 10 identified antibacterial compounds of yerba mate [94], whereas the antibacterial effect of caffeine is first reported here. The previously reported antibacterial effects of artichoke were ascribed to phenolic compounds without further specification [27,105]. Chlorogenic and 3-*O*-feruloyl quinic acids are phenolic compounds, but not cynaratriol, a known sesquiterpene of artichoke, whose antibacterial activity has not yet been reported. Chlorogenic acid and dicaffeoylquinic acid have known anti-α-glucosidase [86,87,95], anti-AChE [88,89,96] and antityrosinase [90,97] activities. No information about a potential effect of caffeine on α-glucosidase, which seems to be new information, was found in the literature. The reported inhibition of AChE [92] and tyrosinase [93] by caffeine was confirmed by our results. The β-glucosidase inhibiting activity for these compounds and the cynaratriol and 3-*O*-feruloyl quinic acid inhibitory effects on the different enzymes are described for the first time. With regard to the β-glucuronidase inhibitory activity, only chlorogenic acid [91] and caffeoylquinic acid [98] have been reported to have an inhibitory effect on this enzyme so far.

Zone **XI** in *Eschscholzia* (**18**) was assigned to coeluting fatty acids, i.e., linoleic acid ($C_{18}H_{32}O_2$, base peak at *m/z* 279.2330 $[M - H]^-$) and linolenic acid ($C_{18}H_{30}O_2$, base peak at *m/z* 277.2175 $[M - H]^-$). This linoleic/linolenic acid zone was active in the tested assays, which is in accordance with a recent study [7]. Our results were confirmed by the reported antibacterial [99] and α-glucosidase inhibiting effects [100] of linoleic and linolenic acids.

2.4. Outlook

It is worth mentioning that the developed effect-directed profiling can also be used for quantitative studies based on the enzymatic or biological response or for the calculation of IC_{50} and MIC values, as recently shown in other studies [106–109]. Planar assays have proven to be just as reliable as current microtiter plate assays, but additionally offer all the advantages of separating a complex sample [110,111]. HPTLC-EDA workflows are highly efficient in time and costs (by a factor of 7 less costs [111]), robust with regard to matrix (minimalistic sample preparation) and reveal new options, such as the simultaneous detection and differentiation of agonistic and antagonistic effects of individual compounds in a mixture [112].

3. Materials and Methods

3.1. Chemicals and Reagents

Fast Blue B salt (95%) was purchased from MP Biomedicals, Eschwege, Germany. We obtained 5-Bromo-4-chloro-3-indolyl β-D-glucuronide sodium salt (X-Gluc, ≥98%) from Carbosynth, Compton–Berkshire, UK. α-Glucosidase (from *Saccharomyces cerevisiae*, 1000 U/vial), acarbose (for pharm.), tyrosinase (from mushroom, ≥1000 U/mg, 25 kU/vial), β-glucuronidase (from *Escherichia coli*, 5000 U/vial), imidazole (≥99.5%), acetylcholinesterase (AChE, from *Electrophorus electricus*, ≥245 U/mg solid, 10 kU/vial), 3-[(3-cholamidopropyl)-dimethylammonio]-1-propanesulfonate (CHAPS, ≥98%), di-ammoniumhydrogen phosphate (99%), pepton from casein (tryptone, for microbio.), cyclohexane (HPLC grade), sodium acetate, monopotassium phosphate, magnesium sulfate heptahydrate and sodium chloride (all p. a. quality and waterfree) were obtained from Fluka or Sigma–Aldrich, Steinheim, Germany. *Aliivibrio fischeri* bacteria (no. 7151) were from the German Collection of Microorganisms and Cell Cultures, Düsseldorf, Germany. 2-Naphthyl-α-D-glucopyranoside (95%) was delivered by Fluorochem, Hadfield Derbyshire, UK. Bovine serum albumin (BSA, fraction V, ≥98%), acetic acid (Ph. Eur.), dipotassium hydrogen phosphate (waterfree, ≥99%), sodium dihydrogen phosphate monohydrate (≥98%), glycerol (86%), potassium dihydrogen phosphate (99%), sodium hydroxide (≥98%), disodium hydrogen phosphate (≥99%), polyethylenglycol (PEG) 8000, koji acid (>98%), acetic acid (Ph. Eur.), hydrochloric acid (HCl, 37%, purest), tris(hydroxymethyl)aminomethane (Tris, ≥99.9%) were purchased from Carl Roth, Karlsruhe, Germany. Methanol, ethanol and cyclohexane (all HPLC grade) as well as formic acid (99%, LC–MS) were obtained from vwr, Darmstadt, Germany. Toluene (HPLC grade) was from Promochem, LGC Standards, Wesel, Germany. HPTLC plates silica gel 60 F_{254} MS grade (20 cm × 10 cm, Lot HX69361434) and citric acid monohydrate were obtained from Merck, Darmstadt, Germany. 2-Naphthyl-β-D-glucopyranoside (95%) and β-glucosidase (from almonds, 3040 U/mg) was purchased from ABCR, Karlsruhe, Germany. 1-Naphthyl acetate (≥98%) was delivered by AppliChem, Darmstadt, Germany. D-Saccharolactone (≥98%) and (2S)-2-amino-3-(3,4-dihydroxyphenyl)propanoic acid (L-DOPA, 96%) was from Santa Cruz Biotechnology, Santa Cruz, CA. Ethyl acetate (≥99.8%) and yeast extract powder (Chemsolute, for microbiol.) were purchased from Th. Geyer, Renningen, Germany. Bidistilled water was prepared using the Destamat Bi 18 E system of Heraeus, Hanau, Germany. For incubation, a polypropylene box (27 cm × 16 cm × 10 cm, KIS, ABM, Wolframs–Eschenbach, Germany) was covered by wetted filter papers at the bottom and sides.

3.2. Production of Fortified Plant Powders

The harvest of each plant was performed at an optimal period as specified in Table S1. From the respective field, each plant was collected and either frozen (grape pomace) or dried within 12 h (roots were washed before treatment to remove soil). Thereby, each plant material was specified (Table S1) and agricultural treatments were documented (data not shown). Dried plants were bagged into paper or mesh bags, frozen grape pomace was stored in plastic containers. Each extract was produced on an industrial scale (0.6–2.4 t per batch), depending on the plant material (PiLeJe Industrie, Saint–Bonnet de Rochefort, France). The same patented ipowder® process, in which a plant extract was added to the same ground plant material [8], was applied to each plant. However, temperature and duration of the extraction varied depending on the plant material (Table S1). Briefly, a finely crushed dry plant material was extracted using water, except for the frozen grape pomace, which was extracted using ethanol–water (3:7, v/v; for stability of the extract). The ratio of plant to first native extract was 4–7 to 1 depending on the plant material. Each plant extract was filtered, concentrated and spray-dried over a new batch of the same ground plant material. For the resulting fortified ipowder® plant extracts, the final ratio of dry plant to final extract was 2–5 to 1, depending on the plant material (listed in detail in Table S1). For evaluation of the fortification success, the green tea raw material was also investigated and compared with the respective ipowder® extract (Table S1, green tea leaf powder ID 5 used to make the fortified extract ID 1). More than 95% of the plant powder particles were less than 500 µm in size. The powders are stable for 5 years when stored at room temperature in a dry and dark place.

3.3. Extraction of the Plant Powders

Each fortified plant powder (1 g each) and the green tea leaf powder were suspended in 10 mL methanol, extracted by ultrasonication for 30 min and centrifuged at $3000 \times g$ for 15 min. Each supernatant, of which an aliquot was transferred to a sampler vial for analysis, was stored in the dark at $-18\ °C$ (stable for 2 years). Note that the analytical extract of green tea leaf powder (the raw material ID 5 used to make the fortified extract ID 1, Table S1) is referred to as nonfortified extract in the text.

3.4. HPTLC–UV/Vis/FLD Method

All methanolic extracts (2 µL/band each, except 3 µL/band for the β-glucuronidase assay and 4 µL/band for the tyrosinase assay) were applied as 8-mm band on the HPTLC plate (distance to lower edge 8 mm, distance to side edge at least 15 mm, dosage speed of 150 nL/s, Automatic TLC Sampler ATS 4, CAMAG, Muttenz, Switzerland). In a twin trough chamber or Automated Development Chamber (ADC 2, CAMAG), the plate was developed using one of the two solvent mixtures (10 mL each) up to a migration distance of 70 mm (measured from the lower plate edge, taking ca. 35 min). Either a more polar solvent mixture MP 1 consisting of ethyl acetate–toluene–formic acid–water 16:4:3:2 ($v/v/v/v$), or a more apolar solvent mixture MP 2 consisting of cyclohexane–ethyl acetate–formic acid 6:3.8:0.2 ($v/v/v$) was used. The chromatogram was dried in the ADC 2 for 20 min (MP 2) or 30 min (MP 1). Documentation was performed at white light illumination (Vis), UV 254 nm and FLD 366 nm (TLC Visualizer, CAMAG).

3.5. Neutralization of Acidic Traces

The formic acid traces which remained on the chromatogram were neutralized. Briefly, 0.75 mL phosphate buffer solution (8 g disodium hydrogen phosphate in 60 mL water, adjusted to pH 7.5–7.8 with 0.1 M citric acid, ad 100 mL) was piezoelectrically sprayed (yellow nozzle, level 6, hood and tray for 20 cm × 10 cm plates, Derivatizer, CAMAG) according to a recently optimized workflow [104]. The plate was dried in a stream of cold air (2 min by hair dryer, then 15 min by ADC 2).

3.6. Effect-Directed Profiling

Six chromatograms were prepared in parallel. For each assay, respective positive controls were applied above the solvent front at the upper plate edge of the neutralized, dried chromatogram [6]. For example, three bands of an aqueous saccharolactone solution (100, 150 and 200 ng/band) were applied for the β-glucuronidase assay. Each plate was subjected to the respective assay solutions or suspensions by piezoelectric spraying (placed on a filter paper sheet, if not stated otherwise, yellow nozzle, level 6, Derivatizer, CAMAG). After the first spraying, the bottom side of the nozzle was manually dried with a lint-free tissue to avoid a dropping on the plate during the second spraying. For incubation (at 37 °C, if not stated otherwise), each plate was placed horizontally in a humid poly-propylene box (premoistened for 30 min at room temperature with 35 mL water spread on filter papers aligned on walls and bottom). Drying was performed in a stream of cold air (hair dryer). If not stated otherwise, documentation was performed at white light illumination in the reflectance mode (TLC Visualizer, CAMAG). Aliquoted enzyme and L-DOPA substrate solutions were stored at −18 °C, whereas other solutions were stored in the dark at 4 °C; all were stable for several months.

3.6.1. HPTLC–*A. fischeri* Bioassay

The HPTLC–*A. fischeri* culture was prepared according to DIN EN ISO 11348–1, 2009. After a day, the progress of the bacterial growth (increasing bioluminescence) was visually controlled by shaking the culture flask in a dark room. When the emitted green-blue light was brilliant, the *A. fischeri* suspension (3.5 mL) was piezoelectrically sprayed on the chromatogram. Thereby, the vapor settling down phase was interrupted to take out the plate (hood was closed again to suck out the rest of the vapor into the trap), which was placed wet into the BioLuminizer cabinet CAMAG). The bioautogram was documented over 30 min (exposure time 1.0 min, trigger interval 3.0 min). Dark or bright zones indicated bioactive compounds acting against Gram–negative bacteria [11].

3.6.2. HPTLC–α-Glucosidase Inhibition Assay

The chromatogram was piezoelectrically sprayed with 1 mL substrate solution (12 mg 2-naphthyl-α-D-glucopyranoside in 9 mL ethanol and 1 mL 0.01 M sodium chloride solution) and dried (1 min). Then, it was sprayed with 0.5 mL sodium acetate buffer (10.3 g of sodium acetate in 250 mL water, adjusted to pH 7.5 with 0.1-M acetic acid) used for prewetting, and 1 mL α-glucosidase solution (10 U/mL sodium acetate buffer, pH 7.5). After incubation (15 min), the plate was sprayed with 0.4 mL aqueous Fast Blue B salt solution (2 mg/mL) used as chromogenic reagent and dried (3 min).

3.6.3. HPTLC–β-Glucosidase Inhibition Assay

The workflow was performed analogously to the HPTLC–α-glucosidase assay, except for using β-glucosidase (1000 U/mL), 2-naphthyl-β-D-glucopyranoside and an incubation of 30 min.

3.6.4. HPTLC–AChE Inhibition Assay

The chromatogram was piezoelectrically sprayed (green nozzle) with 0.5 mL Tris-HCl buffer solution (pH 7.8, 0.05 M) used for prewetting and then 1.5 mL AChE solution (6.66 U/mL). After incubation (25 min), the plate was sprayed with 0.5 mL of the 1:1 substrate/chromogenic reagent mixture (ethanolic 1-naphthyl acetate solution and aqueous Fast Blue B salt solution, 3 mg/mL each) and dried (3 min).

3.6.5. HPTLC–Tyrosinase Inhibition Assay

The chromatogram was piezoelectrically sprayed (blue nozzle) with 1 mL L-DOPA substrate solution (L-DOPA 4.5 mg/mL, CHAPS 2.5 mg/mL and PEG 8000 7.5 mg/mL dissolved in 20 mM phosphate buffer of 0.7 g dipotassium phosphate and 0.84 g disodium phosphate in 0.5 L deionized water, adjusted to pH 6.8 by appropriate salt addition) and dried (1 min). Then, it was sprayed with 1 mL tyrosinase solution (400 U/mL phosphate buffer 20 mM, pH 6.8), incubated at room temperature (20 min) and dried (3 min).

3.6.6. HPTLC–β-Glucuronidase Inhibition Assay

The chromatogram was piezoelectrically sprayed with 0.25 mL phosphate buffer (pH 7.0) used for prewetting, and immediately with 1.5 mL β-glucuronidase solution (25 U/mL phosphate buffer plus 0.25 g BSA). After incubation (15 min), the chromatogram was sprayed with 0.5 mL X-Gluc substrate solution (2 mg/mL in water, red nozzle), incubated (1 h) and dried (50 °C, heating plate, 10 min).

3.7. HPTLC–HRMS Analysis

All methanolic extracts (2 µL or 200 µg/band each) were applied twice as two sets on the same HPTLC plate. The plate was cut (smartCut Plate Cutter, CAMAG) after development, and one sample set was subjected to the bioassay to match/mark (soft pencil) the coordinates of active zones on the other plate part inspected at UV 254 nm and FLD 366 nm. Each marked zone was tightly fixed by the oval elution head (4 mm × 2 mm, 320 N) and online eluted for 30 s with methanol at 100 µL/min via the Plate Express (Advion, Ithaca, NY, USA) or TLC–MS Interface 2 (CAMAG) into the Q Exactive Plus Hybrid Quadrupole-Orbitrap Mass Spectrometer (Thermo Fisher Scientific, Dreieich, Germany). All full scan HPTLC–HRMS spectra (m/z 100–1000) were recorded in the positive and negative heated electrospray ionization (HESI) mode. Sodium diisooctyl phthalate at m/z 413.26623 was used as lock mass. The HRMS parameters were set as follows: spray voltage 3.5 kV, capillary temperature 270 °C and resolution 280,000. Data processing was done with Xcalibur 3.0.63 software (Thermo Fisher Scientific). In between each zone elution, a clean back piece of an aluminum plate was eluted to reduce cross contamination (rinse the outline tube of the interface to HRMS). Several representative plate backgrounds were recorded at a similar hR_F position as the zones of interest. A respective plate background was subtracted from the analyte mass spectrum to reduce the system signals.

4. Conclusions

The developed effect-directed profiling was found to be a generic procedure, as it worked successfully for the assessment of 17 different fortified plant extracts. It pointed directly to important multipotent candidates in the multicomponent mixtures. The initial hypotheses were confirmed. The 17 fortified extracts showed characteristic effects in the six selected assays, new information was obtained by the effect-directed profiles and the bioanalytical hyphenation complemented the toolbox. Multipotent compounds with antibacterial, α-glucosidase, β-glucosidase, AChE, tyrosinase and/or β-glucuronidase-inhibiting effects were detected in most extracts, and characterized further by HPTLC–HESI–HRMS. New activities were observed and first assigned here, such as the antibacterial and enzymatic activities of *Eleutherococcus* and diverse activities of cynaratriol and caffeine. Hence, the profiling not only confirmed what was expected (considered as proof), but also revealed new activities. As the same amounts of all 17 different fortified plant extracts were applied, their comparative evaluation with regard to the effect was a great advantage. An image is worth a thousand words. The top candidates among the plants with activity against Gram–negative bacteria were *Eleutherococcus*, grape pomace, passiflora, *Eschscholzia*, rosemary and hops. Hops and rosemary especially were the top candidates for α-glucosidase inhibition. The catechins in green tea turned out to be multitalented with regard to their intense activities in all six assays. Each of the six effects was higher in the fortified green tea sample when compared to the nonfortified one, which exemplarily confirmed the fortification process. Given the global production chain of plant extracts, nontarget effect-directed profiling is highly attractive for routine use. It added effect information to the present chemical marker-oriented information, and thus provided comprehensive information on the quality and safety of the plant extracts.

Supplementary Materials: The following are available online: Table S1: Data on the raw material and production process of the investigated plant extracts; Figure S1: HPTLC–UV/Vis chromatograms used for the *A. fischeri* bioassay (bioautogram in Figure 2A); Figure S2: HPTLC–UV/FLD chromatograms used for the α-glucosidase inhibition assay (autogram in Figure 2B); Figure S3: Stability check after two years, exemplarily shown for rosemary (ID **3**) via HPTLC–FLD chromatograms and HPTLC–Vis tyrosinase inhibition autograms.

Author Contributions: Conceptualization, resources, G.E.M., V.B., L.L. and M.D.; methodology, supervision, formal analysis, data curation, funding acquisition, project administration, writing—original draft, G.E.M.; investigation, J.H. (HPTLC experiments) and C.C. (powder preparation); writing—review and editing, G.E.M., V.B., L.L. and M.D. All authors have read and agreed to the published version of the manuscript.

Funding: This research was partially funded by PiLeJe. Instrumentation was partially funded by the Deutsche Forschungsgemeinschaft (DFG, German Research Foundation)—INST 162/471-1 FUGG; INST 162/536-1 FUGG.

Institutional Review Board Statement: Not applicable.

Informed Consent Statement: Not applicable.

Data Availability Statement: The data presented in this study are available on request from the corresponding author.

Acknowledgments: Thanks are owed to Claude Blondeau, PiLeJe Laboratoire, for editorial assistance, and to Imanuel Yüce, JLU Giessen, for the recording of the HPTLC–HRMS spectra.

Conflicts of Interest: The authors have no conflict of interest.

Sample Availability: Samples of the extracts are available from the authors.

References

1. Williamson, E. Synergy and other interactions in phytomedicines. *Phytomedicine* **2001**, *8*, 401–409. [CrossRef]
2. Verpoorte, R.; Kim, H.K.; Choi, Y.H. Synergy: Easier to say than to prove. *Synergy* **2018**, *7*, 34–35. [CrossRef]
3. Morlock, G.E.; Heil, J. HI-HPTLC-UV/Vis/FLD-HESI-HRMS and bioprofiling of steviol glycosides, steviol, and isosteviol in *Stevia* leaves and foods. *Anal. Bioanal. Chem.* **2020**, *412*, 6431–6448. [CrossRef]
4. Guinobert, I.; Bardot, V.; Dubourdeaux, M. De la plante aux effets biologiques de l'extrait: Quand la démarche scientifique éclaire les usages. *Phytothérapie* **2019**, *17*, 149–155. [CrossRef]
5. Bardot, V.; Escalon, A.; Ripoche, I.; Denis, S.; Alric, M.; Chalancon, S.; Chalard, P.; Cotte, C.; Berthomier, L.; Leremboure, M.; et al. Benefits of the ipowder® extraction process applied to *Melissa officinalis* L: Improvement of antioxidant activity and in vitro gastro-intestinal release profile of rosmarinic acid. *Food Funct.* **2020**, *11*, 722–729. [CrossRef] [PubMed]
6. Morlock, G.E.; Heil, J.; Inarejos-Garcia, A.M.; Maeder, J. Effect-directed profiling of powdered tea extracts for catechins, theaflavins, flavonols and caffeine. *Antioxidants* **2021**, *10*, 117. [CrossRef] [PubMed]
7. Chandana, N.G.A.S.S.; Morlock, G.E. Eight different bioactivity profiles of 40 cinnamon extracts to discover multipotent compounds by multi-imaging planar chromatography hyphenated with effect-directed analysis and high-resolution mass spectrometry. *Food Chem.* **2021**. in print.
8. Dubourdeaux, M. Procédé de Préparation d'extraits Végétaux Permettant l'obtention d'une Nouvelle Forme Galénique. 14 January 2009. Available online: https://patents.google.com/patent/EP2080436A2/en (accessed on 21 January 2021).
9. Móricz, Á.M.; Ott, P.G.; Morlock, G.E. Discovered acetylcholinesterase inhibition and antibacterial activity of polyacetylenes in tansy root extract via effect-directed chromatographic fingerprints. *J. Chromatogr. A* **2018**, *1543*, 73–80. [CrossRef] [PubMed]
10. Krüger, S.; Bergin, A.; Morlock, G.E. Effect-directed analysis of ginger (*Zingiber officinale*) and its food products, and quantification of bioactive compounds via high-performance thin-layer chromatography and mass spectrometry. *Food Chem.* **2018**, *243*, 258–268. [CrossRef]
11. Klöppel, A.; Grasse, W.; Brümmer, F.; Morlock, G. HPTLC coupled with bioluminescence and mass spectrometry for bioactivity-based analysis of secondary metabolites in marine sponges. *J. Planar Chromatogr. Mod. TLC* **2008**, *21*, 431–436. [CrossRef]
12. Krüger, S.; Hüsken, L.; Fornasari, R.; Scainelli, I.; Morlock, G.E. Effect-directed fingerprints of 77 botanical extracts via a generic high-performance thin-layer chromatography method combined with assays and mass spectrometry. *J. Chromatogr. A* **2017**, *1529*, 93–106. [CrossRef]
13. DIN EN ISO 11348-1. *Water Quality—Determination of the Inhibitory Effect of Water Samples on the Light Emission of Vibrio fischeri (Luminescent bacteria test)—Part 1: Method Using Freshly Prepared Bacteria*; Beuth Verlag: Berlin, Germany, 2009.
14. Bartmańska, A.; Wałecka-Zacharska, E.; Tronina, T.; Popłoński, J.; Sordon, S.; Brzezowska, E.; Bania, J.; Huszcza, E. Antimicrobial Properties of Spent Hops Extracts, Flavonoids Isolated Therefrom, and Their Derivatives. *Molecules* **2018**, *23*, 2059. [CrossRef]

15. Karabín, M.; Hudcová, T.; Jelínek, L.; Dostálek, P. Biologically Active Compounds from Hops and Prospects for Their Use. *Compr. Rev. Food Sci. Food Saf.* **2016**, *15*, 542–567. [CrossRef] [PubMed]

16. Lee, S.; Shin, D.-S.; Oh, K.-B.; Shin, K.H. Antibacterial compounds from the leaves of *Acanthopanax senticosus*. *Arch. Pharm. Res.* **2003**, *26*, 40–42. [CrossRef] [PubMed]

17. Davydov, M.; Krikorian, A.D. *Eleutherococcus senticosus* (Rupr. & Maxim.) Maxim. (Araliaceae) as an adaptogen: A closer look. *J. Ethnopharmacol.* **2000**, *72*, 345–393. [CrossRef] [PubMed]

18. Masadeh, M.M.; Alkofahi, A.S.; Alzoubi, K.H.; Tumah, H.N.; Bani-Hani, K. Anti-*Helicobactor pylori* activity of some Jordanian medicinal plants. *Pharm. Biol.* **2014**, *52*, 566–569. [CrossRef] [PubMed]

19. Singh, S.; Jain, L.; Pandey, M.B.; Singh, U.P.; Pandey, V.B. Antifungal activity of the alkaloids from *Eschscholzia californica*. *Folia Microbiol. (Praha)* **2009**, *54*, 204–206. [CrossRef] [PubMed]

20. Smither-Kopperl, M.L. Plant Guide for California Poppy (*Eschscholzia californica*), Lockeford, CA 95237. 2018. Available online: https://plants.usda.gov/plantguide/pdf/pg_esca2.pdf (accessed on 21 January 2021).

21. Kuete, V. Health Effects of Alkaloids from African Medicinal Plants. In *Toxicological Survey of African Medicinal Plants*; Kuete, V., Ed.; Elsevier: Amsterdam, The Netherlands, 2014; pp. 611–633. ISBN 9780128000182.

22. Taylor, P.W. Interactions of Tea-Derived Catechin Gallates with Bacterial Pathogens. *Molecules* **2020**, *25*, 1986. [CrossRef]

23. Gopal, J.; Muthu, M.; Paul, D.; Kim, D.-H.; Chun, S. Bactericidal activity of green tea extracts: The importance of catechin containing nano particles. *Sci. Rep.* **2016**, *6*, 266. [CrossRef]

24. Oliveira, D.A.; Salvador, A.A.; Smânia, A.; Smânia, E.F.A.; Maraschin, M.; Ferreira, S.R.S. Antimicrobial activity and composition profile of grape (*Vitis vinifera*) pomace extracts obtained by supercritical fluids. *J. Biotechnol.* **2013**, *164*, 423–432. [CrossRef]

25. Ehsani, A.; Alizadeh, O.; Hashemi, M.; Afshari, A.; Aminzare, M. Phytochemical, antioxidant and antibacterial properties of *Melissa officinalis* and *Dracocephalum moldavica* essential oils. *Vet. Res. Forum* **2017**, *8*, 223–229. [PubMed]

26. Thompson, A.; Meah, D.; Ahmed, N.; Conniff-Jenkins, R.; Chileshe, E.; Phillips, C.O.; Claypole, T.C.; Forman, D.W.; Row, P.E. Comparison of the antibacterial activity of essential oils and extracts of medicinal and culinary herbs to investigate potential new treatments for irritable bowel syndrome. *BMC Complement. Altern. Med.* **2013**, *13*, 76. [CrossRef] [PubMed]

27. Pereira, C.; Barros, L.; José Alves, M.; Santos-Buelga, C.; Ferreira, I.C.F.R. Artichoke and milk thistle pills and syrups as sources of phenolic compounds with antimicrobial activity. *Food Funct.* **2016**, *7*, 3083–3090. [CrossRef]

28. Rauha, J.P.; Remes, S.; Heinonen, M.; Hopia, A.; Kähkönen, M.; Kujala, T.; Pihlaja, K.; Vuorela, H.; Vuorela, P. Antimicrobial effects of Finnish plant extracts containing flavonoids and other phenolic compounds. *Int. J. Food Microbiol.* **2000**, *56*, 3–12. [CrossRef]

29. French, K.E.; Harvey, J.; McCullagh, J.S.O. Targeted and Untargeted Metabolic Profiling of Wild Grassland Plants identifies Antibiotic and Anthelmintic Compounds Targeting Pathogen Physiology, Metabolism and Reproduction. *Sci. Rep.* **2018**, *8*, 1695. [CrossRef]

30. Deliorman Orhan, D.; Orhan, N.; Özçelik, B.; Ergun, F. Biological activities of *Vitis vinifera* L. leaves. *Turk. J. Biol.* **2009**, 341–348.

31. Sharifi-Rad, M.; Mnayer, D.; Morais-Braga, M.F.B.; Carneiro, J.N.P.; Bezerra, C.F.; Coutinho, H.D.M.; Salehi, B.; Martorell, M.; Del Mar Contreras, M.; Soltani-Nejad, A.; et al. *Echinacea* plants as antioxidant and antibacterial agents: From traditional medicine to biotechnological applications. *Phytother. Res.* **2018**, *32*, 1653–1663. [CrossRef] [PubMed]

32. Raudsepp, P.; Koskar, J.; Anton, D.; Meremäe, K.; Kapp, K.; Laurson, P.; Bleive, U.; Kaldmäe, H.; Roasto, M.; Püssa, T. Antibacterial and antioxidative properties of different parts of garden rhubarb, blackcurrant, chokeberry and blue honeysuckle. *J. Sci. Food Agric.* **2019**, *99*, 2311–2320. [CrossRef]

33. Kungel, P.T.A.N.; Correa, V.G.; Corrêa, R.C.G.; Peralta, R.A.; Soković, M.; Calhelha, R.C.; Bracht, A.; Ferreira, I.C.F.R.; Peralta, R.M. Antioxidant and antimicrobial activities of a purified polysaccharide from yerba mate (*Ilex paraguariensis*). *Int. J. Biol. Macromol.* **2018**, *114*, 1161–1167. [CrossRef] [PubMed]

34. Wang, J.; Zhao, J.; Liu, H.; Zhou, L.; Liu, Z.; Wang, J.; Han, J.; Yu, Z.; Yang, F. Chemical analysis and biological activity of the essential oils of two valerianaceous species from China: *Nardostachys chinensis* and *Valeriana officinalis*. *Molecules* **2010**, *15*, 6411–6422. [CrossRef]

35. Parizadeh, H.; Garampalli, R.H. Evaluation of Some Lichen Extracts for β-Glucosidase Inhibitory as a Possible Source of Herbal Anti-diabetic Drugs. *Am. J. Biochem.* **2016**, *6*, 46–50.

36. Chokki, M.; Cudălbeanu, M.; Zongo, C.; Dah-Nouvlessounon, D.; Ghinea, I.O.; Furdui, B.; Raclea, R.; Savadogo, A.; Baba-Moussa, L.; Avamescu, S.M.; et al. Exploring Antioxidant and Enzymes (A-Amylase and B-Glucosidase) Inhibitory Activity of *Morinda lucida* and *Momordica charantia* Leaves from Benin. *Foods* **2020**, *9*, 434. [CrossRef]

37. Yang, X.; Kong, F. Evaluation of the in vitro α-glucosidase inhibitory activity of green tea polyphenols and different tea types. *J. Sci. Food Agric.* **2016**, *96*, 777–782. [CrossRef]

38. Yilmazer-Musa, M.; Griffith, A.M.; Michels, A.J.; Schneider, E.; Frei, B. Grape seed and tea extracts and catechin 3-gallates are potent inhibitors of α-amylase and α-glucosidase activity. *J. Agric. Food Chem.* **2012**, *60*, 8924–8929. [CrossRef]

39. Guo, P.-C.; Shen, H.-D.; Fang, J.-J.; Ding, T.-M.; Ding, X.-P.; Liu, J.-F. On-line high-performance liquid chromatography coupled with biochemical detection method for screening of α-glucosidase inhibitors in green tea. *Biomed. Chromatogr.* **2018**, *32*, e4281. [CrossRef] [PubMed]

40. Olennikov, D.N.; Kashchenko, N.I.; Chirikova, N.K. Meadowsweet Teas as New Functional Beverages: Comparative Analysis of Nutrients, Phytochemicals and Biological Effects of Four *Filipendula* Species. *Molecules* **2016**, *22*, 16. [CrossRef] [PubMed]

41. Kwon, Y.I.; Vattem, D.A.; Shetty, K. Evaluation of clonal herbs of Lamiaceae species for management of diabetes and hypertension. *Asia Pac. J. Clin. Nutr.* **2006**, *15*, 107–118. [PubMed]

42. Ranilla, L.G.; Kwon, Y.-I.; Apostolidis, E.; Shetty, K. Phenolic compounds, antioxidant activity and in vitro inhibitory potential against key enzymes relevant for hyperglycemia and hypertension of commonly used medicinal plants, herbs and spices in Latin America. *Bioresour. Technol.* **2010**, *101*, 4676–4689. [CrossRef] [PubMed]

43. Santos, J.S.; Escher, G.B.; Vieira do Carmo, M.; Azevedo, L.; Boscacci Marques, M.; Daguer, H.; Molognoni, L.; Inés Genovese, M.; Wen, M.; Zhang, L.; et al. A new analytical concept based on chemistry and toxicology for herbal extracts analysis: From phenolic composition to bioactivity. *Food Res. Int.* **2020**, *132*, 109090. [CrossRef]

44. Turkiewicz, I.P.; Wojdyło, A.; Tkacz, K.; Nowicka, P.; Hernández, F. Antidiabetic, Anticholinesterase and Antioxidant Activity vs. Terpenoids and Phenolic Compounds in Selected New Cultivars and Hybrids of Artichoke *Cynara scolymus* L. *Molecules* **2019**, *24*, 1222. [CrossRef]

45. Liu, M.; Yin, H.; Liu, G.; Dong, J.; Qian, Z.; Miao, J. Xanthohumol, a prenylated chalcone from beer hops, acts as an α-glucosidase inhibitor in vitro. *J. Agric. Food Chem.* **2014**, *62*, 5548–5554. [CrossRef]

46. Baranowska-Wójcik, E.; Szwajgier, D.; Winiarska-Mieczan, A. Regardless of the Brewing Conditions, Various Types of Tea are a Source of Acetylcholinesterase Inhibitors. *Nutrients* **2020**, *12*, 709. [CrossRef]

47. Dastmalchi, K.; Ollilainen, V.; Lackman, P.; Boije af Gennäs, G.; Dorman, H.J.D.; Järvinen, P.P.; Yli-Kauhaluoma, J.; Hiltunen, R. Acetylcholinesterase inhibitory guided fractionation of *Melissa officinalis* L. *Bioorg. Med. Chem.* **2009**, *17*, 867–871. [CrossRef]

48. Ozarowski, M.; Mikolajczak, P.L.; Bogacz, A.; Gryszczynska, A.; Kujawska, M.; Jodynis-Liebert, J.; Piasecka, A.; Napieczynska, H.; Szulc, M.; Kujawski, R.; et al. *Rosmarinus officinalis* L. leaf extract improves memory impairment and affects acetylcholinesterase and butyrylcholinesterase activities in rat brain. *Fitoterapia* **2013**, *91*, 261–271. [CrossRef]

49. Santos, E.C.S.; Bicca, M.A.; Blum-Silva, C.H.; Costa, A.P.R.; Dos Santos, A.A.; Schenkel, E.P.; Farina, M.; Reginatto, F.H.; de Lima, T.C.M. Anxiolytic-like, stimulant and neuroprotective effects of *Ilex paraguariensis* extracts in mice. *Neuroscience* **2015**, *292*, 13–21. [CrossRef] [PubMed]

50. Neagu, E.; Paun, G.; Albu, C.; Radu, G.-L. Assessment of acetylcholinesterase and tyrosinase inhibitory and antioxidant activity of *Alchemilla vulgaris* and *Filipendula ulmaria* extracts. *J. Taiwan Inst. Chem. Eng.* **2015**, *52*, 1–6. [CrossRef]

51. Salazar, P.B.; Minahk, C.; Rodriguez-Vaquero, M.J. Polyphenols from Cafayate Grape Pomace Display High Inhibitory Activity on Human Acetylcholinesterase. Available online: https://www.asev.org/abstract/polyphenols-cafayate-grape-pomace-display-high-inhibitory-activity-human (accessed on 16 December 2020).

52. Cahlíková, L.; Macáková, K.; Kunes, J.; Kurfürst, M.; Opletal, L.; Cvacka, J.; Chlebek, J.; Blundene, G. Acetylcholinesterase and butyrylcholinesterase inhibitory compounds from *Eschscholzia californica* (Papaveraceae). *Nat. Prod. Commun.* **2010**, *5*, 1035–1038. [CrossRef] [PubMed]

53. No, J.K.; Soung, D.Y.; Kim, Y.J.; Shim, K.H.; Jun, Y.S.; Rhee, S.H.; Yokozawa, T.; Chung, H.Y. Inhibition of tyrosinase by green tea components. *Life Sci.* **1999**, *65*, PL241-6. [CrossRef]

54. Jun, H.-J.; Roh, M.; Kim, H.W.; Houng, S.-J.; Cho, B.; Yun, E.J.; Hossain, M.A.; Lee, H.; Kim, K.H.; Lee, S.-J. Dual inhibitions of lemon balm (*Melissa officinalis*) ethanolic extract on melanogenesis in B16-F1 murine melanocytes: Inhibition of tyrosinase activity and its gene expression. *Food Sci. Biotechnol.* **2011**, *20*, 1051–1059. [CrossRef]

55. Al-Mamary, M. The antioxidant and tyrosinase inhibitory activities of some essential oils obtained from aromatic plants grown and used in Yemen. *Sci. Res. Essays* **2011**, *6*. [CrossRef]

56. Ferri, M.; Rondini, G.; Calabretta, M.M.; Michelini, E.; Vallini, V.; Fava, F.; Roda, A.; Minnucci, G.; Tassoni, A. White grape pomace extracts, obtained by a sequential enzymatic plus ethanol-based extraction, exert antioxidant, anti-tyrosinase and anti-inflammatory activities. *N. Biotechnol.* **2017**, *39*, 51–58. [CrossRef] [PubMed]

57. Molan, A.-L.; Liu, Z.; Tiwari, R. The ability of green tea to positively modulate key markers of gastrointestinal function in rats. *Phytother. Res.* **2010**, *24*, 1614–1619. [CrossRef]

58. Molan, A.-L.; Liu, Z.; Plimmer, G. Evaluation of the effect of blackcurrant products on gut microbiota and on markers of risk for colon cancer in humans. *Phytother. Res.* **2014**, *28*, 416–422. [CrossRef] [PubMed]

59. Yan, S.; Shao, H.; Zhou, Z.; Wang, Q.; Zhao, L.; Yang, X. Non-extractable polyphenols of green tea and their antioxidant, anti-α-glucosidase capacity, and release during in vitro digestion. *J. Funct. Foods* **2018**, *42*, 129–136. [CrossRef]

60. Guyot, S.; Pellerin, P.; Brillouet, J.M.; Cheynier, V. Inhibition of β-Glucosidase (*Amygdalae dulces*) by (+)-Catechin Oxidation Products and Procyanidin Dimers. *Biosci. Biotechnol. Biochem.* **1996**, *60*, 1131–1135. [CrossRef]

61. Wang, W.; Fu, X.-W.; Dai, X.-L.; Hua, F.; Chu, G.-X.; Chu, M.-J.; Hu, F.-L.; Ling, T.-J.; Gao, L.-P.; Xie, Z.-W.; et al. Novel acetylcholinesterase inhibitors from Zijuan tea and biosynthetic pathway of caffeoylated catechin in tea plant. *Food Chem.* **2017**, *237*, 1172–1178. [CrossRef] [PubMed]

62. Kavak, D.D.; Altıok, E.; Bayraktar, O.; Ülkü, S. Pistacia terebinthus extract: As a potential antioxidant, antimicrobial and possible β-glucuronidase inhibitor. *J. Mol. Catal. B Enzym.* **2010**, *64*, 167–171. [CrossRef]

63. Oboh, G.; Ogunsuyi, O.B.; Ogunbadejo, M.D.; Adefegha, S.A. Influence of gallic acid on α-amylase and α-glucosidase inhibitory properties of acarbose. *J. Food Drug Anal.* **2016**, *24*, 627–634. [CrossRef]

64. Li, P.-H.; Lin, Y.-W.; Lu, W.-C.; Hu, J.-M.; Huang, D.-W. In Vitro Hypoglycemic Activity of the Phenolic Compounds in Longan Fruit (*Dimocarpus longan* var. Fen Ke) Shell Against α-Glucosidase and β-Galactosidase. *Int. J. Food Prop.* **2015**, *19*, 1786–1797. [CrossRef]

65. Kaur, A.; Randhawa, K.; Singh, V.; Shri, R. Bioactivity-Guided Isolation of Acetylcholinesterase Inhibitor from *Ganoderma mediosinense* (Agaricomycetes). *Int. J. Med. Mushrooms* **2019**, *21*, 755–763. [CrossRef]

66. Su, T.-R.; Lin, J.-J.; Tsai, C.-C.; Huang, T.-K.; Yang, Z.-Y.; Wu, M.-O.; Zheng, Y.-Q.; Su, C.-C.; Wu, Y.-J. Inhibition of Melanogenesis by Gallic Acid: Possible Involvement of the PI3K/Akt, MEK/ERK and Wnt/β-Catenin Signaling Pathways in B16F10 Cells. *Int. J. Mol. Sci.* **2013**, *14*, 20443–20458. [CrossRef] [PubMed]

67. Zhu, F.; Wang, J.; Takano, H.; Xu, Z.; Nishiwaki, H.; Yonekura, L.; Yang, R.; Tamura, H. Rosmarinic acid and its ester derivatives for enhancing antibacterial, α-glucosidase inhibitory, and lipid accumulation suppression activities. *J. Food Biochem.* **2019**, *43*, e12719. [CrossRef] [PubMed]

68. Akhtar, M.S.; Hossain, M.A.; Said, S.A. Isolation and characterization of antimicrobial compound from the stem-bark of the traditionally used medicinal plant Adenium obesum. *J. Tradit. Complement. Med.* **2017**, *7*, 296–300. [CrossRef]

69. Matejczyk, M.; Świsłocka, R.; Golonko, A.; Lewandowski, W.; Hawrylik, E. Cytotoxic, genotoxic and antimicrobial activity of caffeic and rosmarinic acids and their lithium, sodium and potassium salts as potential anticancer compounds. *Adv. Med. Sci.* **2018**, *63*, 14–21. [CrossRef]

70. Abedini, A.; Roumy, V.; Mahieux, S.; Biabiany, M.; Standaert-Vitse, A.; Rivière, C.; Sahpaz, S.; Bailleul, F.; Neut, C.; Hennebelle, T. Rosmarinic Acid and Its Methyl Ester as Antimicrobial Components of the Hydromethanolic Extract of *Hyptis atrorubens Poit.* (Lamiaceae). *Evid. Based Complement. Altern. Med.* **2013**, *2013*, 1–11. [CrossRef]

71. Jordán, M.J.; Lax, V.; Rota, M.C.; Lorán, S.; Sotomayor, J.A. Relevance of carnosic acid, carnosol, and rosmarinic acid concentrations in the in vitro antioxidant and antimicrobial activities of *Rosmarinus officinalis* (L.) methanolic extracts. *J. Agric. Food Chem.* **2012**, *60*, 9603–9608. [CrossRef]

72. Gülçin, İ.; Scozzafava, A.; Supuran, C.T.; Koksal, Z.; Turkan, F.; Çetinkaya, S.; Bingöl, Z.; Huyut, Z.; Alwasel, S.H. Rosmarinic acid inhibits some metabolic enzymes including glutathione S-transferase, lactoperoxidase, acetylcholinesterase, butyrylcholinesterase and carbonic anhydrase isoenzymes. *J. Enzyme Inhib. Med. Chem.* **2016**, *31*, 1698–1702. [CrossRef]

73. Zuo, A.-R.; Dong, H.-H.; Yu, Y.-Y.; Shu, Q.-L.; Zheng, L.-X.; Yu, X.-Y.; Cao, S.-W. The antityrosinase and antioxidant activities of flavonoids dominated by the number and location of phenolic hydroxyl groups. *Chin. Med.* **2018**, *13*, 51. [CrossRef]

74. Bernardes, W.A.; Lucarini, R.; Tozatti, M.G.; Souza, M.G.M.; Silva, M.L.A.; Filho, A.A.d.S.; Martins, C.H.G.; Crotti, A.E.M.; Pauletti, P.M.; Groppo, M.; et al. Antimicrobial activity of *Rosmarinus officinalis* against oral pathogens: Relevance of carnosic acid and carnosol. *Chem. Biodivers.* **2010**, *7*, 1835–1840. [CrossRef]

75. Collins, M.A.; Charles, H.P. Antimicrobial activity of Carnosol and Ursolic acid: Two anti-oxidant constituents of *Rosmarinus officinalis* L. *Food Microbiol.* **1987**, *4*, 311–315. [CrossRef]

76. Ma, Y.-Y.; Zhao, D.-G.; Zhang, R.; He, X.; Li, B.Q.; Zhang, X.-Z.; Wang, Z.; Zhang, K. Identification of bioactive compounds that contribute to the α-glucosidase inhibitory activity of rosemary. *Food Funct.* **2020**, *11*, 1692–1701. [CrossRef]

77. Shirasugi, I.; Sakakibara, Y.; Yamasaki, M.; Nishiyama, K.; Matsui, T.; Liu, M.-C.; Suiko, M. Novel screening method for potential skin-whitening compounds by a luciferase reporter assay. *Biosci. Biotechnol. Biochem.* **2010**, *74*, 2253–2258. [CrossRef] [PubMed]

78. Jesus, J.A.; Lago, J.H.G.; Laurenti, M.D.; Yamamoto, E.S.; Passero, L.F.D. Antimicrobial activity of oleanolic and ursolic acids: An update. *Evid. Based Complement. Alternat. Med.* **2015**, *2015*, 620472. [CrossRef]

79. Zheng, X.; Wang, H.; Zhang, P.; Gao, L.; Yan, N.; Li, P.; Liu, X.; Du, Y.; Shen, G. Chemical Composition, Antioxidant Activity and α-Glucosidase Inhibitory Activity of Chaenomeles Speciosa from Four Production Areas in China. *Molecules* **2018**, *23*, 2518. [CrossRef] [PubMed]

80. Ding, H.; Hu, X.; Xu, X.; Zhang, G.; Gong, D. Inhibitory mechanism of two allosteric inhibitors, oleanolic acid and ursolic acid on α-glucosidase. *Int. J. Biol. Macromol.* **2018**, *107*, 1844–1855. [CrossRef]

81. Loesche, A.; Köwitsch, A.; Lucas, S.D.; Al-Halabi, Z.; Sippl, W.; Al-Harrasi, A.; Csuk, R. Ursolic and oleanolic acid derivatives with cholinesterase inhibiting potential. *Bioorg. Chem.* **2019**, *85*, 23–32. [CrossRef] [PubMed]

82. Chung, Y.K.; Heo, H.J.; Kim, E.K.; Kim, H.K.; Huh, T.L.; Lim, Y.; Kim, S.K.; Shin, D.H. Inhibitory effect of ursolic acid purified from *Origanum majorana* L. on the acetylcholinesterase. *Mol. Cells* **2001**, *11*, 137–143.

83. Han, S.K.; Kim, Y.G.; Kang, H.C.; Huh, J.R.; Kim, J.Y.; Baek, N.-I.; Lee, D.-K.; Lee, D.-G. Oleanolic acid from *Fragaria ananassa* calyx leads to inhibition of α-MSH-induced melanogenesis in B16-F10 melanoma cells. *J. Korean Soc. Appl. Biol. Chem.* **2014**, *57*, 735–742. [CrossRef]

84. Muñoz, E.; Avila, J.G.; Alarcón, J.; Kubo, I.; Werner, E.; Céspedes, C.L. Tyrosinase inhibitors from *Calceolaria integrifolia* s.l.: *Calceolaria talcana* aerial parts. *J. Agric. Food Chem.* **2013**, *61*, 4336–4343. [CrossRef]

85. Puupponen-Pimia, R.; Nohynek, L.; Meier, C.; Kahkonen, M.; Heinonen, M.; Hopia, A.; Oksman-Caldentey, K.-M. Antimicrobial properties of phenolic compounds from berries. *J. Appl. Microbiol.* **2001**, *90*, 494–507. [CrossRef]

86. Pérez-Nájera, V.C.; Gutiérrez-Uribe, J.A.; Antunes-Ricardo, M.; Hidalgo-Figueroa, S.; Del-Toro-Sánchez, C.L.; Salazar-Olivo, L.A.; Lugo-Cervantes, E. Smilax aristolochiifolia Root Extract and Its Compounds Chlorogenic Acid and Astilbin Inhibit the Activity of α-Amylase and α-Glucosidase Enzymes. *Evid. Based Complement. Altern. Med.* **2018**, *2018*, 6247306. [CrossRef] [PubMed]

87. Oboh, G.; Agunloye, O.M.; Adefegha, S.A.; Akinyemi, A.J.; Ademiluyi, A.O. Caffeic and chlorogenic acids inhibit key enzymes linked to type 2 diabetes (*in vitro*): A comparative study. *J. Basic Clin. Physiol. Pharmacol.* **2015**, *26*, 165–170. [CrossRef] [PubMed]

88. Kwon, S.-H.; Lee, H.-K.; Kim, J.-A.; Hong, S.-I.; Kim, H.-C.; Jo, T.-H.; Park, Y.-I.; Lee, C.-K.; Kim, Y.-B.; Lee, S.-Y.; et al. Neuroprotective effects of chlorogenic acid on scopolamine-induced amnesia via anti-acetylcholinesterase and anti-oxidative activities in mice. *Eur. J. Pharmacol.* **2010**, *649*, 210–217. [CrossRef]

89. Guo, Z.; Li, J. Chlorogenic Acid Prevents Alcohol-induced Brain Damage in Neonatal Rat. *Transl. Neurosci.* **2017**, *8*, 176–181. [CrossRef]

90. Li, H.-R.; Habasi, M.; Xie, L.-Z.; Aisa, H.A. Effect of chlorogenic acid on melanogenesis of B16 melanoma cells. *Molecules* **2014**, *19*, 12940–12948. [CrossRef]

91. Karak, S.; Das, S.; Biswas, M.; Choudhury, A.; Dutta, M.; Chaudhury, K.; De, B. Phytochemical composition, β-glucuronidase inhibition, and antioxidant properties of two fractions of *Piper betle* leaf aqueous extract. *J. Food Biochem.* **2019**, *43*, e13048. [CrossRef] [PubMed]

92. Pohanka, M.; Dobes, P. Caffeine inhibits acetylcholinesterase, but not butyrylcholinesterase. *Int. J. Mol. Sci.* **2013**, *14*, 9873–9882. [CrossRef] [PubMed]

93. Yang, Y.; Sun, X.; Ni, H.; Du, X.; Chen, F.; Jiang, Z.; Li, Q. Identification and Characterization of the Tyrosinase Inhibitory Activity of Caffeine from Camellia Pollen. *J. Agric. Food Chem.* **2019**, *67*, 12741–12751. [CrossRef]

94. Rempe, C.S.; Burris, K.P.; Woo, H.L.; Goodrich, B.; Gosnell, D.K.; Tschaplinski, T.J.; Stewart, C.N.; Schuch, R. Computational Ranking of Yerba Mate Small Molecules Based on Their Predicted Contribution to Antibacterial Activity against Methicillin-Resistant *Staphylococcus aureus*. *PLoS ONE* **2015**, *10*, e0123925. [CrossRef]

95. Tong, J.; Ma, B.; Ge, L.; Mo, Q.; Zhou, G.; He, J.; Wang, Y. Dicaffeoylquinic Acid-Enriched Fraction of *Cichorium glandulosum* Seeds Attenuates Experimental Type 1 Diabetes via Multipathway Protection. *J. Agric. Food Chem.* **2015**, *63*, 10791–10802. [CrossRef]

96. Kang, J.Y.; Park, S.K.; Guo, T.J.; Ha, J.S.; Du Lee, S.; Kim, J.M.; Lee, U.; Kim, D.O.; Heo, H.J. Reversal of Trimethyltin-Induced Learning and Memory Deficits by 3,5-Dicaffeoylquinic Acid. *Oxid. Med. Cell. Longev.* **2016**, *2016*, 6981595. [CrossRef] [PubMed]

97. Iwai, K.; Kishimoto, N.; Kakino, Y.; Mochida, K.; Fujita, T. In vitro antioxidative effects and tyrosinase inhibitory activities of seven hydroxycinnamoyl derivatives in green coffee beans. *J. Agric. Food Chem.* **2004**, *52*, 4893–4898. [CrossRef] [PubMed]

98. Shi, J.; Sun, C.; Huang, H.; Lin, W.; Gao, J.; Lin, Y.; Zhang, Z.; Huo, X.; Tian, X.; Yu, Z.; et al. β-Glucuronidase- and OATP2B1-mediated drug interaction of scutellarin in Dengzhan Xixin Injection: A formulation aspect. *Drug Dev. Res.* **2020**, *81*, 609–619. [CrossRef] [PubMed]

99. Kusumah, D.; Wakui, M.; Murakami, M.; Xie, X.; Yukihito, K.; Maeda, I. Linoleic acid, α-linolenic acid, and monolinolenins as antibacterial substances in the heat-processed soybean fermented with *Rhizopus oligosporus*. *Biosci. Biotechnol. Biochem.* **2020**, *84*, 1285–1290. [CrossRef]

100. Artanti, N.; Tachibana, S.; Kardono, L.B.S. Effect of Media Compositions on α-glucosidase Inhibitory Activity, Growth and Fatty Acid Content in Mycelium Extracts of *Colletotrichum* sp. TSC13 from *Taxus Sumatrana* (Miq.) de Laub. *Pak. J. Biol. Sci.* **2014**, *17*, 884–890. [CrossRef] [PubMed]

101. Cwikla, C.; Schmidt, K.; Matthias, A.; Bone, K.M.; Lehmann, R.; Tiralongo, E. Investigations into the antibacterial activities of phytotherapeutics against *Helicobacter pylori* and *Campylobacter jejuni*. *Phytother. Res.* **2010**, *24*, 649–656. [CrossRef] [PubMed]

102. Denev, P.; Kratchanova, M.; Ciz, M.; Lojek, A.; Vasicek, O.; Blazheva, D.; Nedelcheva, P.; Vojtek, L.; Hyrsl, P. Antioxidant, antimicrobial and neutrophil-modulating activities of herb extracts. *Acta Biochim. Pol.* **2014**, *61*, 359–367. [CrossRef]

103. Adamczak, A.; Ożarowski, M.; Karpiński, T.M. Antibacterial Activity of Some Flavonoids and Organic Acids Widely Distributed in Plants. *J. Clin. Med.* **2019**, *9*, 109. [CrossRef] [PubMed]

104. Azadniya, E.; Goldoni, L.; Bandiera, T.; Morlock, G.E. Same analytical method for both (bio)assay and zone isolation to identify/quantify bioactive compounds by quantitative nuclear magnetic resonance spectroscopy. *J. Chromatogr. A* **2020**, *1628*, 461434. [CrossRef]

105. El Sohaimy, S.A. Chemical Composition, Antioxidant and Antimicrobial Potential of Artichoke. *Open Nutraceuticals J.* **2014**, *1*, 15–20. [CrossRef]

106. Jamshidi-Aidji, M.; Morlock, G.E. Fast equivalency estimation of unknown enzyme inhibitors in situ the effect-directed fingerprint, shown for *Bacillus* lipopeptide extracts. *Anal. Chem.* **2018**, *90*, 14260–14268. [CrossRef]

107. Jamshidi-Aidji, M.; Morlock, G.E. From bioprofiling and characterization to bioquantification of natural antibiotics by direct bioautography linked to high-resolution mass spectrometry: Exemplarily shown for *Salvia miltiorrhiza* root. *Anal. Chem.* **2016**, *88*, 10979–10986. [CrossRef]

108. Klingelhöfer, I.; Morlock, G.E. Bioprofiling of surface/wastewater and bioquantitation of discovered endocrine-active compounds by streamlined direct bioautography. *Anal. Chem.* **2015**, *87*, 11098–11104. [CrossRef] [PubMed]

109. Morlock, G.E.; Klingelhöfer, I. Liquid chromatography-bioassay-mass spectrometry for profiling of physiologically active food. *Anal. Chem.* **2014**, *86*, 8289–8295. [CrossRef] [PubMed]

110. Meyer, D.; Marin-Kuan, M.; Debon, E.; Serrant, P.; Cottet-Fontannaz, C.; Schilter, B.; Morlock, G.E. Detection of low levels of genotoxic compounds in food contact materials using an alternative HPTLC-SOS-Umu-C Assay. *ALTEX Altern. Anim. Exp.* **2020**. [CrossRef]

111. Azadniya, E.; Mollergues, J.; Stroheker, T.; Billerbeck, K.; Morlock, G.E. New incorporation of the S9 system into methods for detecting acetylcholinesterase inhibition, comparison and proof-of-concept using food contact materials. *Anal. Chim. Acta* **2020**, *1129*, 76–84. [CrossRef] [PubMed]

112. Klingelhöfer, I.; Hockamp, N.; Morlock, G.E. Non-targeted detection and differentiation of agonists *versus* antagonists, directly in bioprofiles of everyday products. *Anal. Chim. Acta* **2020**, *1125*, 288–298. [CrossRef]

Article

GC-FID-MS Based Metabolomics to Access Plum Brandy Quality

Stefan Ivanović [1], Katarina Simić [1], Vele Tešević [2], Ljubodrag Vujisić [2], Marko Ljekočević [3] and Dejan Gođevac [1,*]

[1] National Institute of the Republic of Serbia, Institute of Chemistry, Technology and Metallurgy, University of Belgrade, Njegoševa 12, 11000 Belgrade, Serbia; stefan.ivanovic@ihtm.bg.ac.rs (S.I.); katarina.simic@ihtm.bg.ac.rs (K.S.)
[2] Faculty of Chemistry, University of Belgrade, Studentski trg 12-16, 11000 Belgrade, Serbia; vtesevic@chem.bg.ac.rs (V.T.); ljubaw@chem.bg.ac.rs (L.V.)
[3] Zarić Distillery, Maksima Markovića 42, 31260 Kosjerić, Serbia; marko.lekocevic@gmail.com
* Correspondence: dgodjev@chem.bg.ac.rs

Abstract: Plum brandy (Slivovitz (en); Šljivovica(sr)) is an alcoholic beverage that is increasingly consumed all over the world. Its quality assessment has become of great importance. In our study, the main volatiles and aroma compounds of 108 non-aged plum brandies originating from three plum cultivars, and fermented using different conditions, were investigated. The chemical profiles obtained after two-step GC-FID-MS analysis were subjected to multivariate data analysis to reveal the peculiarity in different cultivars and fermentation process. Correlation of plum brandy chemical composition with its sensory characteristics obtained by expert commission was also performed. The utilization of PCA and OPLS-DA multivariate analysis methods on GC-FID-MS, enabled discrimination of brandy samples based on differences in plum varieties, pH of plum mash, and addition of selected yeast or enzymes during fermentation. The correlation of brandy GC-FID-MS profiles with their sensory properties was achieved by OPLS multivariate analysis. Proposed workflow confirmed the potential of GC-FID-MS in combination with multivariate data analysis that can be applied to assess the plum brandy quality.

Keywords: plum brandy; metabolomics; GC-FID-MS

Citation: Ivanović, S.; Simić, K.; Tešević, V.; Vujisić, L.; Ljekočević, M.; Gođevac, D. GC-FID-MS Based Metabolomics to Access Plum Brandy Quality. *Molecules* **2021**, *26*, 1391. https://doi.org/10.3390/molecules26051391

Academic Editors: Young Hae Choi, Young Pyo Jang, Yuntao Dai and Luis Francisco Salomé-Abarca

Received: 5 February 2021
Accepted: 27 February 2021
Published: 5 March 2021

Publisher's Note: MDPI stays neutral with regard to jurisdictional claims in published maps and institutional affiliations.

1. Introduction

Plum brandy (Slivovitz (en); Šljivovica(sr)) is produced in Eastern and Central Europe, and consumed worldwide. The quality of plum brandy depends on many factors, such as: environmental characteristics of plum cultivation, characteristics of plum cultivars, and technological characteristics of the production process [1]. Revenue in the brandy segment was USD 63,620 million in 2020 worldwide, and the market is expected to exhibit a strong 6.9% compound annual growth rate over the forecast period from 2020 to 2023 [2]. Beside ethanol, plum brandy contains a complex mixture of chemical compounds formed by plum and yeast metabolism during the distillation process.

Domestic or European plum (*Prunus domestica* L.) is the most represented and traditionally the most important fruit species in Serbia. In the world production of plums, the Republic of Serbia was ranked second in Europe in 2019 and third in the world, behind China and Romania, with a production of 558,930 t [3]. The most popular plum cultivars in Serbia are Požegača, Čačanska rodna, Stanley, Čačanska lepotica, Čačanska najbolja, and Čačanska rana [4].

Plum fruits may be consumed fresh, dried, or prepared into preserves, compotes, mousse, pulp, candied fruit, frozen fruit, jams, and jelly products [5]. The largest amount of plum fruits produced in Serbia (more than 75%) is processed into brandy [6].

Serbia has a long tradition of manufacturing natural fruit brandies and plum spirits called Rakija or Šljivovica are produced as a national drink.

In the production of plum spirits with distinctive aromatic characteristics, old widespread plum cultivars (such as plum cultivars Požegača (PZ)) have been traditionally used. Additionally, some previously rarely used and autochthonous cultivars have been used more intensively as a raw material in the production of spirits. There are a number of old native cultivars, such as Crvena ranka (CR) (*P. domestica* L.) and Trnovača (TR) (*Prunus insititia* L.), which are suitable for high quality brandies. CR plum (also known as Darosavka, Šumadinka, Crvenjača, or Ranošljiva) is an autochthonous variety of Serbia [7].

Traditional fruit brandy production still mostly uses a simple distillation pot. In recent years, column distillation has slowly entered the production of spirits in small distilleries and two different types of distillation equipment are now commonly used in the production of fruit spirits: the Charentais alembic pot (French style) and the batch distillation column (German style) [8].

Apart from water and ethanol as the main constituents, brandy also contains a number of other components, the concentration of which is mostly dependent upon the cultivar, production still, and differences in microflora during fermentation. The quality of the fruit spirits depends on the concentration and the relationship of the individual volatile compounds. He et al. analyzed a traditional Chinese brandy using solid head microextraction and GC-MS. The obtained volatile profiles were then analyzed by OPLS-DA for classification of the samples according to their aroma and region [9]. A number of studies on the compositions of volatile compounds in both stone and pome fruit spirits, such as plum brandy, apricot brandy, mirabelle brandy, and Cornelian cherry brandy have been reported [9–16].

Volatile compounds are one of the main characteristics that determine a brandy's organoleptic quality and style. This is the result of the contribution of hundreds of volatile compounds, including higher alcohols, esters, acids, aldehydes, ketones, terpenes, norisoprenoids, and volatile phenols. These compounds can be classified into four groups: primary aromatic compounds (whose entire aroma appears exactly as in the fruit during ripening); secondary aromatic components (formed during alcoholic fermentation); tertiary aromatic compounds (formed during the distillation process); and quaternary aromatic compounds (formed during the maturation process) [11–16].

In most cases, the producers make fruit brandies in the traditional way without using selected yeast strains, enzymes, or other agents. However, in order to obtain good quality fruit brandy, many producers use selected yeast strains.

The work reported by Satora and Tuszyński [17] showed that the samples obtained after spontaneous fermentation were distinguished by a high content of acetoin, ethyl acetate, and total esters, accompanied by a low level of methanol and fusel alcohols. They have also found that non-*Saccharomyces* yeasts were responsible for higher concentrations of esters and methanol, while *S. cerevisiae* strains resulted in increased levels of higher alcohols. In comparison with isolated indigenous strains of *S. cerevisiae* synthesized relatively low amounts of higher alcohols in regard to commercial cultures [11].

The quality of plum brandy is a complex concept. In addition to its chemical composition, the utmost importance is attributed to its sensory characteristics as one of the most important factors of quality. The results of sensory evaluations together with the results of chemical analysis may be used to make a final judgement on the quality of plum brandy only after their processing by adequate mathematical-statistical methods [18].

M. Jakubíková et al. used synchronous fluorescence, UV−Vis, and near infrared (NIR) spectroscopy coupled with chemometric methods to discriminate samples of high-quality plum brandies of different varietal origins (*Prunus domestica* L.) [19]. Bajer et al. analysed volatile compounds profiles of various fruit spirits (plum, mirabelle, apricot, pear, and apple spirits) by method of headspace solid-phase microextraction coupled to gas chromatography with flame ionization detection (HS-SPME/GC-FID) and obtained chromatographic data used for authentication of fruit spirits [20]. Based on the results obtained by HPLC–FLD analysis and using chemometric methods (principal component analysis (PCA) and linear discriminant analysis LDA), Jakubíková et al. detected and quantified two volatile phenols (eugenol, 4-ethylphenol) and three anisoles (4-vinylanisole, 4-allylanisole, 4-propenylanisole) in plum brandies with different origins. Moreover, quantitative profiles of the eugenol, 4-vinylanisole, and 4-ethylphenol showed high diversity between summer and autumn plum brandy samples [21].

The chemical complexity and high alcohol content of fruit brandy make sensory evaluation of the product very challenging. Moreover, sensory evaluation of brandy must be strictly controlled to ensure valid results.

In our study, GC-FID-MS matabolomics was employed to discover fine differences in non-aged plum brandies chemical composition establishing desirable and undesirable metabolites that make differences between plum brandies of different quality. The main volatiles and aroma compounds of 108 plum brandies originating from three plum cultivars, and fermented using different conditions, were investigated. The chemical profiles obtained after two-step GC-FID-MS analysis were subjected to multivariate data analysis to reveal the peculiarity in different cultivars and fermentation process.

Finally, correlation of plum brandy chemical composition with its sensory characteristics obtained by expert commission was performed.

2. Results and Discussion

2.1. GC-FID-MS Profiling of Plum Brandies

To follow the variability of plum brandies composition in regard to plum cultivars and using different technological process prior fermentation, the qualitatively and quantitatively measurements of all molecules present therein is needed. This approach is well known in plant or human metabolomics studies, where GC-FID-MS in combination with multivariate data analysis is the method of choice in case of volatile compounds [22]. In case of plum brandies, it is an even more suitable method because both brandies and gas chromatography share the same technology—the distillation process.

The 108 plum brandies were produced from three plum cultivars using different fermentation conditions: two different pH adjustment of plum mash, addition of yeast Lalvin QA23, addition of enzymes Lallzyme BETA™ and Lallzyme CUVÉE BLANC™.

In total, 89 compounds were identified and taken into account for aroma metabolites profiling.

2.2. Multivariate Models Creation

All the 89 volatiles and aroma metabolites concentrations obtained from flame ionization detector (FID) areas for 108 samples were subjected to multivariate data analysis. Firstly, principal component analysis (PCA) as an unsupervised variable reduction technique to develop smaller number of novel variables that will account for most of the variation in the observed variables was performed. It resulted in 13 principal components (PCs) model explaining 83.1% of the total data variance. The optimal model dimensionality is obtained after 7-fold internal cross-validation. If a new PC improved the predictive power compared with preceding PC, the new PC is retained in the model. Based on PCA score plot of the first two PCs (Figure 1a), three groups of samples were separated according to the plum cultivars. Interestingly, the samples were also separated along PC5 due to different pH value of plum mash prior the fermentation (Figure 1b).

Next, orthogonal partial least squares to latent structures-discriminant analysis (OPLS-DA) was performed. In this supervised technique, novel variables will account for maximum separation between predefined classes. An additional advantage of the orthogonal model is the facilitated interpretation due to separation of the systematic variation of the variables into two parts: one linearly related to class information and one orthogonal to the class information [23]. Thus, OPLS-DA is suitable for finding variables having the greatest discriminatory power between classes. Two OPLS-DA models were created containing CG-MS-FID data of brandies produced from different cultivars: CR versus PZ and TR versus PZ. To investigate the influence of different technological process prior to fermentation, more OPLS-DA was performed: pH value of plum mash (pH 3 versus pH 3.5), addition of yeast Lalvin QA23 versus natural occurring yeast, addition of enzyme Lallzyme BETA™ versus no addition of enzyme, and addition of enzyme Lallzyme CUVÉE BLANC™ versus no addition of enzyme.

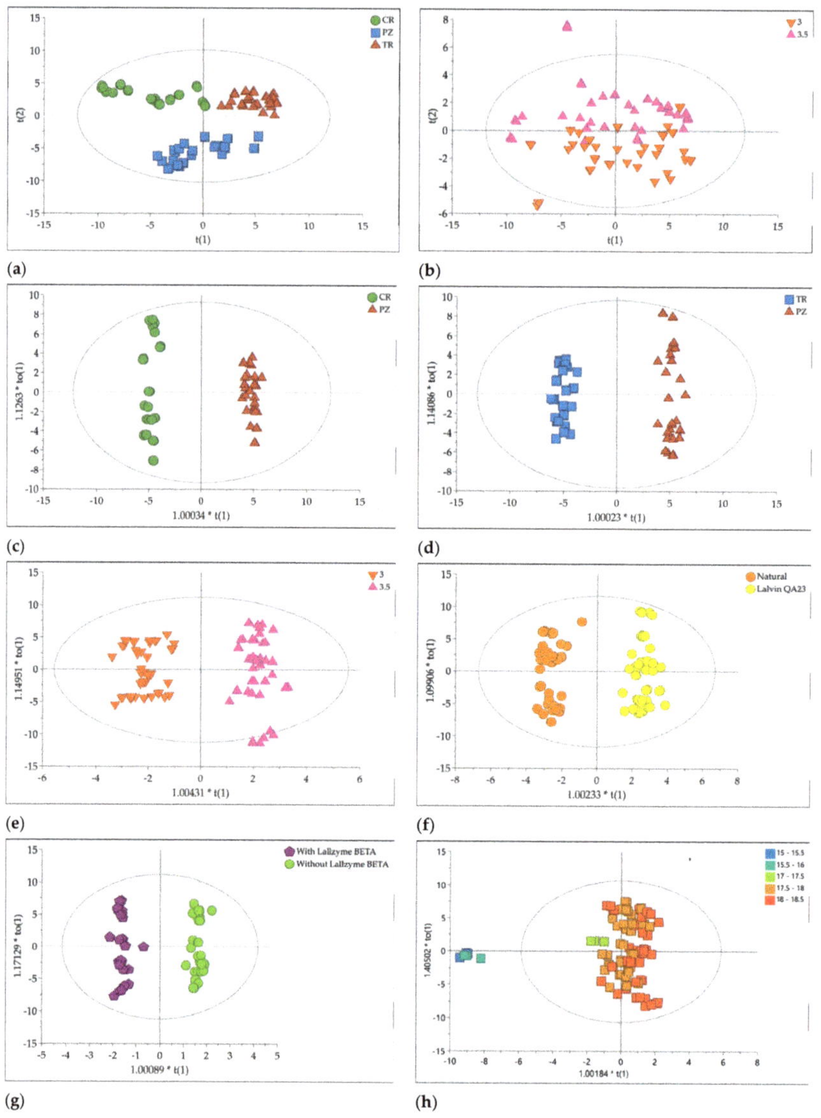

Figure 1. (**a**) PCA score plot (PC1 versus PC2) of all studied samples. The scores are colored according to the cultivars: CR—Crvena ranka, PZ—Požegača, TR—Trnovača; (**b**) PCA score plot (PC1 versus PC5) of all studied samples The scores are colored according to the pH of plum mash; (**c**) OPLS-DA score plot comprising cultivars CR versus PZ; (**d**) OPLS-DA score plot comprising cultivars TR versus PZ; (**e**) OPLS-DA score plot comprising different pH of plum mash, pH 3.0 versus pH 3.5; (**f**) OPLS-DA score plot comprising different yeast, natural versus Lalvin QA23™; (**g**) OPLS-DA score plot comprising enzyme addition, no addition versus Lallzyme BETA™; (**h**) OPLS score plot comprising correlation of brandy composition with its sensory properties. The scores are colored according to the grades of sensory evaluation.

The quality of the obtained models was assessed by goodness of fit (R^2) indicating how well the variation of variables is explained using the predictive components and predictive ability of the model (Q^2) indicating how well the model predicts new data, as estimated by cross validation. In the five OPLS-DA models concerning different cultivars, pH of plum mash, yeast and Lallzyme BETA™ addition, R^2 and Q^2 values over 0.9 and close to 1 (maximum value) indicated remarkable goodness of fit, and predictivity (Table 1).

Table 1. Parameters of the multivariate analysis models.

Model No.	Model Name	No. of Samples	No. of Components	R^2	Q^2[1]	p (CV-ANOVA)	F (CV-ANOVA)
M1	PCA	108	13	0.831	0.540	-	-
M2	OPLS-DA cultivars CR/PZ[2]	72	1 + 2	0.991	0.985	$<1 \times 10^{-6}$	710
M3	OPLS-DA cultivars TR/PZ	72	1 + 2	0.988	0.981	$<1 \times 10^{-6}$	567
M4	OPLS-DA pH of plum mash	108	1 + 4	0.947	0.914	$<1 \times 10^{-6}$	103
M5	OPLS-DA natural/selected yeast	108	1 + 4	0.966	0.943	$<1 \times 10^{-6}$	159
M6	OPLS-DA with/without Beta	72	1 + 8	0.983	0.907	$<1 \times 10^{-6}$	29
M7	OPLS-DA with/without Cuvee	72	1 + 0	0.228	-0.091	1	0
M8	OPLS sensory evaluation	108	1 + 7	0.923	0.809	$<1 \times 10^{-6}$	24

[1] Q^2 values were determined by 7-fold internal cross-validation. Only the model related to Lallzyme CUVÉE BLANC™ addition with R^2 around 0.2 and negative Q^2 value cannot be considered valid for further analysis. [2] CR—Crvena ranka; PZ—Požegača; TR—Trnovača.

The OPLS-DA models were validated in two ways. Firstly, in permutation tests the R^2 and Q^2 values of the original models were compared with the R^2 and Q^2 values of several models based on data where the order of class information has been randomly permuted. The adequate results were obtained for all the five OPLS-DA models since regressions of Q^2 lines intersected the vertical axis at below zero, and all Q^2 and R^2 values of permuted Y vectors were lower than the original ones. Further validation was performed by CV-ANOVA, where the significance of the five OPLS-DA models was clearly shown with p values far less than 0.05, except the model related to addition of Lallzyme CUVÉE BLANC™ (Table 1). In the misclassification tables (Table S4), performance metrics for classifications of the OPLS-DA models are summarized.

The score plots of five OPLS-DA models are depicted in Figure 1c–g. This indicated good separation between classes along to the predictive components.

Finally, orthogonal partial least squares to latent structures (OPLS) analysis was applied to correlate brandy composition with its sensory properties. The jointed volatiles and aroma metabolites concentrations obtained from FID areas were used as X variables, while average grades from sensory analysis performed by expert commission were used as Y variables. The goodness of fit and predictive ability were fairly good according to Q^2 and R^2 values. According to CV-ANOVA, the OPLS model was significant with $p < 0.05$ (Table 1). This is also in accordance with the results of permutation test. According to the score plots of the OPLS model, brandy samples with lower grades (15−16) were clearly separated from those with higher grades (17–18.5). The area of grades 17–18.5 were slightly separated (Figure 1h).

The selection of the most influential variables in OPLS and OPLS-DA models was based on variable influence on projection scores of the predictive components (VIPpred). Variables with the VIPpred score above 1.4 were considered as important for the correlation or separation.

2.3. Cultivar Influence on Brandy Profiles

Since two OPLS-DA models containing data of brandies produced from three different cultivars, shared and unique structure plot (SUS-plot) was used to reveal the metabolites specific for each cultivar. The SUS-Plot represents a scatter plot of the loadings scaled as a correlation coefficient (p(corr) (1)) from two separate OPLS-DA models [23] (Figure 2). The negative values of loadings coresponded to the most influential variables responsible for separation of CR and TR in models M2 and M3, respectively. Such loadings for PZ important variables exhibited positive values in both models.

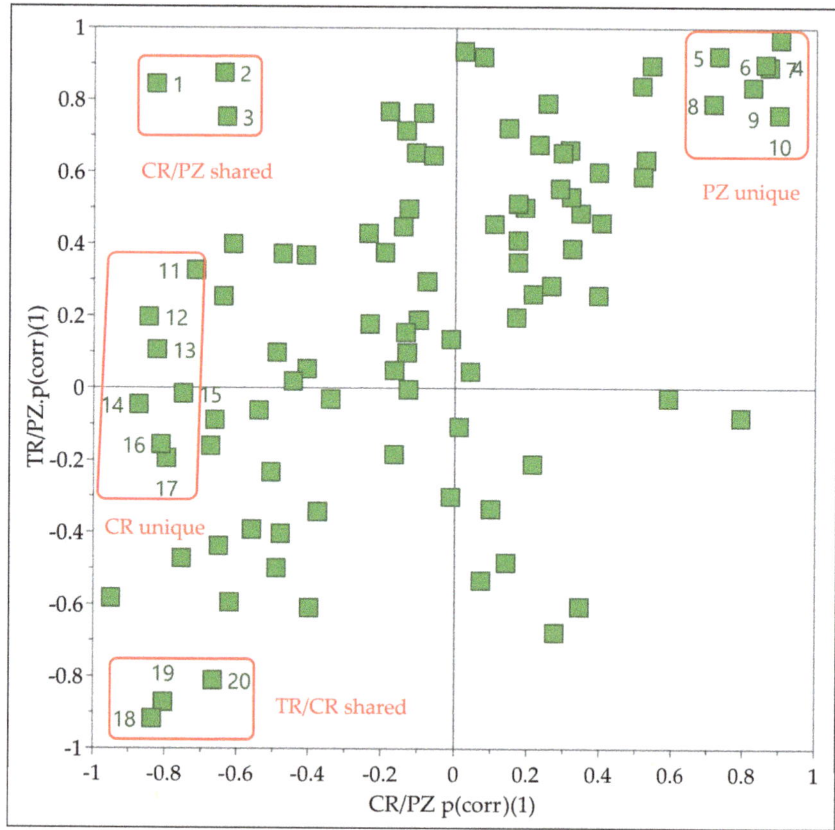

Figure 2. Shared and unique structure plot (SUS-plot) connecting two OPLS-DA models comprising cultivars CR versus PZ (M2) and TR versus PZ (M3). The most influential variables are marked with numbers corresponding to the metabolites depicted in Figure 3.

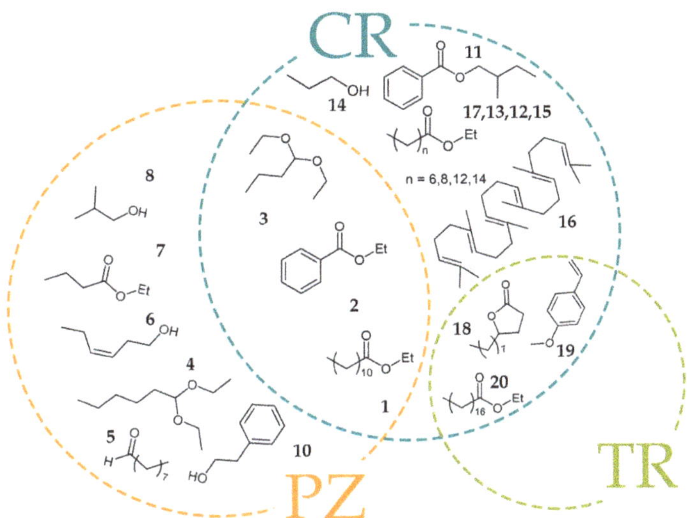

Figure 3. Metabolites important for discrimination between brandies produced from different cultivars.

Thus, metabolites **11–17** found close to the extreme negative end of the X axis in the SUS-plot having CR/PZ p(corr)(1) value close to −1 and TR/PZ p(corr)(1) value close to zero were discriminating for CR cultivar. Similarly, the metabolites **4–10** with both p(corr)(1) values close to 1 were discriminating for PZ cultivar. The metabolites (**1–3**) found in the upper left corner were discriminating for CR and PZ relative to TR. In the same manner, metabolites (**18–20**) in the lower left corner of the SUS-Plot were discriminating for TR and CR relative to PZ.

The impact of cultivars on the volatile organic compounds composition of plum brandy has been studied by Vyviurska et al. It was found that plum brandies distilled by the same technology from 25 plum cultivars could be distinguished by two-dimensional gas chromatographic analysis and sensory evaluation. The main differences were observed in the presence of unsaturated fusel alcohols (3-methyl-3-buten-1-ol, trans-3-hexenol), unsaturated aldehydes (2-butenal, 2-nonenal), monoterpene derivatives (linalool acetate, geraniol acetate) and lactones, which were mainly detected at the trace level [24].

Ethyl esters present quantitatively the largest group of the aroma compounds in the plum distillates. Esters are formed during alcoholic fermentation via yeast metabolism and qualitatively present the major class of flavor compounds in distillates [12].

Our results showed that fatty acid ethyl esters **6**, **8**, **12**, and **14** discriminated the plum brandies obtained from cultivar CR. Ethyl butyrate (**7**) was discriminating ester for plum brandies obtained from cultivar PZ. Ethyl stearate discriminated plum brandies obtained from cultivar TR and CR, while ethyl laurate and ethyl benzoate discriminated brandies from CR and PZ. The esters contents in the plum brandy were in agreement with the previously reported data [10].

Higher alcohols are common constituents in alcoholic beverages and are formed in small amounts by yeast, from sugars and from amino acids metabolism during the alcoholic fermentation process. The most common fusel alcohols in distilled spirits include: 2-butanol, *i*-butanol, 1-butanol, 1-propanol, and amyl alcohols [17]. In our work, 1-butanol discriminating brandy was obtained from CR, while *i*-butanol and (Z)-3-Hexen-1-ol were discriminating in the distillates obtained from PZ cultivar.

Diethyl acetals derived from the reaction between the corresponding aldehydes and ethanol, probably during the distillation process [25].

1,1-Diethoxyhexane was a discriminating component for the cultivar PZ and 1,1-diethoxybutane for PZ and CV cultivar.

2.4. Fermentation Process Influence on Brandy Profiles

The volatile profile, taste and flavor as well as yield of plum brandy is strongly dependent on the unique aromatic profile of plum and the yeast present on the surface of the fruit, as well as the yeast used for nutrition [26]. The majority flavor compounds are formed during fermentation by the yeast. These compounds include volatile organic acids, alcohols, aldehydes, and esters. The production and the amount of these compounds found in the fruit brandy are yeast strain dependent.

Spontaneous fermentation of plum musts involves a sequential succession of yeasts. During the first phase of fermentation *Kloeckera apiculata* and *Candida pulcherrima* are dominated strains. However, as fermentation progresses, these non-*Saccharomyces* species are naturally substituted by *Saccharomyces* ones, because the latter are good fermenters and possess high alcohol tolerance [17].

In our study, the samples after spontaneous fermentation were distinguished by a high content of fatty acid ethyl esters (**32**, **34–38**) and nonyl acetate (**33**). Selected yeast was responsible for higher concentrations of metabolites **23**, **27** and **39–44** (Figure 4).

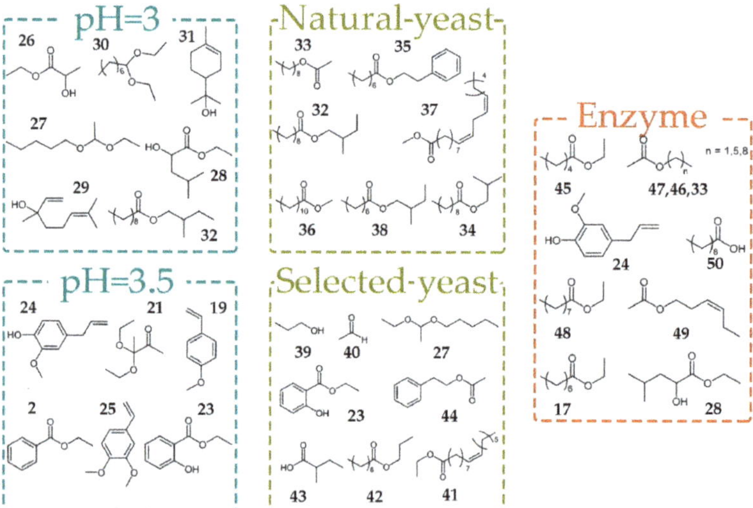

Figure 4. Metabolites important for discrimination between brandies produced under different fermentation conditions.

Satora and Tuszyński have described plum must fermentations by different yeasts isolated from fresh blue plum fruits (*Aureobasidium* sp.) and spontaneously fermenting plum musts (*Kloeckera apiculata* and *Saccharomyces cerevisiae*), as well as commercial wine and distillery strains non-*Saccharomyces* yeasts. They found that samples after spontaneous fermentation were distinguished by a high content of acetoin, ethyl acetate, and total esters, accompanied by a low level of methanol and fusel alcohols. Non-*Saccharomyces* yeasts were responsible for higher concentrations of esters and methanol, while *S. cerevisiae* strains resulted in increased levels of higher alcohols. It was also found that isolated indigenous strains of *S. cerevisiae* synthesized relatively low amounts of higher alcohols compared to commercial cultures [17]. Urošević et al. evaluated the influence of yeast and nutrients on the quality of apricot brandy using five yeast strains *S. cerevisiae* and *S. bayanus*. The best results were obtained with yeast strain *S. cerevisiae* ex r.f. *bayanus* and yeast strain *S. cerevisiae* ex ph.r. *bayanus*, which gave a high content of linalool. The control sample with no nutrients and selected yeast gave a distillate that was evaluated as having the worst quality with higher concentration of ethyl acetate, 1-butanol, 1-hexanol, and amyl alcohols [27].

Among many constituent of plum are organic acids, which are of great importance for alcohol fermentation chemical processes. Plum fruits do not contain higher amounts of acids and their pH is over 3.5, in most cases [28]. pH value affects the growth and fermentation rate of yeast. Our results have shown that pH value considerably influenced the composition of fermentation products. Benzene derivatives (**2, 19, 21, 23–25**) were a dominate group of volatile compounds in brandies obtained on pH = 3.5. Decreasing of the pH value to pH = 3.0 led to increased levels of acetals (**27, 30**), monoterpenes (**29, 31**) and esters **26, 28**, and **32**.

In alcoholic fermentation, enzymes act as catalysts for carbohydrate decomposition reactions and the formation of specific compounds. The use of enzyme preparations also has a significant impact on the quality of plum brandy [27].

Treatment of fermentation broths with pectic enzyme complex Lallzyme BETA™ containing pectinases, beta-glucosidase, rhamnosidase, apiosidase, and arabinofuranosidase provoked increased yield of ethyl esters (**17, 28, 45, 48**), acetates (**33, 46, 47, 49**), eugenole (**24**) and propionic acid (**50**). On the other hand, the treatment with Lallzyme CUVÉE BLANC™, which contains mainly pectinases, did not show any effect on brandy composition.

2.5. Correlation of Brandies Profiles with Their Sensory Properties

In general, the most common volatiles that contribute to the flavor and/or aroma profile of fruit brandies can be broadly categorized into several main groups: higher alcohols, esters, acetates, organic acids, etc. Nevertheless, there are many minor volatile and non-volatile components such as aldehydes, ketones, acetals, lactones, terpenes, phenols, and glycerol, which contribute to the full bouquet of flavors, aroma, and mouthfeel of brandies.

The most of the straight-chain alcohols have a strong pungent smell. At low concentrations, they contribute to the aromatic complexity but at higher levels are characterized by penetrating odors, which mask the aromatic finesse. In distilled spirits, such as brandies, rum, and whisky, fusel alcohols provide most of their common aromatic character [29]. As seen in Figure 5, higher alcohols propanol (**39**), amyl alcohols (**51**) and 1-hexanol (**52**) are metabolites that contribute favorable sensorial properties of plum brandy. 1-Hexanol (**52**) usually has a positive influence on the aroma of a distillate when its concentration reaches 20 mg/L [30], but it also will negatively affect the flavor of brandies by imparting a grass-like note, when the concentration exceeds 100 mg/L [11]. The amyl alcohols/isobutanol and isobutanol/propanol ratios are used in some alcoholic beverages such as whisky as a quality index and should be greater than unity [31]. Studies by Satora et al. found that the plum brandies contained more propanol than isobutanol, without any influence on quality of the beverage [13].

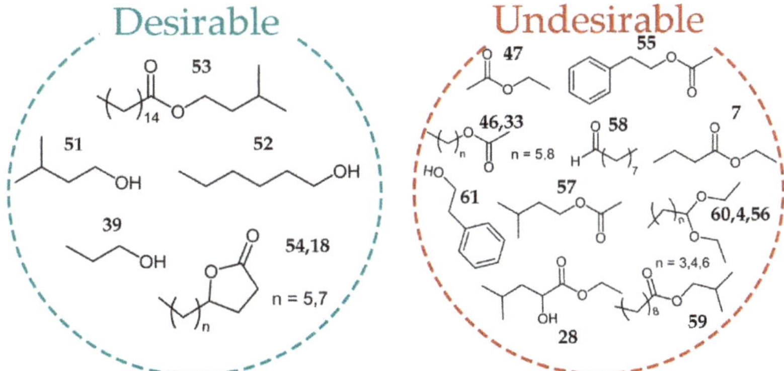

Figure 5. Desirable and undesirable metabolites as determined from correlation of brandies profiles with their sensory properties.

Lactones (**18**, **54**) were the second dominant class of desirable compounds in plum brandies. These compounds, particularly γ-lactones, are important compounds in terms of their contribution to the aroma and, in general, present fruity odor descriptors. γ-Lactones are important flavor and aroma constituents and occur in fruits such as apricot, peach, plum, and strawberry. Horvat et al. [14] found that lactones were also a major class of volatile compounds in southeastern American plums. Studies reported by Gomez et al. designate γ-dodecalactone as the major lactone in Japanese plum (*P. salicina*) [15].

Isopentenyl palmitate (**53**) was also obtained in plum brandy. This ester is rare in fruit brandy, but may be responsible for the apple, pineapple, and banana whiskey odor [32].

Acetates (**33**, **46**, **47**, **55**, and **57**), esters (**7**, **28**, **59**), acetals (**4**, **60**, **56**), octanal (**58**), and phenyl ethyl alcohol were marked as undesirable molecules, but this should be understood conditionally, because for such a complex aroma mixture, the question of exact concentration and synergistic effect are of great importance.

Ethyl acetate (**47**) is the most common and typical ester in fruit brandy. In small concentrations, they have floral notes but at high concentration, they can be very repulsive, exhibiting the odor of solvent [33].

3. Materials and Methods

3.1. Materials

Methanol, ethanol, 1-propanol, acetaldehyde, ethyl-acetate, *iso*-butanol, 1-butanol, isoamyl alcohol, 1-hexanol, dichloromethane, 4-methyl-1-pentanol and methyl 10-undecenoate and formic acid of were purchased from Sigma–Aldrich (Steinheim, Germany). Anhydrous magnesium-sulfate was obtained from Merck (Darmstadt, Germany). All chemicals were p.a. purity. Commercial enzymes β Lallzyme™, Lallzyme Cuvee Blanc™, and yeast Lalvin QA23™ were supplied from Lallemand (Montreal, QC, Canada).

3.2. Plum Sample Collection

The plum fruits were collected at the stage of technological maturity near the town Kosjerić (Western Serbia) in autumn 2017. Three cultivars were collected: Trnovača, Crvena ranka, and Požegača; taking care that the fruits did not contain any impurities (e.g., soil, leaves, branches) and without rotten or damaged fruits in order to avoid the influence of factors that are not the subject of these experiments. Each variety was harvested from the same localities and plantations, which eliminates the influence of the difference in results due to different microclimates, soil composition, or agrotechnics applied in orchards.

3.3. Brandy Production

The amount of mashed fruits without stones in each experiment was 45 kg. A total of twelve experiments in triplicate were set for each cultivar (Supplementary Materials, Table S1). The addition of enzymes β Lallzyme™, Lallzyme Cuvee Blanc™, yeast Lalvin QA23™, and pH value was varied. Concentration of enzyme β Lallazyme™ was 0.05 g/kg in experiments II, V, VIII, and XI. The concentration of Lallzyme Cuvee Blanc™ was 0.02 g/kg in experiments III, VI, IX, XII; and concentration of yeast Lalvin QA23™ was 0.22 g/kg in experiments VII–XII. In experiments IV–VI, and X–XII the pH value was adjusted with metatartaric acid to 3.0, or 3.5. Experiment I was control without the addition of enzymes or yeast. Fermentations were conducted in 60 L plastic containers. When fermentation was done all samples were distilled. Double distillation was conducted in a 15 L copper pot. The first distillate was subjected to a second distillation, and three fractions were collected: the first ("head"), the second ("heart"), and the third ("tail"). The "head" fraction was collected until volume reach 1.5% of total volume of the first distillate. The "heart" fraction was collected until the distillate reached a value of 65% (*v/v*) ethanol. The obtained distillates were diluted with demineralized water to the strength of 45% (*v/v*). Total of 108 plum brandies, originated from three plum cultivars, were manufactured using different fermentation conditions.

3.4. Sample Preparation

A total of 108 plum brandies, originated from three plum cultivars, were manufactured using different fermentation conditions. Main volatiles and aroma compounds in the 108 manufactured plum brandies were determined by GC-FID and GC-FID-MS analysis, respectively.

Sample preparation for determination of main volatiles (acetaldehyde, ethyl-acetate, methanol, 1-propanol, *i*-butanol, 1-butanol, isoamyl alcohol, and 1-hexanol) in the plum brandies by GC-FID was carried out by adding 3 μL of 4-methyl-1-pentanol as an internal standard to 5 mL of brandy. A mixture of authentic standards is made by adding 3 μL of each compound to 5 mL of 40% ethanol [12].

The extraction of aroma compounds for GC-FID–MS analysis was started by diluting an aliquot of 100 mL of plum beverage with 100 mL distilled water, followed by adding of 15 mL of dichloromethane, 1 mL of internal standard (methyl ester of 10-undecenoic acid, 1.8 mg/mL in dichloromethane), and continuously extracted on vortex for 3 min. The dichloromethane extract was dried over anhydrous magnesium sulfate, and concentrated under nitrogen flow to final volume of 1.5 mL.

3.5. GC-FID and GC-FID-MS Analysis

The GC–FID analysis of main volatiles were conducting on Agilent 7890A gas chromatograph (Agilent Technologies, Santa Clara, CA, USA) equipped with polar HP-INNOWax capillary column (30 m × 0.32 mm, 0.25 μm film thickness). The analyses were performed in split mode 15:1 with helium as carrier gas at 15.74 psi in constant pressure mode. The injection volume was 1 μL, the injector temperature was 220 °C, column temperature was programmed linearly in the range of 40–90 °C at rate of 5 °C/min with initial 10 min hold, then temperature increased to 240 °C at rate of 50 °C with final 10 min hold. The FID temperature was 300 °C.

The GC–FID-MS analysis of aroma compounds were analyzed on Agilent 7890A GC system (Agilent Technologies, Santa Clara, CA, USA) equipped with a 5975C mass selective detector (MSD) and a FID connected by capillary flow technology through a two-way splitter. A non-polar HP-5MSI capillary column (30 m × 0.25 mm, 0.25 μm film thickness) was used. Column temperature was programmed linearly in the range of 60 °C to 270 °C at a rate of 3 °C/min, then at a rate of 20 °C/min to 310 °C with final 8 min hold. Helium was used as carrier, auxiliary and make up gas; inlet pressure was constant at 19.7 psi (flow 1.0 mL/min at 210 °C), auxiliary pressure was 3.8 psi and FID make up flow was 25 mL/min. FID temperature was 300 °C, split ratio was 5:1 and injection volume was 1 μL for all analysis. Mass spectra obtained by electron ionization with 70 eV at 200 °C. Quadrupole temperature was set to 150 °C and MS range was 40–550 amu. Transfer line temperature was 315 °C.

3.6. Data Processing

Identification of main volatiles were done by comparison of retention times of authentic standards in GC-FID chromatograms [10]. Concentrations of these components were calculated, using the following equation:

$$C_i = (A_i/A_s) \times C_s \times RF_i \tag{1}$$

where:

C_i is the concentration of the component to be determined

A_i—the area below the peak of the component of unknown concentration

A_s—area below peak standard

C_s—concentration standard

RF_i—relative response factor

Library search and mass spectral deconvolution and extraction of aroma compounds were performed using the MSD ChemStation software, version E02.02 (Agilent Technologies, Santa Clara, CA, USA), the NIST AMDIS (Automated Mass Spectral Deconvolution and Identification System) software, version 2.70, and the commercially available Adams04, NIST17, and Wiley07 libraries containing approximately 500,000 spectra. The concentration of the aroma compounds was determined using the peak area of internal standard methyl ester of 10-undecenoic acid, and calculated using the following equation:

$$C_i = (A_i / A_s) \times C_s \times DF_i \qquad (2)$$

where:

C_i is the concentration of the component to be determined

A_i—the area below the peak of the component of unknown concentration

A_s—area below peak standard

C_s—concentration standard

DF_i—dilution factor

The obtained concentrations of main volatiles and aroma compounds were merged in one table, and subjected to multivariate data analysis. Principal component analysis (PCA), orthogonal partial least squares to latent structures (OPLS), and orthogonal partial least squares to latent structures-discriminant analysis (OPLS-DA) methods were performed with SIMCA software (version 15, Sartorius, Göttingen, Germany). The data were mean centered, and scaled to unit variance.

3.7. Sensory Analysis of Brandies

Sensory analysis was performed by Buxbaum modified method with a maximum of 20 points [18,28].

The sensory evaluation of brandies was performed in three repetitions. The evaluations were performed by three verified evaluators. The evaluation was anonymous, according to the so-called point-type system. The main quality parameters were evaluated: color, clarity, typicality, smell, and taste.

4. Conclusions

The GC-FID-MS-based metabolomics approach was used for the detection of peculiarities in different cultivars and fermentation process, as well as in brandy sensory characteristics. Furthermore, utilization of PCA and OPLS-DA multivariate analysis methods on GC-FID-MS data revealed metabolites important for discrimination between brandies produced under different fermentation conditions (differences in plum varieties, pH of plum mash, and addition of selected yeast or enzymes during fermentation).

Correlation of brandy GC-FID-MS profiles with their sensory properties achieved by OPLS multivariate data analysis. In total, six desirable and 13 undesirable metabolites revealed to support fine tuning between non-aged plum brandies of different quality.

Proposed workflow confirmed the potential of GC-FID-MS in combination with multivariate data analysis that can be applied to assess brandy quality and to understand the chemistry behind the extraordinary plum brandy quality.

Though the authors studied brandies from three local plum cultivars, it is expected that the results and multivariate models could be extended to other stone fruit spirits.

Supplementary Materials: The following are available online, Table S1: The plum mash experiments of Požegača, Crvena ranka and Trnovača cultivars depending on the added of enzyme, yeast (0-not added. 1-added) and pH values, Table S2: The main volatile compounds from plum brandy (GC-FID), Table S3: Aroma compounds from plum brandy (GC-FID/MSA), Table S4: Misclassification tables of the OPLS-DA models.

Author Contributions: Conceptualization, D.G. and M.L.; methodology, D.G. and L.V.; validation, S.I.; formal analysis, S.I. and K.S.; investigation, S.I.; resources, M.L.; data curation, S.I.; writing—original draft preparation, S.I., V.T. and D.G.; writing—review and editing, L.V.; visualization, S.I.;

supervision, D.G. and L.V.; project administration, D.G. and V.T.; funding acquisition, D.G. and M.L. All authors have read and agreed to the published version of the manuscript.

Funding: This research was funded by Ministry of Education, Science and Technological Development (Serbia), grant numbers 451-03-9/2021-14/200026 and 451-03-9/2021-14/200168.

Institutional Review Board Statement: Not applicable.

Informed Consent Statement: Not applicable.

Data Availability Statement: Not applicable.

Acknowledgments: The authors thank Accreditet Laboratory CH IHTM, Belgrade, Serbia and Zarić Distillery, Kosjerić, Serbia for financial support.

Conflicts of Interest: The authors declare no conflict of interest.

Sample Availability: Not applicable.

References

1. Miličević, B.; Lukić, I.; Babic, J.; Šubarić, D.; Miličević, R.; Ackar, D.; Miličević, D. Aroma and sensory characteristics of Slavonian plum brandy. *Technol. Acta* **2012**, *5*, 1–7.
2. Statista Brandy. Available online: https://www.statista.com/outlook/10020500/100/brandy/worldwide (accessed on 18 January 2021).
3. Food and Agriculture Organisation of the United Nations Food and Agriculture Data. Available online: http://www.fao.org/faostat (accessed on 18 January 2021).
4. Milošević, T.; Milošević, N. Quantitative analysis of the main biological and fruit quality traits of F1 plum genotypes (*Prunus domestica* L.). *Acta Sci. Pol. Hortorum Cultus Ogrod.* **2011**, *10*, 95–107.
5. Milosevic, T.; Glisic, I. Agronomic properties and nutritional status of plum trees (Prunus domestica L.) influenced by different cultivars. *J. Soil Sci. Plant Nutr.* **2013**, *13*, 706–714. [CrossRef]
6. Mišić, D.P. *Šljiva*; Partenon: Belgrade, Serbia, 2006; ISBN 978-86-7157-289-7.
7. Ljekocevic, M.; Jadranin, M.; Stankovic, J.; Popovic, B.; Nikicevic, N.; Petrovic, A.; Tesevic, V. Phenolic composition and anti-DPPH radical scavenging activity of plum wine produced from three plum cultivars. *J. Serbian Chem. Soc.* **2019**, *84*, 141–151. [CrossRef]
8. Spaho, N.; Đukic-Ratković, D.; Nikićević, N.; Blesić, M.; Tešević, V.; Mijatović, B.; Murtić, M.S. Aroma compounds in barrel aged apple distillates from two different distillation techniques. *J. Inst. Brew.* **2019**, *125*, 389–397. [CrossRef]
9. He, X.; Yangming, H.; Górska-Horczyczak, E.; Wierzbicka, A.; Jeleń, H.H. Rapid analysis of Baijiu volatile compounds fingerprint for their aroma and regional origin authenticity assessment. *Food Chem.* **2021**, *337*, 128002. [CrossRef] [PubMed]
10. Tešević, V.; Nikićević, N.; Jovanović, A.; Djoković, D.; Vujisić, L.; Vučković, I.; Bonić, M. Volatile Components from Old Plum Brandies. *Food Technol. Biotechnol.* **2005**, *43*, 367–372.
11. Satora, P.; Tuszyński, T. Influence of indigenous yeasts on the fermentation and volatile profile of plum brandies. *Food Microbiol.* **2010**, *27*, 418–424. [CrossRef]
12. Spaho, N.; Dürr, P.; Grba, S.; Velagić-Habul, E.; Blesić, M. Effects of distillation cut on the distribution of higher alcohols and esters in brandy produced from three plum varieties. *J. Inst. Brew.* **2013**, *119*, 48–56. [CrossRef]
13. Satora, P.; Tuszyński, T. Chemical characteristics of Śliwowica Łącka and other plum brandies. *J. Sci. Food Agric.* **2008**, *88*, 167–174. [CrossRef]
14. Horvat, R.J.; Chapman, G.W., Jr.; Senter, S.D.; Norton, J.D.; Okie, W.R.; Robertson, J.A. Comparison of the volatile compounds from several commercial plum cultivars. *J. Sci. Food Agric.* **1992**, *60*, 21–23. [CrossRef]
15. Gomez, E.; Ledbetter, C.A.; Hartsell, P.L. Volatile compounds in apricot, plum, and their interspecific hybrids. *J. Agric. Food Chem.* **1993**, *41*, 1669–1676. [CrossRef]
16. Ledauphin, J.; Le Milbeau, C.; Barillier, D.; Hennequin, D. Differences in the Volatile Compositions of French Labeled Brandies (Armagnac, Calvados, Cognac, and Mirabelle) Using GC-MS and PLS-DA. *J. Agric. Food Chem.* **2010**, *58*, 7782–7793. [CrossRef] [PubMed]
17. Satora, P.; Tuszynski, T. Biodiversity of Yeasts During Plum Wegierka Zwykla Spontaneous Fermentation. *Food Technol. Biotechnol.* **2005**, *43*.
18. Nikicevic, N. Terminology used in sensory evaluation of plum brandy sljivovica quality. *J. Agric. Sci. Belgrade* **2005**, *50*, 89–99. [CrossRef]
19. Jakubíková, M.; Sádecká, J.; Kleinová, A. On the use of the fluorescence, ultraviolet–visible and near infrared spectroscopy with chemometrics for the discrimination between plum brandies of different varietal origins. *Food Chem.* **2018**, *239*, 889–897. [CrossRef] [PubMed]
20. Bajer, T.; Hill, M.; Ventura, K.; Bajerová, P. Authentification of fruit spirits using HS-SPME/GC-FID and OPLS methods. *Sci. Rep.* **2020**, *10*, 18965. [CrossRef] [PubMed]

21. Jakubíková, M.; Sádecká, J.; Hroboňová, K. Classification of plum brandies based on phenol and anisole compounds using HPLC. *Eur. Food Res. Technol.* **2019**, *245*, 1709–1717. [CrossRef]

22. Beale, D.J.; Pinu, F.R.; Kouremenos, K.A.; Poojary, M.M.; Narayana, V.K.; Boughton, B.A.; Kanojia, K.; Dayalan, S.; Jones, O.A.H.; Dias, D.A. Review of recent developments in GC–MS approaches to metabolomics-based research. *Metabolomics* **2018**, *14*, 152. [CrossRef]

23. Wiklund, S.; Johansson, E.; Sjöström, L.; Mellerowicz, E.J.; Edlund, U.; Shockcor, J.P.; Gottfries, J.; Moritz, A.T.; Trygg, J. Visualization of GC/TOF-MS-Based Metabolomics Data for Identification of Biochemically Interesting Compounds Using OPLS Class Models. *Anal. Chem.* **2008**, *80*, 115–122. [CrossRef] [PubMed]

24. Vyviurska, O.; Matura, F.; Furdíková, K.; Špánik, I. Volatile fingerprinting of the plum brandies produced from different fruit varieties. *J. Food Sci. Technol.* **2017**, *54*, 4284–4301. [CrossRef] [PubMed]

25. Awad, P.; Athès, V.; Decloux, M.E.; Ferrari, G.; Snakkers, G.; Raguenaud, P.; Giampaoli, P. Evolution of Volatile Compounds during the Distillation of Cognac Spirit. *J. Agric. Food Chem.* **2017**, *65*, 7736–7748. [CrossRef] [PubMed]

26. Pielech-Przybylska, K.; Balcerek, M.; Nowak, A.; Patelski, P.; Dziekońska-Kubczak, U. Influence of yeast on the yield of fermentation and volatile profile of 'Węgierka Zwykła' plum distillates. *J. Inst. Brew.* **2016**, *122*, 612–623. [CrossRef]

27. Urošević, I. Uticaj Sojeva Selekcionisanog Kvasca i Hraniva u Fermentaciji na Hemijski Sastav i Senzorne Karakteristike Voćnih Rakija. Ph.D. Thesis, University of Belgrade, Faculty of Agriculture, Belgrade, Serbia, 2015.

28. Nikićević, N.; Tešević, V. Possibilities for methanol content reduction in plum brandy. *J. Agric. Sci. Belgrade* **2005**, *50*, 49–60. [CrossRef]

29. Tsakiris, A.; Kallithraka, S.; Kourkoutas, Y. Grape brandy production, composition and sensory evaluation. *J. Sci. Food Agric.* **2014**, *94*, 404–414. [CrossRef]

30. Apostolopoulou, A.; Flouros, A.; Demertzis, P.; Akrida-Demertzi, K. Differences in concentration of principal volatile constituents in traditional Greek distillates. *Food Control.* **2005**, *16*, 157–164. [CrossRef]

31. Cortés, S.; Gil, M.L.; Fernández, E. Volatile composition of traditional and industrial Orujo spirits. *Food Control.* **2005**, *16*, 383–388. [CrossRef]

32. FlavorDb Whisky. Available online: https://cosylab.iiitd.edu.in/flavordb/entity_details?id=25 (accessed on 18 January 2021).

33. Urosevic, I.; Nikicevic, N.; Stankovic, L.; Andjelkovic, B.; Urosevic, T.; Krstic, G.; Tesevic, V. Influence of yeast and nutrients on quality of apricot brandy. *J. Serbian Chem. Soc.* **2014**, *79*, 1223–1234. [CrossRef]

molecules

MDPI

Article

NMR and LC-MS-Based Metabolomics to Study Osmotic Stress in Lignan-Deficient Flax

Kamar Hamade [1], Ophélie Fliniaux [1], Jean-Xavier Fontaine [1], Roland Molinié [1], Elvis Otogo Nnang [1], Solène Bassard [1], Stéphanie Guénin [2], Laurent Gutierrez [2], Eric Lainé [3], Christophe Hano [3], Serge Pilard [4], Akram Hijazi [5], Assem El Kak [6] and François Mesnard [1,*]

[1]   UMRT INRAE 1158 BioEcoAgro, Laboratoire BIOPI, University of Picardie Jules Verne, 80000 Amiens, France; kamar.hamade@u-picardie.fr (K.H.); ophelie.fliniaux@u-picardie.fr (O.F.); jean-xavier.fontaine@u-picardie.fr (J.-X.F.); roland.molinie@u-picardie.fr (R.M.); elvisotogonnang@gmail.com (E.O.N.); solene.bassard@u-picardie.fr (S.B.)
[2]   CRRBM, University of Picardie Jules Verne, 80000 Amiens, France; stephanie.vandecasteele@u-picardie.fr (S.G.); laurent.gutierrez@u-picardie.fr (L.G.)
[3]   USC INRAE 1328, Laboratoire LBLGC, Antenne Scientifique Universitaire de Chartres, University of Orleans, 28000 Chartres, France; eric.laine@univ-orleans.fr (E.L.); christophe.hano@univ-orleans.fr (C.H.)
[4]   Plateforme Analytique, University of Picardie Jules Verne, 80000 Amiens, France; serge.pilard@u-picardie.fr
[5]   Platform for Research and Analysis in Environmental Sciences (PRASE), Lebanese University, Beirut 6573, Lebanon; akram.hijazi@ul.edu.lb
[6]   Laboratoire de Biotechnologie des Substances Naturelles et Produits de Santé (BSNPS), Lebanese University, Beirut 6573, Lebanon; aelkak@ul.edu.lb
*   Correspondence: francois.mesnard@u-picardie.fr; Tel.: +33-322-82-7787

Citation: Hamade, K.; Fliniaux, O.; Fontaine, J.-X.; Molinié, R.; Otogo Nnang, E.; Bassard, S.; Guénin, S.; Gutierrez, L.; Lainé, E.; Hano, C.; et al. NMR and LC-MS-Based Metabolomics to Study Osmotic Stress in Lignan-Deficient Flax. *Molecules* **2021**, *26*, 767. https://doi.org/10.3390/molecules26030767

Academic Editors: Young Hae Choi, Young Pyo Jang, Yuntao Dai and Luis Francisco Salomé-Abarca

Received: 10 December 2020
Accepted: 28 January 2021
Published: 2 February 2021

Publisher's Note: MDPI stays neutral with regard to jurisdictional claims in published maps and institutional affiliations.

Abstract: Lignans, phenolic plant secondary metabolites, are derived from the phenylpropanoid biosynthetic pathway. Although, being investigated for their health benefits in terms of antioxidant, antitumor, anti-inflammatory and antiviral properties, the role of these molecules in plants remains incompletely elucidated; a potential role in stress response mechanisms has been, however, proposed. In this study, a non-targeted metabolomic analysis of the roots, stems, and leaves of wild-type and PLR1-RNAi transgenic flax, devoid of (+) secoisolariciresinol diglucoside ((+) SDG)—the main flaxseed lignan, was performed using ^{1}H-NMR and LC-MS, in order to obtain further insight into the involvement of lignan in the response of plant to osmotic stress. Results showed that wild-type and lignan-deficient flax plants have different metabolic responses after being exposed to osmotic stress conditions, but they both showed the capacity to induce an adaptive response to osmotic stress. These findings suggest the indirect involvement of lignans in osmotic stress response.

Keywords: lignans; secoisolariciresinol; RNA interference; flax; osmotic stress; metabolomics; ^{1}H-NMR; LC-MS

1. Introduction

Flax (*Linum usitatissimum*) has been extensively used as a plant model for lignan biosynthesis studies. Flaxseed is a rich source of lignans and neolignans and contains about 75–800 times more lignans than other plant-derived food [1]. Secoisolariciresinol diglucoside (SDG) is the major lignan present in flaxseed and exists as a macromolecule in the flaxseed hull. This polymer complex is composed of SDG, which are held together by 3-hydroxy-3-methylglutaric acid (HMGA) residues; The hydroxycinnamic acids, 4-*O*-β-D-glucopyranosyl-*p*-coumaric glucoside (CouAG), and 4-*O*-β-D-glucopyranosyl ferulic acid glucoside (FAG) are ester-linked to the glucosyl moiety of SDG as end units. Stereochemistry and enantiomeric composition of lignans vary between plant species, but also between different organs and/or different developmental stages within the same species [2]. For example, flax seeds contain almost 99% (+)- and 1% (−)-SDG, whereas its aerial parts during flowering stage accumulate only (−)-enantiomers of lignans [3].

71

In flax, the lignan biosynthesis pathway starts with the stereospecific coupling of two coniferyl alcohol residues, in the presence of dirigent proteins (DIR), leading to the formation of (−)-pinoresinol, which is step-wisely converted into (−)-lariciresinol and then into (+)-secoisolariciresinol through the action of a NADPH-dependent pinoresinol-lariciresinol reductase (PLR). In *L. usitatissimum*, five PLR genes are present in the genome that encodes full PLR proteins, but only two have been widely studied: *LuPLR1* and *LuPLR2*. Both *LuPLR1* and *LuPLR2* transcripts were detected in the seeds and roots but only the *LuPLR2* transcript is detected in leaves and stems [4,5]. PLR1 and PLR2 showed opposite enantiospecificity towards pinoresinol and lariciresinol. Further on, PLR1 catalyzes the conversion of (−)-pinoresinol into (−)-lariciresinol and contributes to the synthesis of (+)-secoisolariciresinol, whereas PLR2 catalyzes the conversion of (+)-pinoresinol into (+)-lariciresinol and contribute to the synthesis of (−)-secoisolariciresinol [5,6].

Lignans are well recognized for their antioxidative, anti-inflammatory, and antiestrogenic activities, key properties that contribute to their ability to reduce risk and protect against various types of cancer, especially hormone-dependent cancers, like breast, colon, and prostate cancer [7–9]. Lignans also exhibit an important role in the reduction of hypercholesterolemia, atherosclerosis, hypertension and diabetes [10–13]. Furthermore, in the last years, several research works demonstrated that dietary lignans could have beneficial effects in the prevention of neurological disorders [14].

However, in plants, the exact biological role of the lignans still remains unclear. They have potent defensive properties, including antiviral [15], antimicrobial [16], antifungal [17,18], and antifeedant activities [19]. Along with these, lignans can act as phytoalexins: they are synthesized in various plant species in response to fungal infections [20]. A possible phytoanticipin activity of lignans is also reported in heartwood where they act as a protective post-lignification infusion [21]. In addition, due to their antioxidant activity, lignans can protect the seed oil against oxidation by scavenging reactive oxygen species (ROS) or maintaining seed dormancy [22–24]. Lignans have been proposed to have a role in defense against oxidative stress during the process of lignification as it is the case in poplar [25]. Additionally, a study by Corbin et al. [26] has shown a regulation of the transcription of the *LuPLR1* gene in flax by the hormone abscisic acid (ABA) suggesting a role of lignan in stress response.

In a previous study, Renouard et al. [27], successfully used the RNAi-mediated gene silencing approach to knock-down the *LuPLR1* gene in flax. As a consequence, the accumulation of the major lignan SDG was nearly suppressed in PLR1-RNAi seeds. These plants, devoid of lignan, provide a tool to allow a comparison between wild-type and PLR1-RNAi transgenic plants, under stress condition and give better insight into the role of lignan in plants.

In the present work, we used the PLR1-RNAi transgenic flax to elucidate whether lignan is involved in the response to osmotic stress (simulating drought stress). For this purpose, we performed a comparative non-targeted metabolic analysis of the roots, stems, and leaves of wild-type and PLR1-RNAi transgenic *L. usitatissimum*, using proton-nuclear magnetic resonance spectroscopy (^{1}H-NMR) and liquid chromatography-mass spectrometry (LC-MS), to identify the common and specific metabolic responses in plants under osmotic stress conditions induced by polyethylene glycol (PEG-6000).

2. Results and Discussion

2.1. Phenotype Analysis

The presence of the selectable marker gene *NPTII* was confirmed in the samples of all homozygous Pi1AM plants selected for the study. Figure 1A shows the PCR analysis of genomic DNA of wild-type (LuT) and transgenic plants (Pi1AM). Plants expressing the *NPTII* marker gene showed no significant morphological differences from the non-transgenic plants, when they were grown under control conditions (Figure 1B). In addition, after harvest, root, stem, and leaf dry weights were determined, and no significant differences were shown between wild-type and PLR1-RNAi transgenic lines (Supplementary Data 1).

These observations suggest that the lignans have no essential role in plant growth under the present conditions.

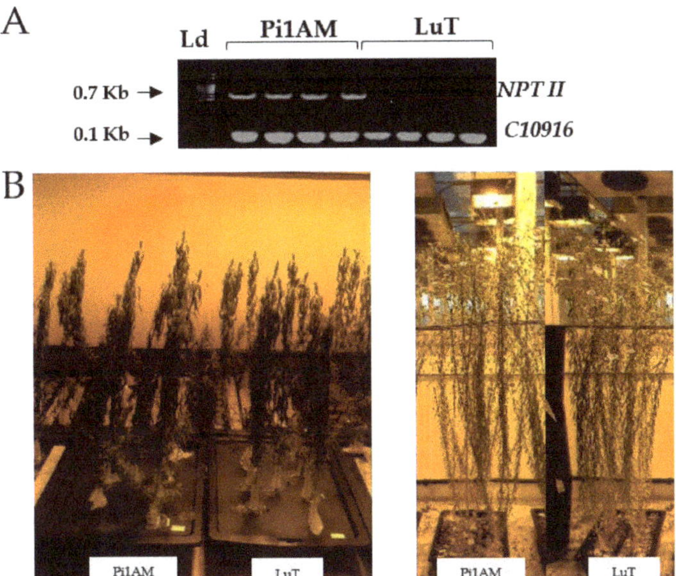

Figure 1. (**A**) Amplification of *NPTII* sequence from genomic DNA of wild-type (LuT) and transgenic plants transformed with pART-RNAiPLR binary vector (Pi1AM). PCR products obtained with a specific set of primers designed for the amplification of a 0.7 Kb *NPTII* fragment were separated by 2.0% agarose gel electrophoresis and visualized with GelRed, under UV light. Ld, a 100 pb DNA Ladder (Solis BioDyne). Flax housekeeping gene was used as a reference gene to normalize the PCR data. (**B**) Phenotypes of wild-type and transgenic plants under control conditions. Pictures were taken at 25 and 75 days after sowing (left and right respectively).

2.2. Metabolite Profiling

2.2.1. Difference of Metabolic Profiles between Wild-Type and PLR1-RNAi Transgenic Lines under Control Conditions

Before studying the impact of osmotic stress on the *L. usitatissimum* metabolome, we firstly compared the metabolite composition of each of the three parts (root, stem, and leaf), of wild-type (LuT) and transgenic (Pi1AM) plants, in control conditions at different time points (day 1 (D1) and day 5 (D5)).

Metabolomic profiles analysis was performed using LC-MS and NMR. While NMR is less sensitive than LC-MS, it can be used for quantification and structural identification of unknown compounds. NMR is, therefore, ideal for untargeted metabolomic studies. However, LC-MS is more sensitive than NMR, and can detect a higher number of metabolites, particularly less abundant compounds such as secondary metabolites. Thus, applying both NMR and LC-MS improves the coverage of the metabolome and enhances both metabolite resolution and annotation.

The LC-MS and NMR spectral data obtained for each plant part, were subjected separately to a multivariate non-targeted analysis-PLS-DA. The score plot of PLS-DA, shown in Figure 2, shows that transgenic LuPLR1-RNAi samples (Pi1AM) were grouped away from the wild-type samples, in various plant parts. These results highlight the metabolomic variability between LuT and Pi1AM. The next step to compare LuT and Pi1AM consisted of identifying the discriminant metabolites in each plant part.

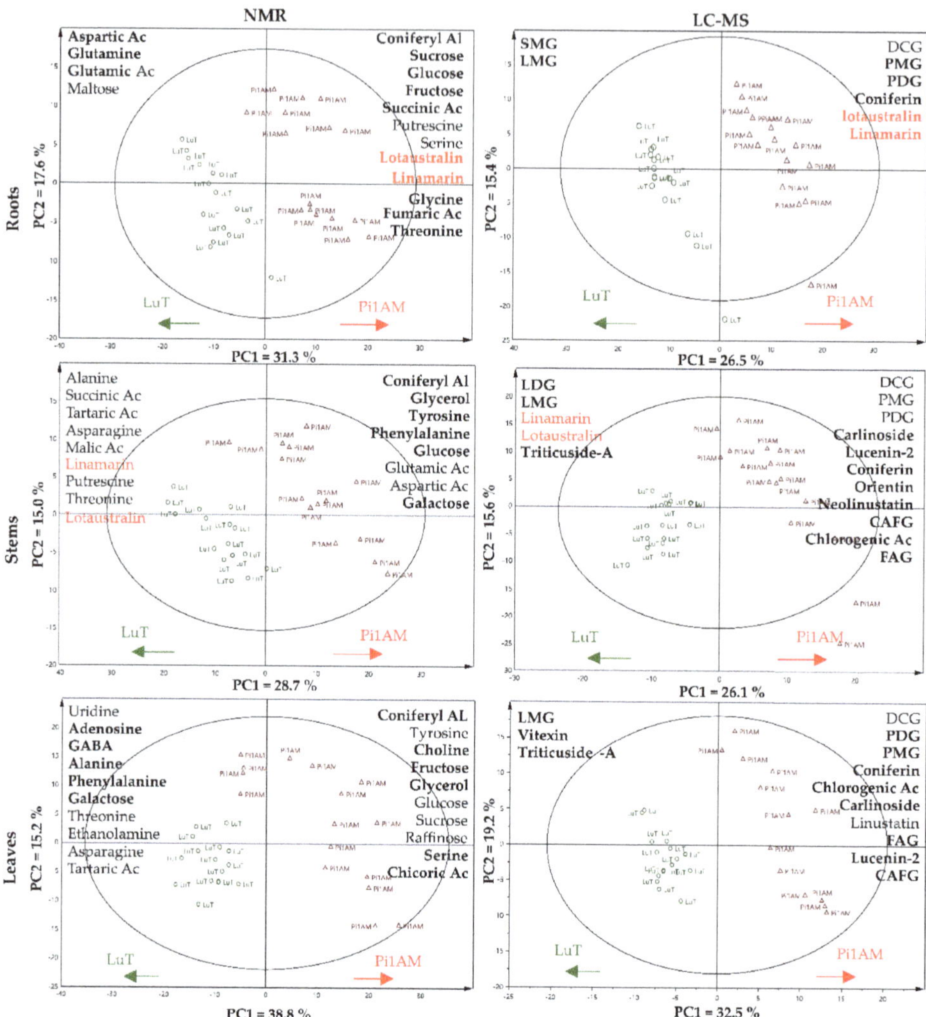

Figure 2. Score plot of partial least squares discriminant analysis (PLS-DA) of metabolite content in flax roots, stems, and leaves, based on ^{1}H-NMR and LC-MS data for LuT and Pi1AM, in control conditions. Samples represent LuT plants (circle) and Pi1AM plants (triangle). Arrows indicate the increases in relative metabolite content in various parts of the wild-type (LuT) or transgenic (Pi1AM) plants. Metabolites are listed in descending order of discrimination. Boldface text indicates discriminant metabolites of the two lines in both control and stress conditions. Red text indicates metabolites measured in common with LC-MS and ^{1}H-NMR. Variable Importance in Projection (VIP) score and the Log10 (Ratio Pi1AM/LuT) of the main discriminant metabolites for each PLS-DA model, are shown in supplementary data 2.

The results of statistical analysis (Wilcoxon Rank Sum test) indicate a significant difference in 22 metabolites in roots, 32 in stems, and 33 in leaves, between LuT and Pi1AM. Differences were observed in primary metabolites as well as in secondary metabolites in different plant parts.

Changes in Primary Metabolism

The changes in primary metabolites are more evident by the PLS-DA model obtained with NMR spectral data. Differences in the accumulation of these metabolites between

LuT and Pi1AM in different plant parts indicate that the change is not restricted to a few chemical classes or categories, and involves different metabolic pathways.

Among primary metabolites, coniferyl alcohol displayed similar trends in different plant parts: The amount of this compound although, exhibited to be 7.91, 8.65, and 10.32 times higher in Pi1AM than in LuT in roots, stems and leaves, respectively.

In addition, when compared to LuT roots, those of Pi1AM, the transgenic line, had significantly higher contents of nine metabolites but significantly lower contents of four metabolites. These up-accumulated metabolites were mainly glucose, fructose, succinic acid, putrescine, serine, sucrose, glycine, fumaric acid and threonine. In contrast, LuT roots were significantly higher in aspartic acid, glutamine, glutamic acid, and maltose contents.

Moreover, Pi1AM stems presented an up-accumulation of seven metabolites, which include glycerol, tyrosine, phenylalanine, glucose, glutamic acid, aspartic acid, and galactose, when compared to LuT stems while these latter showed an up-accumulation of seven different metabolites identified as alanine, succinic acid, tartaric acid, asparagine, malic acid, putrescine, and threonine.

In leaves, the Pi1AM line had significantly higher contents of nine metabolites than LuT including many sugars such as fructose, glucose, sucrose, raffinose, but also tyrosine, choline, glycerol, serine, and chicoric acid. Despite, LuT displayed a higher content in 10 metabolites including amino acids such as GABA, alanine, phenylalanine, threonine, asparagine, and other compounds, such as uridine, adenosine, galactose, ethanolamine, and tartaric acid.

Despite the genetic modification is in a part of the genome coding for compounds that are involved in metabolic pathways away from the biosynthetic pathway of amino acids, sugars, and organic acids, we observed major changes for these latter. These primary metabolites are central molecules. Changes at these levels can therefore affect a large part of the other metabolic pathways of the plant.

In the different plant parts, nitrogen metabolism showed changes in the content of amino acids such as glutamic acid, glutamine, aspartic acid, and asparagine, in the transgenic line compared to the wild-type. These amino acids are commonly involved in nitrogen recycling and transport.

These observations lead to hypothesize that these amino acids can be affected by the disruption of the activity of phenylalanine ammonia-lyase (PAL) which can be caused by the accumulation of coniferyl alcohol, in the transgenic line.

PAL constitutes the point of connection between the primary metabolism of shikimate, which leads to aromatic amino acids and the secondary metabolism of phenylpropanoids. Generally, phenylpropanoid biosynthesis begins with the deamination of phenylalanine by phenylalanine ammonia-lyase (PAL) to yield cinnamic acid [28]. In the second step, the action of cinnamate 4-hydrolase (C4H) transforms the cinnamic acid into *p*-coumaric acid. Then, 4-coumaroyl CoA ligase (4CL) catalyzes *p*-coumaric acid into *p*-coumaroyl-CoA. This latter is a crucial branch point leading to the generation of various phenolic compounds such as flavonoids, lignins and lignans [29]. In transgenic plants, coniferyl alcohol can accumulate eventually to levels that could be toxic to the plant cells. Therefore, cells can put in place a negative feedback to limit this accumulation. This regulation must act on the pathway of phenylpropanoid, responsible for the production of coniferyl alcohol. PAL are the most potential candidates for this process, since the regulation of PALs can also affect nitrogen metabolism.

In fact, PALs are responsible for the release of NH^{4+} from phenylalanine which is then incorporated into glutamate to form glutamine. Then, amino and amide groups from glutamate and glutamine are transferred to other molecules by transamination and transamidation reactions for the synthesis of amino acids [30].

Changes in Secondary Metabolism

On the other hand, and with regard to secondary metabolites, the loading plots obtained from LC-MS data indicate that, the major change can be observed in the lignan metabolite group in the different plant parts. An 18.1, 5.31, and 13.78-fold increase in the content of pinoresinol mono-glucoside (PMG), was observed in roots, stems, and leaves of the Pi1AM line, respectively. The diglucosylated forms of this metabolite (PDG), also showed a 12.27, 4.73 and approximately 15-folds increase in roots, stems, and leaves of Pi1AM, respectively. The drastic accumulation of pinoresinol may initiate a downregulation on the dimerization process of coniferyl alcohol that yield pinoresinol, as a negative feedback, and that, as a consequence, can explain the coniferyl alcohol accumulation in the transgenic line.

However, the accumulation level of pinoresinol in both its mono and diglucoside forms, was higher than that of the coniferyl alcohol, in the transgenic line. This observation suggests that coniferyl alcohol can be accumulated in a different form other than its free form. This hypothesis was supported by an accumulation of coniferin observed in the three parts of the transgenic plant compared to the wild-type.

Coniferin is a glycoside of coniferyl alcohol, and might serve as an alternative precursor in lignan production. Coniferin is accumulated in Pi1AM compared to the LuT line, with 2.32, 2.4, and 7.68-fold increases in roots, stems, and leaves, respectively. This accumulation is also in accordance with the increase of coniferyl alcohol. Thus, it is likely that, the coniferyl alcohol accumulates in both free and glycosylated forms in the transgenic line.

Lariciresinol mono-glucoside (LMG) showed to be lower in Pi1AM when compared to the LuT line, with a Pi1AM/LuT ratios (calculated by the formula: ratio Pi1AM/LuT = content Pi1AM plant/content LuT plant) of 0.31, 0.73, and 0.50 for roots, stems, and leaves, respectively.

In our study, lariciresinol in di-glycosilated form (LDG) is not detected, both in roots and leaves of *L. usitatissimum*. However, LDG is detected in stems in a lower amount in Pi1AM when compared to the LuT (Pi1AM/LuT ratio: 0.17)

The negative correlation between the accumulation of pinoresinol and lariciresinol with the PLR activity confirmed that the accumulation of theses lignans occurred due to the reduction of the PLR enzyme activity.

A remarkable reduction of secoisolariciresinol mono-glucoside (SMG) was observed only in roots of the Pi1AM line compared with that in roots of LuT. Secoisolariciresinol are usually found in a glycosylated form (SDG), its intermediate monoglucoside (SMG) forms not being accumulated in the seed. In this study, the presence of SMG was detected in roots. The decreased amount of SMG in the transgenic line is then, a consequence of the PLR enzyme suppression. PLR, in fact, a bifunctional enzyme which catalyze the reduction of pinoresinol to lariciresinol and then lariciresinol to secoisolaricirésinol [27]. By suppression of *LuPLR1* expression, the lignan biosynthesis pathway from pinoresinol to the end products would be blocked, leading to an accumulation of pinoresinol and a decrease of both lariciresinol and secoisolariciresinol within the transgenic line.

Both PLR1 and PLR2 enzymes were detected in the seeds and roots. However, in stems and leaves, only PLR2 was expressed. However, our results showed that the integration of PLR1-RNAi in the *L. usitatissimum* genome leads to a different metabolome in roots, but also in stems and leaves of wild-type and transgenic lines. Thus, it could be suggested that the suppression of *LuPLR1* expression is accompanied by the *LuPLR2* cosuppression. This can be theoretically, possible due to a major similarity of *LuPLR1* sequence used to synthetize PLR1-RNAi construct, with *LuPLR2* gene (60.3% of sequence homology) as already observed by Corbin et al. [6].

Herein it should be noted that, the decreased rate in both LMG and SMG is not proportionally equivalent to the increased rate of pinoresinol, in the transgenic line. This led us to hypothesize that, in the transgenic line, a compensation can occur with an increase in the production of pinoresinol in order to recover the loss of LMG and SMG. This compensation could be carried out by the cell to maintain the role of lignan.

Because lignans play an important role in plant defense, their content should be maintained in the transgenic line. The low amounts of larciresinol in the cell can therefore activate regulators in order to transmit a signal into the cell to continue producing pinoresinol. A possible regulation of dirigent proteins by abscisic acid ABA could be suggested, since this phytohormone appears to be involved in the regulation of lignan biosynthesis as previously described in *L. usitatissimum* [31], as well as in the regulation of dirigent proteins in response to abiotic stress as previously described in *B. hygrometrica* [32].

Furthermore, the presence of neolignan dehydrodiconiferyl alcohol glucoside (DCG) was detected in roots, stems and leaves of the transgenic lines, but it was not found in the wild-type plant. A similar result was seen in PLR1-RNAi transgenic flax seeds [27]. The aforementioned results can be attributed to a down-regulation of the *LuPLR1* gene expression, leading to the synthesis of the neolignans. Thus, the DCG formation could result from the coupling of two monolignol moieties without the intervention of a dirigent protein, a consequence of coniferyl alcohol accumulation or the dirigent protein retroinhibition as a result of the pinoresinol accumulation caused by the *LuPLR1* suppression.

Moreover, further change in secondary metabolites was also observed between LuT and Pi1AM lines in each plant part.

In roots, cyanogenic compounds such as linamarin and lotaustralin showed significantly higher contents in Pi1AM than those in LuT. While in stems, Pi1AM showed a lower content than LuT in linamarin and lotaustralin and a higher content in neolinustatin. On the other hand, in leaves, these cyanogenic compounds did not show any statistically significant changes, in the comparison between Pi1AM and LuT but linustatin, showed a higher content in Pi1AM line. A previous study on NMR metabolomics performed on flax seeds with different ω-3 fatty acid contents has shown that a decrease in secoisolariciresinol glucoside (+)-SDG was accompanied with a decrease in linustatin [33]. Another study on spatiotemporal distribution of lignans and cyanogenic glucosides in flaxseed revealed that seeds devoid of (+)-SDG did not accumulate linustatin [34]. However, so far, no specific and precise relation between lignans and cyanogenic compounds has been reported.

It should be noted that, although the majority of the metabolites listed above were measured by one of the two analytical techniques. The cyanogenic glycosides including linamarin, lotaustralin, linustatin, and neolinustatin, were detected in common with both ^{1}H-NMR and LC-MS. Interestingly, for each of these metabolites, LC-MS data showed almost a similar relative value (Pi1AM/LuT), when compared to those obtained with ^{1}H-NMR, in each of the plant parts. This similarity confirms the robustness of these two analytical methods. For example, for linamarin and lotaustralin: NMR data showed that Pi1AM/LuT root ratios were 1.44 and 1.5, respectively. LC-MS analysis, showed very similar values for Pi1AM/LuT ratios (1.41 for linamarin and 1.49 for lotaustralin). In stems, according to NMR data, Pi1AM/LuT ratio was 0.82 for linamarin and 0.88 for lotaustralin. While, using LC-MS, Pi1AM/LuT ratio were approximately 0.76 for linamarin and 0.76 for lotaustralin. In the case of leaves, both NMR and LC-MS data revealed no significant difference in lotaustralin and linamarin contents between Pi1AM and LuT.

In addition, in stems, Pi1AM showed a higher content of carlinoside, lucenin-2, orientin, caffeic acid glucoside (CAFG), chlorogenic acid, and ferulic acid glucoside (FAG) than in LuT. However, a lower content than LuT in triticuside-A.

Leaves of Pi1AM line showed a higher content in chlorogenic acid, carlinoside, FAG, lucenin-2 and CAFG, than shows the LuT. This latter showed, however, a higher content in vitexin and vicenin-2.

Carlinoside, lucenin-2, vitexin, vicenin-2, orientin and triticuside-A, are flavone C-glycosides derived from the same precursor of lignan, p-coumaroyl CoA, which in turn are derived from the phenylpropanoid pathway. Chalcone synthase (CHS) uses p-coumaroyl-CoA as a starter molecule to form the naringenin chalcone (4,2′,4′,6′-tetrahydroxychalcone), which serves as precursor for a different class of flavonoids [35]. Thus, the changes observed for these flavone C-glycosides in the transgenic line compared to the wild-type, could be explained by the disruption of their precursors p-coumaroyl CoA, which might be caused by the accumulation of coniferyl alcohol.

Ferulic and caffeic acids are also involved in phenylpropanoid metabolism. These compounds are hydroxylated derivatives of cinnamic acid. For monolignol biosynthesis, cinnamic acid is first hydroxylated by cinnamate 4-hydroxylase to yield *p*-coumaric acid.

Then, *p*-coumaric acid and their derivates (caffeic and ferulic acid) are catalyzed by the action of 4-coumaroyl CoA ligase (4CL) to form the p-coumaroyl CoA ester, which is a point of divergence of different pathway leading to the biosynthesis of other phenylpropanoids.

The methylation of caffeoyl-CoA (activated form of caffeic acid) to feruloyl-CoA (activated form of ferulic acid) is catalyzed by caffeoyl-CoA *O*-methyltransferase. Then, feruloyl-CoA, is reduced by cinnamoyl-CoA reductase to form the coniferaldehyde and further by cinnamyl alcohol dehydrogenase to yield coniferyl alcohol [36].

In plant cells, ferulic and caffeic acids rarely appear in their free form. Generally, they are mostly found in glycosylated form, such as FAG and CAFG, or in esterified form, such as chlorogenic acid and chicoric acid [37]. The accumulation of these acids in glucoside and esterified forms in the transgenic line, can be due to the accumulation of coniferyl alcohol caused by the disruption of lignan synthesis. Thus, there might be a negative feedback on the activity of the coumaroyl CoA ligase (4CL), in order to reduce the biosynthesis of CoA esters which represent the activated forms of the metabolism of phenylpropanoids.

Based on the above mentioned, we can hypothesize that, the integration of PLR1-RNAi transgene in *L. usitatissimum* genome, affects flax metabolic activity and leads to a different metabolome in various plant parts of wild-type and transgenic lines.

2.2.2. Metabolites Change in Response to Osmotic Stress in Wild-Type and PLR1-RNAi Transgenic Line

Drought is becoming one of the major limiting factors in agriculture worldwide leading to vast reductions in crop yield. It affects physiological and molecular processes and disrupt metabolic homeostasis, causing significant changes in the chemical composition of many plants resulting from the adjustment of metabolic pathways for adaptation. Hence, studying the metabolic response of transgenic line deprived for lignan production compared to wild-type of flax under osmotic stress conditions seems crucial and will provide important information about whether lignan play a role in the mechanism of the plants against this abiotic stress.

The PCA score plot of wild-type (LuT) and transgenic lines (Pi1AM) of each of the three parts (root, stem, and leaf), shows a characteristic grouping for control and stressed samples (Supplementary Data 3). The effect of the kinetics of stressed samples in different plant part was investigated. A Wilcoxon Rank Sum test was performed on the datasets for LuT and Pi1AM stressed sample, to verify whether there was a significant difference between the two periods of osmotic stress application. The test highlighted differences in metabolite content during the osmotic stress period (Data not shown). Thus, D1 and D5 stressed samples could be considered as a separate group for both lines.

Metabolic Profiles in Response to Osmotic Stress in Roots

In roots, the metabolic response to osmotic stress differed between LuT and Pi1AM. Statistical analysis highlighted 31 metabolite contents that change under osmotic stress for LuT and 31 for Pi1AM (Figure 3). Among these, some responded similarly to osmotic stress in both LuT and Pi1AM, some were specific to LuT and others to Pi1AM.

Thirteen amino acid (AAs) that increased significantly during osmotic stress application period were common to both lines: proline, asparagine, leucine, GABA, tryptophane, alanine, phenylalanine, valine, tyrosine, threonine, isoleucine, glycine, and serine. The reorganization of almost all these AAs appeared after one-day post-treatment with PEG and becomes more visible after five days, which indicates that metabolite content variation in roots, occurred early in response to osmotic stress in both lines. The largest fold increase in AAs content was observed for proline. This compound accumulated in a gradual manner with stress period, and at D5, its content was multiplied by 4.99- and 4.52-fold (calculated by the formula: ratio S/C = content plant treated by PEG-6000/content control plant), for LuT and Pi1AM respectively. There was also an increase in the amount of putrescine, a diamine derived from amino acid, in stressed samples compared to control samples in both lines.

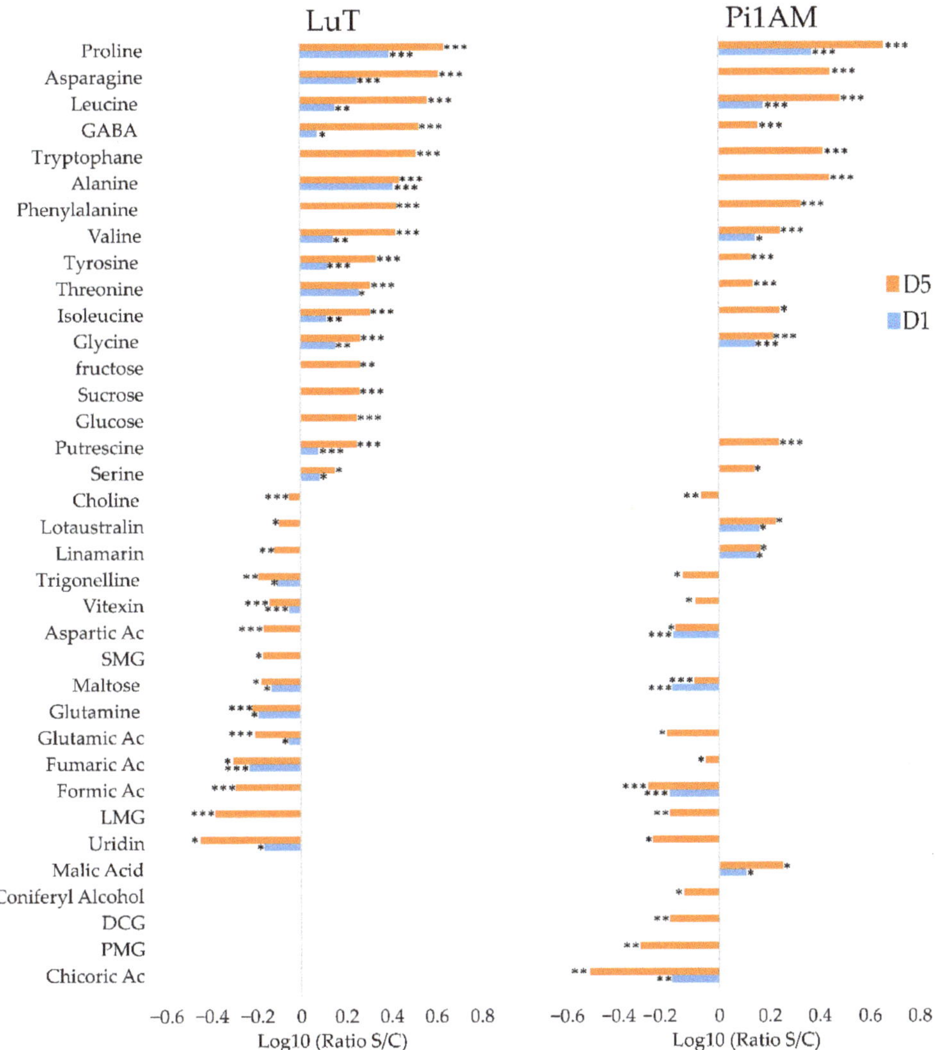

Figure 3. Comparison of metabolite content reorganization in wild-type (LuT) and transgenic (Pi1AM) roots after osmotic stress at different time-points: one and five days. Bars on logarithmic scale (log10) represent the mean relative response ratio S/C (content plant treated by PEG 6000/content control plant). Negative values represent a lower content and positive values a higher content of metabolites in stressed plants. Different time points are indicated by blue bars for D1 and red bars for D5. Values are significantly different at: *** $p < 0.001$; ** $p < 0.01$; * $p < 0.05$ after the Wilcoxon Rank Sum test ($n = 8$).

Decreased aspartic and glutamic acid contents were also observed in both lines, after PEG treatment. Only one AA changes specifically in LuT after PEG treatment: glutamine decreased in stressed roots samples compared to control samples in LuT, while the content of this metabolite was not affected in Pi1AM under osmotic stress.

In addition, LuT and Pi1AM showed different others kinds of primary metabolites that decreased during osmotic stress, such as choline, fumaric acid, formic acid, and uridine, as well as of secondary metabolites, such as trigonelline, vitexin, and LMG.

A significant accumulation of non-structural carbohydrates (NSCs) such as glucose (ratio S/C: 1.77), sucrose (ratio S/C: 1.82) and fructose (ratio S/C: 1.84), was observed specifically for LuT stressed samples, compared with LuT control samples, after five days of stress. Additionally, five days of treatment with PEG, led to a significant decrease in the content of secoisolariciresinol monoglucoside (SMG) (ratio S/C: 0.68) on LuT roots. However, this latter compound was not detected in Pi1AM roots.

For Pi1AM, five days of osmotic stress, induced a decrease in content of both lignans such as PMG (ratio S/C: 0.49) and neolignans such as DCG (ratio S/C: 0.64). A decrease in a similar manner was also observed for coniferyl alcohol content (ratio S/C: 0.7).

There were other metabolites that changed specifically in roots of Pi1AM, under osmotic stress. This was the case for chicoric acid that decreased and for malic acid that increased under osmotic stress conditions.

The content of linamarin and lotaustralin decreased in LuT, at D5 of stress conditions, with an S/C ratios of 0.8 and 0.76, respectively, while they accumulated more in Pi1AM in stress conditions with S/C ratios increased 1.45 and 1.67 folds respectively, after five days of osmotic stress, which was observed on both ^{1}H-NMR and LC-MS.

Metabolic Profiles in Response to Osmotic Stress in Stems

In stems, comparisons between the two datasets therefore reveal common responses between LuT and Pi1AM, and responses specific. A total of 43 metabolites were observed significantly influenced in the stressed LuT samples, compared to 29 in Pi1AM (Figure 4).

It appears that AAs content variation occurred early in stem of both lines (after one-day post treatment with PEG) relatively compared to other metabolites, and becomes more visible after five days post treatment with PEG.

Content of 12 among of these AAs increased (proline, isoleucine, asparagine, leucine, valine, threonine, glycine, phenylalanine, tryptophane, tyrosine, alanine, GABA, and serine), and for two of them it decreased (aspartic acid and glutamic acid). These reorganization in AAs content are similar to those observed in roots under osmotic stress. The highest increase also concerned the well-known osmorotectant proline that the content increased by 5.8- and 4.59-fold in the stems of LuT and Pi1AM, respectively, after five days of osmotic stress.

Other metabolites responded similarly to osmotic stress in the two lines. The contents of caffeic acid, tartaric acid and putrescine, were significantly accumulated under osmotic stress in both of the LuT and Pi1AM lines. In contrast, their contents in linamarin, lucenin-2, carlinoside, orientin, and fumaric acid were significantly decreased in stems, after osmotic stress.

The content of some kinds of sugar, including sucrose, and raffinose, was decreased in stems, for both Pi1AM and LuT lines, under osmotic stress.

Furthermore, glucose, fructose and galactose, decreased specifically in stems of LuT, after five days of osmotic stress.

The specific response to osmotic stress in LuT stems, was associated to the accumulation of ferulic acid and glycerol, and the decrease of the accumulation of primary metabolites such as succinic acid, choline, malic acid, uridine and ethanolamine, but also phenolic acid such as chlorogenic acid, flavonoids, such as isovitexin, vitexin, and vicenin-2, and cyanogenic glycosides, such as linustatin, lotaustralin, and neolinustatin, as well as lignans, such as LDG.

While the specific metabolic response to stress in Pi1AM, mainly consisted in a decrease of the content of metabolites involved in lignan biosynthetic pathway including DCG, coniferyl alcohol, PDG, and PMG.

Metabolic Profiles in Response to Osmotic Stress in Leaves

The response of metabolites to osmotic stress in leaves differed between LuT and Pi1AM lines. There were 38 metabolites with a significant change in leaves content under osmotic stress for each of LuT and Pi1AM (Figure 5). Most of up-regulated metabolites in both lines belonged to AAs, including proline, leucine, tryptophane, valine, isoleucine, phenylalanine, threonine, tyrosine, GABA, serine, asparagine, alanine, and glycine. In both lines it seems that the intensity of change of these AAs is dependent of stress period. The up-regulation had significantly been noticed after a single day of treatment with PEG and become more visible after five days. The largest significant increase in LuT and Pi1AM was proline with an S/C ratios of 10.29 and 6.8 respectively, after five days of osmotic stress. The AAs reorganization in leaves under osmotic stress is also reflected by the decrease in the content of aspartic acid in both lines. These trends in AAs changes are similar to what has been observed previously in roots and stems of both lines, under osmotic stress. A similar change in these AAs contents was reported by Quéro et al. in wild-type flax (*L. usitatissimum*) leaves, under osmotic stress conditions [38].

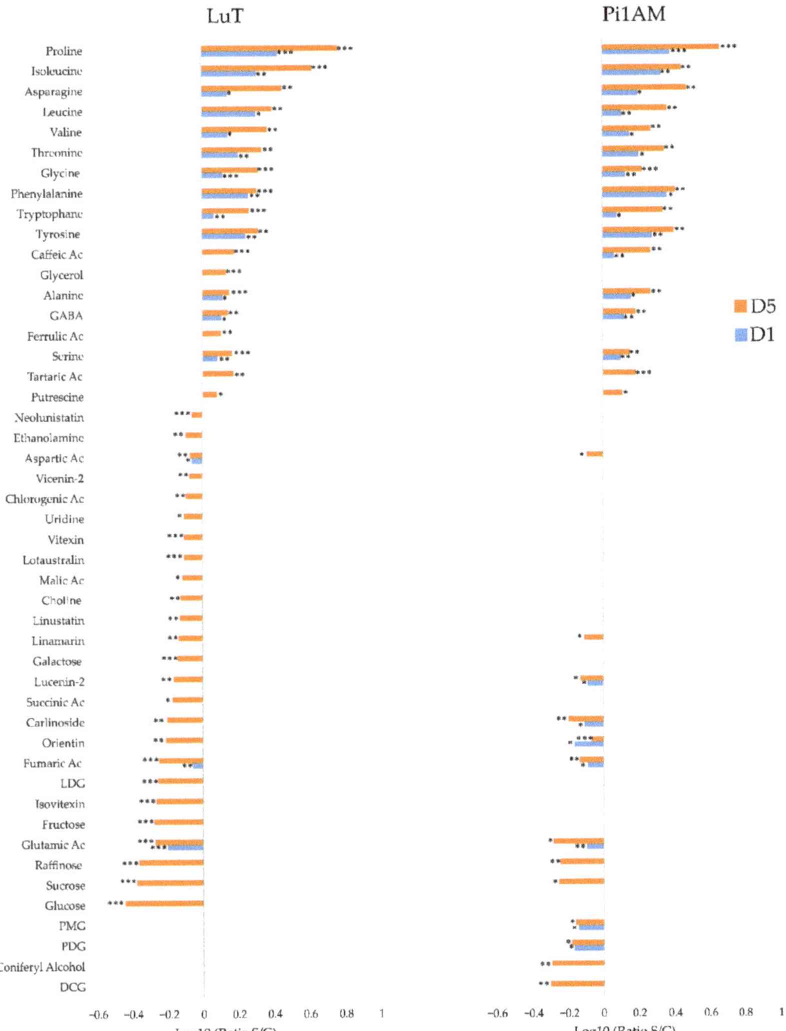

Figure 4. Comparison of metabolite content reorganization in wild-type (LuT) and transgenic (Pi1AM) stems after osmotic stress at different time-points: one and five days. Bars on logarithmic scale (log10) represent the mean relative response ratio S/C (content plant treated by PEG 6000/content control plant). Negative values represent a lower content and positive values a higher content of metabolites in stressed plants. Different time points are indicated by blue bars for D1 and red bars for D5. Values are significantly different at: *** $p < 0.001$; ** $p < 0.01$; * $p < 0.05$ after the Wilcoxon Rank Sum test ($n = 8$).

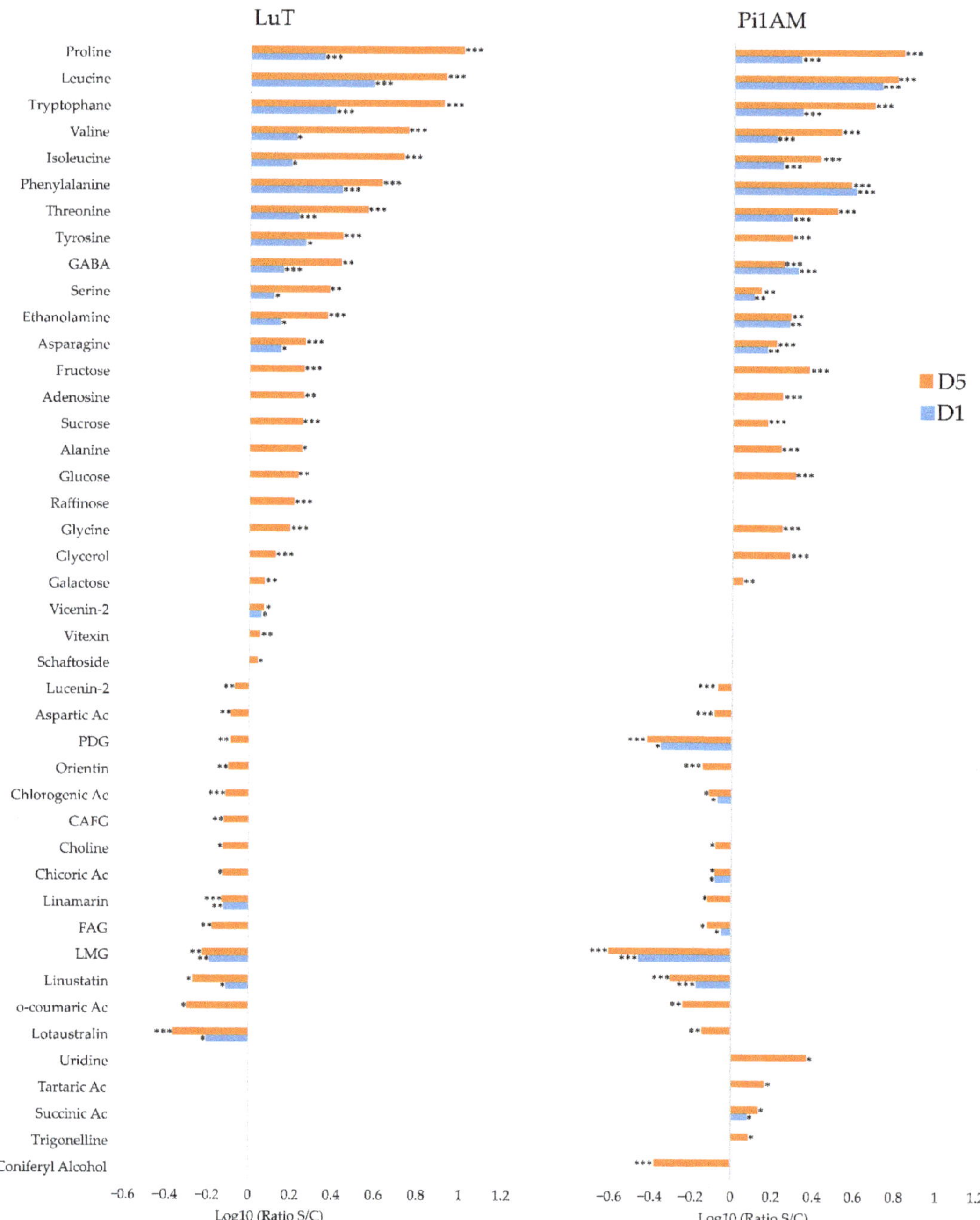

Figure 5. Comparison of metabolite content reorganization in wild-type (LuT) and transgenic (Pi1AM) leaves after osmotic stress at different time-points: one and five days. Bars on logarithmic scale (log10) represent the mean relative response ratio of S/C (content plant treated by PEG 6000/content control plant). Negative values represent a lower content and positive values a higher content of metabolites in stressed plants. Different time points are indicated by blue bars for D1 and red bars for D5. Values are significantly different at: *** $p < 0.001$; ** $p < 0.01$; * $p < 0.05$ after the Wilcoxon Rank Sum test ($n = 8$).

Other compounds accumulated in the same way in leaves of both lines when plants were submitted to osmotic stress. This was the case of ethanolamine, adenosine, and glycerol and as well as sugars, including fructose, sucrose, glucose, and galactose.

In addition to aspartic acid, there were 12 metabolites that the content decreased in leaves, under osmotic stress, which were common to both lines. This decrease was observed for lucenin-2, PDG, orientin, chlorogenic acid, choline, chicoric acid, FAG, LMG, and coumaric acid, but also for cyanogenic compounds including linamarin, linustatin, and lotaustralin. However, LuT showed specifically, significant increase of the accumulation of raffinose, vicenin-2, vitexin and schaftoside, and a decrease of the accumulation of CAFG in leaves under osmotic stress.

Conversely, some metabolites such as uridine, tartaric acid, succinic acid, trigonelline, and coniferyl alcohol were not affected by osmotic stress in LuT, but were significantly involved in the Pi1AM osmotic stress response. Among these metabolites, the content of coniferyl alcohol decreased while for the others, metabolites content in leaves increased under osmotic stress.

In the present work, an evident difference in the metabolic response to osmotic stress between different plant parts, as well as lines, was observed. These observations suggest that the mechanism of osmotic stress response of transgenic line is somehow different of that of wild-type, leading to the hypothesis that balanced lignan content may be important for proper stress response.

Major changes in the primary and secondary metabolic pathways of roots, stems and leaves of both wild-type and transgenic line after PEG treatment, as well as the proposed relations between these metabolic pathways are presented in balance sheet shown in Figure 6.

2.2.3. Comparison of the Metabolites in Wild-Type and PLR1-RNAi Transgenic Line under Osmotic Stress Conditions

PLS-DA was performed with both NMR and LC-MS data corresponding to stressed samples for LuT and Pi1AM for each of the three parts of the plant (Figure 7). The representation reveals that, in roots, stems, and leaves, LuT and Pi1AM lines are still separated because they still have a different metabolome under osmotic stress conditions. The metabolites that discriminated LuT and Pi1AM under osmotic stress in each of the three organs of these two lines, are represented in Figure 7, and their respective ratios, and significance are reported in Table 1.

In roots, some metabolites were discriminant of the two lines in control conditions and remained so, in the same range, in osmotic stress, due to a comparable increase or decrease in these metabolites after osmotic stress for both lines. This was the case for glutamic acid, aspartic acid, glutamine, SMG, and LMG that accumulated more in LuT, whereas glycine, succinic acid, threonine, fructose, sucrose, fumaric acid, glucose, lotaustralin, linamarin, coniferin, coniferyl alcohol, PDG, PMG, and DCG accumulated more in Pi1AM.

Some metabolites appeared to be discriminant for both lines only in stress conditions while they were not discriminant in control conditions. This was the case for GABA, tyrosine, alanine, uridine, formic acid, malic acid, chicoric acid, and trigonelline.

Some metabolites were discriminant for both lines in control conditions but no longer discriminant in stress conditions. This was the case for maltose, serine, and putrescine.

In stems, glycerol, galactose, tyrosine, glucose, phenylalanine, LMG, LDG, triticuside-A, CAFG, chlorogenic acid, FAG, neolinustatin, orientine, lucenin-2, carlinoside, DCG, PDG, PMG, coniferin, and coniferyl alcohol were discriminant of the two lines in control conditions and still discriminant under osmotic stress. All these metabolites are more accumulated in Pi1AM than in LuT, except LMG, LDG, and triticuside-A, which are accumulated more in LuT than in Pi1AM under stress conditions.

Ethanolamine, fumaric acid, uridine, fructose, vicenin-2 and vitexin appeared to be discriminant for both lines only in stress conditions. The amount of these metabolites was the same for both lines in control conditions:

Alanine, succinic acid, tartaric acid, asparagine, malic acid, putrescine, threonine, aspartic acid, glutamic acid, linamarin, and lotaustralin were discriminant for the two lines, in control conditions, but they did not interact in the discrimination of both lines, in stress conditions.

In leaves, some metabolites were discriminant of the two lines in both control and stress conditions. This was the case for phenylalanine, GABA, alanine, serine, galactose, LMG, vitexin, and vicenin-2 that were more accumulated in LuT and for glycerol, fructose, choline, glucose, lucenin-2, chicoric acid, CAFG, carlinoside, FAG, chlorogenic acid, coniferyl alcohol, coniferin, PMG, PDG, and DCG that were more accumulated in Pi1AM.

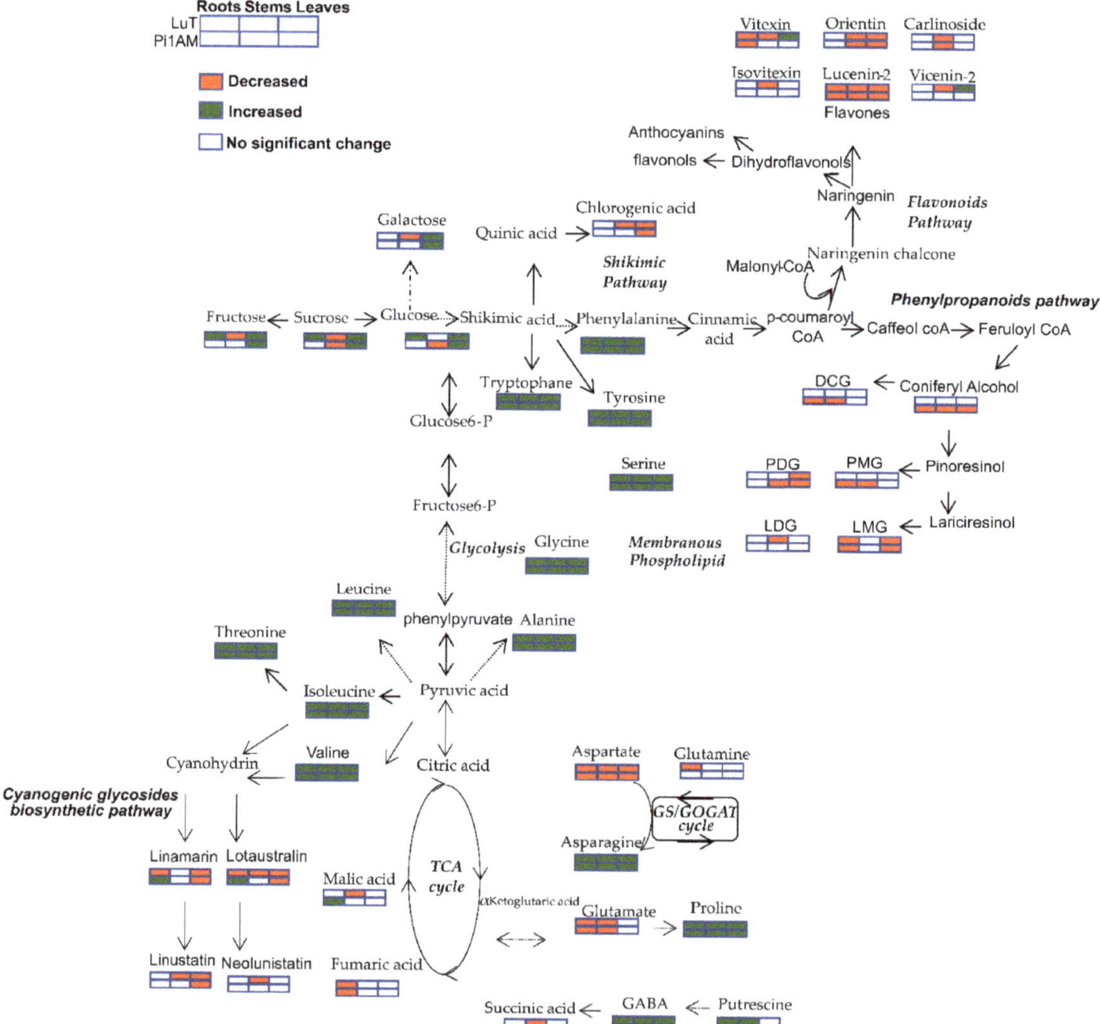

Figure 6. Major changes in the primary and secondary metabolic pathways of roots, stems and leaves of LuT and Pi1AM, after PEG treatment. The proposed metabolic pathways were based on the literature. Metabolites with red boxes represent significant decreases, with green ones represent significant increases, and with empty ones represent no significant change. The level of significance was set at $p < 0.001$, $p < 0.01$ and $p < 0.05$ after the Wilcoxon Rank Sum test.

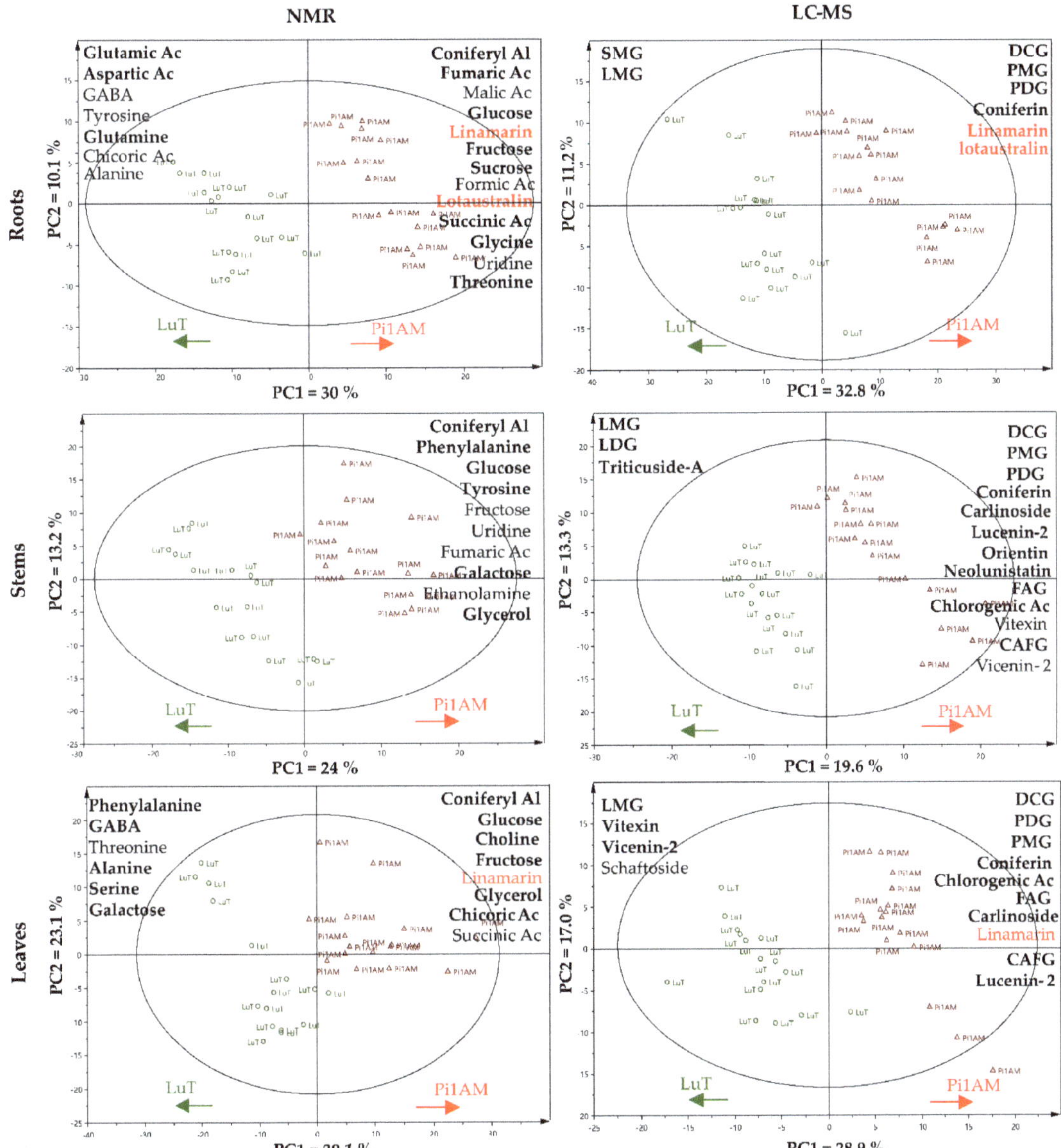

Figure 7. Score plot of partial least squares discriminant analysis (PLS-DA) of metabolite content in flax roots, stems and leaves, based on ^{1}H-NMR and LC-MS data for LuT and Pi1AM, in stress conditions. Samples represent LuT plants (circle) and Pi1AM plants (triangle). Arrows indicate the increases in relative metabolite content in various parts of the wild-type (LuT) or transgenic (Pi1AM) plants. Metabolites are listed in descending order of discrimination. Boldface text indicates discriminant metabolites of the two lines in control and stress conditions. Red text indicates metabolites measured in common with LC-MS and ^{1}H-NMR. Importance in Projection (VIP) score and the Log10 (ratio Pi1AM/LuT) of the main discriminant metabolites for each PLS-DA model, are shown in supplementary data 4.

Table 1. Comparison of discriminant metabolite contents for LuT and Pi1AM in stress versus control conditions.

Roots			Stems			Leaves		
	Ratio Pi1AM/LuT			Ratio Pi1AM/LuT			Ratio Pi1AM/LuT	
Metabolite	Stress	Control	Metabolite	Stress	Control	Metabolite	Stress	Control
Primary metabolites			**Primary metabolites**			**Primary metabolites**		
Glutamic Acid	0.5 ***	0.65 **	Glycerol	1.3 *	5.56 **	Phenylalanine	0.42 **	0.7 **
Aspartic Acid	0.53 ***	0.55 ***	Ethanolamine	1.38 **	_	GABA	0.47 *	0.63 *
GABA	0.53 ***	_	Galactose	1.47 *	1.19 **	Threonine	0.47 **	_
Tyrosine	0.54 ***	_	Fumaric Acid	1.5 **	_	Alanine	0.62 *	0.64 ***
Glutamine	0.56 ***	0.63 ***	Uridine	1.68 **	_	Serine	0.64 *	1.13 *
Alanine	0.85 ***	_	Fructose	1.87 *	_	Galactose	0.66 **	0.75 ***
Threonine	1.2 ***	1.23 *	Tyrosine	2.01 ***	1.88 ***	Succinic Acid	1.2 *	_
Uridine	1.34 ***	_	Glucose	2.31 **	1.42 **	Glycerol	1.36 *	1.5 **
Glycine	1.38 ***	1.41 ***	Phenylalanine	3.05 ***	1.6 **	Fructose	2.00 *	1.72 ***
Succinic Acid	1.42 ***	1.64 **	Coniferyl Alcohol	7.72 ***	8.65 ***	Choline	2.05 **	1.75 ***
Formic Acid	1.43 ***	_	Alanine	_	0.7 **	Glucose	2.15 **	1.34 *
Sucrose	1.45 ***	2.07 ***	Succinic Acid	_	0.73 ***	Coniferyl Alcohol	4.79 ***	10.32 ***
Fructose	1.69 ***	1.73 ***	Tartaric Acid	_	0.74 **	Uridine	_	0.55 *
Glucose	1.79 ***	1.85 ***	Asparagine	_	0.76 *	Adenosine	_	0.6 ***
Malic Acid	2.03 ***	_	Malic Acid	_	0.82 ***	Threonine	_	0.76 **
Fumaric Acid	2.11 ***	1.34 *	Putrescine	_	0.84 **	Ethanolamine	_	0.79 **
Coniferyl Alcohol	5.51 ***	7.91 ***	Threonine	_	0.87 *	Asparagine	_	0.85 *
Maltose	_	0.87 *	Aspartic Acid	_	1.27 *	Tartaric Acid	_	0.91 *
Serine	_	1.48 ***	Glutamic Acid	_	1.39 *	Raffinose	_	1.28 **
Putrescine	_	1.6 ***	**Secondary metabolites**			Sucrose	_	1.32 **
Secondary metabolites			LMG	0.46 ***	0.73 ***	Tyrosine	_	3.73 ***
SMG	0.06 ***	0.15 ***	LDG	0.64 ***	0.17 ***	**Secondary metabolites**		
LMG	0.28 ***	0.31 **	Triticuside-A	0.8 *	0.8 **	LMG	0.27 ***	0.50 ***
Chicoric Acid	0.71 ***	_	Vicenin-2	1.36 ***	_	Vitexin	0.65 ***	0.69 ***
Trigonelline	1.21 ***	_	CAFG	1.45 **	1.42 **	Vicenin-2	0.7 **	0.77 **
Lotaustralin	1.81 **	1.49 **	Vitexin	1.54 ***	_	Schaftoside	0.9 ***	_
Linamarin	1.75 **	1.41 **	Chlorogenic Acid	1.63 **	1.32 **	Lucenin-2	1.27 *	1.21 **
Coniferin	1.83 ***	2.32 ***	FAG	1.81 ***	1.21 *	Chicoric Ac	1.33 ***	1.23 ***
PDG	12.66 ***	12.27 ***	Neolunistatin	2.04 **	1.84 **	CAFG	1.42 *	1.11 **
PMG	18.63 ***	18.13 ***	Orientin	2.67 ***	1.92 ***	linamarin	1.45 ***	_
			Lucenin-2	3.07 ***	2.57 ***	Carlinoide	1.5 ***	1.45 ***
			Carlinoside	3.51 **	2.8 ***	FAG	1.66 ***	1.24 *
			Coniferin	3.63 ***	2.4 ***	Chlorogenic Acid	2.19 ***	1.51 **
			PDG	3.83 ***	4.73 ***	Coniferin	6.45	7.68
			PMG	4.25 ***	5.31 ***	PMG	14.1	13.78 ***
			Linamarin	_	0.76 ***	PDG	14.5 ***	15.03 ***
			Lotaustralin	_	0.76*	Linustatin	_	1.27 **

Note: Results of the Wilcoxon Rank Sum test and ratios are reported for each discriminant metabolite; Values are significantly different at: *** $p < 0.001$; ** $p < 0.01$; * $p < 0.05$. After the Wilcoxon Rank Sum test.

As for, threonine, succinic acid, schaftoside and linamarin, they appeared to be discriminant for LuT and Pi1AM, only in osmotic stress conditions. The amount of these metabolites was similar in LuT and Pi1AM, in control conditions.

In addition, uridine, adenosine, threonine, ethanolamine, asparagine, tartaric acid, raffinose, sucrose, tyrosine, and linustatin were discriminant for both lines, in control conditions whereas they are not discriminant in stress conditions.

Herein, osmotic stress appears to affect the whole plant metabolome in the different analyzed parts. Roots, stems, and leaves, display different responses in order to preserve water status for survival.

However, several organic molecules, known as osmolytes, play a crucial role during osmotic adjustment, including AAs, sugars, and organic acids, which potentially help in maintaining osmotic balance within the plant cells.

Amino acids protect the plants cell membranes, stabilize the structure of biomolecules, play a role as a scavenger of reactive oxygen species (ROS), and provide a reserve of nitrogen and carbon, mainly for the synthesis of specific enzymes and precursors for secondary metabolites such as flavonoids and lignins, during stress conditions [39–41]. Thus, an early accumulation of amino acids contributes to a greater level of stress tolerance [42].

In the present work, for the most of the amino acids that were measured, no significant differences in their concentrations have been reported between both lines, in control conditions. Furthermore, under osmotic stress, they showed a rapid transient increase in content, after one day of PEG treatment. This increase followed almost the same trend in LuT and Pi1AM lines.

Proline has been repeatedly reported as one of the most important components of osmotic adjustment in different plant species including flax [38,43]. A large variability exists between and within plant species for their capacity to accumulate proline [44]. Genotypes with high capacity to accumulate proline under osmotic stress are generally considered to be tolerant to this type of stress [45]. In this study, no significant differences in proline concentrations have been reported between both lines, in control conditions as well as after exposure to osmotic stress.

In particular, branched-chain amino acids (BCAAs), such as leucine, valine, and isoleucine, increase in response to osmotic stress, to enhance stress resistance. These compounds can serve as precursors for cyanogenic glycosides and other secondary metabolites, to acquire a defense response against abiotic stress [46]. In this study, the amount of leucine, valine and isoleucine, in roots, stems, and leaves, was the same for LuT and Pi1AM lines in control condition and remained so in stress conditions, due to a comparable increase in these metabolites during osmotic stress for both lines. We can thus assume that both LuT and Pi1AM showed the same capacity in producing AAs in their early response that enhances tolerance and/or resistance to osmotic stress, in various flax parts.

In addition to amino acid, the breakdown of homeostasis caused by water deficit induces the accumulation of carbohydrates. The accumulation of sugars mainly fructose and glucose, is indeed considered as an adaptive response that may also contribute to enhance plant stress tolerance [47]. Soluble sugar plays a major role in osmotic adjustment of plants facing the conditions of drought stress [48]. It also maintains the turgidity of leaves and prevent dehydration of membranes and proteins. Furthermore, sugar accumulation reduces leaf photosynthetic rate, under drought stress [49]. However, glucose enhances the plant adaptability under drought stress by inducing stomatal closure in leaves [50]. In another study, Quéro et al. have shown that glucose and fructose increased in response to osmotic stress in leaves of wild-type *L. usitatissimum* [38]. In this study, the glucose and fructose content of leaves increased significantly, in the same manner in both LuT and Pi1AM, in response to osmotic stress.

On the basis of the comparison of osmolyte contents between the two lines under control and osmotic stress, we may conclude that the transgenic line has a similar capacity with the wild-type line, in regulating osmotic stress by producing soluble sugars and AAs. Additionally, no visible differences were detected in terms of leaf area, plant height and plant dry weight in Pi1AM compared to LuT, when subjected to PEG-induced osmotic stress. Overall, it may be assumed that the capacity of plant to induce an adaptative response that improves osmotic stress tolerance was similar in both LuT and Pi1AM.

Moreover, our results showed that the levels of some secondary metabolites, including cyanogen, lignan, and flavonoids, differ between wild-type and transgenic plants in both control and stress conditions.

Interestingly, a significant common decrease in the content of metabolites involved in lignan biosynthetic pathway was observed at least in one part of plants subjected to osmotic stress in comparison to control plants. This was the case for coniferyl alcohol, PMG and DCG that decreased specifically in Pi1AM, for LMG that decreased in both lines, for SMG and LDG that decreased only in LuT, and for PDG that decreased in both lines in leaves, but only in Pi1AM in stems, under stress conditions. These results indicate that osmotic stress response and lignan biosynthesis pathways may have a crosstalk with each other in *L. usitatissimum*.

3. Materials and Methods

3.1. Plant Material

Flax seeds of the variety Barbara, for PLR1-RNAi transgenic (Pi1AM, homozygous), and wild-type (LuT) lines, have been provided by the Laboratory of Woody Plants and Crops Biology (LBLGC, INRAE USC 1328) in Chartres, France. Plants were grown in a hydroponic cultivation system on Hoagland solution [51,52], in a growth chamber with a hemeroperiod of 16 h at 21 °C, 70% relative humidity and a light intensity of 90 μmol·m^{-1}·s^{-1}. After 30 days, half of each line was grown on Hoagland medium supplemented with PEG-6000 (30 mM) (stressed plants), while the half remaining plants still grown on Hoagland medium (control plants). Then, control and treated plants were harvested at different time-points: one and five days (D1, D5) after osmotic stress application. During the harvest, the aerial and root parts from each plant, were separated, immersed in liquid nitrogen, stored at −80 °C, and freeze-dried. Leaves, stems, and roots of each plant, were then separately ground to a fine powder. For each condition, eight independent samples coming from different plants have been collected (n = 8) and were subjected to further analysis.

3.2. PCR Analysis of Transgenic Plants

Genomic DNA was extracted from leaves (10 mg dried weight) of wild-type and transgenic plants, using the E.Z.N.A.® Plant DNA DS Kit (Omega, Bio-teK (Norcross, GA USA)), according to the manufacturer's instructions, and quantified using a Nanodrop® spectrophotometer. To confirm their transgenic nature, the extracted DNA from all isolates were tested for the presence of *NPTII* selectable marker gene. Primers used for the amplification of a 0.7-Kb *NPTII* fragment were: P2, 5′-ATCGGGAGCGGCGATACCGTA-3′ (position 201) as 5′ primer and P1, 5′-GAGGCTATTCGGCTATGA CTG-3′ (position 900) as 3′ primer [25]. PCR amplification was carried out in a total volume of 25 μL containing 12.5 μL of Quick-Load® Taq 2X Master Mix, BioLabs, 0.5 μL of each primer (0.25 μM), 2 μL of extracted DNA and 10 μl ultrapure water. Amplifications were performed with initial denaturation step of 5 min at 95 °C, followed by 40 cycles of 10 s denaturation at 95 °C, 30 s annealing at 55 °C for the *NPTII* gene, and a final extension step of 1 min at 68 °C. PCR products were run in 2% agarose gel electrophoresis, with 0.5X TBE (Tris, borate, EDTA), at 100 V for 45 min, and bands were visualized with GelRed, under UV light, using a ChemiDoc Imaging system.

3.3. Metabolite Wxtractions

Metabolites were separately extracted from 100 mg of powdered leaves or stems and 60 mg of powdered roots. A total volume of 800, 700 and 600 μL Water/methanol (1:1), used as extracting solvent, was added to leaves, stems and root powder respectively, and the samples were mixed for 10 min at 60 °C, using a ThermoMixer® (Eppendorf AG, Germany) at 2000 rpm, followed by 30 min of sonication at 60 °C using ultrasonic bath at 35 kHz. The samples were centrifuged at 4 °C for 10 min at 12,000 rpm. A total of 400 μL of the supernatant were collected. The described extraction process was repeated two times by adding 400 µL of extraction solvent in the beginning of each new extraction. A final volume of 1.2 mL was collected, of which 800µL were used for NMR analysis. The remaining 400 µL were then used for LC-MS analysis.

3.4. Metabolite Analysis by NMR

3.4.1. Sample Preparation

The pH of each sample prepared for NMR analysis was adjusted to 6.00 ± 0.02, and the samples were dried under vacuum and then dissolved in 800 µL of deuterated solvent prepared in a mixture of (1:1) Methanol-d4: KH_2PO_4 buffer (0.1 M) in D_2O at pH 6.0 with TMSP (0.0125%), NaN_3 (0.6 mg/mL), and 15 mM EDTA-d12. Then, samples were briefly vortexed, sonicated, and centrifuged. The supernatant was placed in 5-mm NMR tubes and then used for NMR analysis.

3.4.2. NMR Data Acquisition

All NMR spectra were acquired at 300 K with a Bruker Avance III 600 MHz spectrometer operating at 600.13 MHz for ^{1}H, and 150.91 MHz for ^{13}C, using a 5-mm multinuclear broadband, equipped with z-gradient (TXI 5 mm tube). CD_3OD was used as the internal lock. All 1D ^{1}H-NMR spectra were collected using 32 scans of 131 K data points and a spectral width of 8417 Hz with a relaxation delay of 13 s, and a water suppression pulse sequence. The resulting spectra were automatically phased and baseline corrected, using Bruker Topspin software (version 3.5) [51]. 2D J-resolved NMR spectra were processed with a 2 s relaxation delay using 16 scans per 64 increments collected into 64K data points, with spectral widths of 8417.5 Hz in F2 and 50 Hz in F1. 2D HSQC spectra were recorded with a 2 s relaxation delay, using 64 scans per 256 increments that were collected into 4 K data points using spectral widths of 8417.5 Hz in F2 and 50 Hz in F1. All spectra were calibrated to TMSP at 0.0 ppm by the Topspin v3.5 (Bruker) software. Each of the 2 D acquisitions were performed by analyzing one control and one stressed samples of each part of plant corresponding to each of the two lines (Pi1AM and LuT).

3.4.3. NMR Data Treatment

^{1}H-NMR spectra were automatically converted into ASCII format and the data from each part of plant (roots, stems and leaves) were collected and imported into Matlab software (version 2018a, the Mathworks Inc, Natick, MA, USA) where baseline correction was performed with the algorithm "airPLS 2.0" [53–55]. All spectra were aligned using the icoshift algorithm (v 1.2.1) with manually defined alignment bins. Then specific integration intervals of the spectra "buckets" were defined manually, and each bucket was integrated. Regions of the NMR spectra corresponding to the methanol-d4 (3.34–3.30 ppm), to residual water (4.85–4.70 ppm), to TMSP (0.01–0 ppm) and to PEG-6000 (3.68–3.64 ppm), were removed. All signals with intensities lower than 3.3 times the mean variance from such a noise region were considered to be noise and were also removed. The obtained datasets were then used for statistical analysis. Metabolites in the 1D and 2D NMR spectra of flax extracts were identified based on comparison with spectra and chemical shifts of reference compounds of database previously developed in the laboratory.

3.4.4. Metabolite Analysis by LC/MS

Sample Preparation

Extracts of stems and leaves were diluted 5 and 10 times, respectively, with methanol/water (50/50), while no dilution was performed for root extracts. All samples were filtered through 0.22 µm PTFE membrane filters and placed in glass vials for further LC-MS analysis. For each plant part, a QC sample were prepared; 10 µL from 20 different samples were taken and thoroughly mixed, reaching a total volume of 200 µL. For roots, stems, and leaves, analysis samples were run in a randomized order.

LC-MS Data Acquisition

The metabolomics analysis was performed using ACQUITY UPLC H-Class system (Waters Micromass, UK), coupled to a SYNAPT G2-Si Q-TOF mass spectrometer (Waters-Micromass, Manchester, UK), which was equipped with an electrospray ion source (ESI). UPLC separation was performed using a Kinetex C18 (1.7 μm, 100 mm × 2.1 mm, Phenomenex, Torrance, CA, USA) column. The column temperature was maintained at 50 °C. The injected volume was 1 μL. Water (A) and methanol (B), both supplemented with 0.1% formic acid, were used as mobile phases. A stepwise gradient method was used for elution at a flow rate of 0.4 mL/min, with the following conditions: 5–95% B (0–7 min), followed by 3 min of isocratic 95% solvent B, and 1 min gradient to 95% (A), followed by 5 min of re-equilibrium at 100% A. MS data was collected in the negative ion mode, over a m/z range of 50 to 1150. The parameters of electrospray ionization (ESI) source were set as follows: capillary voltage at 3 keV, the cone voltage at 3 V, source temperature at 120 °C, desolvation at 450 °C, the cone gas flow 6.5 bar and the desolvation gas flow 800 L/h. Analyses of the samples were carried out in a mode of a full MS survey, at a resolution of 2000 (FWHM). MSMS scans for most intense peaks were performed to produce high resolution MSMS spectra, with a collision energy of 30 eV. Data acquisition was performed by MasslynxTM v4.1 software (Waters, Milford, MA, USA). The QC samples were injected at the beginning of the run to set up the system and then every eight samples, so they were used to ensure system conditioning within the analytical sequence.

LC-MS/MS Data Processing

The acquired spectral data were converted to mzXML format using the Proteowizard MSConvert tool from 0 to 10 min RT in order to avoid features coming from cleaning step of the gradient.

Then, the mzXML files data of each part of the plant were loaded and pre-processed with the XCMS package (v3.0.2) in the open-source R software (v3.2.2).

The centWave algorithm was used for peak detection with the following optimized parameters: minimum peak width = 3 s, maximum peak width = 15 s, ppm = 5, threshold = 2, mzdiff = 0.005 and prefilter: (4,100,000), and noise filter = 1000. Peaks were well aligned by XCMS, using the following parameters: bw = 5.0 and mzwid = 0.025. Retention time correction was performed by the Obiwarp algorithm, which aligns multiple samples by using a center sample.

A filling step was included to reduce the number of missing peaks, using the fillPeaks tool. For each sample, Peak area, retention time and peak widths, calculated as the difference between the end and start of the integration points, were extracted from XCMS data.

After the whole data processing, a table (matrix) including retention time and m/z, sample names, and ion intensities, was obtained for each sample set. Then, before the statistical analysis, these matrices were prepared in order to remove the features before the injection peak (less than 1 min).

The percentage of relative standard deviation (%RSD) was calculated for all metabolic features in QC samples and the features with %RSD greater than 30% were removed due to its variability.

Metabolites were principally identified by matching masses, retention times and fragment patterns of pure standards.

Statistical Treatment of Data

The obtained datasets for NMR and LC-MS were imported into SIMCA-P software (version 15.0; Umetrics, Umea, Sweden), and applied to principal component analysis (PCA) and partial least squares discriminant analysis (PLS-DA) for multivariate statistical analysis. The UV scaling method was used.

The Wilcoxon Rank Sum test was used in R's statistics base-package in order to test the significant difference in metabolite content, between analyzed groups, with different *p*-values ($p < 0.05$; $p < 0.01$; $p < 0.001$).

4. Conclusions

In the present study, the wild-type and transgenic flax plants displayed different metabolic behaviors after being subjected to osmotic stress conditions, which could be due to their different metabolic backgrounds but also to different metabolic rearrangements. In the last case, we can suggest that an altered lignan composition in transgenic plant changes the metabolic response of the plant to osmotic stress.

In addition, our experimental observations revealed that wild-type and transgenic line showed the same morphological response to osmotic stress. Moreover, they showed the same capacity to induce an adaptive response that improves osmotic stress tolerance. Thus, we can assume that lignan perturbations did not affect plant stress response leading to osmotic stress resistance. Taken together, our findings indicate that the lignans could not be directly involved in osmotic stress resistance, but in the metabolic pathways responsible of the osmotic stress response.

Supplementary Materials: The following are available online, Table S1. Comparison of dry weight of different part (roots, stems and leaves) of wild-type (LuT) and transgenic (Pi1AM) flax lines, at D1 and D5 in control conditions ($n = 8$). Table S2. The VIP score and Log10 (Ratio Pi1AM/LuT) of the discriminant metabolites between LuT and Pi1AM of different plant part, in control conditions. Figure S1. Score plot of principal component analysis (PCA) based on ^{1}H-NMR and LC-MS data for LuT control or stressed samples in flax roots, stems and leaves. Figure S2. Score plot of principal component analysis (PCA) based on 1H-NMR and LC-MS data for Pi1AM control or stressed samples in flax roots, stems and leaves. Table S3. The VIP score and Log10 (Ratio Pi1AM/LuT) of the discriminant metabolites between LuT and Pi1AM of different plant part, in stress conditions.

Author Contributions: Conceptualization: F.M. and A.E.K.; methodology: K.H., S.B., and S.G.; software: K.H. and J.-X.F.; validation: L.G., O.F., C.H., and F.M.; formal analysis: K.H., R.M., E.O.N., and S.P.; investigation: K.H. and O.F.; resources: C.H. and E.L.; writing—original draft preparation: K.H.; writing—review and editing: K.H., O.F., and F.M.; supervision: F.M., O.F., A.H., and A.E.K.; project administration: F.M.; funding acquisition: F.M. and A.E.K. All authors have read and agreed to the published version of the manuscript.

Funding: This research received no external funding.

Institutional Review Board Statement: Not applicable.

Informed Consent Statement: Not applicable.

Data Availability Statement: The data presented in this study are available on request from the corresponding author.

Acknowledgments: The authors wish to thank the University of Picardie Jules Verne for its financial support towards the publication of this work.

Conflicts of Interest: The authors declare no conflict of interest.

Sample Availability: Samples of the compounds are available on request from the corresponding author.

References

1. Imran, M.; Ahmad, N.; Anjum, F.M.; Khan, M.K.; Mushtaq, Z.; Nadeem, M.; Hussain, S. Potential Protective Properties of Flax Lignan Secoisolariciresinol Diglucoside. *Nutr. J.* **2015**, *14*, 71. [CrossRef] [PubMed]
2. Umezawa, T. Diversity in Lignan Biosynthesis. *Phytochem. Rev.* **2003**, *2*, 371–390. [CrossRef]
3. Schmidt, T.; Hemmati, S.; Fuss, E.; Alfermann, A. A Combined HPLC-UV and HPLC-MS Method for the Identification of Lignans and Its Application to the Lignans of Linum Usitatissimum L. and L. Bienne Mill. *Phytochem. Anal. Phytochem Anal.* **2006**, *17*, 299–311. [CrossRef] [PubMed]
4. Hano, C.; Martin, I.; Fliniaux, O.; Legrand, B.; Gutierrez, L.; Arroo, R.R.J.; Mesnard, F.; Lamblin, F.; Lainé, E. Pinoresinol–Lariciresinol Reductase Gene Expression and Secoisolariciresinol Diglucoside Accumulation in Developing Flax (Linum Usitatissimum) Seeds. *Planta* **2006**, *224*, 1291–1301. [CrossRef]

5. Hemmati, S.; Heimendahl, C.; Klaes, M.; Alfermann, A.; Schmidt, T.; Fuss, E. Pinoresinol-Lariciresinol Reductases with Opposite Enantiospecificity Determine the Enantiomeric Composition of Lignans in the Different Organs of *Linum Usitatissimum* L. *Planta Med.* **2010**, *76*, 928–934. [CrossRef]

6. Corbin, C.; Drouet, S.; Mateljak, I.; Markulin, L.; Decourtil, C.; Renouard, S.; Lopez, T.; Doussot, J.; Lamblin, F.; Auguin, D.; et al. Functional Characterization of the Pinoresinol-Lariciresinol Reductase-2 Gene Reveals Its Roles in Yatein Biosynthesis and Flax Defense Response. *Planta* **2017**, *246*, 405–420. [CrossRef]

7. Lamblin, F.; Hano, C.; Fliniaux, O.; Mesnard, F.; Fliniaux, M.-A.; Lainé, E. [Interest of lignans in prevention and treatment of cancers]. *Med. Sci. MS* **2008**, *24*, 511–519. [CrossRef]

8. Dikshit, A.; Gao, C.; Small, C.; Hales, K.; Hales, D.B. Flaxseed and Its Components Differentially Affect Estrogen Targets in Pre-Neoplastic Hen Ovaries. *J. Steroid Biochem. Mol. Biol.* **2016**, *159*, 73–85. [CrossRef]

9. Hano, C.; Corbin, C.; Drouet, S.; Quéro, A.; Rombaut, N.; Savoire, R.; Molinié, R.; Thomasset, B.; Mesnard, F.; Lainé, E. The Lignan (+)-Secoisolariciresinol Extracted from Flax Hulls Is an Effective Protectant of Linseed Oil and Its Emulsion against Oxidative Damage. *Eur. J. Lipid Sci. Technol.* **2017**, *119*, 1600219. [CrossRef]

10. Kezimana, P.; Dmitriev, A.A.; Kudryavtseva, A.V.; Romanova, E.V.; Melnikova, N.V. Secoisolariciresinol Diglucoside of Flaxseed and Its Metabolites: Biosynthesis and Potential for Nutraceuticals. *Front. Genet.* **2018**, *9*. [CrossRef]

11. Zhang, W.; Wang, X.; Liu, Y.; Tian, H.; Flickinger, B.; Empie, M.W.; Sun, S.Z. Dietary Flaxseed Lignan Extract Lowers Plasma Cholesterol and Glucose Concentrations in Hypercholesterolaemic Subjects. *Br. J. Nutr.* **2008**, *99*, 1301–1309. [CrossRef] [PubMed]

12. Rodríguez-García, C.; Sánchez-Quesada, C.; Toledo, E.; Delgado-Rodríguez, M.; Gaforio, J. Naturally Lignan-Rich Foods: A Dietary Tool for Health Promotion? *Molecules* **2019**, *24*, 917. [CrossRef] [PubMed]

13. Pan, A.; Sun, J.; Chen, Y.; Ye, X.; Li, H.; Yu, Z.; Wang, Y.; Gu, W.; Zhang, X.; Chen, X.; et al. Effects of a Flaxseed-Derived Lignan Supplement in Type 2 Diabetic Patients: A Randomized, Double-Blind, Cross-Over Trial. *PLoS ONE* **2007**, *2*, e1148. [CrossRef] [PubMed]

14. Senizza, A.; Rocchetti, G.; Mosele, J.I.; Patrone, V.; Callegari, M.L.; Morelli, L.; Lucini, L. Lignans and Gut Microbiota: An Interplay Revealing Potential Health Implications. *Molecules* **2020**, *25*, 5709. [CrossRef]

15. Chang, C.W.; Lin, M.T.; Lee, S.S.; Liu, K.C.; Hsu, F.L.; Lin, J.Y. Differential Inhibition of Reverse Transcriptase and Cellular DNA Polymerase-Alpha Activities by Lignans Isolated from Chinese Herbs, Phyllanthus Myrtifolius Moon, and Tannins from Lonicera Japonica Thunb and Castanopsis Hystrix. *Antivir. Res.* **1995**, *27*, 367–374. [CrossRef]

16. Ekalu, A.; Ayo, R.G.-O.; Habila, J.D.; Hamisu, I. In Vitro Antimicrobial Activity of Lignan from the Stem Bark of Strombosia Grandifolia Hook.f. Ex Benth. *Bull. Natl. Res. Cent.* **2019**, *43*, 115. [CrossRef]

17. Akiyama, K.; Yamauchi, S.; Nakato, T.; Maruyama, M.; Sugahara, T.; Kishida, T. Antifungal Activity of Tetra-Substituted Tetrahydrofuran Lignan, (−)-Virgatusin, and Its Structure-Activity Relationship. *Biosci. Biotechnol. Biochem.* **2007**, *71*, 1028–1035. [CrossRef]

18. Cho, J.Y.; Choi, G.J.; Son, S.W.; Jang, K.S.; Lim, H.K.; Lee, S.O.; Sung, N.D.; Cho, K.Y.; Kim, J.-C. Isolation and Antifungal Activity of Lignans from Myristica Fragrans against Various Plant Pathogenic Fungi. *Pest. Manag. Sci.* **2007**, *63*, 935–940. [CrossRef]

19. Zhang, W.; Wang, Y.; Geng, Z.; Guo, S.; Cao, J.; Zhang, Z.; Pang, X.; Chen, Z.; Du, S.; Deng, Z. Antifeedant Activities of Lignans from Stem Bark of Zanthoxylum Armatum DC. against Tribolium Castaneum. *Molecules* **2018**, *23*, 617. [CrossRef]

20. Gang, D.R.; Kasahara, H.; Xia, Z.Q.; Vander Mijnsbrugge, K.; Bauw, G.; Boerjan, W.; Van Montagu, M.; Davin, L.B.; Lewis, N.G. Evolution of Plant Defense Mechanisms. Relationships of Phenylcoumaran Benzylic Ether Reductases to Pinoresinol-Lariciresinol and Isoflavone Reductases. *J. Biol. Chem.* **1999**, *274*, 7516–7527. [CrossRef]

21. Gang, D.R.; Fujita, M.; Davin, L.B.; Lewis, N.G. The "Abnormal Lignins": Mapping Heartwood Formation through the Lignan Biosynthetic Pathway. In *Lignin and Lignan Biosynthesis*; ACS Symposium Series; American Chemical Society: Washington, DC, USA, 1998; Volume 697, pp. 389–421. ISBN 978-0-8412-3566-3.

22. Thongphasuk, P.; Suttisri, R.; Bavovada, R.; Verpoorte, R. Antioxidant Lignan Glucosides from Strychnos Vanprukii. *Fitoterapia* **2004**, *75*, 623–628. [CrossRef] [PubMed]

23. Mahendra Kumar, C.; Singh, S.A. Bioactive Lignans from Sesame (Sesamum Indicum L.): Evaluation of Their Antioxidant and Antibacterial Effects for Food Applications. *J. Food Sci. Technol.* **2015**, *52*, 2934–2941. [CrossRef]

24. Li, Y.; Wei, J.; Fang, J.; Lv, W.; Ji, Y.; Aioub, A.A.A.; Zhang, J.; Hu, Z. Insecticidal Activity of Four Lignans Isolated from Phryma Leptostachya. *Molecules* **2019**, *24*, 1976. [CrossRef] [PubMed]

25. Niculaes, C.; Morreel, K.; Kim, H.; Lu, F.; McKee, L.S.; Ivens, B.; Haustraete, J.; Vanholme, B.; Rycke, R.D.; Hertzberg, M.; et al. Phenylcoumaran Benzylic Ether Reductase Prevents Accumulation of Compounds Formed under Oxidative Conditions in Poplar Xylem. *Plant. Cell* **2014**, *26*, 3775–3791. [CrossRef]

26. Corbin, C.; Decourtil, C.; Marosevic, D.; Bailly, M.; Lopez, T.; Renouard, S.; Doussot, J.; Dutilleul, C.; Auguin, D.; Giglioli-Guivarc'h, N.; et al. Role of Protein Farnesylation Events in the ABA-Mediated Regulation of the Pinoresinol-Lariciresinol Reductase 1 (LuPLR1) Gene Expression and Lignan Biosynthesis in Flax (Linum Usitatissimum L.). *Plant Physiol. Biochem. PPB* **2013**, *72*, 96–111. [CrossRef]

27. Renouard, S.; Tribalatc, M.-A.; Lamblin, F.; Mongelard, G.; Fliniaux, O.; Corbin, C.; Marosevic, D.; Pilard, S.; Demailly, H.; Gutierrez, L.; et al. RNAi-Mediated Pinoresinol Lariciresinol Reductase Gene Silencing in Flax (Linum Usitatissimum L.) Seed Coat: Consequences on Lignans and Neolignans Accumulation. *J. Plant. Physiol.* **2014**, *171*, 1372–1377. [CrossRef]

28. Barros-Rios, J.; Serrani-Yarce, J.; Chen, F.; Baxter, D.; Venables, B.; Dixon, R. Role of Bifunctional Ammonia-Lyase in Grass Cell Wall Biosynthesis. *Nat. Plants* **2016**, *2*, 16050. [CrossRef]
29. Vogt, T. Phenylpropanoid Biosynthesis. *Mol. Plant* **2010**, *3*, 2–20. [CrossRef]
30. Bernard, S.M.; Habash, D.Z. The Importance of Cytosolic Glutamine Synthetase in Nitrogen Assimilation and Recycling. *New Phytol.* **2009**, *182*, 608–620. [CrossRef]
31. Corbin, C.; Renouard, S.; Lopez, T.; Lamblin, F.; Lainé, E.; Hano, C. Identification and Characterization of Cis-Acting Elements Involved in the Regulation of ABA- and/or GA-Mediated LuPLR1 Gene Expression and Lignan Biosynthesis in Flax (Linum Usitatissimum L.) Cell Cultures. *J. Plant. Physiol.* **2013**, *170*, 516–522. [CrossRef]
32. Wu, R.; Wang, L.; Wang, Z.; Shang, H.; Liu, X.; Zhu, Y.; Qi, D.; Deng, X. Cloning and Expression Analysis of a Dirigent Protein Gene from the Resurrection Plant Boea Hygrometrica. *Prog. Nat. Sci.* **2009**, *19*, 347–352. [CrossRef]
33. Ramsay, A.; Fliniaux, O.; Fang, J.; Molinié, R.; Roscher, A.; Grand, E.; Guillot, X.; Kovensky, J.; Fliniaux, M.-A.; Schneider, B.; et al. Development of an NMR Metabolomics-Based Tool for Selection of Flaxseed Varieties. *Metabolomics* **2014**, *10*. [CrossRef]
34. Dalisay, D.S.; Kim, K.W.; Lee, C.; Yang, H.; Rübel, O.; Bowen, B.P.; Davin, L.B.; Lewis, N.G. Dirigent Protein-Mediated Lignan and Cyanogenic Glucoside Formation in Flax Seed: Integrated Omics and MALDI Mass Spectrometry Imaging. *J. Nat. Prod.* **2015**, *78*, 1231–1242. [CrossRef]
35. Jez, J.; Ferrer, J.-L.; Bowman, M.; Austin, M.; Schröder, J.; Dixon, R.; Noel, J. Structure and Mechanism of Chalcone Synthase-like Polyketide Synthases. *J. Ind. Microbiol. Biotechnol.* **2002**, *27*, 393–398. [CrossRef] [PubMed]
36. Marchiosi, R.; dos Santos, W.D.; Constantin, R.P.; de Lima, R.B.; Soares, A.R.; Finger-Teixeira, A.; Mota, T.R.; de Oliveira, D.M.; de Pavia Foletto-Felipe, M.; Abrahão, J.; et al. Biosynthesis and Metabolic Actions of Simple Phenolic Acids in Plants. *Phytochem. Rev.* **2020**, *19*, 865–906. [CrossRef]
37. Shahidi, F.; Chandrasekara, A. Hydroxycinnamates and Their in Vitro and in Vivo Antioxidant Activities. *Phytochem. Rev.* **2009**, *9*, 147–170. [CrossRef]
38. Quéro, A.; Molinié, R.; Elboutachfaiti, R.; Petit, E.; Pau-Roblot, C.; Guillot, X.; Mesnard, F.; Courtois, J. Osmotic Stress Alters the Balance between Organic and Inorganic Solutes in Flax (Linum Usitatissimum). *J. Plant. Physiol.* **2014**, *171*, 55–64. [CrossRef]
39. Delauney, A.J.; Verma, D.P.S. Proline Biosynthesis and Osmoregulation in Plants. *Plant. J.* **1993**, *4*, 215–223. [CrossRef]
40. Hare, P.D.; Natal, P.U.; Cress, W.A. Metabolic Implications of Stress-Induced Proline Accumulation in Plants. *Plant Growth Regul. Neth.* **1997**, *21*, 79–102. [CrossRef]
41. Munns, R. Comparative Physiology of Salt and Water Stress. *Plant Cell Environ.* **2002**, *25*, 239–250. [CrossRef]
42. Kovács, Z.; Simon-Sarkadi, L.; Vashegyi, I.; Kocsy, G. Different Accumulation of Free Amino Acids during Short- and Long-Term Osmotic Stress in Wheat. Available online: https://www.hindawi.com/journals/tswj/2012/216521/ (accessed on 25 November 2020).
43. Ashraf, M.; Foolad, M.R. Roles of Glycine Betaine and Proline in Improving Plant Abiotic Stress Resistance. *Environ. Exp. Bot.* **2007**, *59*, 206–216. [CrossRef]
44. Sharma, S.; Verslues, P.E. Mechanisms Independent of Abscisic Acid (ABA) or Proline Feedback Have a Predominant Role in Transcriptional Regulation of Proline Metabolism during Low Water Potential and Stress Recovery. *Plant Cell Environ.* **2010**, *33*, 1838–1851. [CrossRef] [PubMed]
45. Abbas, S.R.; Ahmad, S.D.; Sabir, S.M.; Shah, A.H. Detection of Drought Tolerant Sugarcane Genotypes (Saccharum Officinarum) Using Lipid Peroxidation, Antioxidant Activity, Glycine-Betaine and Proline Contents. *J. Soil Sci. Plant Nutr.* **2014**, *14*, 233–243. [CrossRef]
46. Vetter, J. Plant Cyanogenic Glycosides. *Toxicon Off. J. Int. Soc. Toxinol.* **2000**, *38*, 11–36. [CrossRef]
47. Bartels, D. Desiccation Tolerance Studied in the Resurrection Plant Craterostigma Plantagineum. *Integr. Comp. Biol.* **2005**, *45*, 696–701. [CrossRef] [PubMed]
48. Koster, K.L.; Leopold, A.C. Sugars and Desiccation Tolerance in Seeds. *Plant. Physiol.* **1988**, *88*, 829–832. [CrossRef]
49. Liu, F.; Jensen, C.R.; Andersen, M.N. Drought Stress Effect on Carbohydrate Concentration in Soybean Leaves and Pods during Early Reproductive Development: Its Implication in Altering Pod Set. *Field Crop. Res.* **2004**, *86*, 1–13. [CrossRef]
50. Osakabe, Y.; Yamaguchi-Shinozaki, K.; Shinozaki, K.; Tran, L.-S.P. ABA Control of Plant Macroelement Membrane Transport Systems in Response to Water Deficit and High Salinity. *New Phytol.* **2014**, *202*, 35–49. [CrossRef]
51. Hoagland, D.R.; Arnon, D.I. The Water-Culture Method for Growing Plants without Soil. *Circ. Calif. Agric. Exp. Stn.* **1950**, *347*, 32.
52. Tocquin, P.; Corbesier, L.; Havelange, A.; Pieltain, A.; Kurtem, E.; Bernier, G.; Périlleux, C. A Novel High Efficiency, Low Maintenance, Hydroponic System for Synchronous Growth and Flowering of Arabidopsis Thaliana. *BMC Plant Biol.* **2003**, *3*, 2. [CrossRef]
53. Deborde, C.; Fontaine, J.-X.; Jacob, D.; Botana, A.; Nicaise, V.; Richard-Forget, F.; Lecomte, S.; Decourtil, C.; Hamade, K.; Mesnard, F.; et al. Optimizing 1D 1H-NMR Profiling of Plant Samples for High Throughput Analysis: Extract Preparation, Standardization, Automation and Spectra Processing. *Metabolomics* **2019**, *15*, 28. [CrossRef] [PubMed]
54. Gan, F.; Ruan, G.; Mo, J. Baseline Correction by Improved Iterative Polynomial Fitting with Automatic Threshold. *Chemom. Intell. Lab. Syst.* **2006**, *82*, 59–65. [CrossRef]
55. Zhang, Z.-M.; Chen, S.; Liang, Y.-Z. Baseline Correction Using Adaptive Iteratively Reweighted Penalized Least Squares. *Analyst* **2010**, *135*, 1138–1146. [CrossRef] [PubMed]

molecules

 MDPI

Article

Localization of Major Ephedra Alkaloids in Whole Aerial Parts of Ephedrae Herba Using Direct Analysis in Real Time-Time of Flight-Mass Spectrometry

Nayoung Yun [1,†], Hye Jin Kim [2,‡], Sang Cheol Park [1], Geonha Park [1], Min Kyoung Kim [1], Young Hae Choi [2,3] and Young Pyo Jang [1,2,*]

[1] Department of Life and Nanopharmaceutical Sciences, College of Pharmacy, Kyung Hee University, Hoegi-dong 1, Dongdaemun-gu, Seoul 02447, Korea; nyyun@helixmith.com (N.Y.); sancho@khu.ac.kr (S.C.P.); ginapark0326@khu.ac.kr (G.P.); mindung3@khu.ac.kr (M.K.K.)

[2] Division of Pharmacognosy, College of Pharmacy, Kyung Hee University, Hoegi-dong 1, Dongdaemun-gu, Seoul 02447, Korea; hk2798@cumc.columbia.edu (H.J.K.); y.h.choi@biology.leidenuniv.nl (Y.H.C.)

[3] Natural Products Laboratory, Institute of Biology, Leiden University, Sylviusweg 72, 2333 BE Leiden, The Netherlands

* Correspondence: ypjang@khu.ac.kr; Tel.: +82-2-961-9421

† Current address: Helixmith Co., Ltd. 5th Fl. Bldg. 203, SNU, 1, Gwanak-ro, Gawanak-gu, Seoul 08826, Korea.

‡ Current address: Department of Ophthalmology, Columbia University Medical Center, New York, NY 10032, USA.

Citation: Yun, N.; Kim, H.J.; Park, S.C.; Park, G.; Kim, M.K.; Choi, Y.H.; Jang, Y.P. Localization of Major Ephedra Alkaloids in Whole Aerial Parts of Ephedrae Herba Using Direct Analysis in Real Time-Time of Flight-Mass Spectrometry. *Molecules* **2021**, *26*, 580. https://doi.org/10.3390/molecules26030580

Academic Editor: Jane Ward

Received: 25 December 2020
Accepted: 20 January 2021
Published: 22 January 2021

Publisher's Note: MDPI stays neutral with regard to jurisdictional claims in published maps and institutional affiliations.

Abstract: Mass spectrometry-based molecular imaging has been utilized to map the spatial distribution of target metabolites in various matrixes. Among the diverse mass spectrometry techniques, matrix-assisted laser desorption/ionization-mass spectrometry (MALDI-MS) is the most popular for molecular imaging due to its powerful spatial resolution. This unparalleled high resolution, however, can paradoxically act as a bottleneck when the bio-imaging of large areas, such as a whole plant, is required. To address this issue and provide a more versatile tool for large scale bio-imaging, direct analysis in real-time-time of flight-mass spectrometry (DART-TOF-MS), an ambient ionization MS, was applied to whole plant bio-imaging of a medicinal plant, Ephedrae Herba. The whole aerial part of the plant was cut into 10–20 cm long pieces, and each part was further cut longitudinally to compare the contents of major ephedra alkaloids between the outer surface and inner part of the stem. Using optimized DART-TOF-MS conditions, molecular imaging of major ephedra alkaloids of the whole aerial part of a single plant was successfully achieved. The concentration of alkaloids analyzed in this study was found to be higher on the inner section than the outer surface of stems. Moreover, side branches, which are used in traditional medicine, represented a far higher concentration of alkaloids than the main stem. In terms of the spatial metabolic distribution, the contents of alkaloids gradually decreased towards the end of branch tips. In this study, a fast and simple macro-scale MS imaging of the whole plant was successfully developed using DART-TOF-MS. This application on the localization of secondary metabolites in whole plants can provide an area of new research using ambient ionization mass spectroscopy and an unprecedented macro-scale view of the biosynthesis and distribution of active components in medicinal plants.

Keywords: molecular imaging; direct analysis in real-time mass spectrometry; ephedra alkaloids; ephedrine; methylephedrine; *Ephedra sinica*

1. Introduction

Mass spectrometry-based molecular imaging has been used as a powerful analytical technique to visualize the distribution of endogenous and/or exogenous biomolecules in diverse tissues. This kind of image analysis can provide some information on the spatial location of metabolites, increasing the quality of the data obtained with conventional

95

metabolomics. Using this technique, various molecules, such as lipids [1], proteins [2], peptides [2,3], and many specialized metabolites, e.g., flavonoids and glucosinolates [4], have been successfully imaged directly in the animal or plant tissue. It is Matrix-Assisted Laser Desorption/Ionization (MALDI) that, among many ionization techniques of mass spectroscopy, is undoubtedly one of the most popular ionization methods used in molecular imaging due to its efficiency in the ionization of large molecules, such as proteins and peptides. The popularity of MALDI for molecular imaging is due mainly to its uniquely high spatial resolution that makes micro-scale tissue imaging possible. Its use has been reported in many previous studies, such as the successful identification of candidates of tumoral markers of prostate cancer [5], the characterization of the spatial distribution of lens proteins and their modifications in lens sections [6], the knowledge of the distribution of neuropeptides in mouse pituitary glands [7], and understanding the distribution of specific drugs in tissue [8]. On the other hand, the limitations of MALDI imaging are well known. It requires a series of sample preparation steps, including embedding, section, mounting, matrix coating, and each step has a significant impact on the quality of the results [9].

Notwithstanding the potential of MALDI-based molecular imaging in mammalian tissues, there are only a few applications involving plants, and if any, these are limited to specific organs, such as leaves, stem, and grain tissue, which have been analyzed, for example, for their content of metabolites, such as flavonoid and saponins [10,11]. Thus, the use of MALDI techniques in plant imaging is mainly limited to a small area of the sample. For plant studies, particularly when molecular imaging is applied at a macro-scale to determine the distribution of metabolites at an organ level, the high spatial resolution power of MALDI limits the area of analysis due to the unacceptably lengthy sample preparation and analysis time needed for large-surface samples. Regarding sample preparation, MALDI analysis requires a step in which the tissue has to be covered with a matrix solution for the ionization and co-crystallized with this matrix [12]. The ionization efficiency of the molecules, however, varies according to the type of matrix [8].

As explained, the extension of the imaging area is an important criterion in plant research applications. This is especially crucial considering that mapping the distribution of compounds in a whole plant is necessary to obtain significant information about the biosynthesis, transportation, and biological functions of metabolites. With this in mind, different types of ionization sources were tested for the macro-scale molecular imaging to widen the analytical range.

A number of MS-methods based on ambient desorption ionization techniques have been developed to circumvent these limitations and increase the efficiency in terms of the time involved in the analysis, namely, direct analysis in real-time (DART), desorption electrospray ionization (DESI) [13], and atmospheric solids analysis probe (ASAP) [14]. Especially, DART has the edge over other techniques as the coverage of area and ease of sample preparation results in reduced analysis time. In the DART ion source, metastable helium gas reacts with atmospheric water to produce protonated water clusters, and the molecules on the sample surface are directly ionized by the transfer of protons from these water clusters [15]. Since the analysis takes place in open-air conditions, samples can be analyzed in diverse states or formats, i.e., as gases, liquids, solids, or powdered, without any specific sample pre-treatment. The ion source of DART can ionize various types of compounds directly from the surface of samples with little fragmentation, allowing molecular ion peaks to be efficiently detected from diverse surfaces, such as human skin, paper money, glass, thin layer chromatography (TLC) plates, clothing, and metals [16–18]. The DART ion source has, however, a relatively low spatial resolution of 1–2 mm as it is limited by the width of the helium gas beam [19]. Nevertheless, as an image screening tool, it is considered to have a high throughput capacity of analysis, which can be efficiently adapted to the large-scale analysis of samples, such as whole plants.

In this paper, we reported an application of the DART ion source to image contents of active components in a medicinal plant with the purpose of testing the feasibility of the DART ion source as a tool for the macro-scale localization of plants' metabolites. The whole aerial part of a single *Ephedra sinica* (Ephedraceae), a plant traditionally used for weight loss and for its diaphoretic, stimulant, and antiasthmatic properties, was selected for this study. The major ephedra alkaloids—ephedrine/pseudoephedrine and methylephedrine/methylpseudoephedrine (Figure 1)—were imaged throughout the whole aerial part of this plant, and their spatial distributions were determined.

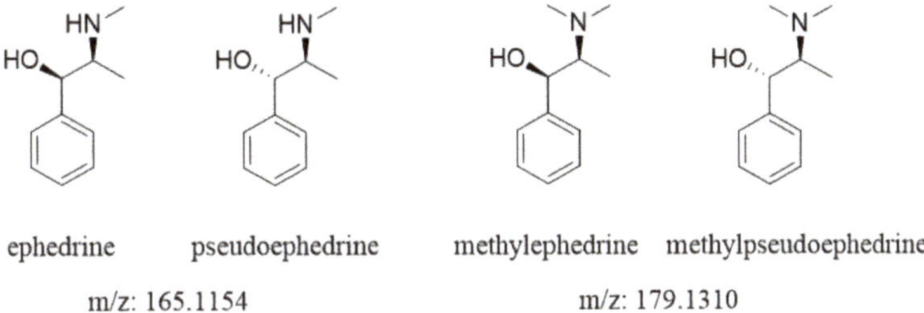

Figure 1. Major alkaloids of ephedra plants. Ephedrine and its diastereomer pseudoephedrine have the same molecular weight, so it is not differentiated by a mass spectrometer, and the same is true in the case of methylephedrine and methylpseudoephedrine. Ephedrine/pseudoephedrine: $C_{10}H_{15}NO$, methylephedrine/methylpseudoephedrine: $C_{11}H_{17}NO$.

2. Results

In DART analysis, the temperature of helium gas largely affects the ionization sensitivity and data quality. Therefore, prior to the analysis of ephedra alkaloids in the samples, a number of experiments were implemented to determine its optimum temperature. The two targeted ephedra alkaloids were not clearly detected at 250 °C, and when the temperature was increased to 350 °C, some surface of the sample started to burn; thus, the temperature of helium gas was set to 300 °C for all the data measurement. The spatial resolution of the system employed in this study was evaluated by separate experiments using a standard solution on a TLC plate, and the maximum resolution was determined to be approximately 1.0 mm. The total experimental time of plant analysis was around 10 h that was still reasonable for the image analysis of the whole aerial part of a single ephedra plant.

Representative DART-TOF-MS spectra obtained from the outer and inner surface of both ephedra side branches and the main stem can be observed in Figure 2. The outer surface of the branch showed the strong peak of the protonated molecule $[M + H]^+$ of methylephedrine and/or methylpseudoephedrine (m/z 180.1323) with a very weak peak of protonated ephedrine (Figure 2A). The inner section of the branch produced both peaks of protonated ephedrine/pseudoephedrine (m/z 166.1126) and protonated methylephedrine and methylpseudoephedrine (Figure 2B). The mass accuracies of observed mass values for the alkaloids are represented in Table 1. Interestingly, there was a large variation of the alkaloid concentrations according to the location since these alkaloids were barely detectable on both the outer and inner surfaces of the main stem, and the MS spectra showed a completely different pattern to those of the side branch (Figure 2C,D).

Molecular imaging of active components in a whole, single ephedra plant was successfully carried out by the integration of imaging data of the parts. Being diastereoisomers, ephedrine and pseudoephedrine and methylephedrine and methylpseudoephedrine have the same molecular mass. It was thus impossible to distinguish ephedrine and methylephedrine from their diastereomers only by MS analysis. Therefore, the total contents of the diastereomers pairs of ephedra alkaloids were used for molecular imaging. Imaging for total contents of ephedrine and pseudoephedrine (m/z 166) is presented in Figure 3. The green color (ion intensity of ca. 15×10^3 and less) represented the lowest detectable intensity of alkaloids, and the color gradually changed to blue (ion intensity of 50×10^3–90×10^3), purple (ion intensity of 90×10^3–110×10^3), and red (ion intensity of ca. 120×10^3 and more) as the ion intensity increased. Most parts of the outer surface of ephedra aerial parts appeared as green, while major parts of the inner surface were colored blue or red. These results mean that the content of ephedrine and pseudoephedrine in the inner section surface was much higher than on the outer surface. It was also observed that the side branches of the plant had a much higher content of ephedrine and pseudoephedrine than the main stem of the ephedra plant. Interestingly, it is not a woody stem but a side branch that is medicinally used in Korea [20]. The total content of ephedrine and pseudoephedrine decreased towards the end tip of the branch, where primary metabolism generating essential primary metabolites is believed to dominate over secondary metabolism, synthesizing ephedra alkaloids. As N-methyltransferase has been known as the key enzyme family in the biosynthesis of ephedra alkaloids, spatial existence and/or activities of these enzymes need to be evaluated to unveil the reason for the differential location of these alkaloids [21]. Molecular imaging for the total content of methylephedrine and methylpseudoephedrine (m/z 180) is presented in Figure 4. The molecular imaging of m/z 180 showed a similar pattern to that observed for ephedrine and pseudoephedrine. The content of methylephedrine and methylpseudoephedrine also showed higher intensity on the inner surface, decreasing towards the tip of the branch. In contrast to the images for ephedrine and pseudoephedrine, some ion peaks of methylephedrine and methylpseudoephedrine were observed on the outer surface, as represented by blue and red color. While the maximum ion intensity for ephedrine/pseudoephedrine was around 150,000, the maximum ion intensity for methylated ephedrine/pseudoephedrine was significantly higher (ca. 250,000), and this represented that the content of methylephedrine and methylpseudoephedrine was higher than the sum of ephedrine and pseudoephedrine. This result seems to be opposite to that reported in a previous study using HPLC in which the content of ephedrine and pseudoephedrine in Ephedrae Herba was reported to be about 10 times that of methylephedrine and methylpseudoephedrine [22]. Due to the characteristic of DART-MS, only the compounds present on the surface are ionized and detected, so the quantitative results cannot be compared to those from the whole extract of the sample. In this study, it is not possible to generalize this to the amount contained in the whole plant, as only the relative amount of alkaloids present in some particular areas can be obtained. Nevertheless, it is significant that the distribution of some alkaloids present on the outer and inner surfaces of the entire plant can be seen as a whole. It is a great advantage that the distribution of the compounds of interest in the whole plant can be mapped as an overall picture, even if the accuracy is compromised.

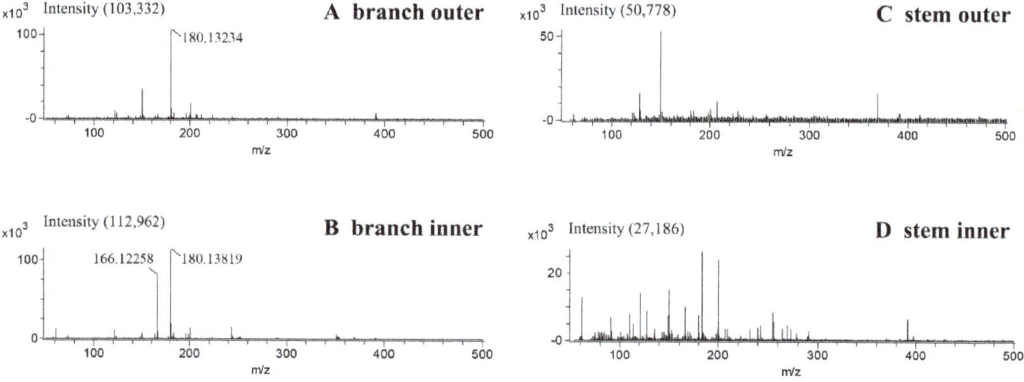

Figure 2. Representative direct analysis in real time-time of flight-mass spectrometry (DART-TOF-MS) spectra of the outer surface (**A**) and an inner surface (**B**) of a branch and outer surface (**C**) and an inner surface (**D**) of the main stem (woody stem) of ephedra aerial parts.

Table 1. Molecular formula, theoretical mass, and observed mass of major ephedra alkaloids and their protonated molecules. The mass difference was represented as millidalton (mDa).

Alkaloid	Molecular Formula	Theoretical Mass (Da)	Observed Mass (Da)	Mass Difference (mDa)
Ephedrine/Pseudoephedrine	$C_{10}H_{15}NO$	165.1154		
Protonated ephedrine	$[C_{10}H_{15}NO + H]^+$	166.1227	166.1226	−0.1
Methylephedrine/ Methylpseudoephedrine	$C_{11}H_{17}NO$	179.1310		
Protonated methylephedrine	$[C_{11}H_{17}NO + H]^+$	180.1383	180.1323	−6.0

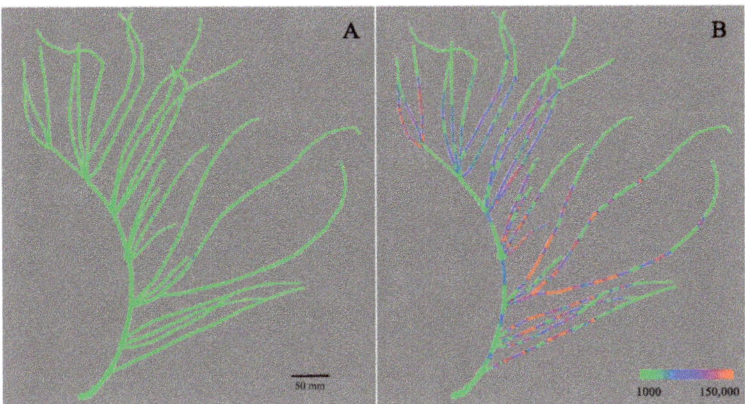

Figure 3. DART-MS images of total contents of ephedrine/pseudoephedrine (m/z 166) on the outer surface (**A**) and an inner surface (**B**) of ephedra aerial part. The color range is from 1000 to 150,000 ion intensity.

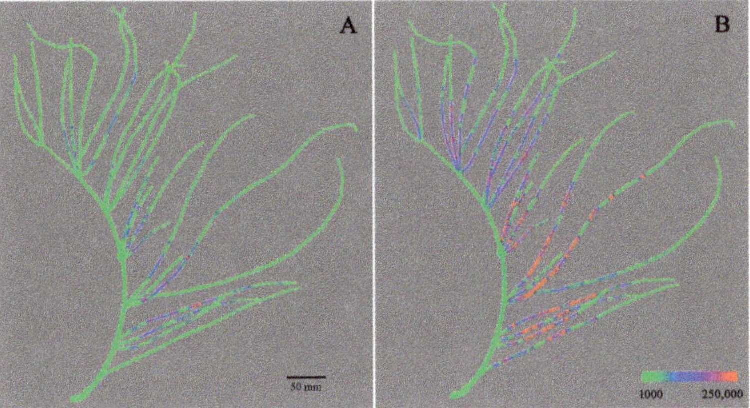

Figure 4. DART-MS molecular images for total contents of methylephedrine/methylpseudoephedrine (m/z 180) on the outer surface (**A**) and an inner surface (**B**) of ephedra aerial part. The color range is from 1000 to 250,000 ion intensity.

3. Discussion

In this work, we showed that rapid and simple macro-scale molecular imaging could be achieved using DART-TOF-MS. This could constitute an improvement in MALDI-MS bio-imaging, which requires a long time to scan large surfaces of a whole plant. On the contrary, this method could be used effectively to image the target compounds in whole plants in a relatively short time without the need of a specific chemical matrix nor a specific probe. This experiment broadens the area of applications of DART-MS, allowing the bio-imaging of the whole plant, thus providing an overall view of the distribution of metabolites in whole plants. The results of the implementation of this technique can also be used to gain a deeper understanding of biological or physiological functions of plant secondary metabolites, the identification of biological markers, biosynthesis, and the transportation of metabolites upon biotic and/or abiotic stress, among many other applications.

4. Materials and Methods

4.1. DART-TOF-MS Measurement of Ephedrae Herba

A single *Ephedra sinica* was harvested from the herbal garden of the College of Pharmacy, Kyung Hee University, and stored in a deep freezer (Thermo Fisher Scientific, Asheville, NC, USA) at $-70\ °C$. A voucher specimen (KHUP-0801) was deposited in the Museum of Korean Traditional Herbal Medicines located in the College of Pharmacy, Kyung Hee University. The aerial part, branch, and woody stem were cut into 10–20 cm pieces, and each piece (diameter range: 2–5 mm) was cut in half vertically so that similar parts of the center were exposed. The measurement was carried out using a DART ion-source (IonSense, Sangus, MA, USA) coupled to an Accu-TOF-MS (JMS-T100TD, JEOL, Tokyo, Japan). The optimized MS operating conditions were as follows: samples were analyzed in the positive ion mode; voltage of first orifice lens was 15 V, ring lens voltage was 5 V, and helium gas flow rate was 3 L/min. Mass scale calibration was accomplished by introducing a glass capillary with polyethylene glycol 600 (PEG 600, Sigma-Aldrich, St. Louis, MO, USA) in the DART. The analyzer was set with a peak voltage of 1000 V, a bias voltage of 28 V, a pusher bias voltage of -0.55 V, and a detector voltage of 2200 V. Mass detection range was m/z 50 to 1000.

The helium gas (Shinyang, Seoul, Korea) temperature and sample module were optimized before molecular imaging. The temperature of the helium gas in the ionization source was chosen according to the ionization efficiency of the target alkaloids among six different temperatures, i.e., 100 °C, 150 °C, 200 °C, 250 °C, 300 °C, and 350 °C. Since there was no specific sample module for molecular imaging with the DART-MS, a house-made module was used for the analysis. A piece of dissected branch and stem of ephedra plant was fixed on a glass plate with double-sided adhesive tape and introduced into the ion source using the appropriate sample module. In order to acquire accurate and reliable MS data, the glass plate bearing the plant sample was introduced using a sample module on a rail under the helium gas flow. Among the three different sample modules available, the 12 DIP-it holder module (IonSense, Sangus, MA, USA) was able to move the plate consistently, exposing the surface of the plant sample just beneath the helium gas jet. Three slide glass plates had to be stacked and glued to align them exactly with the helium gas jet (Figure 5). The analysis was conducted on both the outer surface and inner surface of Ephedrae Herba to compare its contents in ephedra alkaloids. The rail speed was set to 0.2 mm/s.

4.2. Data Processing

The raw data resulting from the spectral analysis were processed for molecular imaging. For this, selected ion chromatograms for the targeted ephedra alkaloids were extracted. As protonated molecular ion ($[M + H]^+$) is mainly generated in DART ionization, selected ion chromatograms of m/z 166 and m/z 180 were extracted for ephedrine/pseudoephedrine and methylephedrine/methylpseudoephedrine, respectively. The identities of alkaloids were confirmed by direct comparison of the experimental high-resolution mass number with their theoretical mass numbers. The extracted chromatograms were converted to an

ASCII file table in which the intensity of specific ions was recorded at 0.4 s intervals. The average of ten data points, which corresponded to 0.8 mm of the sample, was used for the imaging and assigned different colors using a specific function in the Excel program known as 'conditional formatting', which assigns different colors according to their numerical value. Whole plant imaging was completed by applying the corresponding colors to the photo of the plant using Adobe Photoshop (Adobe CS6). Each herb part had a length of 10~20 cm, and 1250~2500 data points were acquired and processed for the imaging.

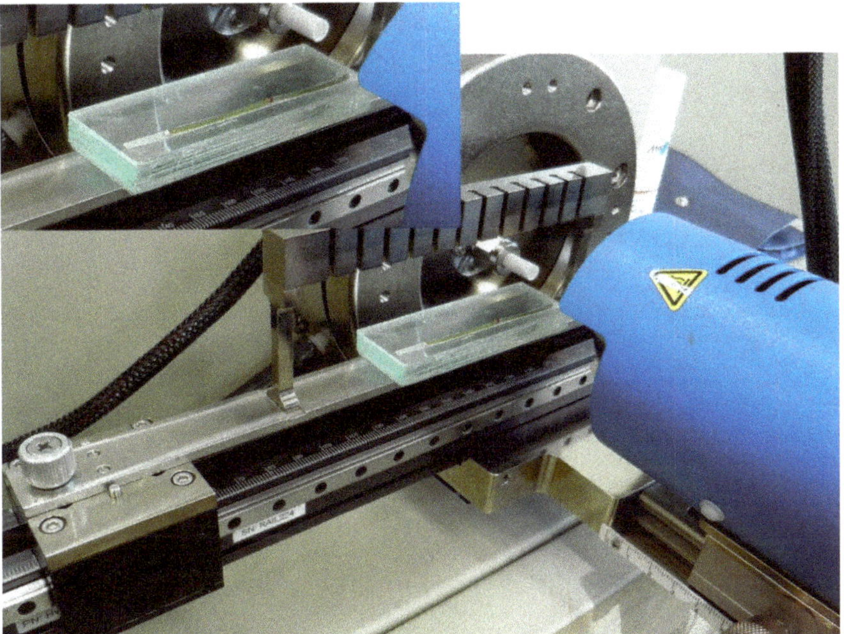

Figure 5. House-made sample module for DART-TOF-MS imaging system.

Author Contributions: Conceptualization, Y.P.J.; methodology, N.Y.; software, S.C.P.; validation, H.J.K. and G.P.; formal analysis, N.Y. and H.J.K.; data curation, S.C.P.; writing—original draft preparation, N.Y.; writing—review and editing, Y.P.J. and Y.H.C.; visualization, N.Y., G.P., and M.K.K.; supervision, Y.P.J. All authors have read and agreed to the published version of the manuscript.

Funding: This research received no external funding.

Data Availability Statement: The data presented in this study are available on request from the corresponding author.

Conflicts of Interest: The authors declare no conflict of interest.

Sample Availability: Not available.

References

1. Murphy, R.C.; Hankin, J.A.; Barkley, R.M. Imaging of lipid species by MALDI mass spectrometry. *J. Lipid Res.* **2009**, *50*, S317–S322. [CrossRef]
2. Caprioli, R.M.; Farmer, T.B.; Gile, J. Molecular imaging of biological samples: Localization of peptides and proteins using MALDI-TOF MS. *Anal. Chem.* **1997**, *69*, 4751–4760. [CrossRef]
3. Altelaar, A.F.; Taban, I.M.; McDonnell, L.A.; Verhaert, P.D.E.M.; de Lange, R.P.J.; Adan, R.A.H.; Mooi, W.J.; Heeren, R.; Piersma, S.R. High-resolution MALDI imaging mass spectrometry allows localization of peptide distributions at cellular length scales in pituitary tissue sections. *Int. J. Mass Spectrom.* **2007**, *260*, 203–211. [CrossRef]

4. Shroff, R.; Vergara, F.; Muck, A.; Svatoš, A.; Gershenzon, J. Nonuniform distribution of glucosinolates in Arabidopsis thaliana leaves has important consequences for plant defense. *Proc. Natl. Acad. Sci. USA* **2008**, *105*, 6196–6201. [CrossRef] [PubMed]

5. Schwamborn, K.; Krieg, R.C.; Reska, M.; Jakse, G.; Knuechel, R.; Wellmann, A. Identifying prostate carcinoma by MALDI-Imaging. *Int. J. Mol. Med.* **2007**, *20*, 155. [CrossRef] [PubMed]

6. Han, J.; Schey, K.L. MALDI tissue imaging of ocular lens α-crystallin. *Investig. Ophthalm. Vis. Sci.* **2006**, *47*, 2990–2996. [CrossRef] [PubMed]

7. Römpp, A.; Guenther, S.; Schober, Y.; Schulz, O.; Takats, Z.; Kummer, W.; Spengler, B. Histology by Mass Spectrometry: Label-Free Tissue Characterization Obtained from High-Accuracy Bioanalytical Imaging. *Angew. Chem. Int. Ed.* **2010**, *49*, 3834–3838. [CrossRef]

8. Castellino, S.; Groseclose, M.R.; Wagner, D. MALDI imaging mass spectrometry: Bridging biology and chemistry in drug development. *Bioanalysis* **2011**, *3*, 2427–2441. [CrossRef]

9. Gemperline, E.; Chen, B.; Li, L. Challenges and recent advances in mass spectrometric imaging of neurotransmitters. *Bioanalysis* **2014**, *6*, 525–540. [CrossRef]

10. Enomoto, H.; Takahashi, S.; Takeda, S.; Hatta, H. Distribution of Flavan-3-ol Species in Ripe Strawberry Fruit Revealed by Matrix-Assisted Laser Desorption/Ionization-Mass Spectrometry Imaging. *Molecules* **2019**, *25*, 103. [CrossRef]

11. Montini, L.; Crocoll, C.; Gleadow, R.M.; Motawia, M.S.; Janfelt, C.; Bjarnholt, N. Matrix-Assisted Laser Desorption/Ionization-Mass Spectrometry Imaging of Metabolites during Sorghum Germination. *Plant. Physiol.* **2020**, *183*, 925–942. [CrossRef] [PubMed]

12. Eberlin, L.S.; Liu, X.; Ferreira, C.R.; Santagata, S.; Agar, N.Y.R.; Cooks, R.G. Desorption electrospray ionization then MALDI mass spectrometry imaging of lipid and protein distributions in single tissue sections. *Anal. Chem.* **2011**, *83*, 8366–8371. [CrossRef] [PubMed]

13. Takats, Z.; Wiseman, J.M.; Cooks, R.G. Ambient mass spectrometry using desorption electrospray ionization (DESI): Instrumentation, mechanisms and applications in forensics, chemistry, and biology. *J. Mass Spectrom.* **2005**, *40*, 1261–1275. [CrossRef]

14. Petucci, C.; Diffendal, J. Atmospheric solids analysis probe: A rapid ionization technique for small molecule drugs. *J. Mass. Spectrom.* **2008**, *43*, 1565–1568. [CrossRef] [PubMed]

15. Cody, R.B.; Laramée, J.A.; Durst, H.D. Versatile new ion source for the analysis of materials in open air under ambient conditions. *Anal. Chem.* **2005**, *77*, 2297–2302. [CrossRef] [PubMed]

16. Kim, H.J.; Jang, Y.P. Direct analysis of curcumin in turmeric by DART-MS. *Phytochem. Anal.* **2009**, *20*, 372–377. [CrossRef]

17. Kpegba, K.; Spadaro, T.; Cody, R.B.; Nesnas, N.; Olson, J.A. Analysis of self-assembled monolayers on gold surfaces using direct analysis in real time mass spectrometry. *Anal. Chem.* **2007**, *79*, 5479–5483. [CrossRef]

18. McEwen, C.N.; McKay, R.G.; Larsen, B.S. Analysis of solids, liquids, and biological tissues using solids probe introduction at atmospheric pressure on commercial LC/MS instruments. *Anal. Chem.* **2005**, *77*, 7826–7831. [CrossRef]

19. Chernetsova, E.S.; Morlock, G.E.; Revelsky, I.A. DART mass spectrometry and its applications in chemical analysis. *Russ. Chem. Rev.* **2011**, *80*, 235–255. [CrossRef]

20. Ministry of Food and Drug Safety. *Korean Pharmacopoeia Ver. X*; Ministry of Food and Drug Safety: Chungcheongbuk-do, Korea, 2017; pp. 1291–1292.

21. Krizevski, R.; Bar, E.; Shalit, O.; Sitrit, Y.; Ben-Shabat, S.; Lewinsohn, E. Composition and stereochemistry of ephedrine alkaloids accumulation in *Ephedra sinica* Stapf. *Phytochemistry* **2010**, *71*, 895–903. [CrossRef]

22. Hong, H.; Chen, H.; Yang, D.; Shang, M.; Wang, X.; Cai, S.; Mikage, M. Comparison of contents of five ephedrine alkaloids in three official origins of Ephedra Herb in China by high-performance liquid chromatography. *J. Nat. Med.* **2011**, *65*, 623–628. [CrossRef] [PubMed]

Article

NMR Profiling of *Ononis diffusa* Identifies Cytotoxic Compounds against Cetuximab-Resistant Colon Cancer Cell Lines

Vittoria Graziani [1,†], Nicoletta Potenza [1], Brigida D'Abrosca [1], Teresa Troiani [2], Stefania Napolitano [2], Antonio Fiorentino [1,*] and Monica Scognamiglio [1,*]

[1] Dipartimento di Scienze e Tecnologie Ambientali, Biologiche e Farmaceutiche, Università degli Studi della Campania "Luigi Vanvitelli", Via Vivaldi 43, 81100 Caserta, Italy; v.graziani@qmul.ac.uk (V.G.); nicoletta.potenza@unicampania.it (N.P.); brigida.dabrosca@unicampania.it (B.D.)

[2] Oncologia medica, Dipartimento di Medicina di precisione, Università degli Studi della Campania "Luigi Vanvitelli", S. Andrea delle Dame, Via L. De Crecchio 7, 80138 Napoli, Italy; teresa.troiani@unicampania.it (T.T.); stefania.napolitano@unicampania.it (S.N.)

[*] Correspondence: antonio.fiorentino@unicampania.it (A.F.); monica.scognamiglio@unicampania.it (M.S.); Tel.: +39-0823274576 (A.F.)

[†] Current address: Barts Cancer Institute, Queen Mary University of London, John Vane Science Building, Charterhouse Square, London EC1M 6BQ, UK.

Citation: Graziani, V.; Potenza, N.; D'Abrosca, B.; Troiani, T.; Napolitano, S.; Fiorentino, A.; Scognamiglio, M. NMR Profiling of *Ononis diffusa* Identifies Cytotoxic Compounds against Cetuximab-Resistant Colon Cancer Cell Lines. *Molecules* **2021**, *26*, 3266. https://doi.org/10.3390/molecules26113266

Academic Editors: Young Hae Choi, Young Pyo Jang, Yuntao Dai and Luis Francisco Salomé-Abarca

Received: 26 April 2021
Accepted: 26 May 2021
Published: 28 May 2021

Publisher's Note: MDPI stays neutral with regard to jurisdictional claims in published maps and institutional affiliations.

Abstract: In the search of new natural products to be explored as possible anticancer drugs, two plant species, namely *Ononis diffusa* and *Ononis variegata*, were screened against colorectal cancer cell lines. The cytotoxic activity of the crude extracts was tested on a panel of colon cancer cell models including cetuximab-sensitive (Caco-2, GEO, SW48), intrinsic (HT-29 and HCT-116), and acquired (GEO-CR, SW48-CR) cetuximab-resistant cell lines. *Ononis diffusa* showed remarkable cytotoxic activity, especially on the cetuximab-resistant cell lines. The active extract composition was determined by NMR analysis. Given its complexity, a partial purification was then carried out. The fractions obtained were again tested for their biological activity and their metabolite content was determined by 1D and 2D NMR analysis. The study led to the identification of a fraction enriched in oxylipins that showed a 92% growth inhibition of the HT-29 cell line at a concentration of 50 μg/mL.

Keywords: cytotoxic activity; natural products; NMR; *Ononis diffusa*; *Ononis variegata*; oxylipins

1. Introduction

Natural products play a crucial role in the discovery of anticancer compounds even today, when new technologies are available to easily obtain a wide range of drug candidates. The success of natural products is inherent to their structure: they possess a well-defined three-dimensional scaffold with functional groups precisely oriented in space. Furthermore, they are characterized by an enormous structural and functional diversity [1,2]. These features confer them with extraordinary selectivity and specificity, when compared to artificially designed molecules [3]. Plants, in particular, are a unique source of chemicals since they use specialized metabolites to interact with the environment and in response to several biotic and abiotic stresses [4,5]. Humankind has, therefore, always exploited this ability to produce a wide and diversified array of chemicals [6].

In recent years, the need for anticancer compounds has become particularly pressing and, in this context, colorectal cancer is definitely under the spotlight. This cancer is indeed one of the leading causes of cancer-related death worldwide and one of the most frequently diagnosed malignant diseases in Europe [7]. Although the outcomes of patients with metastatic colorectal cancer (mCRC) has improved in recent years [8], new challenges are on the horizon. Nowadays, resistance to both chemotherapy and molecularly targeted therapies represents a major problem for setting up effective treatments. Specifically,

103

clinical data highlight the emergence of acquired resistance to anti-EGFR therapies [9]. EGFR (Epithelial Growth Factor Receptor) is a transmembrane tyrosine kinase receptor that, once activated, triggers two main signaling pathways that are involved in cell proliferation, survival and motility [10]. Drugs like the monoclonal antibody cetuximab are effective at blocking the EGFR receptor. However, patients carrying mutations at some of the intracellular effectors of EGFR activation have intrinsic resistance to cetuximab treatment [11]. Furthermore, around 25% of cetuximab-sensitive mCRC patients develop, after an initial response, secondary resistance to this drug [11,12]. Therefore, it is urgent to find chemotherapeutic agents to prevent or overcome the limit imposed by these resistances to the effectiveness of the anticancer drug.

In the search for compounds with anticancer activity in plants of Mediterranean region, we recently screened a set of plants of the Fabaceae family with a metabolomics-based approach [13]. This study led to the identification of candidate molecules that are currently under further investigation. In fact, there appears to be a link between members of the Fabaceae family and colon cancer prevention and therapy [13].

Herewith, we undertook the study of two further Fabaceae plants which were not analyzed in our previous study: *Ononis diffusa* Ten. and *Ononis variegata* L. *Ononis* is a large genus of perennial herbs and shrubs, including food, foraging and medicinal plants [14]. Several *Ononis* species have been studied, and bioactive compounds including flavonoids, isoflavonoids, alkylresorcinols, coumarin derivatives, and terpenes have been reported from them [14–17]. However, no phytochemical studies have been previously carried out on *O. diffusa* and *O. variegata*.

Extracts from the two species which are the objects of this study were first screened for their antiproliferative activity on colorectal cancer cell lines (Caco-2, HT-29, and HCT-116) characterized by different genetic profiles. Specifically, the Caco-2 cell line is sensitive to cetuximab as it has no genetic defects associated with anti-EGFR therapy resistance. Meanwhile, the HT-29 and HCT-116 cell lines are intrinsically resistant to cetuximab since they harbor BRAF V600E and KRAS/PIK3CA mutations, respectively [18]. *Ononis diffusa* showed a marked cytotoxic effect in all the cell lines and especially in those exhibiting drug resistance. Thus, *O. diffusa* crude extract was further tested on cetuximab-sensitive (GEO and SW48) and on acquired cetuximab-resistant (GEO-CR and SW48-CR) human colon cancer cell lines. GEO-CR and SW48-CR represent a preclinical model for the study of acquired resistance to cetuximab. The potential cytotoxicity of the extracts was evaluated through MTT assays. In the attempt to identify the metabolites responsible for the activity, NMR profiles of the extracts were obtained. Furthermore, a partial purification, always paired with the NMR analysis, and biological tests led to the identification of a mixture of oxylipins as putative bioactive compounds responsible for the antiproliferative properties of the crude extract.

2. Results

2.1. Cytotoxic Activity

2.1.1. Cytotoxic Activity of Crude Extracts

Plant extracts were obtained with a mixture of methanol and water (1:1). The potential antiproliferative activity of the plant crude extracts was evaluated on a panel of colon cancer cell lines (Caco-2, HT-29 and HCT-116). While *O. variegata* showed no inhibition of cell proliferation at any of the tested concentrations, *O. diffusa* showed a significant inhibition against all of the cell lines, beginning at a concentration of 50 µg/mL (Figure 1). *Ononis diffusa* extract exhibited a more pronounced effect on HT-29 cell line.

2.1.2. Cytotoxic Activity of the Crude Extract against Colon Cancer Cell Lines with Acquired Resistance to Cetuximab

Based on the significant activity shown by *O. diffusa* extract on the three colon cancer cell lines, including the intrinsically cetuximab-resistant ones, this extract was further explored for its activity against both cetuximab-sensitive (GEO and SW48) and secondary cetuximab-resistant cancer cell lines (GEO-CR and SW48-CR). The extract showed a remarkable inhibition of the proliferation of the acquired resistant cell lines (Figure 2). Indeed, while a growth inhibition of 50% on SW48 cells was observed at the highest tested concentration (250 µg/mL), a 50% inhibition of the corresponding secondary resistant cell line was observed at a dose as low as 50 µg/mL.

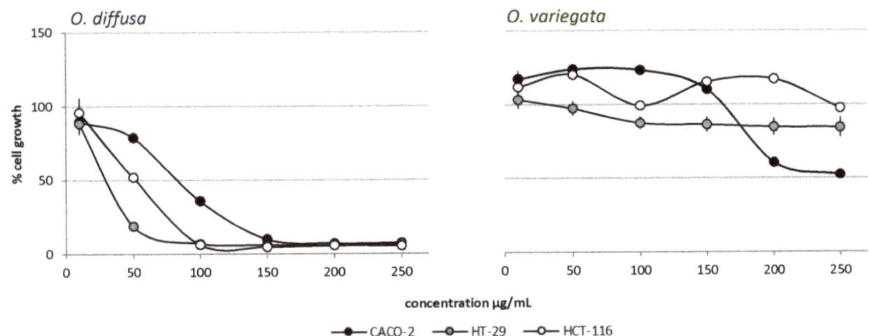

Figure 1. Antiproliferative activity evaluation of *O. diffusa* and *O. variegata* on the Caco-2, HT-29, and HCT-116 cell lines. Cell growth is expressed as percentage from control, and is plotted on the vertical axis, while doses of plant extracts are reported on the horizontal axis.

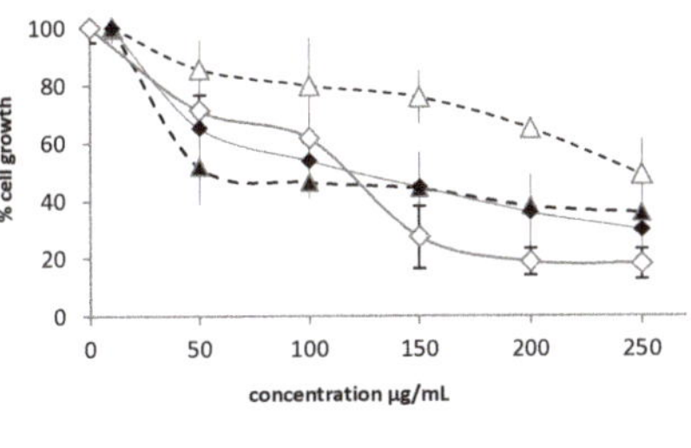

Figure 2. Antiproliferative activity evaluation of *O. diffusa* against the SW48, SW48-CR, GEO, and GEO-CR cell lines. Cell growth is expressed as percentage from control, and it is plotted on the vertical axis, while doses of plant extracts are reported on the horizontal axis.

On the other hand, the activity on the GEO and GEO-CR cell lines was comparable; important effects were observed starting from a concentration of 150 µg/mL of the extract.

2.2. NMR Profiling of the Extracts

NMR spectra were obtained for the extracts of the two Ononis species. The plant material was extracted with a mixture of methanol-d_4 and phosphate buffer in D_2O (1:1) and the solution thereby obtained was analyzed by NMR.

Although *O. variegata* showed no activity in the biological tests, a comparison of the profiles of the two species helped us narrow down the set of the possible bioactive compounds. It was clear, indeed, from the comparison of the ^{1}H-NMR spectra (Figure 3), that there were metabolites present in both species that, furthermore, were the main components of *O. variegata*. These metabolites could not be considered responsible for the activity observed for *O. diffusa* extract. Besides several sugars, amino acids, and organic acids identified on the basis of literature data [19,20], *O. variegata* also showed signals of caffeic acid and of trigonelline (Figure 3) [20,21].

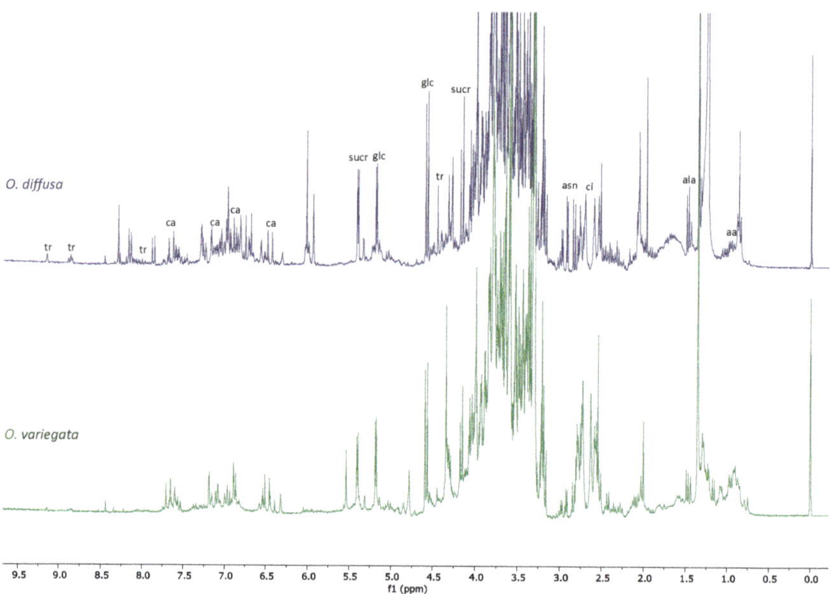

Figure 3. Stacked ^{1}H-NMR spectra of *O. variegata* (green) and *O. diffusa* (blue). Spectra were acquired at 300 MHz, in 1:1 methanol-d_4: buffer. Diagnostic signals of the main metabolites detected in both extracts are indicated on *O. variegata* spectrum by the following abbreviations: aa = amino acids, ala = alanine, asn = asparagine, ca = caffeic acid, ci = citric acid, glc = glucose, sucr = sucrose, tr = trigonelline.

All these metabolites were also detected in the *O. diffusa* extract. Other signals only present in *O. variegata* were very likely attributable to caffeoylquinic acids. However, as these signals were not detected in the active extract, the identification of the metabolites generating them was beyond the scope of the present work.

The aromatic and olefinic region of *O. diffusa* ^{1}H-NMR spectrum showed several peaks that were not detected in *O. variegata* (Figure 3). Further differences were observed in the aliphatic region; protons at δ_H 1.23 and a triplet at δ_H 0.83 indicated the presence of alkyl chains. These signals are usually attributed to fatty acids. However, the solvent mixture herewith used would not explain the extraction of these compounds, which seem to be rather abundant in the extract.

In the attempt to assign signals to metabolites only present in *O. diffusa*, and therefore very likely responsible for the activity, an extensive 2D-NMR study of the extract was carried out. However, the main correlations detected for the signals in the aromatic region did not allow us to further identify the unknown metabolites. While the HSQC and COSY (Figures S1 and S2) experiments suggested the presence of further 1,2,4-trisubstituted ring systems, the long-range correlation experiments were not determining, since the main correlations detected were those of the already-known compounds (whose identity was therefore confirmed, Figure S3). Concerning the signals in the aliphatic region, the presence of an alkyl chain was undoubtful, based on the NMR data previously discussed. However, long-range correlations of protons resonating in this region with carbon signals in the range of 60–80 ppm suggested the possibility that the fatty acids present in the extracts could be oxygenated, as in the case of oxylipins.

2.3. Partial Purification of the Extract and Biological Activity

Since the 2D NMR of *O. diffusa* extract was dominated by the signals of the already-known compounds, a partial purification of a larger quantity of extract was carried out. The extract was first partitioned with water and ethyl acetate. The water fraction obtained was chromatographed on amberlite (XAD-4 and XAD-7). The columns were eluted with methanol first, and then with water. The methanolic fractions were joined together due to their very similar TLC profiles. The fraction (OdM) thereby obtained was analyzed by NMR and the ^{1}H-NMR spectrum, along with the HSCQ experiment, clearly showed that the aim was achieved (Figure 4). The OdM spectra did not show signals of primary metabolites, caffeic acid, or trigonelline (eluting in the water fraction).

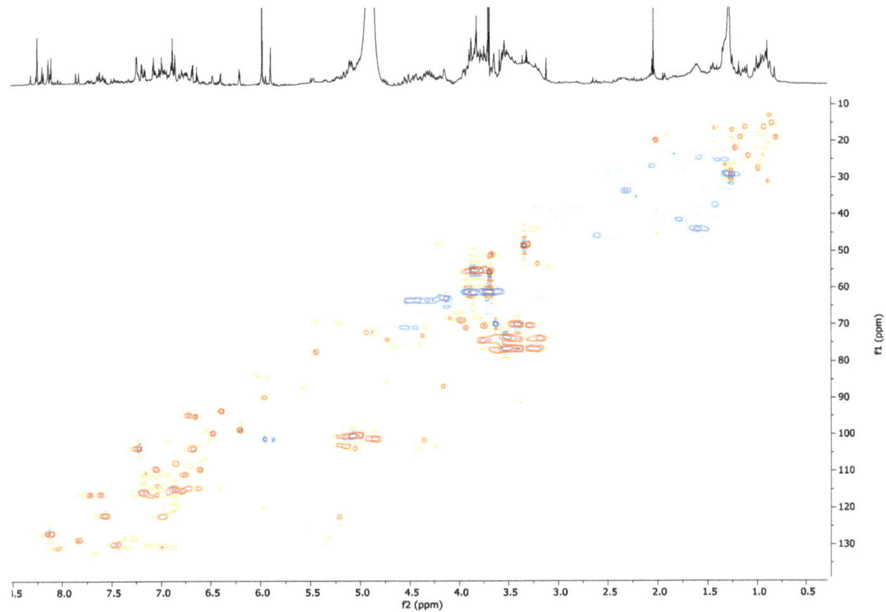

Figure 4. HSQC and ^{1}H-NMR spectra of the OdM fraction.

However, the spectrum was still rich in signals, especially in the aromatic region. The HSQC experiment (Figure 4) and the other 2D NMR experiments revealed several correlations. In particular, it was observed that the singlet proton at δ_H 8.14 was not bound to a carbon atom; this value was in good agreement with a hydrogen atom of an amidic functional group. This signal heterocorrelated, in the CIGAR-HMBC experiment (Figure S4), with a carbonyl carbon at δ_C 176.5. Furthermore, it also correlated, in the same experiment, with the quaternary carbon at δ_C 124.2, which was in turn correlated with two protons of a 1,2,4-tribubstituted aromatic system resonating at δ_H 6.90 (dd, J = 8.2; 2.0 Hz) and 6.97 (d, J = 2.0 Hz). The latter signal showed cross peaks with two oxygenated aromatic carbons at δ_C 147.8 and 146.1, both of which correlated with a third aromatic proton at δ_H 6.76 (d, J = 8.2 Hz) and with a singlet methylene at δ_H 5.87 (δ_C 101.2), indicating the presence of a 3,4-dioxymethylenephenyl group bound to the amide nitrogen. In addition, in the CIGAR experiment, the amidic proton correlated with a further aromatic carbon at δ_C 157.9, which in turn displayed cross-peaks with a proton at δ_H 8.01 (d, J = 9.0 Hz), as an indicator of a second 1,2,4-tribubstituted aromatic system. This proton correlated in the long-range experiment with the amide carbonyl and with a quaternary aromatic carbon at δ_C 118.8, which in turn correlated with other two aromatic protons resonating at δ_H 7.07 (d, J = 2.4 Hz) and 7.22 (dd, J = 9.0; 2.4 Hz), showing cross-peaks with an oxygenated aromatic carbon at δ_C 162.2. This carbon correlated with an anomeric proton at δ_H 4.98, therefore suggesting the presence of a sugar moiety, putatively identified, on the basis of an HSQC-TOCSY experiment, as glucose. These data were in agreement with the presence of a glucopyranosyl-2-hydroxy-*N*-(3,4-dioxymethylenephenyl)benzamide. However, the isolation and complete structural elucidation of this compound will be needed to confirm the hypothesized structure.

More signals were detected in the aromatic region. Signals belonging to the A-ring of flavonoids were observed as two COSY-correlating doublets (J = 2.0 Hz) at δ_H 6.20 and 6.39. These protons were correlated in the HSQC experiment to the carbons at δ_C 93.1 and 98.4, respectively. Furthermore, the former proton also showed long-range correlations with the two carbons resonating at δ_C 161.2 and 165.0. This second carbon was also correlated to the δ_H 6.39, which showed further long-range correlations with the carbons at δ_C 156.9 and 104.0. It was not possible to unambiguously identify these metabolites, due to the lack of further correlations, but the NMR data here described prompted us to hypothesize that the flavonoids were all characterized by a hydroxy function bound to the C-3 [21].

Finally, the signals in the range of 6.8–7.4 ppm were correlated, in the HSQC experiment, with the carbons in the range of 125–145 ppm, suggesting the presence of olefinic protons. These protons also showed COSY correlations with protons resonating in the range of 5.9–6.1 ppm bound to carbons in the range of 120–125 ppm (Figure S5). The proton signals at 5.9–6.1 ppm showed further COSY correlations with signals in the aliphatic region (2.4–2.8 ppm), in turn correlating with protons geminal to oxygen. Since the mixture was complex and these compounds were minor components, it was not possible to definitely identify the compounds. However, it is plausible that these signals belonged to oxylipins.

The OdM fraction was tested on HT-29 cells (Figure 5), but no activity was observed. Two possible explanations were then taken into consideration: (i) the occurrence of synergistic effects in the crude extract or (ii) the fact that the OdM fraction obtained from the amberlite chromatography, although enriched in several specialized metabolites, did not contain a significant amount of the active compounds. These active metabolites could be then found either in the water fraction obtained from the same chromatography, or in the ethyl acetate fraction (OdE) obtained in the previous step. By comparing the NMR profiles of the OdM and the crude extract (Figures 3 and 4), it was clear that, besides the signals belonging to metabolites also detected in the crude extract of the inactive species *O. variegata*, there was also a significant decrease in the intensity of signals so far putatively attributed to oxylipins. These oxylipins could have been in the OdE fraction, which was therefore assayed. The ethyl acetate fraction showed activity on HT-29 cells (Figure 5).

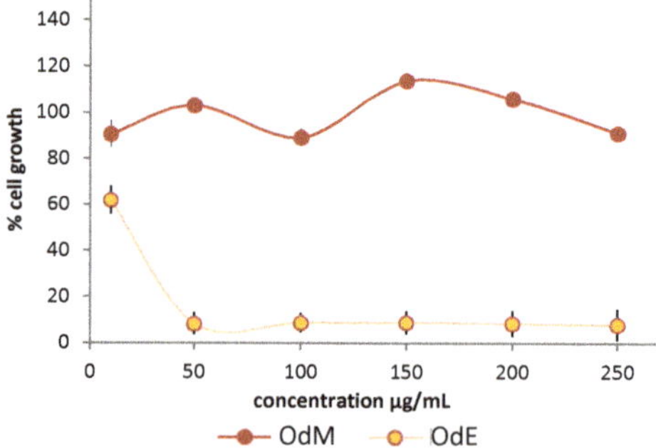

Figure 5. Antiproliferative activity evaluation of the *O. diffusa* partially purified fractions on HT-29 cell lines. Cell growth is expressed as percentage from control, and it is plotted on the vertical axis, while doses of plant extracts are reported on the horizontal axis.

The inhibition was of 92% at a concentration of 50 µg/mL, and around 40% at the concentration of 10 µg/mL. The ^{1}H-NMR of the ethyl acetate extract confirmed that this fraction was enriched in signals that were only detected in traces in the methanol fraction, and that could be attributed to oxylipins (Figure 6).

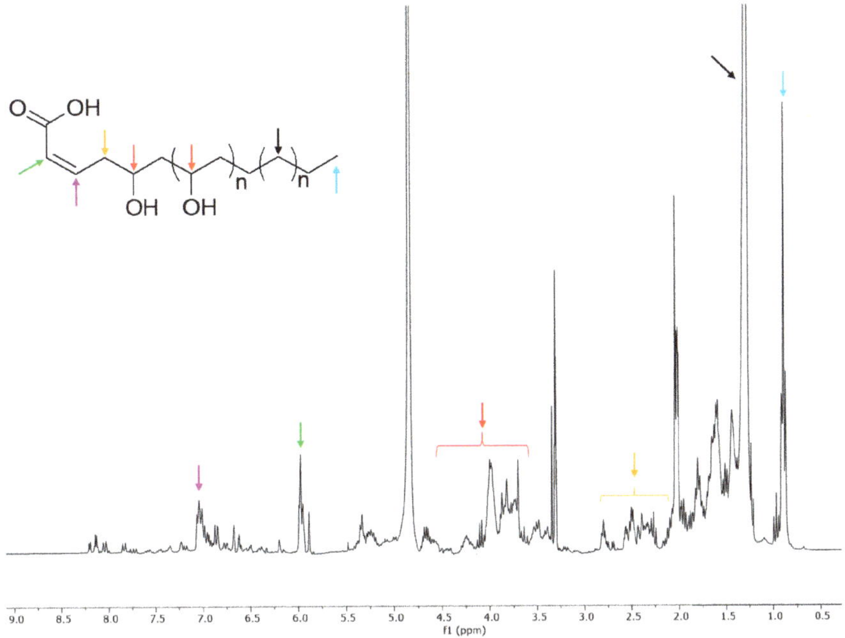

Figure 6. ^{1}H-NMR spectrum of the OdE fraction. A tentative general structure for the oxylipins is proposed, and diagnostic signals are indicated by colored arrows.

Besides the triplet at δ_H 0.83 and the methylene signals at δ_H 1.23, signals belonging to two olefinic protons conjugated to a carboxylic function were detected at δ_H 5.90 and δ_H 7.00. Allylic protons were instead resonating in the range 2–3 ppm, while signals in the region between 3 and 5 ppm were attributable to protons geminal to hydroxyl groups. The hypothesized structure is reported in Figure 6. This hypothesis was drawn based on the analysis of the 2D-NMR spectra of the extract previously discussed.

A complete structural elucidation of these metabolites was not possible in mixture, because it is very likely to have been made up of compounds with variable chain lengths. However, it is also clear that compounds with different substitution patterns were present in the mixture, including compounds that might bear acetyl groups (as shown by singlets at around 2 ppm).

3. Discussion

The present study demonstrates that *O. diffusa* is a source of potential anticancer compounds acting on drug-resistant colon cancer cell lines. The aim of the present work was not only to identify cytotoxic extracts, but particularly sources of compounds able to overcome anti-EGFR therapy resistance in mCRC.

EGFR is a very important target in cancer therapy [11], being a central regulator of tumor progression in a variety of human cancers, including mCRC [22], one of the leading causes of cancer-related death worldwide. One way to inhibit the activation of EGFR is with monoclonal antibodies like cetuximab [23]. Unfortunately, primary resistance (due to specific mutations) to cetuximab in mCRC patients, who therefore do not respond to this treatment, has been reported [24]. Furthermore, it has been shown that one fourth of cetuximab-sensitive mCRC patients develop secondary resistance [11,12]. The emergence of secondary resistance might be due to the selection of drug-insensitive subclones imposed by the continuous EGFR blockade [24]. Finding drugs acting with alternative modes of action, able to overcome or bypass these innate and acquired resistances is therefore crucial. In this study, we used Caco-2, GEO, and SW48 cells as cetuximab-sensitive models, HT-29 and HCT-116 as intrinsically cetuximab-resistant models and GEO-CR and SW48-CR as cell models with acquired resistance to cetuximab. Although the data here reported are preliminary and extensive tests on pure compounds are needed to assess their activity, toxicity, and modes of action, the extract herewith analyzed showed a strong inhibition of cell growth on all the cell lines. Of note, the activity on cetuximab-resistant human cancer cell lines was remarkable. In particular, the extract strongly inhibited the growth of HT-29, a cell line harboring a BRAF mutation, which is a strong negative prognostic biomarker for patients suffering from mCRC [12].

Based on the partial purification and on the fraction testing and profiling, we could identify the class of compounds potentially responsible for the activity exerted by the extracts. It was therefore suggested that the oxylipin components of the extract could be the compounds responsible for the biological activity. The crude extract demonstrated strong cytotoxicity (Figures 1 and 2). Based on a comparison to the inactive extract of the related plant *O. variegata* (Figure 3), which was also tested and analyzed, it was possible to exclude some of the metabolites from the list of candidate bioactive compounds. These compounds (caffeic acid, caffeoyl derivatives, trigonelline, and several primary metabolites) were, indeed, the main components of the *O. variegata* extract (Figure 3), which however showed no activity even at the highest tested concentration. A second level of selection was obtained after the partial purification of the crude extract: OdE and OdM fractions were obtained when the extract was partitioned with ethyl acetate/water and then the aqueous fraction therefrom was further purified on amberlite. Although the OdM fraction was particularly enriched in phenolic compounds, it did not show activity against HT-29 cells (Figures 4 and 5). On the other hand, the OdE fraction was very active and even more strongly inhibiting of HT-29 cell growth than the crude extract. Based on the NMR analysis, it was possible to tentatively identify the main compounds in this fraction as oxylipins (Figure 6 and Figures S1–S3).

Oxylipins are oxidized fatty acids, used by plants mostly as signaling molecules [25,26], to the best of our knowledge. Jasmonic acid is by far the most studied oxylipin, due to its central role as a plant hormone involved in the regulation of developmental and defense-related processes [27]. Given these roles in signaling and as plant hormones, oxylipins are usually present at low concentrations [25] and therefore have also been seldom studied for further biological activities. It has been suggested that these molecules promote apoptosis in animal cells by altering the intracellular calcium signaling and inducing cytoskeletal instability [28], although the molecular mechanism is not yet known. Besides jasmonates, different substitute oxylipins have been reported, including oxylipins with cytotoxic, anti-inflammatory, and potential anticancer properties [29,30]. The oxylipins in *O. diffusa* extract seemed to be polyoxygenated compounds. This hypothesis was supported by their polarity and chromatographic behavior.

Although the oxylipins were the most abundant compounds in the active fraction, we cannot completely exclude the possibility that less-abundant compounds could have been responsible for the observed activity. However, a further effort to perform a complete isolation and structural elucidation of the pure compounds will be carried out as the object of future studies. The observed biological activity makes the OdE fraction a potential source of compounds that could be further explored with the aim of finding drug candidates able to overcome either intrinsic or acquired drug resistance in colorectal cancer cells.

4. Materials and Methods

4.1. Plant Sampling

Plant samples belonging to the two species, *O. diffusa* and *O. variegata* were harvested in the spring at the "Castel Volturno" Nature Reserve (40°57.587′ N, 14°00.105′ E; southern Italy) and identified by Prof. Assunta Esposito. Voucher specimens were deposited at the Herbarium of DiSTABiF of Università degli Studi della Campania "Luigi Vanvitelli." Leaf samples were collected and immediately frozen in liquid N_2 to avoid unwanted enzymatic reactions, and stored at −80 °C before the freeze-drying process. Lyophilized samples were powdered in liquid nitrogen and stored at −20 °C.

4.2. Plant Chemical Profile

4.2.1. Extraction

An aliquot (50 mg) of freeze-dried and powdered plant material extracted with 1.5 mL of phosphate buffer (Fluka Chemika, Buchs, Switzerland; 90 mM; pH 6.0) in D_2O (Cambridge Isotope Laboratories, Andover, MA, USA) containing 0.1% *w/w* trimethylsilylpropionic-2,2,3,3-d_4 acid sodium salt (TMSP, Sigma-Aldrich, St. Louis, MO, USA) and CD_3OD (Sigma-Aldrich, St. Louis, MO, USA) (1:1). The mixture was vortexed at room temperature for 1 min, ultrasonicated (Elma Transsonic Digital, Hohentwiel, Germany) for 40 min, and centrifuged (Beckman Allegra™ 64R, F2402H rotor; Beckman Coulter, Fullerton, CA, USA) at 13,000 rpm for 10 min. A volume of 0.65 mL was transferred to a 5 mm NMR tube and analyzed by NMR [31].

4.2.2. NMR Analysis

NMR spectra were recorded at 25 °C on a Varian Mercury Plus 300 Fourier transform NMR operating at 300.03 MHz for ^{1}H and 75.45 MHz for ^{13}C. CD_3OD was used as the internal lock. 1D and 2D NMR spectra were acquired using Varian standard pulse sequences and as previously described [13].

4.3. Partial Purification of O. diffusa Extract

Dried leaves (60 g) of *O. diffusa* were powdered and underwent three cycles of an ultrasound-assisted extraction with a $MeOH:H_2O$ (1:1) solution (1.8 L), obtaining a crude extract (16 g), which was dissolved in H_2O and separated by liquid–liquid extraction using EtOAc as an extracting solvent, obtaining an ethyl acetate fraction (OdE; 3 g) and a water fraction. The aqueous fraction was chromatographed on Amberlite XAD-4 and XAD-7 with water and then with methanol. The two alcoholic eluates were joined together to give the OdM fraction (1.7 g). Aliquots were used for bioassays and NMR analyses.

4.4. Cell Lines

4.4.1. Cell Cultures

Human HCT-116, HT-29, Caco-2, and SW48 colorectal cancer cell lines were obtained from the American Type Culture Collection (ATCC) (Manassas, VA, USA). The human GEO colon cancer cell was a gift of Dr. N. Normanno (National Cancer Institute, Naples, Italy). The HCT-116 and HT-29 cancer cells were cultured in RPMI 1640 medium (Lonza, Cologne, Germany), and supplemented with 10% FBS, 2 mM L-glutamine, 50 U/mL penicillin, and 100 µg/mL streptomycin (Lonza, Cologne, Germany). Caco-2 cell line was cultured in DMEM medium (Lonza, Cologne, Germany), and supplemented with 10% FBS, 2 mM L-glutamine, 1% non-essential amino acid, 50 U/mL penicillin, and 100 µg/mL streptomycin (Lonza, Cologne, Germany). The GEO and GEO-CR clones were grown on DMEM medium supplemented with 20% FBS, 1% penicillin/streptomycin (Lonza, Cologne, Germany). The SW48 and SW48-CR cells were cultured in RPMI 1640 medium supplemented with 10% FBS and 1% penicillin/streptomycin. All cell lines were maintained in a humified atmosphere of 95% air and 5% CO_2 at 37 °C, and routinely screened for the presence of mycoplasma (Mycoplasma Detection Kit, Roche Diagnostics, Basel, Switzerland).

4.4.2. Proliferation Assay

Cell proliferation was measured with 3-(4,5-dimethylthiazol-2-yl)-2,5-diphenyltetrazolium bromide (MTT) assay [32]. Briefly, cells in the logarithmic growth phase were plated in 96-well plates and incubated for 24 h before exposure to increasing doses of plant extracts (10, 50, 100, 150, 200 and 250 µg/mL). At 48 h after treatment, 50 µL of 1 mg/mL (MTT) were mixed with 200 µL of medium and added to the well. At 1 h after incubation at 37 °C, the medium was removed, and the purple formazan crystals produced in the viable cells were solubilized in 100 µL of dimethyl sulfoxide and quantitated by measurement of absorbance at 570 nm with a plate reader. Results were reported as mean +SD of % of cell growth respect to the control, from six replicates. The control was represented by 0.25% DMSO treatment, corresponding to the higher amount of DMSO used for the tests.

4.4.3. Statistical Analyses

Bioassays were carried out in six replicates. Statistical analyses were performed using Excel 2010 (Microsoft Corporation; Redmond, WA, USA). A Student's *t* test ($p < 0.001$) was used to determine the statistical significance of the experimental results.

Supplementary Materials: The following are available online. Figure S1: HSQC of *O. diffusa* extract. Figure S2: COSY of *O. diffusa* extract. Figure S3: HMBC of *O. diffusa* extract. Figure S4: CIGAR-HMBC of OdM fraction. Figure S5: COSY of OdM fraction.

Author Contributions: Conceptualization, V.G., A.F. and M.S.; methodology, V.G.; investigation, V.G., N.P., T.T. and S.N.; resources, A.F.; data curation, M.S.; writing—original draft preparation, V.G. and M.S.; writing—review and editing, N.P., B.D., T.T. and S.N.; supervision, A.F. and M.S.; funding acquisition, A.F. and M.S. All authors have read and agreed to the published version of the manuscript.

Funding: This work was funded by Regione Campania, "Technology Platform per la Lotta alle Patologie Oncologiche" Grant: iCURE (CUP B21C17000030007–SURF17061BP00000000. MS acknowledges V:ALERE 2020 Program of University of Campania "Luigi Vanvitelli" for funding.

Data Availability Statement: The raw data supporting the conclusions of this article will be made available by the authors upon request, without undue reservation.

Acknowledgments: This work is dedicated to Robert Verpoorte in occasion of his 75th Birthday.

Conflicts of Interest: The authors declare no conflict of interest. The funders had no role in the design of the study; in the collection, analyses, or interpretation of data; in the writing of the manuscript, or in the decision to publish the results.

Sample Availability: Samples are available from the authors.

References

1. Yuliana, N.D.; Jahangir, M.; Verpoorte, R.; Choi, Y.H. Metabolomics for the rapid dereplication of bioactive compounds from natural sources. *Phytochem. Rev.* **2013**, *12*, 293–304. [CrossRef]
2. Verpoorte, R. Exploration of nature's chemodiversity: The role of secondary metabolites as leads in drug development. *Drug Discov. Today* **1998**, *3*, 232–238. [CrossRef]
3. Paterson, I.; Anderson, E.A. Chemistry. The renaissance of natural products as drug candidates. *Science* **2005**, *310*, 451–453. [CrossRef] [PubMed]
4. Pichersky, E.; Lewinsohn, E. Convergent evolution in plant specialized metabolism. *Annu. Rev. Plant Biol.* **2011**, *62*, 549–566. [CrossRef]
5. Nakabayashi, R.; Saito, K. Integrated metabolomics for abiotic stress responses in plants. *Curr. Opin. Plant Biol.* **2015**, *24*, 10–16. [CrossRef]
6. Verpoorte, R.; Choi, Y.H.; Kim, H.K. Ethnopharmacology and systems biology: A perfect holistic match. *J. Ethnopharmacol.* **2005**, *100*, 53–56. [CrossRef] [PubMed]
7. Malvezzi, M.; Carioli, G.; Bertuccio, P.; Rosso, T.; Boffetta, P.; Levi, F.; La Vecchia, C.; Negri, E. European cancer mortality predictions for the year 2016 with focus on leukaemias. *Ann. Oncol.* **2016**, *27*, 725–731. [CrossRef]
8. Hashim, D.; Boffetta, P.; La Vecchia, C.; Rota, M.; Bertuccio, P.; Malvezzi, M.; Negri, E. The global decrease in cancer mortality: Trends and disparities. *Ann. Oncol.* **2016**, *27*, 926–933. [CrossRef]
9. Giordano, G.; Remo, A.; Porras, A.; Pancione, M. Immune Resistance and EGFR Antagonists in Colorectal Cancer. *Cancers* **2019**, *11*, 1089. [CrossRef] [PubMed]
10. Ciardiello, F.; Tortora, G. EGFR antagonists in cancer treatment. *N. Engl. J. Med.* **2008**, *358*, 1160–1174. [CrossRef]
11. Napolitano, S.; Martini, G.; Rinaldi, B.; Martinelli, E.; Donniacuo, M.; Berrino, L.; Vitagliano, D.; Morgillo, F.; Barra, G.; De Palma, R.; et al. Primary and Acquired Resistance of Colorectal Cancer to Anti-EGFR Monoclonal Antibody Can Be Overcome by Combined Treatment of Regorafenib with Cetuximab. *Clin. Cancer Res.* **2015**, *21*, 2975–2983. [CrossRef] [PubMed]
12. Van Cutsem, E.; Kohne, C.H.; Lang, I.; Folprecht, G.; Nowacki, M.P.; Cascinu, S.; Shchepotin, I.; Maurel, J.; Cunningham, D.; Tejpar, S.; et al. Cetuximab plus irinotecan, fluorouracil, and leucovorin as first-line treatment for metastatic colorectal cancer: Updated analysis of overall survival according to tumor KRAS and BRAF mutation status. *J. Clin. Oncol.* **2011**, *29*, 2011–2019. [CrossRef] [PubMed]
13. Graziani, V.; Scognamiglio, M.; Belli, V.; Esposito, A.; D'Abrosca, B.; Chambery, A.; Russo, R.; Panella, M.; Russo, A.; Ciardiello, F.; et al. Metabolomic approach for a rapid identification of natural products with cytotoxic activity against human colorectal cancer cells. *Sci. Rep.* **2018**, *8*, 5309. [CrossRef] [PubMed]

14. Mezrag, A.; Malafronte, N.; Bouheroum, M.; Travaglino, C.; Russo, D.; Milella, L.; Severino, L.; De Tommasi, N.; Braca, A.; Dal Piaz, F. Phytochemical and antioxidant activity studies on Ononis angustissima L. aerial parts: Isolation of two new flavonoids. *Nat. Prod. Res.* **2017**, *31*, 507–514. [CrossRef]
15. Gampe, N.; Darcsi, A.; Nagyne Nedves, A.; Boldizsar, I.; Kursinszki, L.; Beni, S. Phytochemical analysis of Ononis arvensis L. by liquid chromatography coupled with mass spectrometry. *J. Mass Spectrom.* **2019**, *54*, 121–133. [CrossRef] [PubMed]
16. Stojkovic, D.; Dias, M.I.; Drakulic, D.; Barros, L.; Stevanovic, M.; Ferreira, I.C.F.R.; Sokovic, M.D. Methanolic Extract of the Herb Ononis spinosa L. Is an Antifungal Agent with no Cytotoxicity to Primary Human Cells. *Pharmaceuticals* **2020**, *13*, 78. [CrossRef]
17. Ghribi, L.; Waffo-Teguo, P.; Cluzet, S.; Marchal, A.; Marques, J.; Merillon, J.M.; Ben Jannet, H. Isolation and structure elucidation of bioactive compounds from the roots of the Tunisian Ononis angustissima L. *Bioorg. Med. Chem. Lett.* **2015**, *25*, 3825–3830. [CrossRef]
18. Wang, G.; Huang, Y.; Wu, Z.; Zhao, C.; Cong, H.; Ju, S.; Wang, X. KRAS-mutant colon cancer cells respond to combined treatment of ABT263 and axitinib. *Biosci. Rep.* **2019**, *39*, 3. [CrossRef]
19. Scognamiglio, M.; Fiumano, V.; D'Abrosca, B.; Esposito, A.; Choi, Y.H.; Verpoorte, R.; Fiorentino, A. Chemical interactions between plants in Mediterranean vegetation: The influence of selected plant extracts on Aegilops geniculata metabolome. *Phytochemistry* **2014**, *106*, 69–85. [CrossRef]
20. Scognamiglio, M.; D'Abrosca, B.; Esposito, A.; Fiorentino, A. Chemical Composition and Seasonality of Aromatic Mediterranean Plant Species by NMR-Based Metabolomics. *J. Anal. Methods Chem.* **2015**, *2015*, 258570. [CrossRef]
21. Scognamiglio, M.; Schneider, B. Identification of Potential Allelochemicals From Donor Plants and Their Synergistic Effects on the Metabolome of Aegilops geniculata. *Front. Plant Sci.* **2020**, *11*, 1046. [CrossRef]
22. Sigismund, S.; Avanzato, D.; Lanzetti, L. Emerging functions of the EGFR in cancer. *Mol. Oncol.* **2018**, *12*, 3–20. [CrossRef]
23. Blick, S.K.; Scott, L.J. Cetuximab: A review of its use in squamous cell carcinoma of the head and neck and metastatic colorectal cancer. *Drugs* **2007**, *67*, 2585–2607. [CrossRef] [PubMed]
24. Leto, S.M.; Trusolino, L. Primary and acquired resistance to EGFR-targeted therapies in colorectal cancer: Impact on future treatment strategies. *J. Mol. Med.* **2014**, *92*, 709–722. [CrossRef] [PubMed]
25. Blee, E. Impact of phyto-oxylipins in plant defense. *Trends Plant Sci.* **2002**, *7*, 315–322. [CrossRef]
26. Howe, G.A.; Schilmiller, A.L. Oxylipin metabolism in response to stress. *Curr. Opin. Plant Biol.* **2002**, *5*, 230–236. [CrossRef]
27. Farmer, E.E.; Almeras, E.; Krishnamurthy, V. Jasmonates and related oxylipins in plant responses to pathogenesis and herbivory. *Curr. Opin. Plant Biol.* **2003**, *6*, 372–378. [CrossRef]
28. Caldwell, G.S. The influence of bioactive oxylipins from marine diatoms on invertebrate reproduction and development. *Mar. Drugs* **2009**, *7*, 367–400. [CrossRef]
29. Christensen, L.P. Bioactive C17 and C18 Acetylenic Oxylipins from Terrestrial Plants as Potential Lead Compounds for Anticancer Drug Development. *Molecules* **2020**, *25*, 2568. [CrossRef]
30. D'Abrosca, B.; Ciaramella, V.; Graziani, V.; Papaccio, F.; Della Corte, C.M.; Potenza, N.; Fiorentino, A.; Ciardiello, F.; Morgillo, F. Urtica dioica L. inhibits proliferation and enhances cisplatin cytotoxicity in NSCLC cells via Endoplasmic Reticulum-stress mediated apoptosis. *Sci. Rep.* **2019**, *9*, 4986. [CrossRef]
31. Kim, H.K.; Verpoorte, R. Sample preparation for plant metabolomics. *Phytochem. Anal.* **2010**, *21*, 4–13. [CrossRef] [PubMed]
32. Tun, J.O.; Salvador-Reyes, L.A.; Velarde, M.C.; Saito, N.; Suwanborirux, K.; Concepcion, G.P. Synergistic cytotoxicity of renieramycin M and doxorubicin in MCF-7 breast cancer cells. *Mar. Drugs* **2019**, *17*, 536. [CrossRef] [PubMed]

Article

Characterization of *α*-Glucosidase Inhibitors from *Psychotria malayana* Jack Leaves Extract Using LC-MS-Based Multivariate Data Analysis and In-Silico Molecular Docking

Tanzina Sharmin Nipun [1,2], Alfi Khatib [1,3,*], Zalikha Ibrahim [1], Qamar Uddin Ahmed [1], Irna Elina Redzwan [1], Mohd Zuwairi Saiman [4,5,*], Farahaniza Supandi [4], Riesta Primaharinastiti [3] and Hesham R. El-Seedi [6,7]

[1] Pharmacognosy Research Group, Department of Pharmaceutical Chemistry, Kulliyyah of Pharmacy, International Islamic University Malaysia, Kuantan 25200, Malaysia; tsn.np99@gmail.com (T.S.N.); zalikha@iium.edu.my (Z.I.); quahmed@iium.edu.my (Q.U.A.); elina@iium.edu.my (I.E.R.)

[2] Department of Pharmacy, Faculty of Biological Sciences, University of Chittagong, Chittagong 4331, Bangladesh

[3] Faculty of Pharmacy, Airlangga University, Surabaya 60155, Indonesia; r_nastiti@gmail.com

[4] Institute of Biological Sciences, Faculty of Science, University of Malaya, Kuala Lumpur 50603, Malaysia; farahaniza@um.edu.my

[5] Center for Research in Biotechnology for Agriculture (CEBAR), Faculty of Science, University of Malaya, Kuala Lumpur 50603, Malaysia

[6] Pharmacognosy Group, Department of Pharmaceutical Biosciences, BMC, Uppsala University, SE-751 23 Uppsala, Sweden; hesham.el-seedi@farmbio.uu.se

[7] International Research Center for Food Nutrition and Safety, Jiangsu University, Zhenjiang 212013, China

[*] Correspondence: alfikhatib@iium.edu.my (A.K.); zuwairi@um.edu.my (M.Z.S.)

Academic Editors: Young Hae Choi, Young Pyo Jang, Yuntao Dai and Luis Francisco Salomé-Abarca
Received: 6 November 2020; Accepted: 10 December 2020; Published: 12 December 2020

Abstract: *Psychotria malayana* Jack has traditionally been used to treat diabetes. Despite its potential, the scientific proof in relation to this plant is still lacking. Thus, the present study aimed to investigate the *α*-glucosidase inhibitors in *P. malayana* leaf extracts using a metabolomics approach and to elucidate the ligand–protein interactions through in silico techniques. The plant leaves were extracted with methanol and water at five various ratios (100, 75, 50, 25 and 0% *v/v*; water–methanol). Each extract was tested for *α*-glucosidase inhibition, followed by analysis using liquid chromatography tandem to mass spectrometry. The data were further subjected to multivariate data analysis by means of an orthogonal partial least square in order to correlate the chemical profile and the bioactivity. The loading plots revealed that the *m/z* signals correspond to the activity of *α*-glucosidase inhibitors, which led to the identification of three putative bioactive compounds, namely 5′-hydroxymethyl-1′-(1, 2, 3, 9-tetrahydro-pyrrolo (2, 1-*b*) quinazolin-1-yl)-heptan-1′-one (**1**), *α*-terpinyl-*β*-glucoside (**2**), and machaeridiol-A (**3**). Molecular docking of the identified inhibitors was performed using Auto Dock Vina software against the crystal structure of *Saccharomyces cerevisiae* isomaltase (Protein Data Bank code: 3A4A). Four hydrogen bonds were detected in the docked complex, involving several residues, namely ASP352, ARG213, ARG442, GLU277, GLN279, HIE280, and GLU411. Compound **1**, **2**, and **3** showed binding affinity values of −8.3, −7.6, and −10.0 kcal/mol, respectively, which indicate the good binding ability of the compounds towards the enzyme when compared to that of quercetin, a known *α*-glucosidase inhibitor. The three identified compounds that showed potential binding affinity towards the enzymatic protein in molecular docking interactions could be the bioactive compounds associated with the traditional use of this plant.

Keywords: *Psychotria malayana* Jack; type 2 diabetes; *α*-glucosidase inhibitors; LC-MS; metabolomics; molecular docking

1. Introduction

Diabetes mellitus (DM) is the most common chronic and metabolic disease, which affects a large number of populations in the world. Currently, more than 150 million people are suffering from diabetes, and this number is expected to reach 300 million by 2025 [1]. It is characterized by an elevated blood glucose level due to defects in insulin secretion and insulin resistance, or both. DM is associated with eye, renal, cardiovascular, and neurological complications in the long term and is also associated with symptoms such as polyuria, fatigue, weight loss, delayed wound healing, blurred vision, increases in plasma and urine glucose levels [2–4]. Generally, DM can be categorized into two types; type 1 and type 2. Many synthetic medicines like insulin and oral hypoglycemic drugs available on the market are used to treat type 2 diabetes, but the long term use of insulin therapy and oral synthetic medicines may cause severe side effects [5]. Furthermore, oral anti-diabetic drugs are costly; hence, it is very difficult for low and middle-income people to continually use these medications [6]. Taking into account the side effects of synthetic anti-diabetic drugs, the interest in herbal medicinal products for DM treatment is growing [7]. Several phytochemical constituents, including flavonoids, phenolic, coumarins, and other compounds that are plentiful in medicinal herbs have been evaluated as being capable of lowering blood glucose levels [8]. The capability of medicinal plants in treating DM relates to their ability to improve the action of pancreatic tissue by increasing insulin secretion or reducing intestinal glucose absorption [9]. Decreasing postprandial hyperglycemia is a therapeutic approach used for the treatment of DM. This goal can be achieved by inhibiting the carbohydrate hydrolyzing enzyme, α-glucosidase (AG). The AG enzyme plays a vital role in carbohydrate metabolism by breaking starch and disaccharides into glucose [10]. Therefore, inhibition of AG is an effective therapeutic approach for the management of DM.

Psychotria malayana Jack plant, which belongs to the Rubiaceae family, is widespread throughout tropical and subtropical countries. This plant is largely distributed in the west Indonesian archipelago and known to Lombok people as "lolon jarum" [11]. Lombok people traditionally used this plant for the treatment of wounds, skin infections and other skin diseases [11]. People in India, Indonesia and Brazil, have conventionally used this plant to treat digestive problems, stomach pain and infections of the female reproductive system [12]. Karo people in North Sumatra used the species of *Psychotria*, which is locally known as "loning" for the treatment of diabetes [13]. Khaled et al. [12] showed the anti-diabetic property of this plant on diabetic-induced zebrafish. Although several alkaloids in this plant have been reported, none of these alkaloids have been found to exert anti-diabetic activity. More research should be performed to identify bioactive compounds related to the anti-diabetic properties of this plant.

LC-MS-based metabolomics has been used to identify bioactive compounds in medicinal herbs [14]. Condensed tannins and flavonoid derivatives have been identified as antioxidant metabolites from the leaves of *Fragaria vesca* by using LC-MS based metabolomics [15]. Furthermore, Kamalrul et al. [16] identified eight compounds from *Wedelia trilobata* leaves that exhibited potential allelopathic effects. Liquid chromatography (LC) combined with quadrupole time-of-flight (Q-TOF) mass spectrometry (MS) was used to annotate metabolites that could act as α-glucosidase (AG) inhibitors [17]. Suganya et al. [17] reported AG inhibitors from *Clinacanthus nutans* leaves using a similar technology. Additionally, Gilda et al. [18] identified 35 metabolites, along with antioxidant and AG inhibitory effects of the fruits of both *Morus alba* and *Morus nigra*. Similar technology has been applied to evaluate the activity of *Clinacanthus nutans* leaves against cisplatin toxicity [19]. In addition, potential thrombin/factor Xa inhibitors were reported from *Salvia miltiorrhiza* Bunge and *Ligusticum chuanxiong* Hort using the LC-MS-based multivariate data analysis technique [20,21]. Chromatographic technique-based metabolomics was also applied to compare and reveal the chemical profiles of raw and fermented Chinese Ge-Gen-Qin-Lian decoction (a traditional chinese medicine formula) during investigation of its anti-diabetic effects on high-fat diet and streptozotocin-induced rats [22]. The choice of using

LC-MS is due to the technique having a high mass resolution and detection sensitivity in regard to mass detection, thus, it can precisely measure the mass [23,24]. LC-Q-TOF-MS provides improved detection for fragmented ions with a high resolution and mass precision in comparison to many other tandem mass spectrometries [23].

Molecular docking plays an important role in predicting the binding modes and binding abilities of small molecules towards the targeted proteins, which is crucial in designing potential drugs [25]. In the present study, the activity of *P. malayana* Jack leaf extracts were investigated through an AG inhibition assay. Subsequently, the AG inhibitors in this herb were identified using LC-MS-based metabolomics. An in silico study was carried out to predict the molecular interaction between the AG enzyme (protein) and the tentative AG inhibitors (ligands) that were identified in *P. malayana* leaf extracts.

2. Results

2.1. α-Glucosidase Inhibition (AGI) of the Plant Extract

Figure 1 shows the α-glucosidase inhibition (AGI) activity of five extracts of *P. malayana* leaves at the assay concentration of 10 µg/mL. The highest AGI activity (93.1% inhibition) was shown by 100% methanol, followed by the 25% methanol extract with an AGI activity of 73.5% inhibition. However, the AGI activity of the 25% methanol was not statistically significant ($p < 0.05$) from 50% methanol (61.6% inhibition). The AGI activity of 50% methanol and water extract (61.2% inhibition) was not significantly different ($p < 0.05$), but was lower than that of the 25% methanol extract. The lowest AGI activity was shown by the 75% methanol extract (48.5% inhibition). The AGI activity of the different *P. malayana* leaf extracts exhibited the following trend: 100 > 25 > 50 > 0 > 75% *v/v* methanol extract.

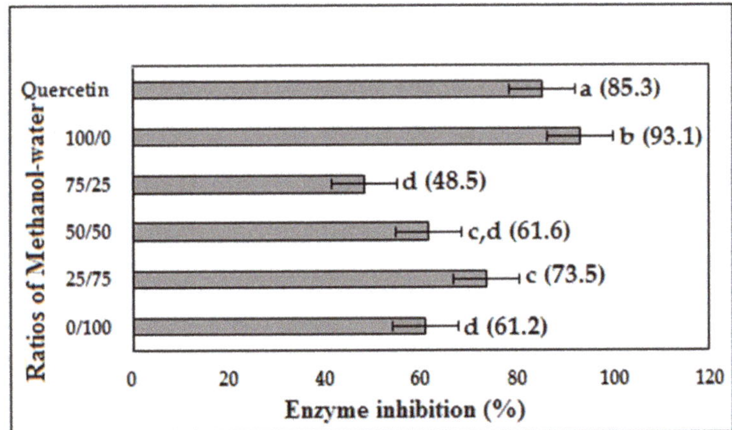

Figure 1. Percentage of α-glucosidase inhibition at the concentration of 10 µg/mL of *P. malayana* leaves extracts. The values that do not share the same letter are significantly different ($p < 0.05$), as measured by Tukey's comparison test.

2.2. Multivariate Data Analysis

Supervised multivariate data analysis, by means of orthogonal partial least square (OPLS), was used to correlate the x variables (LC-MS signals) and the y variable (AGI activity) of each sample. The mass to charge ratio (*m/z*) and retention time were considered as the x variables and resulted from the pre-processing of LCMS data using an online Mzmine platform. Pareto scaling was used in order to suppress the effect of *m/z* noise in the statistical calculation, thus avoiding a biased result.

The Y-axis in Figure 2A represents the variation explained by the calibration model, while the R^2(cum)progression is the degree of variation that can be explained by each of the components. The variation was calculated as a cumulative score and thus, described in a progression manner. The OPLS model produced four principal components (PC), explaining 91% of the total variation (see Figure 2A). PC1 explained the largest variation of the samples (58%), while PC2, PC3, and PC4 represented 16%, 29%, and 33% of the sample variation, respectively. The difference between the R^2Y and Q^2Y of all the PCs in this study was not larger than 0.3, confirming the validation of this multivariable calibration model, according to Eriksson et al. [26]. Another validity evaluation of this model was performed through the analysis of the regression coefficient (R^2) between the predicted and actual AGI activities of the samples (Figure 2B). The R^2 value obtained was 0.91, indicating the validity of this model since it was higher than 0.90 [26]. The permutation test showed that the R^2Y and Q^2Y intercept values of this model were lower than 0.4 and 0.05, respectively, confirming that this calibration model was valid [26].

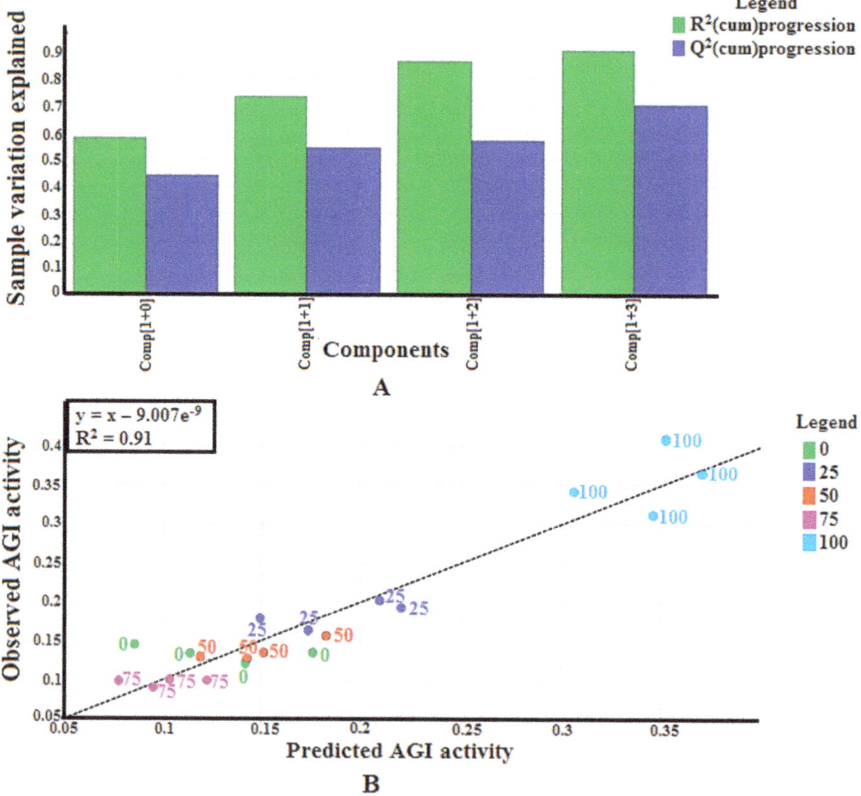

Figure 2. (**A**) Summary of fit of established orthogonal partial least square (OPLS) model for *P. malayana* leaf extracts; (**B**) Observed vs. predicted *α*-glucosidase inhibition (AGI) activity with R^2 value of 0.91 from 20 extracts of *P. malayana* leaves.

Figure 3A shows a score scatter plot that displays the separation among the samples. The sample separation can be viewed well through a combination of OPLS component one and two. The most active extract (100% methanol) was separated in the positive side, while the others—except part of the 25% methanol extract—were separated in the negative side of OPLS component one. While OPLS component two could not separate the most active extract from the less active extracts.

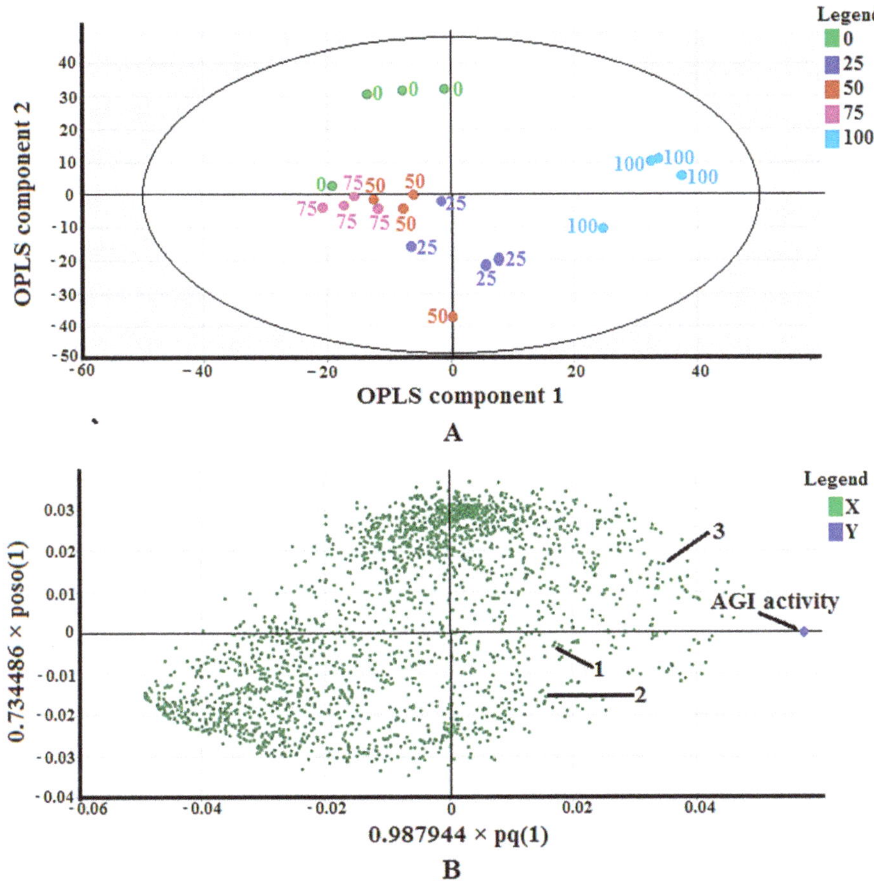

Figure 3. (**A**) Score scatter plot of validated OPLS model for 20 extracts of *P. malayana* leaves; (**B**) The loading scatter plot of OPLS model.

The loading scatter plot, as shown in Figure 3B, maps out the correlation of the x variables (*m/z*) to the y variable (AGI activity). The *m/z* values being closer to the AGI activity indicates their positive correlation to the bioactivity. Plenty of *m/z* vindicated this property. However, only three of them (*m/z* of 315.2, 317.2, and 349.2) could be identified after comparison with the available databases and publications. The detailed explanation of compound identification is described in Section 2.3. The three identified putative compounds are (5′-hydroxymethyl-1′-(1, 2, 3, 9-tetrahydro-pyrrolo [2, 1-*b*] quinazolin-1-yl)-heptan-1-one) (**1**), *α*-terpinyl-*β*-glucoside (**2**), and machaeridiol-A (**3**).

2.3. Identification of Putative Compounds

Identification of putative compounds was performed by analyzing the *m/z* spectrum of fragmented ions of each compound which was obtained from the LC-MS/MS instrument with a positive ionization mode (see Table 1). The fragmentation pathway of three compounds is shown in Figure S1, the supplementary file. The chemical structures of the identified compounds are shown in Figure 4.

Table 1. Tentative metabolites identified in the *P. malayana* Jack leaves through LC-MS/MS fragmentation using positive ionization.

Compound	M + H	MS^2 Fragments Ion	Tentative Metabolites	Reference
1	315.1775 ($C_{19}H_{27}N_2O_2$)	$[M\text{-}CH_2]^+$ at *m/z* 300, $[M\text{-}CH_3]^+$ at *m/z* 299, $[M\text{-}CH_5]^+$ at *m/z* 297, $[M\text{-}C_2H_4]^+$ at *m/z* 286, $[M\text{-}CH_3O]^+$ at *m/z* 283, $[M\text{-}CH_4O]^+$ at *m/z* 282, $[M\text{-}C_4H_2]^+$ at *m/z* 264, $[M\text{-}C_4H_9]^+$ at *m/z* 257, $[M\text{-}C_4H_9O]^+$ at *m/z* 241, $[M\text{-}C_6H_3]^+$ at *m/z* 239, $[M\text{-}C_5H_{11}O]^+$ at *m/z* 227, $[M\text{-}C_7H_{15}O]^+$ at *m/z* 199, $[M\text{-}C_7H_{13}O_2]^+$ at *m/z* 185, $[M\text{-}C_9H_{19}O]^+$ at *m/z* 171, $[M\text{-}C_{10}H_{21}O]^+$ at *m/z* 157, $[M\text{-}C_{12}H_{15}O]^+$ at *m/z* 139, $[M\text{-}C_{11}H_{18}O_2]^+$ at *m/z* 132, $[M\text{-}C_{12}H_{17}O_2]^+$ at *m/z* 121, $[M\text{-}C_{14}H_{21}O_2]^+$ at *m/z* 93, $[M\text{-}C_{13}H_{19}N_2O_2]^+$ at *m/z* 79, $[M\text{-}C_{17}H_{23}N_2O_2]^+$ at *m/z* 27	5′-hydroxymethyl-1′-(1, 2, 3, 9-tetrahydro-pyrrolo (2, 1-*b*) quinazolin-1-yl)-heptan-1′-one	[27,28]
2	317.2882 ($C_{16}H_{29}O_6$)	$[M\text{-}HO]^+$ at *m/z* 299, $[M\text{-}H_3O_2]^+$ at *m/z* 281, $[M\text{-}CH_3O_2]^+$ at *m/z* 269, $[M\text{-}CH_5O_3]^+$ at *m/z* 251, $[M\text{-}C_4H_5]^+$ at *m/z* 263, $[M\text{-}C_9H_{17}O_6]^+$ at *m/z* 95, $[M\text{-}C_{10}H_{19}O_6]^+$ at *m/z* 81, $[M\text{-}C_{11}H_{19}O_6]^+$ at *m/z* 69, $[M\text{-}C_{11}H_{21}O_6]^+$ at *m/z* 67, $[M\text{-}C_{14}H_{23}O_4]^+$ at *m/z* 61, $[M\text{-}C_{14}H_{24}O_4]^+$ at *m/z* 60	α-terpinyl-β-glucoside	[29–33]
3	349.2127 ($C_{24}H_{29}O_2$)	$[M\text{-}HO]^+$ at *m/z* 331, $[M\text{-}H_3O_2]^+$ at *m/z* 313, $[M\text{-}C_3H_3]^+$ at *m/z* 309, $[M\text{-}C_4H_5]^+$ at *m/z* 295, $[M\text{-}C_5H_7]^+$ at *m/z* 281, $[M\text{-}C_6H_9]^+$ at *m/z* 267, $[M\text{-}C_{10}H_{17}]^+$ at *m/z* 211, $[M\text{-}C_{17}H_{14}O]^+$ at *m/z* 114	machaeridiol-A	[34,35]

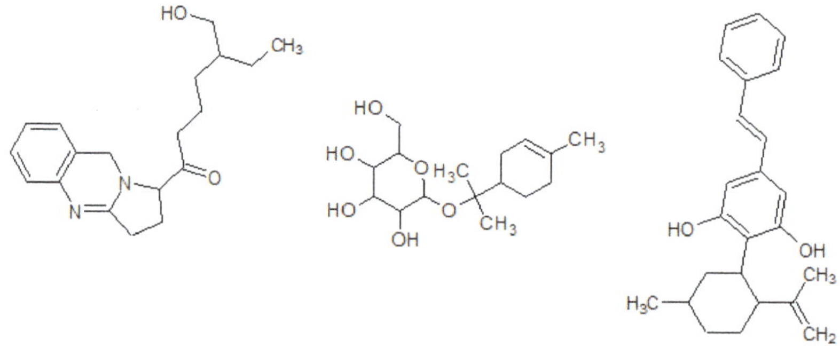

Figure 4. Chemical structures of compound **1**, **2**, and **3**.

Compound **1**, an alkaloid, consists of a quinazoline ring with an aliphatic side chain containing carbonyl and hydroxyl groups. Compound **1** was protonated at the oxygen attached to the O-H functional groups to form the parent ion $[M + H]^+$ having an *m/z* of 315. Subsequent removal of C_2H_5, C_4H_3, C_4H_{10}, $C_4H_{10}O$, $C_{11}H_{19}O$, $C_{13}H_{20}N_2O_2$ and $C_{17}H_{24}N_2O_2$ from the parent ion resulted in ions with *m/z* values of 286, 264, 257, 241, 132, 79, and 27, respectively. Removal of a methyl group from the parent ion produced an ion with an *m/z* of 300. The loss of methane (CH_4) from the parent ion results in the formation of an ion with an *m/z* of 299 from which dihydrogen was removed to produce an ion with an *m/z* of 297. Removal of CH_4O from the parent ion produced an ion with an *m/z* of 283, and the further loss of hydrogen caused the formation of an ion with an *m/z* of 282. In another pathway, the parent ion lost pentanol ($C_5H_{12}O$) to form an ion with an *m/z* of 227, which was further fragmented to produce an ion with an *m/z* of 185 by removal of ethenone (C_2H_2O). The loss of heptanol ($C_7H_{16}O$) from the parent ion leads to the formation of an ion with an *m/z* of 199, from which, further loss of ethylene (C_2H_4)

forms an ion with an *m/z* of 171. Further fragmentation of this ion leads to the formation of an ion with an *m/z* of 157 by removal of methylene (CH_2). The parent ion lost benzene (C_6H_4) to produce an ion with an *m/z* of 239 that was further fragmented to produce an ion with an *m/z* of 139 by removal of hexanal ($C_6H_{12}O$). Later on, a water molecule was removed to produce an ion with an *m/z* of 121 that led to the production of another ion with an *m/z* of 93 by removal of ethylene (C_2H_4).

Compound **2** is a diterpenoid glucoside which was protonated at the oxygen atom of one of the hydroxyl groups of glycoside during positive Electrospray Ionisation (ESI) to form the parent ion $[M + H]^+$ with an *m/z* of 317. The removal of C_4H_6, $C_{11}H_{20}O_6$, $C_{14}H_{24}O_4$, and $C_{14}H_{25}O_4$ from this ion led to the formation of ions with *m/z* values of 263, 69, 61, and 60, respectively. The loss of $C_9H_{18}O_6$ from the parent ion produced an ion with an *m/z* of 95, which was further fragmented to form an ion with an *m/z* of 81 by removal of methylene (CH_2). In another pathway, the loss of $C_9H_{18}O_6$ led to the formation of an ion with an *m/z* of 95. While an ion with an *m/z* of 67 was formed from the ion with an *m/z* of 95 by removal of C_2H_4. Removal of water from the parent ion produced an ion with an *m/z* of 299, this was followed by the loss of another water molecule to form an ion with an *m/z* of 281. The removal of CH_4O_2 from the parent ion formed an ion with an *m/z* of 263, which was further fragmented to produce an ion with an *m/z* of 251 by the removal of water.

Compound **3**, a phenolic compound with two hydroxyl groups in its structure, contains cyclic hydrocarbon with methylene and methyl. In the ESI positive ionization mode, compound **3** was protonated at the oxygen atom of one of the hydroxyl groups to form the parent ion $[M + H]^+$ with an *m/z* of 349. The cyclic rearrangement and removal of C_4H_6, C_5H_8, C_6H_{10}, and $C_{10}H_{18}$ from the parent ion produced ions with an *m/z* of 295, 281, 267, and 211, respectively. Removal of a water molecule from the parent ion produced an ion with an *m/z* of 331, which was further fragmented to produce an ion with an *m/z* of 313 by removal of another water molecule. The loss of C_3H_4 from the parent ion led to the formation of an ion with an *m/z* of 309, this was followed by the loss of $C_{14}H_{11}O$ to form an ion with an *m/z* of 114.

2.4. Molecular Docking

Molecular docking was performed on the three compounds that were identified using LC-MS based metabolomics to identify a potential binding mode of the compounds that could justify their inhibitory activity. Table 2 shows the binding affinities of the control ligand, the detected bioactive compounds and the standard competitive inhibitor (quercetin) towards the AG enzyme. A control docking procedure was performed using the co-crystallized control ligand, alpha-D-glucose (ADG) in order to validate the docking parameters, while quercetin was used as a comparison with the identified compounds. In this work, the complex with a more negative value (i.e., stronger binding affinity) is considered as the best-docked complex. The re-docked ADG was found to bind with 3A4A in a manner identitical to its crystallographic configuration. The value of the root mean square deviation (RMSD) of the re-docked ADG was found to be 0.633 Å, suggesting that the chosen docking parameters are able to reproduce the crystallised conformation. The docking parameters are considered to be acceptable if the RMSD value of the redocked ligand, with respect to the crystallised one, is less than 1.5 Å [36]. The control docking showed that the control ligand exerts a binding affinity of −6.0 kcal.mol towards the enzyme. Seven amino acid residues (ASP352, GLH277, ASH215, HIE112, ASH69, ARG442, HIE351) were involved in the hydrogen bond interactions with ADG. For quercetin, it exerts the binding affinity of −8.4 kcal/mol. The binding is assisted by two residues (ASH215, GLH277) via a hydrogen bond and three residues (PHE303, ASP352, and ARG442) via other types of interactions.

All of the three identified compounds exerted greater binding affinity towards the enzyme than that of the control ligand. Among the three identified compounds, compound **3** showed the highest binding affinity towards the enzyme, with -10.00 kcal/mol. It also had a higher binding affinity value when compared to quercetin (−8.4 kcal/mol). In contrast, compounds **1** and **2** showed binding affinities of −8.3 and −7.6 kcal/mol, respectively, which are lower than that of quercetin. The superimposed 3D docking visualization (Figure 5A) describes the predicted binding site of the three identified compounds, quercetin, and the control ligand (ADG) in the enzymatic protein. According to the figure, all compounds are predicted to bind at domain A of the enzyme, where the catalytic site is located. The same site is also occupied by ADG and quercetin, suggesting that a similar inhibition mechanism might be followed by the identified compounds.

Table 2. Binding affinities of the α-glucosidase enzyme (3A4A) with control ligand, the known competitive inhibitor (quercetin), and the identified active compounds.

Compound	Binding Affinity, kcal/mol
Control ligand (ADG)	−6.0
Quercetin	−8.4
1	−8.3
2	−7.6
3	−10.0

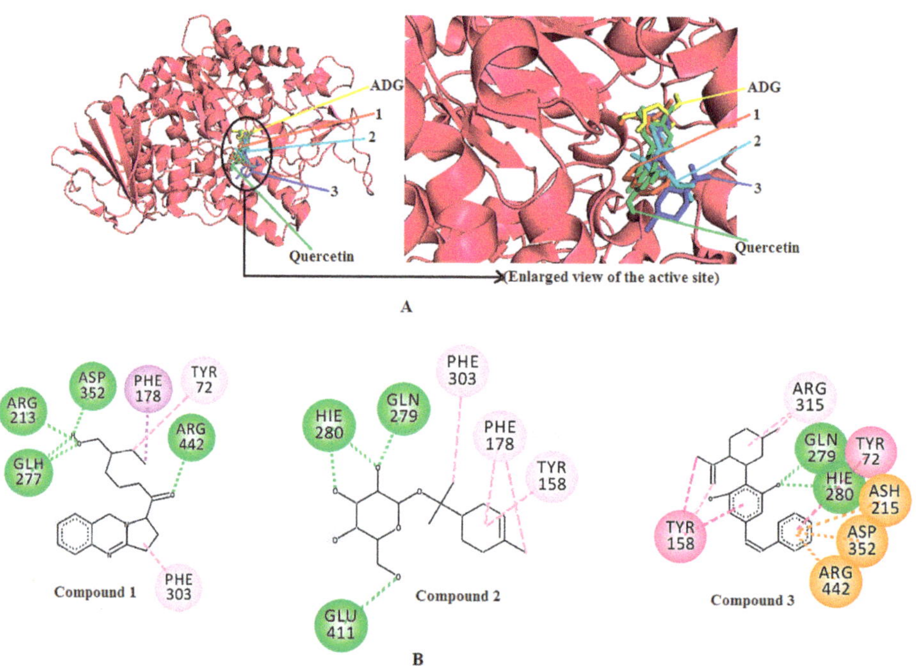

Figure 5. (**A**) Superimposed 3D diagram of control ligand (ADG), positive control (quercetin), the docked compounds **1**, **2**, and **3**. (**B**) 2D binding interactions of the docked compound **1**, **2**, and **3** with α-glucosidase (3A4A).

Table 3 shows the binding interactions of compounds **1**, **2**, and **3**. Along with the H-bond interaction, other interactions such as pi-sigma, pi-alkyl, pi-pi t-shaped were also involved in the docked complexes. The binding interaction analysis showed that the hydroxyl moiety of compound **1** interacts with ARG213, ASP352 and protonated GLU277 residues, whereas the carbonyl moiety of compound **1** interacts with the ARG442 residue through hydrogen bond interaction. Among the four hydrogen bonds, the strongest hydrogen bond is the one involving ASP352 with a distance of 2.08 Å. Meanhile, TYR72 and PHE178 interact with the aliphatic chain of compound **1** through pi-alkyl and pi-sigma interactions, respectively. Whereas, the pyrrolidines moiety of compound **1** exhibited a pi-alkyl interaction with the PHE303 residue of α-glucosidase.

Table 3. Data for the molecular docking of compounds **1**, **2**, and **3** in 3A4A.

Compound Structure	Interacting Amino Acid Residues	Bond Type	Bond Distance(Å)
Compound 1	ASP352	Hydrogen bonding	2.08
	ARG213	Hydrogen bonding	2.82
	GLH277	Hydrogen bonding	2.60
	ARG442	Hydrogen bonding	2.64
	TYR72	Pi-Alkyl	4.49
	PHE178	Pi-Sigma	3.68
	PHE303	Pi-Alkyl	4.76
Compound 2	GLN279	Hydrogen bonding	2.36
	HIE280	Hydrogen bonding	2.28 2.45
	GLU411	Hydrogen bonding	3.39
	PHE303	Pi-Alkyl	4.59
	PHE178	Pi-Alkyl	4.82 5.25
	TYR158	Pi-Alkyl	5.24
Compound 3	GLN279	Hydrogen bonding	2.58
	HIE280	Hydrogen bonding	2.78
	ASP352	Pi-Anion	4.50
	ASH215	Pi-Anion	4.80
	ARG442	Pi-Cation	3.66
	TYR72	Pi-Pi T-shaped	5.30
	TYR158	Pi-Pi T-shaped Pi-Alkyl Pi-Alkyl	4.85 4.16 4.09
	ARG315	Alkyl	4.50 3.84

In the complex containing compound **2**, the tetrahydronpyran moeity of the compound was observed to form two hydrogen bonds with HIE280, with distances of 2.28 Å and 2.45 Å, respectively. The other two hydrogen bonds were established between the tetrahydropyran moiety of compound **2** and GLU411, and GLN279. Whereas, the alicyclic moiety of compound **2** interacted with TYR158 and PHE178 via pi-alkyl interactions. Furthermore, the aliphatic moiety of compound **2** interacted via a pi-alkyl bond with the PHE303 residue at a distance of 4.59 Å.

In the docked complex of compound **3**–AG enzyme, the hydroxyl moiety of the aromatic ring of compound **3** showed interactions with HIS280 and GLU279 residues through hydrogen bonding at distances of 2.78 Å and 2.58 Å, respectively. The aromatic ring of the phenolic and the aliphatic moieties of compound **3** were observed to involve pi–pi t-shaped and pi–alkyl interactions with TYR158. On the other hand, ASP352 and protonated ASP215 showed a pi–anion, and ARG442 showed a pi–cation interaction with the aromatic segment of compound **3**. Furthermore, the aromatic moiety of compound **3** also interacted with the TYR72 residue through a pi–pi t-shaped interaction. ARG315 interacted with the alicyclic moiety of compound **3** through two alkyl interactions at a distance of 4.50 Å and 3.84 Å (Table 3). The two-dimensional (2D) binding interactions of compounds **1**, **2**, and **3** with the active site of the 3A4A enzyme are depicted in Figure 5B.

3. Discussion

Various extraction methods using a methanolic solvent have been applied to enhance the content of the bioactive compounds possessing anti-diabetic and antioxidant activities in medicinal herbal extracts. Kidane et al. [37] reported that the extraction of *Psiadia punctulata* leaves in a methanolic solvent produced an extract with high AGI activity. A similar result was found by Bhatia et al. [38] who reported the high AGI activity of a methanolic extract of *Cornus capitata* Wall leaves. Extraction with aqueous methanol was shown to be better for the recovery of the largest amount of phenolic compounds from *Moringa oleifera* and rice bran, which are related to antioxidant activity [39,40]. In another study, 80% methanol solvent was used to extract antioxidants from numerous natural sources such as rice bran, wheat bran, oat groats and hull, coffee beans, citrus peel and guava leaves [41]. In this study, different ratios of aqueous methanolic solvent were investigated, which led to obtaining an extract with high AGI activity in 100% methanol. The trend of AGI activity (100 > 25 > 50 > 0 > 75% *v/v* methanol–water) is caused by the difference of metabolite profiling among the extracts which were affected by the solvent used during extraction [42–45].

The loading scatter plot obtained using OPLS (Figure 3B) pinpointed three putative compounds correlating to the AGI activity of *P. malayana* leaves. The presence of these compounds in this plant had not been recorded elsewhere; thus, this study reports it for the first time. The presence of compound **1** has been reported by Sutradhar et al. [28], who isolated it from the methanol and water extract of *Sida cordifolia*. Linn. aerial sections. This compound was reported to possess analgesic and anti-inflammatory effects in an animal model [27]. Although no research has been performed on the anti-diabetic effect of this compound, the plant (*S. cordifolia*) that is rich in this compound has been shown to ellicit both hypoglycemic and antioxidant activities [46,47].

Compound **2** is commonly known as α-terpineol-β-D-O-glucopyranoside or α-terpinyl-β-glucoside. Nhiem et al. [31] isolated this compound from a methanol extract of *Acanthopanax koreanum* leaves, although it did not possess antioxidant activity. This compound is also found in other plant parts such as apricot fruits [29], sweet potatoes [30], the aerial parts of *Ligularia alticola* [32], and an ethyl acetate extract of *Saccocalyx satureioides* [33].

Compound **3** was isolated from the stem bark of *Machaerium multiflorum* [34]. However, no anti-diabetic activity has been reported so far for this compound. Most of the reports indicate its antimicrobial activity against *Candida albicans*, *Cryptococcus neoformans*, *Mycobacterium intracellularae*, *Aspergillus fumigatus* [35], and *Staphylococcus aureus* [34].

This present study appears to be the first to investigate the potential anti-diabetic effect of these compounds through an in silico approach. The best method to confirm the AGI activity of these compounds is through an in vitro AGI test on each of the pure compounds. However, these compounds are not commercially available, and the isolation of these compounds is difficult due to their low abundnace in this plant. Thus, in silico computational analysis, evaluating the binding characteristic between the identified compounds and the active site of the enzyme, was carried out.

The *S. cerevisiae* isomaltase is made up of 589 amino acids. It consists of three domains, namely domain A (1–113 and 190–512), domain B (114–189) and domain C (513–589) [48]. There are three catalytic residues ASP215, GLU277 and ASP352, which are situated on the side of the C-terminal of domain A [17,48]. The catalytic residues indicate the active site of the enzymatic protein and all the residues involved in the molecular interaction during re-dock of ADG with 3A4A, contributing the actual catalytic reaction [17]. Based on the results obtained, all compounds demonstrated binding affinities comparable to that of the known AG inhibitor, quercetin, suggesting their potential to inhibit the enzyme.

Compounds **1** and **2** showed higher binding affinities than ADG, but slightly lower binding affinities compared to that of quercetin. Good binding affinity of these compounds to the active site of the enzyme was indicated, which is the same site used by ADG [17]. Among the seven residues that are involved in the binding of ADG towards the AG enzyme, three residues (ASP352, GLH277, and ARG442) were involved in the interactions with compound **1**, however, no active site residues of the control ligand interacted with compound **2**. This might explain the difference in the binding affinity (0.7 kca/mol) between compound **1** and compound **2**. It is anticipated that once the enzyme's active site is blocked, the enzyme is unable to break down carbohydrates, resulting in lower quantities of glucose absorption and eventually lower blood glucose levels [10].

Compound **3** showed the highest binding affinity among the three compounds, quercetin, and ADG. This high binding affinity is attributed to its chemical structure which contains three main parts (aromatic, alicyclic, and aliphatic parts). This allows the compound to interact with three catalytic residues (ASP352, ARG442, and ASH215) via pi–anion and pi–cation interactions, in addition to hydrogen bond interactions. The higher number of hydrophobic interactions between compound **3** and the enzyme may lead to a high affinity towards the active site of the enzyme, which is in line with previous research [49]. Ahmed et al. [49] reported a higher number of hydrophobic interactions between questin (an identified α-glucosidase inhibitor) and 3A4A that indicated a high affinity towards the enzyme's active site by exhibiting a more negative binding affinity value of -8.3 kcal/mol. Three residues (ASP352, ASH215, and ARG442) that are present in the ADG-AG enzyme complex are also involved in the docked complex of compound **3**–AG enzyme, supporting the high binding affinity exerted by compound **3**. Given the above findings, it is anticipated that compound **3** may serve as potential lead compound for the development of an AG inhibitor.

4. Materials and Methods

4.1. Materials

Organic chemicals (methanol, dimethyl sulfoxide) of analytical and chromatography grade and the AG (yeast maltase) enzyme were obtained from Merck (Darmstadt, Germany) and Megazyme (Megazyme, Bray, Ireland), respectively. Quercetin as the standard and ρ-nitrophenyl-ρ-D-glucopyranosidase (PNPG) as the substrate were procured from Sigma-Aldrich (St. Louis, MO, USA).

4.2. Plant Sample Collection

The *P. malayana* Jack leaves were collected from Cermin Nan Gedang at Sarolangun district, Jambi, Indonesia. These were identified by a taxonomist, Dr Shamsul Khamis from the University of Putra, Malaysia. The specimen was deposited at the Kulliyyah of Pharmacy (KOP) Herbarium, IIUM, Kuantan (voucher specimen #PIIUM008-2). The plant samples were allowed to dry for seven days at room temperature (25 ± 5 °C) in the shade and were converted into powder form using a Universal cutting mill (Fritsch, Idar-Oberstein, Germany). The powdered leaves were stored at -80 °C until further analysis [50].

4.3. Preparation of P. malayana Leaves Extracts

Crushed powdered leaves of *P. malayana* (1 g) were subjected to extraction. The extraction was conducted by sonication. The powdered leaves were immersed in 30 mL of methanol–water at various ratios (0, 25, 50, 75, and 100%, *v/v*) and allowed to sonicate for 30 min. Subsequently, the extracts were filtered using Whatman filter paper No.1. A rotary evaporator was used at 40 °C to remove the solvent. To clear any residual solvents, the extracts were then freeze-dried and stored at -80 °C. For this study, a total of 20 samples were prepared (4 replicates for each of the 5 separate extracts).

4.4. AGI Assay

The method reported by Javadi et al. [50] was followed with minor modifications. Quercetin (2 mg in 1 mL DMSO) and ρ-nitrophenyl-ρ-D-glucopyranosidase (PNPG, 6 mg in 20 mL of 50 mM phosphate buffer, pH 6.5) were used as the positive control and substrate, respectively. The yeast, glucosidase, was used in this study under the consideration that its substrate specificity, inhibitor sensitivity, and optimum pH are similar to the mammalian glucosidase [51].

The sample was prepared following the same preparation protocol for quercetin. While the negative control was prepared by replacing the sample with DMSO. A total of 10 μL of the samples, quercetin and DMSO; 100 µL of 30 mM phosphate buffer and 15 µL of 0.02 U/µL AG enzyme were transferred into 96-well plate. A blank was prepared following the same protocol but without the enzyme. The samples and the blank mixture were both incubated at room temperature for 5 min and then treated with 75 µL of PNPG, followed by addition of glycine (pH 10) and another incubation time of 15 min at room temperature to stop the reaction. The absorbance reading was recorded at 405 nm by the microplate reader (Tecan Nanoquant Infinite M200, Männedorf, Switzerland). The inhibition percentage was obtained from the following equation:

$$\text{Inhibition activity (\%)} = [(A_{control} - A_{sample})/A_{control}] \times 100\% \tag{1}$$

where, the absorbance of the negative control is $A_{control}$, and the absorbance of a sample or positive control is A_{sample}.

4.5. LCMS-QTOF Analysis

The method used by Murugesu et al. [17] was followed in this study with a slight modification. LCMS-QTOF instrument (Agilent 1290 Infinity and 6550 iFunnel, Santa Clara, CA, USA) fitted with an electrospray interface (ESI) with positive ion modes was used to analyze the sample. The sample was prepared using 250 µL of methanol to dissolve 1 mg of each plant extract. The mixture was vortexed and sonicated for 15 min, followed by the addition of 250 µL of water. Subsequently, it was centrifuged for 15 min to obtain a clear supernatant which was transferred into an insert glass vial through syringe filtration. Phenomenex Kinetex C18 core-shell technology 100 Å (250 mm × 4.6 mm, 5 µm, California, United States) column was used for sample injection. The column temperature was 27 °C. The column was initially washed with methanol before sample injection. A total of 10 µL sample was injected through an autosampling system. Samples were eluted with a gradient system from 5% methanol in water with 0.1% formic acid to absolute methanol in 20 min. The machine was then operated for another 10 min with absolute methanol at a flow rate of 0.7 mL/min for a total running time of 30 min. The MS/MS data were recorded between *m/z* values of 50 to 1500 with a scanning rate of 1 spectrum per scan, and the MS/MS spectra were collected by a collision energy ramp of 35 eV. The source parameter was set to the following: gas temperature = 200 °C, gas flow = 14 L/min, nebulizer= 35 psig, sheat gas temperature = 350 °C, sheat gas flow = 11 L/min.

ACD/Spec Manager v.12.00 (Advanced Chemistry Development, Inc., ACD/Labs Toronto, ON, Canada) software was used for analyzing the acquired LCMS-QTOF data and for in silico analysis of fragmention pathway. The raw (*.xms) data were converted into CDF (*.cdf) format by ACD/Spec manager. The data were subsequently pre-processed using MZmine software (VTT Technical Research Centre, Oulu, Finland) for baseline correction, peak detection, peak filtering, alignment, smoothing and gap filling and saved as a (*.csv) format [52] following the parameters shown in Table 4. The results were analyzed with SIMCA-P$^+$ 14.0 (Umetrics, Umeå, Sweden) software for multivariate data analysis.

4.6. Molecular Docking

Docking was carried out using AutoDock Vina (version 1.1.2) to predict the binding affinity of the identified compounds to the active site of the enzyme. From Protein Data Bank (PDB), the AG crystal structure was combined (PDB code: 3A4A) [17,53–57]. The 3D structures of the identified compound were collected from the Super Natural II database as *.mol files and quercetin was collected from Pub Chem in *.sdf format. All the structures of the identified compounds were changed to *.pdb format by Open Babel (version 2.3.1) [17]. Then AutoDock Tools (version 1.5.6) was used to convert the compounds in *.pdb format to *. Pdbqt, prior to docking. Gasteiger charges and hydrogen atoms were added [53]. The rotatable bonds were set using AutoDock Tools. The co-crystallized ligand, alpha-D-glucose (ADG), was re-docked into the enzyme (PDB ID: 3A4A). The top-levelled docking conformations were compared with the actual crystallographic conformation based on the values of root mean square deviation (RMSD). The same parameters as in the control docking were applied for the docking of quercetin, as well as compounds **1**, **2** and **3** against the enzyme [36,58]. The grid box parameters were centered at the coordinate of 21.272, −0.751 and 18.634. The sizes of the grid box for X, Y, and Z were 20 Å, 26 Å and 22 Å, respectively, while the exhaustiveness was set as 16. The docking was performed in triplicate. The enzyme crystal structure was protonated at a pH of 6.5 with the PDB2PQR Server, version 2.0.0, to simulate the real state of the bioassay [17,59]. The 3D superimposed diagram of the standard and the detected compounds were rendered using PyMOL TM 1.7.4.5 (Schrödinger, LLC, New York, NY, USA) [17]. All docking outcomes were analyzed by Biovia Discovery Studio Visualizer (San Diego, CA, USA) [17,36,58].

Table 4. Parameter setting for pre-processing of LC-MS data using MZmine software (VTT Technical Research Centre, Finland).

Parameters	Details
Baseline correction	Mass Spectrometry (MS) level = 1 m/z bin width = 1 Asymmetric baseline corrector (smoothing = 100,000, asymmetry = 0.5)
Mass detection	Mass detector = centroid (noise level = 200)
Chromatogram builder	Min time span = 0.2 min Min height = 200 m/z tolerance = 1.0 mz or 2500 ppm
Peak detection	Filter width = 11
Isotopic peaks grouper	m/z tolerance = 1.0 mz or 2500 ppm Retention time (RT) tolerance = 0.2 min Monotonic shape Maximum charge = 1
Duplicate peak filter	m/z tolerance = 1.0 mz or 2500 ppm RT tolerance = 0.2 min
Normalization	Linear Normalization Peak measurement type = peak area
Alignment (Join aligner)	m/z tolerance = 1.0 or 2500 ppm Weight for m/z = 85 RT tolerance = 0.2 min Weight for RT = 15 Require same charge state
Gap filling	Same RT and m/z range gap filler m/z tolerance = 1.0 or 2500 ppm

4.7. Statistical Analysis

Minitab 16 (Minitab Inc., State College, PA, USA) was used for the statistical analysis of the obtained data. One-way variances (ANOVA) and Tukey's test were used to analyze the data with a significance level of $p < 0.05$. The data are interpreted as a mean ± standard deviation (SD). Multivariate data analysis was carried out using SIMCA P + 14.0 (Umetrics, Umeå, Sweden) software where the pareto scaling method was applied.

5. Conclusions

LC-MS-based multivariate data analysis and molecular docking is a promising technique for investigating bioactive compounds associated with folk medicine, for example those from herbal plants. In the present study, three AG inhibitor compounds were identified from methanolic extracts of *P. malayana* leaves by using this approach. The compounds were identified as 5'-hydroxymethyl-1'-(1, 2, 3, 9-tetrahydro-pyrrolo (2, 1-*b*) quinazolin-1-yl)-heptan-1'-one, α-terpinyl-β-glucoside, and machaeridiol-A, being reported for the first time in this plant in this study. Their AG inhibitory activity was confirmed through in silico molecular studies, which predicted the binding mode and binding interactions between the identified compounds and the binding site on the enzyme. This information is useful for future studies on the bioassay assessment of these compounds for exploration of their potential uses as novel anti-diabetic agents. Purification and identification of the unknown compounds suspected to possess AG inhibitory activities in this study are required to be conducted in the future study.

Supplementary Materials: The following is available online, Figure S1: Fragmentation pathway of three identified compounds.

Author Contributions: Conceptualization, T.S.N. and A.K.; methodology, T.S.N. and A.K.; software, T.S.N. and A.K.; validation, T.S.N. and A.K.; formal analysis, T.S.N. and A.K.; investigation, T.S.N. and A.K.; resources, M.Z.S.; T.S.N. and A.K.; data curation, T.S.N.; M.Z.S. and A.K.; writing—original draft preparation, T.S.N.; writing—review and editing, M.Z.S.; F.S.; H.R.E.-S.; T.S.N.; R.P. and A.K.; visualization, T.S.N. and A.K.; supervision, A.K.; Z.I.; Q.U.A. and I.E.R.; project administration, T.S.N.; M.Z.S. and A.K.; funding acquisition, T.S.N.; M.Z.S. and A.K. All authors have read and agreed to the published version of the manuscript.

Funding: This research was funded by the Ministry of Higher Education of Malaysia (Grant No. FRGS-19-0880697) and University of Malaya (Grant No. RU006-2017).

Acknowledgments: The authors thank Ministry of Higher Education of Malaysia (Grant No. FRGS-19-0880697) and University of Malaya (Grant No. RU006-2017) for financial support. The authors also thank the Bangabandhu Science and Technology Fellowship Trust, Ministry of Science and Technology of Bangladesh for a PhD fellowship to T.S.N.

Conflicts of Interest: The authors declare no conflict of interest.

References

1. Boyle, J.P.; Honeycutt, A.A.; Narayan, K.M.V.; Hoerger, T.J.; Geiss, L.S.; Chen, H.; Thompson, T.J. Projection of Diabetes Burden Through 2050. *Diabetes Care* **2001**, *24*, 1936–1940. [CrossRef]
2. Islam, D.; Huque, A.; Mohanta, L.C.; Das, S.K.; Sultana, A. Hypoglycemic and Hypolipidemic Effects of Nelumbo Nucifera Flower in Long-Evans Rats. *J. Herbmed Pharmacol.* **2018**, *7*, 148–154. [CrossRef]
3. Rahimi-Madiseh, M.; Karimian, P.; Kafeshani, M.; Rafieian-Kopaei, M. The Effects of Ethanol Extract of Berberis Vulgaris Fruit on Histopathological Changes and Biochemical Markers of the Liver Damage in Diabetic Rats. *Iran. J. Basic Med. Sci.* **2017**, *20*, 552–556. [CrossRef] [PubMed]
4. Mohamed, N.; Maideen, P.; Balasubramaniam, R. Pharmacologically Relevant Drug Interactions of Sulfonylurea Antidiabetics with Common Herbs. *J. Herbmed Pharmacol.* **2018**, *7*, 200–210. [CrossRef]
5. Graham, J.E.; Stoebner-May, D.G.; Ostir, G.V.; Al, S.; Peek, M.K.; Markides, K.; Ottenbacher, K.J. Health Related Quality of Life in Older Mexican Americans with Diabetes: A Cross-Sectional Study. *Health Qual. Life Outcomes* **2007**, *5*, 1–7. [CrossRef]
6. Haque, N.; Salma, U.; Nurunnabi, T.R.; Uddin, M.J.; Jahangir, M.F.K.; Islam, S.M.Z.; Kamruzzaman, M. Management of Type 2 Diabetes Mellitus by Lifestyle, Diet and Medicinal Plants. *Pakistan J. Biol. Sci.* **2011**, *14*, 13–24. [CrossRef] [PubMed]
7. Gro, J.K.; Yada, S.; Vats, V. Medicinal Plants of India with Anti-Diabetic Potential. *J. Ethnopharmacol.* **2002**, *81*, 81–100.
8. Jung, M.; Park, M.; Lee, H.C.; Kang, Y.; Kang, E.S.; Kim, K. Antidiabetic Agents from Medicinal Plants. *Curr. Med. Chem.* **2006**, *13*, 1203–1218. [CrossRef]
9. Kooti, W.; Farokhipour, M.; Asadzadeh, Z.; Ashtary-Larky, D.; Asadi-Samani, M. The role of medicinal plants in the treatment of diabetes:a systematic review. *Electron. Physician* **2016**, *8*, 1832–1842. [CrossRef]
10. Nair, S.S.; Kavrekar, V.; Mishra, A. In Vitro Studies on Alpha Amylase and Alpha Glucosidase Inhibitory Activities of Selected Plant Extracts. *Eur. J. Exp. Biol.* **2013**, *3*, 128–132.
11. Hadi, S.; Asnawati, D.; Ersalena, V.F.; Azwari, A.; Rahmawati, K.P. Characterization of Alkaloids From The Leaves of *Psychotria Malayana* Jack of Lombok Island on The Basis of Gas Chromatography-Mass Spectroscopy. *J. Pure Appl. Chem. Res.* **2014**, *3*, 108–113. [CrossRef]

12. Benchoula, K.; Khatib, A.; Quzwain, F.M.C.; Che Mohamad, C.A.; Wan Sulaiman, W.M.A.; Wahab, R.A.; Ahmed, Q.U.; Ghaffar, M.A.; Saiman, M.Z.; Alajmi, M.F.; et al. Optimization of Hyperglycemic Induction in Zebrafish and Evaluation of Its Blood Glucose Level and Metabolite Fingerprint Treated with *Psychotria Malayana* Jack Leaf Extract. *Molecules* **2019**, *24*, 1506. [CrossRef] [PubMed]

13. Situmorang, R.O.P.; Harianja, A.H.; Silalahi, J. Karo's Local Wisdom: The Use of Woody Plants for Traditional Diabetic Medicines. *Indones. J. For. Res.* **2015**, *2*, 121–130. [CrossRef]

14. Zhou, B.; Xiao, J.F.; Tuli, L.; Ressom, H.W. LC-MS-Based Metabolomics. *Mol. Biosyst.* **2012**, *8*, 470–481. [CrossRef]

15. D'Urso, G.; Pizza, C.; Piacente, S.; Montoro, P. Combination of LC–MS Based Metabolomics and Antioxidant Activity for Evaluation of Bioactive Compounds in Fragaria Vesca Leaves from Italy. *J. Pharm. Biomed. Anal.* **2018**, *150*, 233–240. [CrossRef]

16. Azizan, K.A.; Ibrahim, S.; Ghani, N.H.A.; Nawawi, M.F. LC-MS Based Metabolomics Analysis to Identify Potential Allelochemicals in Wedelia Trilobata. *Rec. Nat. Prod.* **2016**, *10*, 788–793.

17. Murugesu, S.; Ibrahim, Z.; Uddin, Q.; Uzir, B.F.; Nik, N.I.; Perumal, V.; Abas, F.; Shaari, K.; Khatib, A. Identification of α-Glucosidase Inhibitors from Clinacanthus Nutans Leaf Extract Using Liquid Chromatography-Mass Spectrometry-Based Metabolomics and Protein-Ligand Interaction with Molecular Docking. *J. Pharm. Anal.* **2018**. [CrossRef]

18. D'urso, G.; Mes, J.J.; Montoro, P.; Hall, R.D.; de Vos, R.C.H. Identification of Bioactive Phytochemicals in Mulberries. *Metabolites* **2020**, *10*, 7. [CrossRef]

19. Mahmod, I.I.; Ismail, I.S.; Alitheen, N.B.; Normi, Y.M.; Abas, F.; Khatib, A.; Rudiyanto; Latip, J. NMR and LCMS Analytical Platforms Exhibited the Nephroprotective Effect of Clinacanthus Nutans in Cisplatin-Induced Nephrotoxicity in the in Vitro Condition. *BMC Complement. Med. Ther.* **2020**, *20*, 1–18. [CrossRef]

20. Yang, Y.Y.; Wu, Z.Y.; Zhang, H.; Yin, S.J.; Xia, F.B.; Zhang, Q.; Wan, J.B.; Gao, J.L.; Yang, F.Q. LC-MS-Based Multivariate Statistical Analysis for the Screening of Potential Thrombin/Factor Xa Inhibitors from Radix Salvia Miltiorrhiza. *Chin. Med.* **2020**, *15*, 1–13. [CrossRef]

21. Yang, Y.Y.; Wu, Z.Y.; Xia, F.B.; Zhang, H.; Wang, X.; Gao, J.L.; Yang, F.Q.; Wan, J.B. Characterization of Thrombin/Factor Xa Inhibitors in Rhizoma Chuanxiong through UPLC-MS-Based Multivariate Statistical Analysis. *Chin. Med.* **2020**, *15*, 1–14. [CrossRef] [PubMed]

22. Yan, Y.; Du, C.; Li, Z.; Zhang, M.; Li, J.; Jia, J.; Li, A.; Qin, X.; Song, Q. Comparing the Antidiabetic Effects and Chemical Profiles of Raw and Fermented Chinese Ge-Gen-Qin-Lian Decoction by Integrating Untargeted Metabolomics and Targeted Analysis. *Chin. Med.* **2018**, *13*, 1–18. [CrossRef] [PubMed]

23. Pang, B.; Zhu, Y.; Lu, L.; Gu, F.; Chen, H. The Applications and Features of Liquid Chromatography-Mass Spectrometry in the Analysis of Traditional Chinese Medicine. *Evid. Based Complement. Altern. Med.* **2016**, *2016*, 1–8. [CrossRef] [PubMed]

24. Kind, T.; Fiehn, O. Advances in Structure Elucidation of Small Molecules Using Mass Spectrometry. *Bioanal. Rev.* **2010**, *2*, 23–60. [CrossRef] [PubMed]

25. Hombalimath, V.S.; Shet, A.R. In-Silico Analysis for Predicting Protein Ligand Interaction for Snake Venom Protein. *J. Adv. Bioinforma. Appl. Res.* **2012**, *3*, 345–356.

26. Eriksson, L.; Byrne, T.; Johansson, E.; Trygg, J.; Vikström, C. *Multi- and Megavariate Data Analysis Basic Principles and Applications*, 2nd ed.; Umetrics Academy: Umea, Sweden, 2006.

27. Sutradhar, R.K.; Matior Rahman, A.K.M.; Ahmad, M.; Bachar, S.C.; Saha, A.; Guha, S.K. Bioactive Alkaloid from *Sida Cordifolia* Linn. with Analgesic and Anti-Inflammatory Activities. *Iran. J. Pharmacol. Ther.* **2006**, *5*, 175–178.

28. Sutradhar, R.K.; Rahman, A.K.M.M.; Ahmad, M.U.; Saha, K. Alkaloids of *Sida cordifolia* L. *Indian J. Chem.* **2007**, *46B*, 1896–1900.

29. Salles, C.; Jallageas, J.C.; Crouzet, J.C. HPLC Separation of Fruit Diastereoisomeric Monoterpenyl Glycosides. *J. Essent. Oil Res.* **1993**, *5*, 381–390. [CrossRef]

30. Ohta, T.; Omori, T.; Shimojo, H.; Hashimoto, K.; Samuta, T.; Ohba, T. Identification of Monoterpene Alcohol β-Glucosides in Sweet Potatoes and Purification of a Shiro-Koji β-Glucosidase. *Agric. Biol. Chem.* **1991**, *55*, 1811–1816. [CrossRef]

31. Nhiem, N.X.; Kim, K.C.; Kim, A.D.; Hyun, J.W.; Kang, H.K.; Van Kiem, P.; Van Minh, C.; Thu, V.K.; Tai, B.H.; Kim, J.A.; et al. Phenylpropanoids from the Leaves of Acanthopanax Koreanum and Their Antioxidant Activity. *J. Asian Nat. Prod. Res.* **2011**, *13*, 56–61. [CrossRef]

32. Silchenko, A.S.; Kalinovsky, A.I.; Ponomarenko, L.P.; Avilov, S.A.; Andryjaschenko, P.V.; Dmitrenok, P.S.; Gorovoy, P.G.; Kim, N.Y.; Stonik, V.A. Structures of Eremophilane-Type Sesquiterpene Glucosides, Alticolosides A-G, from the Far Eastern Endemic Ligularia Alticola Worosch. *Phytochemistry* **2014**, 1–8. [CrossRef] [PubMed]

33. Kherkhache, H.; Benabdelaziz, I.; Silva, A.M.S.; Lahrech, M.B.; Benalia, M.; Haba, H. A New Indole Alkaloid, Antioxidant and Antibacterial Activities of Crude Extracts from Saccocalyx Satureioides. *Nat. Prod. Res.* **2018**, 1–7. [CrossRef] [PubMed]

34. Muhammad, I.; Li, X.; Dunbar, D.C.; Elsohly, M.A.; Khan, I.A. Antimalarial (+)-Trans-Hexahydrodibenzopyran Derivatives from Machaerium Multiflorum. *J. Nat. Prod.* **2001**, *50*, 6–9.

35. Muhammad, I.; Li, X.C.; Jacob, M.R.; Tekwani, B.L.; Dunbar, D.C.; Ferreira, D. Antimicrobial and Antiparasitic (+)-Trans-Hexahydrodibenzopyrans and Analogues from Machaerium Multiflorum. *J. Nat. Prod.* **2003**, *66*, 804–809. [CrossRef]

36. Wei, S.; Abas, F.; Wai, K.; Yusoff, K. In Vitro and in Silico Evaluations of Diarylpentanoid Series as α-Glucosidase Inhibitor. *Bioorg. Med. Chem. Lett.* **2018**, *28*, 302–309. [CrossRef]

37. Kidane, Y.; Bokrezion, T.; Mebrahtu, J.; Mehari, M.; Gebreab, Y.B.; Fessehaye, N.; Achila, O.O. In Vitro Inhibition of α-Amylase and α-Glucosidase by Extracts from Psiadia Punctulata and Meriandra Bengalensis. *Evid. Based Complement. Altern. Med.* **2018**, *2018*, 1–9. [CrossRef]

38. Bhatia, A.; Singh, B.; Arora, R.; Arora, S. In Vitro Evaluation of the α-Glucosidase Inhibitory Potential of Methanolic Extracts of Traditionally Used Anti-diabetic Plants. *BMC Complement. Altern. Med.* **2019**, *19*, 1–9. [CrossRef]

39. Chatha, S.A.S.; Anwar, F.; Manzoor, M.; Bajwa, J.U.R. Evaluation of the Antioxidant Activity of Rice Bran Extracts Using Different Antioxidant Assays. *Grasas Aceites* **2006**, *57*, 328–335. [CrossRef]

40. Siddhuraju, P.; Becker, K. Antioxidant Properties of Various Solvent Extracts of Total Phenolic Constituents from Three Different Agroclimatic Origins of Drumstick Tree (Moringa Oleifera Lam.) Leaves. *J. Agric. Food Chem.* **2003**, *51*, 2144–2155. [CrossRef]

41. Anwar, F.; Jamil, A.; Iqbal, S.; Sheikh, M.A. Antioxidant Activity of Various Plant Extracts under Ambient and Accelerated Storage of Sunflower Oil. *Grasas Aceites* **2006**, *57*, 189–197. [CrossRef]

42. Nipun, T.S.; Khatib, A.; Ahmed, Q.U.; Redzwan, I.E.; Ibrahim, Z.; Khan, A.Y.F.; Primaharinastiti, R.; Khalifa, S.A.M.; El-Seedi, H.R. Alpha-Glucosidase Inhibitory Effect of *Psychotria Malayana* Jack Leaf: A Rapid Analysis Using Infrared Fingerprinting. *Molecules* **2020**, *25*, 4161. [CrossRef] [PubMed]

43. Saleh, M.S.M.; Siddiqui, M.J.; Mat So'ad, S.Z.; Roheem, F.O.; Saidi-Besbes, S.; Khatib, A. Correlation of FT-IR Fingerprint and α-Glucosidase Inhibitory Activity of Salak (Salacca Zalacca) Fruit Extracts Utilizing Orthogonal Partial Least Square. *Molecules* **2018**, *23*, 1434. [CrossRef] [PubMed]

44. Nokhala, A.; Ahmed, Q.U.; Saleh, M.S.M.; Nipun, T.S.; Khan, A.Y.F.; Siddiqui, M.J. Characterization of α-Glucosidase Inhibitory Activity of Tetracera Scandens Leaves by Fourier Transform Infrared Spectroscopy-Based Metabolomics. *Adv. Tradit. Med.* **2020**, *20*, 169–180. [CrossRef]

45. Sharif, M.F.; Bennett, M.T. The Effect of Different Methods and Solvents on the Extraction of of Polyphenols in Ginger (Zingiber Officinale). *J. Teknol.* **2016**, *78*, 49–54. [CrossRef]

46. Jain, A.; Choubey, S.; Singour, P.K.; Rajak, H.; Pawar, R.S. *Sida Cordifolia* (Linn)—An Overview. *J. Appl. Pharm. Sci.* **2011**, *1*, 23–31.

47. Momin, M.A.M.; Bellah, S.F.; Rahman, S.M.R.; Rahman, A.A.; Murshid, G.M.M.; Emran, T.B. Phytopharmacological Evaluation of Ethanol Extract of *Sida Cordifolia* L. Roots. *Asian Pac. J. Trop. Biomed.* **2014**, *4*, 18–24. [CrossRef]

48. Yamamoto, K.; Miyake, H.; Kusunoki, M.; Osaki, S. Crystal Structures of Isomaltase from Saccharomyces Cerevisiae and in Complex with Its Competitive Inhibitor Maltose. *FEBS J.* **2010**, *277*, 4205–4214. [CrossRef]

49. Nokhala, A.; Siddiqui, M.J.; Ahmed, Q.U.; Bustamam, M.S.A.; Zakaria, Z.A. Investigation of A-glucosidase Inhibitory Metabolites from Tetracera Scandens Leaves by GC–MS Metabolite Profiling and Docking Studies. *Biomolecules* **2020**, *10*, 287. [CrossRef]

50. Javadi, N.; Abas, F.; Hamid, A.A.; Simoh, S.; Shaari, K.; Ismail, I.S.; Mediani, A.; Khatib, A. GC-MS-Based Metabolite Profiling of Cosmos Caudatus Leaves Possessing Alpha-Glucosidase Inhibitory Activity. *J. Food Sci.* **2014**, *79*, C1130–C1136. [CrossRef]

51. Dhanawansa, R.; Faridmoayer, A.; van der Merwe, G.; Li, Y.X.; Scaman, C.H. Overexpression, Purification, and Partial Characterization of Saccharomyces Cerevisiae Processing Alpha Glucosidase I. *Glycobiology* **2002**, *12*, 229–234. [CrossRef]

52. Pluskal, T.; Castillo, S.; Villar-briones, A.; Ore, M. MZmine 2: Modular Framework for Processing, Visualizing, and Analyzing Mass Spectrometry- Based Molecular Profile Data. *BMC Bioinform.* **2010**, *11*, 1–11. [CrossRef] [PubMed]

53. Han, L.; Fang, C.; Zhu, R.; Peng, Q.; Li, D.; Wang, M. International Journal of Biological Macromolecules Inhibitory Effect of Phloretin on Glucosidase: Kinetics, Interaction Mechanism and Molecular Docking. *Int. J. Biol. Macromol.* **2017**, *95*, 520–527. [CrossRef] [PubMed]

54. Imran, S.; Taha, M.; Ismail, H.; Kashif, S.M.; Rahim, F.; Jamil, W.; Wahab, H.; Khan, K.M. Synthesis, In Vitro and Docking Studies of New Flavone Ethers as α-Glucosidase Inhibitors. *Chem. Biol. Drug Des.* **2016**, *87*, 361–373. [CrossRef] [PubMed]

55. Taha, M.; Adnan, S.; Shah, A.; Afifi, M.; Imran, S.; Sultan, S.; Rahim, F.; Mohammed, K. Synthesis, α-Glucosidase Inhibition and Molecular Docking Study of Coumarin Based Derivatives. *Bioorg. Chem.* **2018**, *77*, 586–592. [CrossRef] [PubMed]

56. Lestari, W.; Dewi, R.T.; Broto, L.; Kardono, S.; Yanuar, A. Docking Sulochrin and Its Derivative as α-Glucosidase Inhibitors of Saccharomyces Cerevisiae. *Indones. J. Chem.* **2017**, *17*, 144–150. [CrossRef]

57. Yan, J.; Zhang, G.; Pan, J.; Wang, Y. α-Glucosidase Inhibition by Luteolin: Kinetics, Interaction and Molecular Docking. *Int. J. Biol. Macromol.* **2014**, *64*, 213–223. [CrossRef] [PubMed]

58. Leong, S.W.; Awin, T.; Munirah, S.; Faudzi, M.; Shaari, M.M.K.; Abas, F. Synthesis and Biological Evaluation of Asymmetrical Diarylpentanoids as Antiinflammatory, Anti-α-Glucosidase, and Antioxidant Agents. *Med. Chem. Res.* **2019**, *28*, 2002–2009. [CrossRef]

59. Dolinsky, T.J.; Nielsen, J.E.; Mccammon, J.A.; Baker, N.A. PDB2PQR: An Automated Pipeline for the Setup of Poisson—Boltzmann Electrostatics Calculations. *Nucleic Acids Res.* **2004**, *32*, 665–667. [CrossRef]

Sample Availability: Herbal samples are kept in Kulliyyah of Pharmacy's herbarium IIUM Kuantan, and available from the authors.

Publisher's Note: MDPI stays neutral with regard to jurisdictional claims in published maps and institutional affiliations.

Article

Berberine Inhibits Telomerase Activity and Induces Cell Cycle Arrest and Telomere Erosion in Colorectal Cancer Cell Line, HCT 116

Muhammad Azizan Samad [1], **Mohd Zuwairi Saiman** [1,2], **Nazia Abdul Majid** [1,*], **Saiful Anuar Karsani** [1] **and Jamilah Syafawati Yaacob** [1,2,*]

[1] Institute of Biological Sciences, Faculty of Science, Universiti Malaya, Kuala Lumpur 50603, Malaysia; azizan_12@siswa.um.edu.my (M.A.S.); zuwairi@um.edu.my (M.Z.S.); saiful72@um.edu.my (S.A.K.)

[2] Centre for Research in Biotechnology for Agriculture (CEBAR), Faculty of Science, Universiti Malaya, Kuala Lumpur 50603, Malaysia

* Correspondence: nazia@um.edu.my (N.A.M.); jamilahsyafawati@um.edu.my (J.S.Y.); Tel.: +60-3-7967-5833 (N.A.M.); +60-3-7967-4090 (J.S.Y.)

Abstract: Colorectal cancer (CRC) is the most common cancer among males and females, which is associated with the increment of telomerase level and activity. Some plant-derived compounds are telomerase inhibitors that have the potential to decrease telomerase activity and/or level in various cancer cell lines. Unfortunately, a deeper understanding of the effects of telomerase inhibitor compound(s) on CRC cells is still lacking. Therefore, in this study, the aspects of telomerase inhibitors on a CRC cell line (HCT 116) were investigated. Screening on HCT 116 at 48 h showed that berberine (10.30 ± 0.89 μg/mL) is the most effective (lowest IC_{50} value) telomerase inhibitor compared to boldine (37.87 ± 3.12 μg/mL) and silymarin (>200 μg/mL). Further analyses exhibited that berberine treatment caused G_0/G_1 phase arrest at 48 h due to high cyclin D1 (CCND1) and low cyclin-dependent kinase 4 (CDK4) protein and mRNA levels, simultaneous downregulation of human telomerase reverse transcriptase (*TERT*) mRNA and human telomerase RNA component (*TERC*) levels, as well as a decrease in the TERT protein level and telomerase activity. The effect of berberine treatment on the cell cycle was time dependent as it resulted in a delayed cell cycle and doubling time by 2.18-fold. Telomerase activity and level was significantly decreased, and telomere erosion followed suit. In summary, our findings suggested that berberine could decrease telomerase activity and level of HCT 116, which in turn inhibits the proliferative ability of the cells.

Keywords: colorectal cancer; HCT 116; telomerase; telomerase inhibitor; berberine; downregulation; cell cycle arrest; telomere erosion

Citation: Samad, M.A.; Saiman, M.Z.; Abdul Majid, N.; Karsani, S.A.; Yaacob, J.S. Berberine Inhibits Telomerase Activity and Induces Cell Cycle Arrest and Telomere Erosion in Colorectal Cancer Cell Line, HCT 116. *Molecules* **2021**, *26*, 376. https://doi.org/10.3390/molecules26020376

Received: 15 October 2020
Accepted: 18 December 2020
Published: 13 January 2021

Publisher's Note: MDPI stays neutral with regard to jurisdictional claims in published maps and institutional affiliations.

1. Introduction

Colorectal cancer (CRC) (10.2% incidence) is the third most common cancer among males and females worldwide after lung (11.6% incidence) and breast cancers (11.6% incidence) [1]. CRC is often associated with an elevated telomerase level [2] and activity [3]. Telomerase is an enzyme responsible for adding the repetitive sequence (TTAGGG) at telomeric ends for telomere maintenance in humans [4]. Telomerase is a prognostic marker of CRC progression [3,5]. Patients with high telomerase activity have poorer prognosis and disease-free survival rates compared to those with low and moderate telomerase activity.

It has been previously suggested that human telomerase consists of six subunits, namely the human telomerase reverse transcriptase (hTERT or TERT), human telomerase RNA component (*hTR* or *TERC*), telomerase-associated protein 1 (TEP1), heat shock protein 90 (HSP90), prostaglandin E synthase 3 (P23), and dyskerin (DKC) [6]. However, recent review articles discussed that telomerase, which is synthesized during the S phase, has more than just six subunits [7]. It has been shown that telomerase activity changed proportionally with the level of TERT but not with the other five subunits [6]. This could be due to the fact that TERT is the catalytic subunit of telomerase [8]. *TERC* is a functional RNA sequence that is responsible for providing a template for 3′TTAGGG5′ addition at telomere ends [4]. It was also stated that TERT and *TERC* are enough for telomerase to function [6]. Due to this reason, TERT and *TERC* serves as the prime target for telomerase level and activity inhibition. The most recent publication on the telomerase structure of the colorectal cancer cell line, HCT 116, surprisingly revealed that there were no TEP1, P23, and HSP90 subunits present [9]. This evidence agrees with the previous statement, whereby only TERT and *TERC* are required for telomerase to function. It might also suggest that telomerase could adopt different structures.

Since telomerase is the hallmark of CRC, targeting telomerase could be a potential treatment for telomerase-positive cancers, specifically CRC in this case. Many natural products have been listed as telomerase-targeting compounds or, in other words, telomerase inhibitors [10]. Boldine, silymarin, and berberine are some that are included in the list. Boldine is an alkaloid that could be found abundantly in the leaves of *Peumus boldus*, which has been used as a traditional medicine in Chile [11]. Silymarin is a collective name of seven flavonolignans, namely silydianin, isosilchristin, silychristin, silibinin (or silybin) A, isosilibinin (or isosilybin) A, silibinin B, and isosilibinin B, as well as one flavonoid named taxifolin [12,13], which is present in *Silybum marianum* (milk thistle) [14]. On the other hand, berberine is an alkaloid that can be found in the outer bark, roots, and rhizhome of *Berberis* spp, such as *B. aristata*, *B. aquafolium*, and *B. vulgaris* [15]. Besides that, berberine could also be extracted from *Tinospora* spp, such as *T. sinensis* and *T. cordifolia* [16]. These compounds have been reported to be able to downregulate telomerase activity and/or level in various cancer cell lines [10].

Telomerase is highly expressed in proliferative cells, such as cancer, stem, and gamete cells, but not in normal cells [17]. However, telomerase activity is higher in cancer cells compared to most stem cells except for embryonic stem cells [18]. It is challenging to determine whether telomerase inhibition will affect gamete and stem cells as cancer cells are affected; however, due to the shorter length of telomere as well as the difference of the metabolism in cancer cells, it would have undergone apoptosis even before it could adversely affect germ line and stem cells [17]. The aim of this study was to investigate the effects of telomerase inhibitor compound treatment on the cell cycle, doubling time (t_d), telomerase activity, and level as well as on the telomere length of a colorectal cancer cell line, using HCT 116 as a model.

2. Results

2.1. Effect of Different Time Points on the Percentage of Cell Cycle Distribution and Relative Telomerase Activity (RTA)

It has been hypothesized that telomerase activity peaked during S phase [19,20], thus it was necessary to determine the time point that possessed the highest percentage of S phase. Cell cycle analysis was done on HCT 116 cells at different time points to determine the percentage of cell cycle phases (Figure 1). Here, 72 h exhibited the highest percentage of G_0/G_1 phase ($35.13 \pm 0.25\%$) compared to 24 h and 48 h ($27.17 \pm 5.14\%$ and $24.63 \pm 0.67\%$, respectively). Based on the results, the percentage of S phase was significantly highest at 48 h ($50.10 \pm 1.15\%$), followed by 24 h ($43.43 \pm 1.59\%$) and 72 h ($36.70 \pm 0.75\%$). The RTA of HCT 116 indeed peaked at 48 h ($114.26 \pm 1.26\%$) followed by 24 h ($98.69 \pm 0.42\%$) and 72 h ($90.24 \pm 0.91\%$). All three time points gave an insignificant percentage of G_2/M phase. Thus, 48 h was chosen for the treatment duration for subsequent experiments since it showed the highest percentage of S phase and RTA.

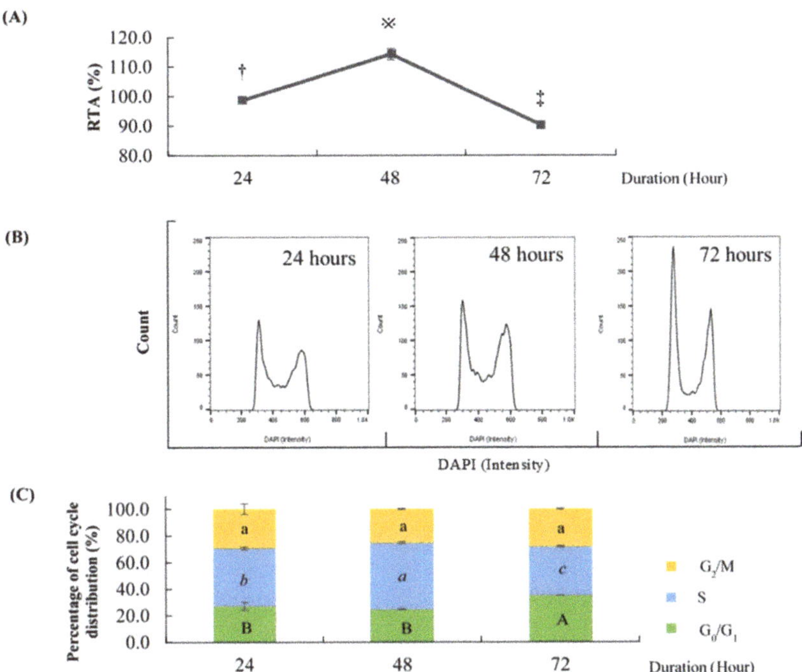

Figure 1. Effects of different time points on relative telomerase activity (RTA) and percentage of HCT 116 cell cycle distribution. (**A**) Telomerase activity of HCT 116 harvested at different time points relative to telomerase-positive control (HEK293). Values are mean ± standard error of triplicate. Mean with different symbols (†, ✻, or ‡) differ significantly at $p < 0.001$. (**B**) Representative cell cycle histogram of HCT 116 harvested at different time points stained with DAPI and analyzed by flow cytometry. (**C**) Percentage of cell cycle distribution of HCT 116 at different time points analyzed by FlowJo version 10. Values are mean ± standard error of triplicate. Mean with different letters differ significantly at $p < 0.05$ (subject to different cell cycle phases). G: Growth, S: Synthesis, M: Mitosis.

2.2. Effect of Different Telomerase Inhibitor Compounds on IC_{50}

HCT 116 cultured in 96-well plates treated with different concentrations of boldine, silymarin, or berberine for 48 h were subjected to SRB assay to determine the percentage of inhibition and the results were interpreted as IC_{50} (compound concentration required to inhibit HCT 116 by 50%). Berberine was found to be the most effective, as it showed the lowest IC_{50} compared to boldine and silymarin (Table 1). Hence, berberine was chosen to be used for subsequent analysis.

Table 1. Effect of different telomerase inhibitor compound treatments on IC_{50}. Values are mean ± standard error of 6 replicates. Mean with different letters differ significantly at $p < 0.05$.

Compound	Boldine	Silymarin	Berberine
IC_{50} (μg/mL)	37.87 ± 3.12 [b]	>200 [c]	10.30 ± 0.89 [a]

2.3. Effect of Berberine Treatment on Cell Cycle Distribution

The cells were treated with berberine for 48 h and were analyzed for cell cycle distribution. Treatment with berberine was observed to cause a cell cycle arrest at G_0/G_1 phase, as shown by the significant increase in the number of cells in G_0/G_1 compared to that of untreated sample (Figure 2). The cells in S phase remained unchanged while the cells also did not proceed to G_2/M, as indicated by a significant decrement of G_2/M.

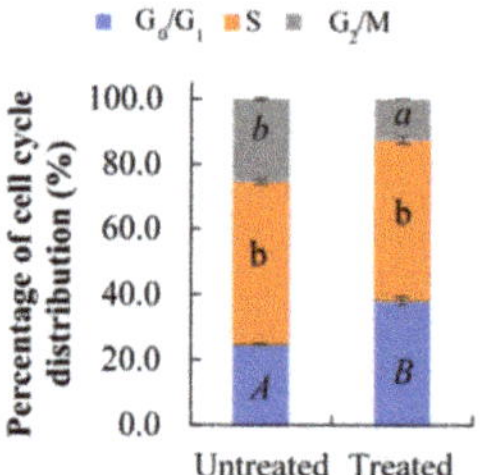

Figure 2. Effect of 10.54 µg/mL berberine treated HCT 116 at 48 h on the percentage of cell cycle distribution. Values are mean ± standard error of triplicate. Mean with different letters differ significantly at $p < 0.001$ (subject to different cell cycle phases).

The cells were also exposed to berberine at various time points, to determine if the effects observed were time dependent. Based on Figure 3, the G_0/G_1 percentage started to increase gradually after 12 h and became constant after 48 h. The percentage of cells in S phase significantly decreased from 0 to 6 h after treatment, then started to increase again from 12 to 24 h, then became stable at 48 h followed by a slight decrement at 72 h; however, the percentages of S phase at 24 and 72 h were not significantly different. G_2/M started to increase from 0 to 6 h after treatment, then decreased from 12 to 24 h, stabilized at 48 h, and then slightly increased at 72 h. This scenario showed that the effect of berberine on cell cycle was time dependent.

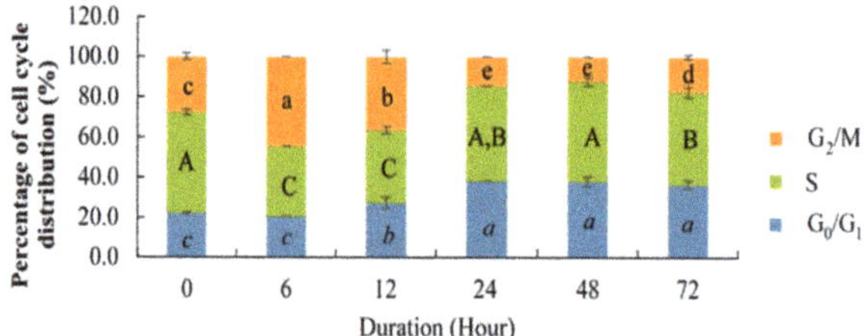

Figure 3. Changes in the percentage of HCT 116 cells in G_0/G_1, S, and G_2/M phases at different durations following treatment with 10.54 µg/mL berberine. Values are mean ± standard error of triplicate. Mean with different letters differ significantly at $p < 0.001$ (subject to different cell cycle phases).

2.4. Effect of Berberine Treatment on Cell Growth and Berberine Cellular Localization

The time-dependent effect of berberine on HCT 116 could also be observed microscopically. Based on Table 2, it was observed that the cell growth was inhibited by berberine treatment. The difference in growth could be clearly observed at 48 and 72 h. Figure 4A also showed that berberine treatment caused the doubling time (t_d) of HCT 116 to be delayed. The green fluorescence intensity observed under the UV channel peaked at 24 h, then decreased at 72 h (see Figure 4B). It was also noted that at 24 h, the nuclei appeared as black round hollows surrounded by green fluorescence, which potentially indicates that berberine had permeated and accumulated in the cytoplasm. The black hollows were no longer observed after 48 and 72 h, suggesting that berberine had permeated and accumulated in the nuclei. Untreated HCT 116 did not fluoresce at all under the UV channel and the image only appeared as black.

Table 2. Effect of 10.54 µg/mL berberine treatment on HCT 116 growth at different durations. Berberine emitted green fluorescence under the ultraviolet (UV) channel of an inverted microscope. White arrows indicate nuclei appeared as black, round hollows surrounded by green fluorescence. Black scale bar = 100 µm. Yellow scale bar = 20 µm.

Duration (Hour)	Cell Growth		Berberine Cellular Localization		
	Untreated	Treated	Phase Contrast	UV Channel	Merged
0					
6					
12					
24					
48					
72					

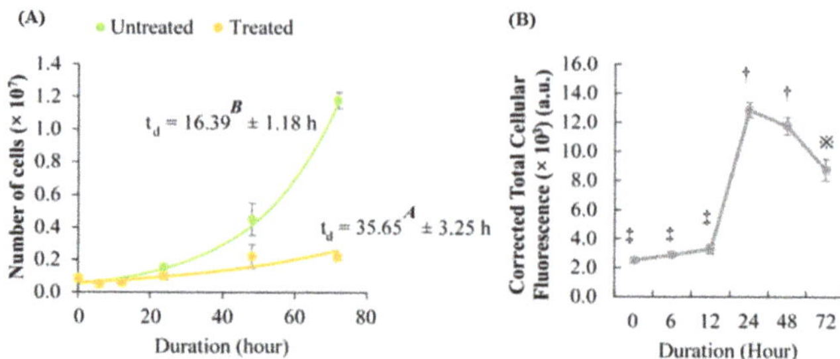

Figure 4. (**A**) Growth curve of untreated and 10.54 μg/mL berberine-treated HCT 116. Values are mean ± standard error of biological quadruplicate. t_d: doubling time. (**B**) Corrected Total Cellular Fluorescence (CTCF) of berberine-treated HCT 116 at different durations. Values are mean ± standard error of biological triplicate. a.u.—arbitrary unit. Mean with different letters or symbols (‡, †, or ※) differ significantly at $p < 0.001$.

2.5. Effect of Berberine Treatment on Relative Telomerase Activity (RTA)

Berberine-treated HCT 116 at 48 h were analyzed by using *TELOTAGGG* Telomerase PCR ELISA. The telomerase activity was observed to significantly decrease with berberine treatment (Figure 5). Compared to untreated HCT 116, the RTA in berberine-treated HCT 116 was decreased by 63.22%. To evaluate cell-free telomerase inhibition, berberine was added into the TRAP reaction mixture to the same final concentration and the results showed that berberine only decreased the RTA by 16.10% (see Figure A1). This might suggest that the inhibition of telomerase–TTAGGG interaction was not the only reason for the decrement in the RTA, whereby the level of telomerase might be affected, which caused the RTA to decrease even further. As such, we are interested in investigating the effect of berberine on TERT and *TERC* levels.

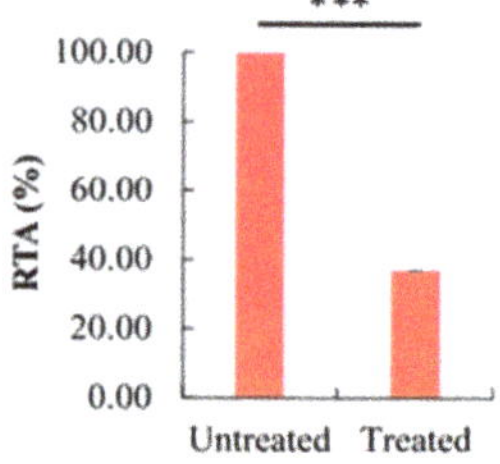

Figure 5. Effect of 10.54 μg/mL berberine-treated HCT 116 at 48 h on relative telomerase activity (RTA). Values are mean ± standard error of triplicate. Means with *** (triple asterisks) differ significantly at $p < 0.001$.

2.6. Effect of Berberine Treatment on CCDN1, CDK4, TERT, and TERC RNA Levels

In cell cycle progression, CCND1 and CDK4 are important regulators for G_1 to S phase transition, whereby the binding of CCND1 to CDK4 is a rate-limiting event [21]. Since G_0/G_1 phase arrest was observed (Figure 2), we are keen to know the effect of berberine on the RNA levels of these regulators. Based on the RT-PCR results (Figure 6), surprisingly, 10.54 μg/mL berberine treatment caused *CCND1* level to be upregulated by 1.71-fold. Besides that, *CDK4*, *TERT*, and *TERC* levels were downregulated by 1.26-, 108.15-, and 2.80-fold, respectively.

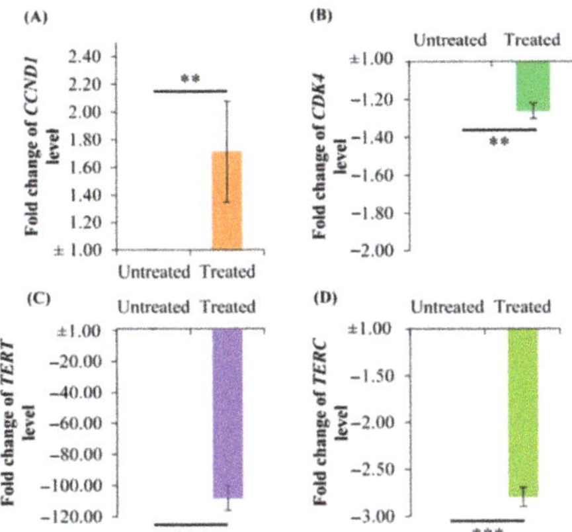

Figure 6. Effect of 10.54 µg/mL berberine-treated HCT 116 at 48 h on (**A**) *CCND1*, (**B**) *CDK4*, (**C**) *TERT*, and (**D**) *TERC* levels. Changes in RNA levels are expressed as fold change in RNA levels relative to untreated HCT 116 at 48 h. Values are mean ± standard error of quadruplicate. Mean with ** (double asterisks) or *** (triple asterisks) differ significantly at $p < 0.01$ or $p < 0.001$, respectively.

2.7. Effect of Berberine Treatment on CCND1, CDK4, and TERT Protein Levels

Parallel to the downregulation of *CDK4* and *TERT* mRNA levels, validation by Western blot results (Figure 7) also showed that berberine treatment at 48 h caused CDK4 and TERT protein levels to decrease significantly by 45.16% and 73.29%, respectively. Besides this, the CCND1 protein level increased significantly by 95.77%.

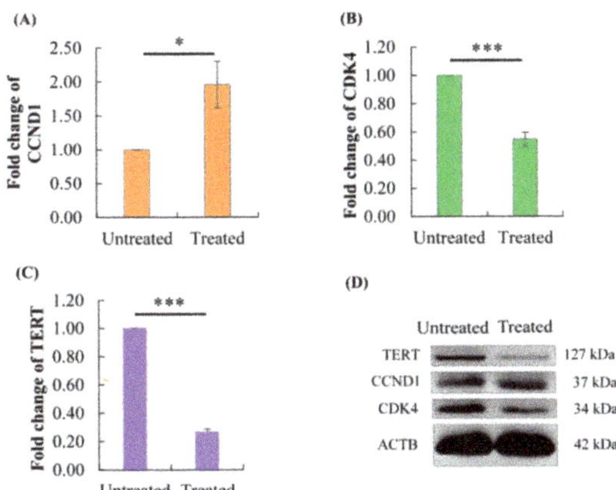

Figure 7. Effect of 10.54 µg/mL berberine-treated HCT 116 at 48 h on (**A**) CCND1, (**B**) CDK4, and (**C**) TERT protein levels. Changes in CCND1, CDK4, and TERT protein levels are expressed as fold change in CCND1, CDK4, and TERT signal intensities, respectively, normalized to ACTB relative to untreated HCT 116 at 48 h. (**D**) Representative Western blot images of TERT, CCND1, CDK4, and ACTB. Values are mean ± standard error of triplicate. Mean with * (single asterisk) or *** (triple asterisks) differ significantly at $p < 0.05$ or $p < 0.001$, respectively.

2.8. Effect of Berberine Treatment on Relative Telomere Length

Consistent with the significantly impaired telomerase activity and level by berberine treatment, telomere attrition also occurred as anticipated (Figure 8), whereby the relative telomere length of berberine-treated HCT 116 after 48 h was 0.81 ± 0.08, compared to untreated HCT 116 (with a relative telomere length of 1.0).

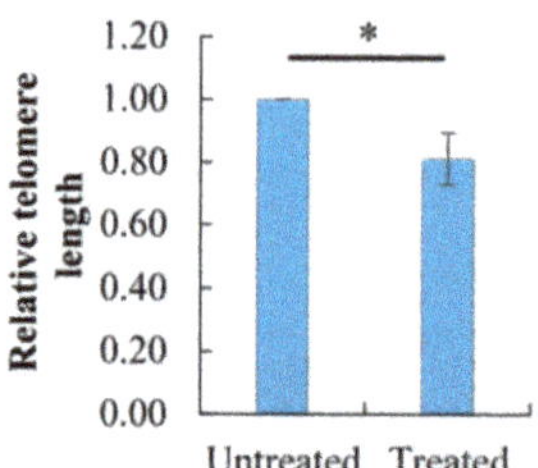

Figure 8. Effect of 10.54 µg/mL berberine-treated HCT 116 at 48 h on the relative telomere length. Changes in telomere length are expressed as fold change relative to untreated HCT 116 at 48 h. Values are mean ± standard error of octuplicate. Mean with * (single asterisk) differ significantly at $p < 0.05$.

3. Discussion

The cytotoxicity of drugs could be classified into cytotoxic (less than 2 µg/mL), moderately cytotoxic (between 2 to 89 µg/mL), and not cytotoxic (more or equal to 90 µg/mL) based on their IC_{50} values [22]. In the present study, berberine and boldine were found to be moderately cytotoxic, while silymarin was not cytotoxic to HCT 116 after 48 h of treatment. Since silymarin did not reach IC_{50}, the value was reported as >200 µg/mL, which was greater than the highest tested concentration as suggested by the National Cancer Institute [23].

In several telomerase-positive cancer cell lines, berberine treatment has been shown to cause G_0/G_1 phase arrest in MCF-7 (breast cancer) after 48 h [24], Huh-7 and HEPG2 (liver cancers) in a dose-dependent manner after 24 h [25], and A2780 (ovarian cancer) after 48 h of treatment [26]. Either too high or too low CCND1 levels would compromise cell cycle progression [27]. In the present study, it was revealed that an abnormal level of cell cycle regulators, in this case high CCND1 levels and low CDK4 levels, would lead to G_0/G_1 phase arrest in HCT 116. Since the binding of CCND1 to CDK4 is a rate-limiting event in G_1 to S phase transition [21], theoretically, the effect of berberine on G_0/G_1 arrest would depend mainly on these two proteins instead of other cyclins and CDKs. Besides that, telomerase is mostly active during S phase, hence the reason the study was focused more on CCND1 and CDK4 instead of other cyclins and CDKs. Due to this reason, the experiments were designed in such flow, so that the study was more focused to the factors that are associated within the telomerase aspect.

Wang, et al. [28] reported that G_0/G_1 arrest in U87 and LN229 (brain cancer) was caused by the downregulation of *miR-21*, which caused a decrement of STAT3 phosphorylation and protein level, thus causing the TERT protein level to decrease. In a separate study, berberine showed the ability to repress the *miR-21* level in oral squamous cell carcinoma cancer stem cell lines (OECM-1 and SAS), which decreased aldehyde dehydrogenase 1 (ALDH1) activity, cell migration, invasion capability, and self-renewal [29]. Moreover, G_0/G_1 phase arrest in HL-60 (leukemia) is also associated with a decrement of telomerase activity as well as *TERT* and *TERC* levels [30]. It was also previously shown that berberine treatment for 24 h induced G_0/G_1 phase arrest in HCT 116 in a dose-dependent manner [31]. Based on these previous findings, it would be fair to remark that the inhibition of cell proliferation (Figure 4) and G_0/G_1 phase arrest in HCT 116 induced by berberine treatment were not exclusively caused by the downregulation of TERT.

Based on various references, the average duration of untreated HCT 116 cell cycle was about 17 h (16.88 ± 2.56 h) [32–36]. Therefore, theoretically, the second round of HCT 116 cell cycle would take place approximately after 34 h. Based on Figure 4A, the average doubling time (t_d) for untreated HCT 116 was 16.39 ± 2.36 h, similar to the average t_d

of previous findings. The average t_d of berberine-treated HCT 116 was 35.65 ± 6.50 h, significantly higher than the average t_d of untreated HCT 116. Data analysis revealed that the t_d of berberine-treated HCT 116 was overdue by approximately two-fold (2.18-fold).

The localization of berberine in nuclei after 48 h further strengthens our justification of choosing this time point for further investigation. This is due to the fact that the transcription, synthesis, and activity of telomerase mainly takes place in the nucleus [7,37]. Since we are targeting the telomerase, berberine should be present in the nucleus for telomerase inhibition to occur. Here, we have demonstrated that berberine cellular localization is time dependent. Similarly, berberine cellular localization has also been shown to be concentration [38] and cell line dependent [39].

It was previously established that berberine treatment on HCT 116 inhibits cell proliferation in a dose- and time-dependent manner [31,40], and induced apoptosis in a dose-dependent manner at 24 [40] and 72 h [31]. Downregulation of TERT levels has been associated with a decrement in telomerase activity [19,30,41] and vice versa [42]. Berberine intervenes in the function of telomerase through binding to the telomeric G-quadruplex structure [43,44], thus preventing the interaction between telomerase and telomere [45]. It was revealed in previous studies that berberine had the potential to regulate gene transcription [46,47]. Other than that, the decrement of the TERT protein level was often associated with downregulation of *TERT* mRNA [28] and vice versa [42]. Berberine could downregulate the TERT protein level in the non-small-cell lung cancer cell line, A549 [48], and cervical cancer cell lines, SiHa and HeLa [15]. Previous studies mostly focused only on the potential of berberine in downregulation of the TERT level; however, in the present study, we demonstrated that berberine could also downregulate the *TERC* level.

Berberine has been shown to decrease the telomerase level and activity, which is significantly followed by telomere erosion. Interestingly, by conducting telomere restriction fragment (TRF) analysis, Xiong et al. [49] revealed that prolonged exposure to berberine for 16 days did not significantly cause telomere shortening in SiHa and HL-60. This could possibly be due to the limitation of TRF, which could potentially overestimate the telomeric length compared to RT-PCR, which is more sensitive and specific [50]. Other than the potential to regulate gene transcription, berberine could interact directly with the telomeric G-quadruplex, whereby the interaction occurred between the negatively charged oxygen of guanine and positively charged nitrogen of berberine [51] with a molar ratio of 2:1 (berberine to G-tetrad) [52], as it stabilized the structure by increasing the melting temperature and as a result, telomerase activity was inhibited [53]. Berberine has been extensively studied in various cancer cell lines; however, in animal models, evidence to support the effects of berberine as a telomerase inhibitor on colorectal cancer is still lacking [54]. Berberine is effective at inhibiting polyps formation [55] and colon tumorigenesis in mice [40], however, currently, there is no in vivo study that validates berberine's presence in the nucleus of in vivo cancer cells of mice after oral administration of berberine. Since berberine is currently used in clinical trials (NCT03281096 and NCT03333265), there is an urge to elucidate berberine mechanisms as an anti-colorectal cancer drug.

In this study, cells from four flasks (biological replicates) were pooled to obtain sufficient material for several different analyses. All extractions and preparations were also performed from the same set of samples for the sake of uniformity. Although pooling of samples may rightfully raise some concerns, it is still widely viewed as a valid method, saves cost, and reduces biological variance, while not seriously affecting the results when compared to un-pooled samples [56–58]. Biological replicates and independent experiments are important to validate the effect of any anticancer drug due to the existence of heterogeneity the within cancer cell population. It was fully acknowledged that the lack of an independent experiment is a limitation in the current study. Though the potential mechanism of telomerase inhibition by berberine using HCT 116 as a model has been shown in the current study, further deeper understanding is vital to elucidate the potential of berberine as a telomerase inhibitor in CRC.

4. Materials and Methods

4.1. Reagents, Primers, and Antibodies

Berberine was diluted in autoclaved Milli-Q® water and sterilized by filtration through a Minisart® 0.22 μm polyethersulfone filter syringe (Sartorius Stedim Biotech, Göttingen, Germany). Boldine and silymarin was dissolved in dimethylsulfoxide (DMSO) and filtered as well. All primers were synthesized by Integrated DNA Technologies (IDT, Coralville, IA, USA). Monoclonal antibodies of CCND1 (A-12), CDK4 (DCS-31), TERT (A-6), and β-actin (ACTB) (AC-15) as well as mouse IgG kappa binding protein conjugated to horseradish peroxidase (m-IgGκ BP-HRP) were obtained from Santa Cruz Biotechnology (Santa Cruz, Dallas, TX, USA). Molecular-grade absolute ethanol (Merck, Darmstadt, Germany) was diluted with autoclaved Milli-Q® water. All reagents were from Sigma-Aldrich unless mentioned otherwise.

4.2. Cell Culture and Harvesting

Colorectal cancer cell line, HCT 116, was cultured in Roswell Memorial Institute (RPMI) medium 1640 (Nacalai Tesque, Kyoto, Japan) supplemented with 10% fetal bovine serum (Tico Europe Ltd., Amstelveen, The Netherlands), 1% Antibiotic-Antimycotic solution (Nacalai Tesque, Kyoto, Japan), maintained at 37 °C in a humidified 5% CO_2-enriched atmosphere (ESCO, Horsham, PA, USA). Cells were rinsed by using $1\times$ phosphate-buffered saline (PBS) (Nacalai Tesque, Kyoto, Japan), repeated three times, and cells were detached by using trypsin-EDTA (Life Technologies, Burlington, Canada), followed by RPMI addition to stop the trypsinization; cells were pelleted by centrifugation at 1000 rpm, resuspended in RPMI for sub-culturing or seeding purpose, or $1\times$ PBS was used to resuspend for harvesting as a washing step, followed by second-time pelleting of the cells. During harvesting, after observation under the microscope and cell count was conducted, cells from four flasks were pooled and pelleted in the amount as required for further analyses. The microcentrifuge tubes containing a pelleted known number of cells were immediately flash frozen in liquid nitrogen before storage in -80 °C until further usage.

4.3. Cell Cycle Analysis

In total, 1×10^6 pelleted cells in 2 mL microcentrifuge tube were washed with 1 mL of sample buffer ($1\times$ PBS containing 1 g/L glucose) once by vortexing briefly, followed by centrifugation at $17,000\times g$ for 5 min. The supernatant was removed and about 0.1 mL was left in the tube. Then, 1 mL of 70% ethanol (-20 °C) was slowly added, 100 μL at a time, drop by drop while vortexing. The suspension was stored at 4 °C for 24 h. After fixation, 1 mL of sample buffer was added while vortexing, followed by centrifugation at $17,000\times g$ for 10 min. Staining solution made up of 0.5 μg/mL 4′,6 -diamidino-2-phenylindole (DAPI) (Miltenyi Biotec, Bergisch Gladbach, Germany), 5 μg/mL RNase A, and 0.1% Tween 20 (Classic Chemicals, Shah Alam, Selangor, Malaysia) in $1\times$ PBS was prepared immediately before incubation. The supernatant was removed, and 1 mL of staining solution was added followed by vortexing and incubation at room temperature for 40 min. Flow cytometry was done by using a MACSQuant® Analyzer (Miltenyi Biotec, Bergisch Gladbach, Germany) following the protocol suggested by Miltenyi Biotec with optimizations. The fluorescence channel was set to V1 and measured by linear acquisition. The V1 voltage was set to 334 V and the trigger was set to 128.00. The number of events was set to 10,000 events for each replicate. The data was analyzed by using FlowJo version 10.

4.4. Sulforhodamine B (SRB) Assay

In total, 5000 cells/100 μL per well were seeded in 96-well plates, incubated at 37 °C in a humidified 5% CO_2-enriched incubator for 24 h before compound treatments. The spent medium was pipetted out and replaced with 150 μL of RPMI containing either boldine, silymarin, or berberine at various concentrations (0.20, 0.39, 0.78, 1.56, 3.13, 6.25, 12.50, 25.00, 50.00, 100.00, and 200.00 μg/mL) achieved by serial dilutions as well as RPMI without any of the named compounds as untreated (control). The treatments were done for 48 h.

After 48 h, each plate was subjected to Sulforhodamine B (SRB) assay [59]. Total protein content of viable cells was assessed in this assay. The plate was shaken on an ELISA microplate reader (Tecan, Männedorf, Switzerland) for 5 min and the absorbance was read at 492 nm. The percentage of inhibition was calculated as the following:

$$\text{Percentage of inhibition (\%)} = \frac{\text{Abs}_{\text{untreated}} - \text{Abs}_{\text{treated}}}{\text{Abs}_{\text{untreated}}} \times 100\% \tag{1}$$

where $\text{Abs}_{\text{untreated}}$ = Absorbance of untreated cells at 492 nm; and $\text{Abs}_{\text{treated}}$ = Absorbance of berberine treated cells at 492 nm.

Graphs of the percentage of inhibition against the concentration were plotted by using GraphPad Prism 7.00 by performing the 4-parameter logistic model (4PL) [60] and the IC_{50} values was determined by using the software. The cells were also cultured in T75 flasks to scale-up the production of HCT116 cells for subsequent analysis and were also used to determine a new IC_{50} value (for berberine).

4.5. Cell Count and Growth Curve Analysis

Cell count was done by using Trypan blue (TB) exclusion assay. In total, 100 µL of the cell suspension were pipetted into a microcentrifuge tube and 100 µL of Trypan blue (Nacalai Tesque, Kyoto, Japan) were gently mixed by pipetting in and out. The Trypan blue-dyed cells were then pipetted to fill up both chambers of a hemocytometer (Electron Microscopy Sciences, Hatfield, PA, USA), followed by observation and cell counting under an inverted microscope. Dead cells were stained blue while living cells were not stained. The average number of cells at each time point was used to plot the exponential growth curve following this equation:

$$y = y_0 e^{\lambda t} \tag{2}$$

where y = Final number of cells; y_0 = Initial number of cells; e = Euler's number; λ = Growth rate; t = Time that has passed; and the t_d was calculated using the following equation:

$$t_d = \ln 2 / \lambda \tag{3}$$

4.6. Berberine Cellular Localization

Cells were observed by using a Leica DMI6000 B inverted microscope (Leica Microsystems, Wetzlar, Germany) under the phase contrast channel. Without changing the position, the channel was changed to UV to observe berberine fluorescence. To avoid cellular damage by UV radiation, the exposure was kept as minimal as possible, whereby the channel was shut off within 30 s (or less) after capturing the image by using a Digital Color Camera Leica DFC310 FX (Leica Microsystems, Wetzlar, Germany). The corrected total cellular fluorescence (CTCF) was determined by using a free software, ImageJ version 1.52a [61], by using the following equation:

$$\text{CTCF} = \text{IntDen}_{\text{cell}} - \left(A_{\text{cell}} \times \overline{X}_{\text{background}} \right) \tag{4}$$

where $\text{IntDen}_{\text{cell}}$ = Integrated density of selected cell; A_{cell} = Area of selected cell; and $\overline{X}_{\text{background}}$ = Average background fluorescence reading.

4.7. TELOTAGGG Telomerase PCR ELISA

Protein extraction was done based on the protocol of the *TELOTAGGG* Telomerase PCR ELISA kit (Roche, Mannheim, Germany) with minor adjustments, whereby 5×10^5 cells were used. The protein content was assessed by Bradford Assay (Bio-Rad Laboratories, Hercules, CA, USA) as suggested by the Bio-Rad protocol. The Telomeric Repeat Amplification Protocol (TRAP) reaction was done using a thermal cycler, StepOnePlus™ Real Time-Polymerase Chain Reaction (RT-PCR) Systems (Applied Biosystems, Foster City, CA, USA). In total, 2 µg of lysate were used in this reaction.

4.8. RT-PCR

RNA extraction was conducted using the protocol of the ReliaPrep™ RNA Cell Miniprep System (Promega, Madison, WI, USA) with minor adjustments, whereby 1×10^6 cells harvested were lysed and DNase I treatment was done for 1 h 30 min at 30 °C. RNA extracted was quantified by using a NanoDrop™ 2000 (Thermo Fischer Scientific, Waltham, MA, USA). The purity was also assessed using the same instrument based on the A260/280 and A260/230 ratios. RNA integrity was evaluated using 1% native agarose gel electrophoresis containing 5 µL of RedSafe™ (iNtRON Biotechnology, Kirkland, WA, USA) conducted in $1\times$ Tris-Borate-EDTA (TBE) (1st BASE Biochemicals, Singapore Science Park II, Singapore) buffer at 70 V until the tracking dye travelled three-quarters of the gel. Gel picture was visualized, and the image was captured using Gel Documentation System Fusion FX7-7026 (Vilber Lourmat, Marne-la-Vallée, France).

The RNA is said to be intact if the band intensity (28S:18S ratio) is 2:1. Complementary DNA (cDNA) was synthesized using the protocol of the GoTaq® 2-Step RT-qPCR System (Promega, Madison, WI, USA) with minor adjustments, whereby 2500 ng of RNA were transcribed in 20 µL of total reaction mixture. In total, 250 ng of RNA equivalent cDNA were used in RT-PCR reaction mixture and 50 nM of primers (Table 3) were used for each gene. The reaction was conducted using StepOnePlus™ RT-PCR Systems (Applied Biosystems, Foster City, CA, USA) following the cycling conditions shown in Table 4. For no template control (NTC), the cDNA template was replaced with nuclease-free water. For no reverse transcriptase control (NORT), RNA mixture without GoScript™ Reverse Transcriptase, random, and oligo(dT)$_{15}$ primers was used in place of the cDNA template. *CCND1*, *CDK4*, *TERT*, and *TERC* levels were normalized to the geomean of glyceraldehyde-3-phosphate dehydrogenase (*GAPDH*) and *ACTB*.

Table 3. Primer sequences used in RT-PCR.

NCBI ID	Gene	Forward (5′ to 3′)	Reverse (5′ to 3′)	Amplicon Size (bp)
7015	*TERT*	ACTGCGTGCGTCGGTATGC	CGGCTGGAGGTCTGTCAAGGTA	97
7012	*TERC*	AGAGGAACGGAGCGAGTC	GCATGTGTGAGCCGAGTC	80
595	*CCND1*	AACACGGCTCACGCTTACC	GCCCCATCACGACAGACAAAG	94
1019	*CDK4*	ATGTGGAGTGTTGGCTGTATC	CTGGTCGGCTTCAGAGTTTC	78
2597	*GAPDH*	TTGGTATCGTGGAAGGACTCA	CCAGTAGAGGCAGGGATGAT	133
60	*ACTB*	CGTCTTCCCCTCCATCGT	GCCTCGTCGCCCACATAG	87

Table 4. Cycling conditions for RT-PCR.

Step	Cycles	Temperature (°C)	Duration
GoTaq® Hot Start Polymerase activation	1	95	2 min
Denaturation		95	15 s
Annealing	40	53.5	30 s
Extension		60	30 s
Dissociation	1	60 to 95	5 min
Hold		10	∞

4.9. Relative Telomere Length Quantification

Genomic DNA (gDNA) extraction was conducted using the protocol of the Quick-DNA™ Miniprep Plus Kit (Zymo Research, Irvine, CA, USA). In total, 1×10^6 harvested cells were used for the gDNA extraction. A NanoDrop™ 2000 was used for gDNA quantification and quality assessment. The quantification of the relative telomere length was done according to Vasilishina et al. [62] with minor adjustments, whereby 10 ng of gDNA and a GoTaq® 2-Step RT-qPCR System were used. The amplification was done for 40 cycles and conducted using StepOnePlus™ RT-PCR Systems. Single copy gene (SCG), interferon beta 1 (*IFNB1*) was used for normalization.

4.10. Western Blot

In total, 1×10^6 harvested cells were lysed in 60 µL Radio Immunoprecipitation Assay (RIPA) lysis buffer containing 1% protease inhibitor cocktail, sonicated on ice bath for 1 min, incubated for 30 min on ice, followed by centrifugation at $14,000 \times g$ for 15 min to pellet the cell debris, and subjected to protein estimation by employing Bradford Assay. Then, 20 µg of total protein denatured at 70 °C were separated on 4% and 12% SDS-polyacrylamide stacking and resolving gels, respectively. Electrophoresis was done at 80 V for 30 min, followed by 100 V for 1 h 40 min using pre-chilled (4 °C) $1\times$ SDS running buffer, followed by protein transferring onto 0.22-µm nitrocellulose membrane (Pall Corp., Morelos, Mexico). Electroblotting was conducted using a Mini-Protean® Tetra System (Bio-Rad Laboratories, Hercules, CA, USA) with pre-chilled (-20 °C) $1\times$ SDS transfer buffer and ice pack (-80 °C) for 1 h 40 min on ice. The membrane was removed and incubated in Blocking One (Nacalai Tesque, Kyoto, Japan) for 1 h at room temperature, on a shaker at 100 rpm. The membrane was incubated for at least 13 h in a cold room (5–10 °C) and was constantly shaken at 90 rpm in monoclonal primary antibody at 1:1000 dilution for ACTB and 1:100 dilution for CCND1, CDK4, and TERT, respectively, in Blocking One. The membrane was then washed 3 times with $1\times$ Tris-buffered saline containing 0.1% Tween 20 (TBST), 5 min each time, and on a shaker at 120 rpm. The membrane was incubated on a shaker at 90 rpm with m-IgGκ BP-HRP at 1:10,000 dilution in Blocking One for 2 h at room temperature. The $1\times$ TBST washing step was repeated as previously stated. The membranes were incubated in chemiluminescence solution of Western Bright Sirius (Advansta, San Jose, CA, USA) following the manufacturer's protocol before the membrane was visualized, and the image was captured using Gel Documentation System Fusion FX7-7026 (Vilber Lourmat, Marne-la-Vallée, France). ACTB was used as the loading control. Protein signal intensities were analyzed by using ImageJ version 1.52a. CCND1, CDK4, and TERT protein signal intensities were normalized to the loading control.

4.11. Statistical Analysis

All data obtained in this study were subjected to either *t*-test or one-way ANOVA. Statistical values were expressed as mean $\pm$ standard error. *p*-values < 0.05, < 0.01, and < 0.001 were considered as statistically significant, very statistically significant, and highly statistically significant, respectively. Mean values were compared using Duncan's Multiple Range Test (DMRT). Most statistical analyses were conducted using either Microsoft® Office Excel version 365 or IBM SPSS Statistics version 23.

5. Conclusions

In summary, our study demonstrated that berberine decreased the telomerase activity in HCT 116 by simultaneous downregulation of *TERT* and *TERC* levels, which resulted in a decrease of the TERT protein level. Berberine induced G_0/G_1 arrest by the upregulation of CCND1 and downregulation of CDK4, whereby the effect of berberine on the cell cycle was time dependent. Telomerase activity of HCT 116 was decreased by berberine treatment and the telomere length was also significantly eroded. Elucidation of berberine's potential as a telomerase inhibitor in CRC would require further extensive and elaborative efforts in the future.

Author Contributions: J.S.Y., N.A.M., M.A.S. and M.Z.S. conceived and designed the experiment(s), M.A.S. conducted the experiment(s) and wrote the manuscript, M.A.S., J.S.Y., M.Z.S. and N.A.M. analyzed the results. J.S.Y., S.A.K., N.A.M. and M.A.S. contributed reagents/materials/experimental tools. All authors have read and agreed to the published version of the manuscript.

Funding: This research was funded by University Malaya (Grant No. RP030C-15AFR, RU004C-2020 and CEBAR RU006-2018).

Acknowledgments: The authors thank Universiti Malaya, Malaysia for experimental facilities and financial support (Grant Number: RP030C-15AFR, RU004C-2020 and CEBAR RU006-2018) provided.

Conflicts of Interest: The authors declare no conflict of interest. The funders had no role in the design of the study; in the collection, analyses, or interpretation of data; in the writing of the manuscript, or in the decision to publish the results.

Sample Availability: Samples of the compounds are not available from the authors.

Appendix A

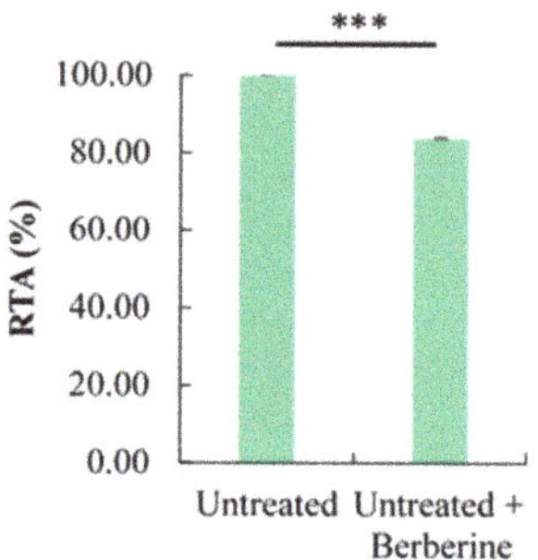

Figure A1. Cell-free telomerase inhibition by direct addition of berberine in the TRAP reaction mixture of untreated HCT 116 cell lysate. The relative telomerase activity (RTA) of untreated HCT 116 cell lysate added with berberine was $83.90 \pm 0.34\%$. Berberine only decreased the RTA of untreated HCT 116 cell lysate by 16.10%. Values are mean $\pm$ standard error of triplicate. Mean with *** (triple asterisks) differ significantly at $p < 0.001$.

References

1. Ferlay, J.; Colombet, M.; Soerjomataram, I.; Mathers, C.; Parkin, D.M.; Piñeros, M.; Znaor, A.; Bray, F. Estimating the global cancer incidence and mortality in 2018: Globocan sources and methods. *Int. J. Cancer* **2019**, *144*, 1941–1953. [CrossRef] [PubMed]
2. Chandrashekar, D.S.; Bashel, B.; Balasubramanya, S.A.H.; Creighton, C.J.; Ponce-Rodriguez, I.; Chakravarthi, B.; Varambally, S. Ualcan: A portal for facilitating tumor subgroup gene expression and survival analyses. *Neoplasia* **2017**, *19*, 649–658. [CrossRef] [PubMed]
3. Tatsumoto, N.; Hiyama, E.; Murakami, Y.; Imamura, Y.; Shay, J.W.; Matsuura, Y.; Yokoyama, T. High telomerase activity is an independent prognostic indicator of poor outcome in colorectal cancer. *Clin. Cancer Res.* **2000**, *6*, 2696–2701. [PubMed]
4. Zvereva, M.I.; Shcherbakova, D.M.; Dontsova, O.A. Telomerase: Structure, functions, and activity regulation. *Biochem. Biokhim.* **2010**, *75*, 1563–1583. [CrossRef] [PubMed]
5. Bertorelle, R.; De Rossi, A. Telomerase as biomarker in colorectal cancer. In *Biomarkers in Cancer*; Preedy, V.R., Patel, V.B., Eds.; Springer: Dordrecht, The Netherlands, 2015; pp. 659–683.
6. Chang, J.T.C.; Chen, Y.L.; Yang, H.T.; Chen, C.Y.; Cheng, A.J. Differential regulation of telomerase activity by six telomerase subunits. *Eur. J. Biochem.* **2002**, *269*, 3442–3450. [CrossRef] [PubMed]
7. MacNeil, D.; Bensoussan, H.; Autexier, C. Telomerase regulation from beginning to the end. *Genes* **2016**, *7*, 64. [CrossRef]
8. Wyatt, H.D.M.; West, S.C.; Beattie, T.L. Intertpreting telomerase structure and function. *Nucleic Acids Res.* **2010**, *38*, 5609–5622. [CrossRef]
9. Nguyen, T.H.D.; Tam, J.; Wu, R.A.; Greber, B.J.; Toso, D.; Nogales, E.; Collins, K. Cryo-em structure of substrate-bound human telomerase holoenzyme. *Nature* **2018**, *557*, 190–195. [CrossRef]
10. Ganesan, K.; Xu, B. Telomerase inhibitors from natural products and their anticancer potential. *Int. J. Mol. Sci.* **2018**, *19*, 13. [CrossRef]
11. O'Brien, P.; Carrasco-Pozo, C.; Speisky, H. Boldine and its antioxidant or health-promoting properties. *Chem. Biol. Interact.* **2006**, *159*, 1–17. [CrossRef]
12. Brantley, S.J.; Oberlies, N.H.; Kroll, D.J.; Paine, M.F. Two flavonolignans from milk thistle (silybum marianum) inhibit cyp2c9-mediated warfarin metabolism at clinically achievable concentrations. *J. Pharmacol. Exp. Ther.* **2010**, *332*, 1081–1087. [CrossRef] [PubMed]
13. Davis-Searles, P.R.; Nakanishi, Y.; Kim, N.-C.; Graf, T.N.; Oberlies, N.H.; Wani, M.C.; Wall, M.E.; Agarwal, R.; Kroll, D.J. Milk thistle and prostate cancer: Differential effects of pure flavonolignans from silybum marianum on antiproliferative end points in human prostate carcinoma cells. *Cancer Res.* **2005**, *65*, 4448–4457. [CrossRef] [PubMed]
14. Surai, P. Silymarin as a natural antioxidant: An overview of the current evidence and perspectives. *Antioxidants* **2015**, *4*, 204–247. [CrossRef]

15. Mahata, S.; Bharti, A.C.; Shukla, S.; Tyagi, A.; Husain, S.A.; Das, B.C. Berberine modulates ap-1 activity to suppress hpv transcription and downstream signaling to induce growth arrest and apoptosis in cervical cancer cells. *Mol. Cancer* **2011**, *10*, 39. [CrossRef]
16. Srinivasan, G.V.; Unnikrishnan, K.P.; Rema Shree, A.B.; Balachandran, I. Hplc estimation of berberine in tinospora cordifolia and tinospora sinensis. *Indian J. Pharm. Sci.* **2008**, *70*, 96–99. [CrossRef] [PubMed]
17. Eskandari-Nasab, E.; Dahmardeh, F.; Rezaeifar, A.; Dahmardeh, T. Telomere and telomerase: From discovery to cancer treatment. *Gene Cell Tissue* **2015**, *2*, e28084. [CrossRef]
18. Hiyama, E.; Hiyama, K. Telomere and telomerase in stem cells. *Br. J. Cancer* **2007**, *96*, 1020–1024. [CrossRef] [PubMed]
19. Noureini, S.K.; Wink, M. Dose-dependent cytotoxic effects of boldine in hepg-2 cells-telomerase inhibition and apoptosis induction. *Molecules* **2015**, *20*, 3730–3743. [CrossRef]
20. Zhu, X.; Kumar, R.; Mandal, M.; Sharma, N.; Sharma, H.W.; Dhingra, U.; Sokoloski, J.A.; Hsiao, R.; Narayanan, R. Cell cycle-dependent modulation of telomerase activity in tumor cells. *Proc. Natl. Acad. Sci. USA* **1996**, *93*, 6091–6095. [CrossRef]
21. Klein, E.A.; Assoian, R.K. Transcriptional regulation of the cyclin d1 gene at a glance. *J. Cell Sci.* **2008**, *121*, 3853–3857. [CrossRef]
22. Ioset, J.-R.; Brun, R.; Wenzler, T.; Kaiser, M.; Yardley, V. Drug screening for kinetoplastids diseases. In *A Training Manual for Screening in Neglected Diseases*; Department for International Development: London, UK, 2009.
23. National Cancer Institute. Nci-60 Screening Methodology. Available online: https://dtp.cancer.gov/discovery_development/nci-60/methodology.htm (accessed on 19 May 2019).
24. Barzegar, E.; Fouladdel, S.; Movahhed, T.K.; Atashpour, S.; Ghahremani, M.H.; Ostad, S.N.; Azizi, E. Effects of berberine on proliferation, cell cycle distribution and apoptosis of human breast cancer t47d and mcf7 cell lines. *Iran. J. Basic Med. Sci.* **2015**, *18*, 334. [PubMed]
25. Li, F.; Dong, X.; Lin, P.; Jiang, J. Regulation of akt/foxo3a/skp2 axis is critically involved in berberine-induced cell cycle arrest in hepatocellular carcinoma cells. *Int. J. Mol. Sci.* **2018**, *19*, 327.
26. Chen, Q.; Qin, R.; Fang, Y.; Li, H. Berberine sensitizes human ovarian cancer cells to cisplatin through mir-93/pten/akt signaling pathway. *Cell. Physiol. Biochem.* **2015**, *36*, 956–965. [CrossRef] [PubMed]
27. Jirawatnotai, S.; Hu, Y.; Livingston, D.M.; Sicinski, P. Proteomic identification of a direct role for cyclin d1 in DNA damage repair. *Cancer Res.* **2012**, *72*, 4289–4293. [CrossRef]
28. Wang, Y.Y.; Sun, G.; Luo, H.; Wang, X.F.; Lan, F.M.; Yue, X.; Fu, L.S.; Pu, P.Y.; Kang, C.S.; Liu, N. Mir-21 modulates h tert through a stat3-dependent manner on glioblastoma cell growth. *CNS Neurosci. Ther* **2012**, *18*, 722–728. [CrossRef]
29. Lin, C.-Y.; Hsieh, P.-L.; Liao, Y.-W.; Peng, C.-Y.; Lu, M.-Y.; Yang, C.-H.; Yu, C.-C.; Liu, C.-M. Correction: Berberine-targeted mir-21 chemosensitizes oral carcinomas stem cells. *Oncotarget* **2018**, *9*, 24870. [CrossRef]
30. Xin, X.; Senthilkumar, P.; Schnoor, J.L.; Ludewig, G. Effects of pcb126 and pcb153 on telomerase activity and telomere length in undifferentiated and differentiated hl-60 cells. *Environ. Sci. Pollut. Res.* **2016**, *23*, 2173–2185. [CrossRef]
31. Wu, K.; Yang, Q.; Mu, Y.; Zhou, L.; Liu, Y.; Zhou, Q.; He, B. Berberine inhibits the proliferation of colon cancer cells by inactivating wnt/β-catenin signaling. *Int. J. Oncol.* **2012**, *41*, 292–298.
32. Brattain, M.G.; Fine, W.D.; Khaled, F.M.; Thompson, J.; Brattain, D.E. Heterogeneity of malignant cells from a human colonic carcinoma. *Cancer Res.* **1981**, *41*, 1751–1756.
33. Owa, T.; Ozawa, Y.; Semba, T. Joint Use of Sulfonamide Based Compound with Angiogenesis Inhibitor. EP Patent 2,364,699,A1, 14 September 2011.
34. Jain, M.; Nilsson, R.; Sharma, S.; Madhusudhan, N.; Kitami, T.; Souza, A.L.; Kafri, R.; Kirschner, M.W.; Clish, C.B.; Mootha, V.K. Metabolite profiling identifies a key role for glycine in rapid cancer cell proliferation. *Science* **2012**, *336*, 1040–1044. [CrossRef]
35. Pereira, P.D.; Serra-Caetano, A.; Cabrita, M.; Bekman, E.; Braga, J.; Rino, J.; Santus, R.; Filipe, P.L.; Sousa, A.E.; Ferreira, J.A. Quantification of cell cycle kinetics by edu (5-ethynyl-2′-deoxyuridine)-coupled-fluorescence-intensity analysis. *Oncotarget* **2017**, *8*, 40514. [CrossRef] [PubMed]
36. National Cancer Institute. Cell Lines in the in vitro screen. Available online: https://dtp.cancer.gov/discovery_development/nci-60/cell_list.htm (accessed on 10 April 2019).
37. Nguyen, D.; Grenier St-Sauveur, V.; Bergeron, D.; Dupuis-Sandoval, F.; Scott, M.S.; Bachand, F. A polyadenylation-dependent 3′ end maturation pathway is required for the synthesis of the human telomerase rna. *Cell Rep.* **2015**, *13*, 2244–2257. [CrossRef] [PubMed]
38. Serafim, T.L.; Oliveira, P.J.; Sardao, V.A.; Perkins, E.; Parke, D.; Holy, J. Different concentrations of berberine result in distinct cellular localization patterns and cell cycle effects in a melanoma cell line. *Cancer Chemther. Pharmacol.* **2008**, *61*, 1007–1018. [CrossRef] [PubMed]
39. Guaman Ortiz, L.M.; Croce, A.L.; Aredia, F.; Sapienza, S.; Fiorillo, G.; Syeda, T.M.; Buzzetti, F.; Lombardi, P.; Scovassi, A.I. Effect of new berberine derivatives on colon cancer cells. *Acta Biochimica Biophysica Sinica* **2015**, *47*, 824–833. [CrossRef]
40. Li, W.; Hua, B.; Saud, S.M.; Lin, H.; Hou, W.; Matter, M.S.; Jia, L.; Colburn, N.H.; Young, M.R. Berberine regulates amp-activated protein kinase signaling pathways and inhibits colon tumorigenesis in mice. *Mol. Carcinogenes.* **2015**, *54*, 1096–1109. [CrossRef]
41. Noureini, S.K.; Tanavar, F. Boldine, a natural aporphine alkaloid, inhibits telomerase at non-toxic concentrations. *Chem.-Biol. Interact.* **2015**, *231*, 27–34. [CrossRef]
42. Ko, E.; Seo, H.W.; Jung, G. Telomere length and reactive oxygen species levels are positively associated with a high risk of mortality and recurrence in hepatocellular carcinoma. *Hepatology* **2018**, *67*, 1378–1391. [CrossRef]

43. Moraca, F.; Amato, J.; Ortuso, F.; Artese, A.; Pagano, B.; Novellino, E.; Alcaro, S.; Parrinello, M.; Limongelli, V. Ligand binding to telomeric g-quadruplex DNA investigated by funnel-metadynamics simulations. *Proc. Natl. Acad. Sci. USA* **2017**, *114*, E2136–E2145. [CrossRef]
44. Bhadra, K.; Kumar, G.S. Interaction of berberine, palmatine, coralyne, and sanguinarine to quadruplex DNA: A comparative spectroscopic and calorimetric study. *Biochimica Biophysica Acta* **2011**, *1810*, 485–496. [CrossRef]
45. Neidle, S. Quadruplex nucleic acids as novel therapeutic targets. *J. Med. Chem.* **2016**, *59*, 5987–6011. [CrossRef]
46. Wang, Y.; Kheir, M.M.; Chai, Y.; Hu, J.; Xing, D.; Lei, F.; Du, L. Comprehensive study in the inhibitory effect of berberine on gene transcription, including tata box. *PLoS ONE* **2011**, *6*, e23495. [CrossRef] [PubMed]
47. Yuan, Z.-Y.; Lu, X.; Lei, F.; Chai, Y.-S.; Wang, Y.-G.; Jiang, J.-F.; Feng, T.-S.; Wang, X.-P.; Yu, X.; Yan, X.-J. Tata boxes in gene transcription and poly (a) tails in mrna stability: New perspective on the effects of berberine. *Sci. Rep.* **2015**, *5*, 18326. [CrossRef]
48. Fu, L.; Chen, W.; Guo, W.; Wang, J.; Tian, Y.; Shi, D.; Zhang, X.; Qiu, H.; Xiao, X.; Kang, T. Berberine targets ap-2/htert, nf-κb/cox-2, hif-1α/vegf and cytochrome-c/caspase signaling to suppress human cancer cell growth. *PLoS ONE* **2013**, *8*, e69240. [CrossRef] [PubMed]
49. Xiong, Y.X.; Su, H.F.; Lv, P.; Ma, Y.; Wang, S.K.; Miao, H.; Liu, H.Y.; Tan, J.H.; Ou, T.M.; Gu, L.Q.; et al. A newly identified berberine derivative induces cancer cell senescence by stabilizing endogenous g-quadruplexes and sparking a DNA damage response at the telomere region. *Oncotarget* **2015**, *6*, 35625–35635. [CrossRef] [PubMed]
50. Gutierrez-Rodrigues, F.; Santana-Lemos, B.A.; Scheucher, P.S.; Alves-Paiva, R.M.; Calado, R.T. Direct comparison of flow-fish and qpcr as diagnostic tests for telomere length measurement in humans. *PLoS ONE* **2014**, *9*, e113747. [CrossRef] [PubMed]
51. Arora, A.; Balasubramanian, C.; Kumar, N.; Agrawal, S.; Ojha, R.P.; Maiti, S. Binding of berberine to human telomeric quadruplex—Spectroscopic, calorimetric and molecular modeling studies. *FEBS J.* **2008**, *275*, 3971–3983. [CrossRef] [PubMed]
52. Bazzicalupi, C.; Ferraroni, M.; Bilia, A.R.; Scheggi, F.; Gratteri, P. The crystal structure of human telomeric DNA complexed with berberine: An interesting case of stacked ligand to g-tetrad ratio higher than 1:1. *Nucleic Acids Res.* **2013**, *41*, 632–638. [CrossRef] [PubMed]
53. Zhang, W.J.; Ou, T.M.; Lu, Y.J.; Huang, Y.Y.; Wu, W.B.; Huang, Z.S.; Zhou, J.L.; Wong, K.Y.; Gu, L.Q. 9-substituted berberine derivatives as g-quadruplex stabilizing ligands in telomeric DNA. *Bioorg. Med. Chem.* **2007**, *15*, 5493–5501. [CrossRef]
54. Xu, J.; Long, Y.; Ni, L.; Yuan, X.; Yu, N.; Wu, R.; Tao, J.; Zhang, Y. Anticancer effect of berberine based on experimental animal models of various cancers: A systematic review and meta-analysis. *BMC Cancer* **2019**, *19*, 589. [CrossRef]
55. Zhang, J.; Cao, H.; Zhang, B.; Cao, H.; Xu, X.; Ruan, H.; Yi, T.; Tan, L.; Qu, R.; Song, G. Berberine potently attenuates intestinal polyps growth in apcmin mice and familial adenomatous polyposis patients through inhibition of wnt signalling. *J. Cell. Mol. Med.* **2013**, *17*, 1484–1493. [CrossRef]
56. Assefa, A.T.; Vandesompele, J.; Thas, O. On the utility of rna sample pooling to optimize cost and statistical power in rna sequencing experiments. *BMC Genom.* **2020**, *21*, 1–14. [CrossRef]
57. Metzger, K.; Tuchscherer, A.; Palin, M.-F.; Ponsuksili, S.; Kalbe, C. Establishment and validation of cell pools using primary muscle cells derived from satellite cells of pig skeletal muscle. *In Vitro Cell. Dev. Biol. Anim.* **2019**, 1–7. [CrossRef] [PubMed]
58. Diz, A.P.; Truebano, M.; Skibinski, D.O. The consequences of sample pooling in proteomics: An empirical study. *Electrophoresis* **2009**, *30*, 2967–2975. [CrossRef] [PubMed]
59. Phang, C.-W.; Karsani, S.A.; Sethi, G.; Malek, S.N.A. Flavokawain c inhibits cell cycle and promotes apoptosis, associated with endoplasmic reticulum stress and regulation of mapks and akt signaling pathways in hct 116 human colon carcinoma cells. *PLoS ONE* **2016**, *11*, e0148775. [CrossRef] [PubMed]
60. Sebaugh, J. Guidelines for accurate ec50/ic50 estimation. *Pharm. Stat.* **2011**, *10*, 128–134. [CrossRef] [PubMed]
61. Schneider, C.A.; Rasband, W.S.; Eliceiri, K.W. Nih image to imagej: 25 years of image analysis. *Nature Methods* **2012**, *9*, 671–675. [CrossRef]
62. Vasilishina, A.; Kropotov, A.; Spivak, I.; Bernadotte, A. Relative human telomere length quantification by real-time pcr. In *Cellular Senescence*; Springer: Berlin/Heidelberg, Germany, 2019; pp. 39–44.

Article

Identification of a Prenyl Chalcone as a Competitive Lipoxygenase Inhibitor: Screening, Biochemical Evaluation and Molecular Modeling Studies

Maria Luiza Zeraik [1,2,*], Ivani Pauli [3], Luiz A. Dutra [2], Raquel S. Cruz [2], Marilia Valli [2,3], Luana C. Paracatu [4], Carolina M. Q. G. de Faria [4], Valdecir F. Ximenes [4], Luis O. Regasini [2,5], Adriano D. Andricopulo [3] and Vanderlan S. Bolzani [2,*]

[1] Laboratório de Fitoquímica e Biomoléculas (LabFitoBio), Departamento de Química, Universidade Estadual de Londrina (UEL), Londrina, PR 86051-990, Brazil

[2] Núcleo de Bioensaios, Biossíntese e Ecofisiologia de Produtos Naturais (NuBBE), Departamento de Química Orgânica, Instituto de Química, Universidade Estadual Paulista (UNESP), Araraquara, SP 14800-060, Brazil; luizdutra_qf@yahoo.com.br (L.A.D.); raqsabara@hotmail.com (R.S.C.); mariliava@gmail.com (M.V.); luisregasini@gmail.com (L.O.R.)

[3] Laboratório de Química Medicinal e Computacional (LQMC), Centro de Pesquisa e Inovação em Biodiversidade e Fármacos (CIBFar), Instituto de Física de São Carlos, Universidade de São Paulo (USP), São Carlos, SP 13563-120, Brazil; ivanipauli@gmail.com (I.P.); aandrico@ifsc.usp.br (A.D.A.)

[4] Departamento de Química, Faculdade de Ciências, Universidade Estadual Paulista (UNESP), Bauru, SP 17020-360, Brazil; luanachiquetto@hotmail.com (L.C.P.); carolina.quinello@gmail.com (C.M.Q.G.d.F.); valdecir.ximenes@unesp.br (V.F.X.)

[5] Instituto de Biociências, Letras e Ciências Exatas, Universidade Estadual Paulista (UNESP), São José do Rio Preto, SP 15054-000, Brazil

* Correspondence: zeraikml@uel.br (M.L.Z.); bolzaniv@iq.unesp.br (V.S.B.)

Citation: Zeraik, M.L.; Pauli, I.; Dutra, L.A.; Cruz, R.S.; Valli, M.; Paracatu, L.C.; de Faria, C.M.Q.G.; Ximenes, V.F.; Regasini, L.O.; Andricopulo, A.D.; et al. Identification of a Prenyl Chalcone as a Competitive Lipoxygenase Inhibitor: Screening, Biochemical Evaluation and Molecular Modeling Studies. *Molecules* **2021**, *26*, 2205. https://doi.org/10.3390/molecules26082205

Academic Editors: Young Hae Choi, Young Pyo Jang, Yuntao Dai and Luis Francisco Salomé-Abarca

Received: 28 February 2021
Accepted: 7 April 2021
Published: 12 April 2021

Abstract: Cyclooxygenase (COX) and lipoxygenase (LOX) are key targets for the development of new anti-inflammatory agents. LOX, which is involved in the biosynthesis of mediators in inflammation and allergic reactions, was selected for a biochemical screening campaign to identify LOX inhibitors by employing the main natural product library of Brazilian biodiversity. Two prenyl chalcones were identified as potent inhibitors of LOX-1 in the screening. The most active compound, (*E*)-2-*O*-farnesyl chalcone, decreased the rate of oxygen consumption to an extent similar to that of the positive control, nordihydroguaiaretic acid. Additionally, studies on the mechanism of the action indicated that (*E*)-2-*O*-farnesyl chalcone is a competitive LOX-1 inhibitor. Molecular modeling studies indicated the importance of the prenyl moieties for the binding of the inhibitors to the LOX binding site, which is related to their pharmacological properties.

Keywords: inflammation; docking; NuBBE database; chalcones

1. Introduction

Lipoxygenases are a family of iron-containing dioxygenases which are widely distributed in both animal and plant species. Human 5-lipoxygenase (5-LOX) is involved in the metabolism of arachidonic acid to leukotrienes, which are potent mediators of inflammation and allergic reactions [1,2]. Plants also contain numerous lipoxygenases, with at least eight having been identified in soybean (*Glycine max*), such as soybean lipoxygenase-1 (LOX-1), all of which use linoleic acid as a substrate [3,4]. Two important approaches for the design and synthesis of anti-inflammatory agents are based on the inhibition of two enzymes, cyclooxygenase (COX) and lipoxygenase, both of which are involved in the metabolism of arachidonic acid. Cyclooxygenase (COX) catalyzes the conversion of arachidonic acid to prostaglandins, with COX-1 being the constitutive isoform that is expressed in most tissues, while COX-2 is only expressed when induced by the response to inflammation [5].

Selective inhibitors of the COX-2 isoform have been developed as anti-inflammatory drugs, such as celecoxib (Celebrex) and rofecoxib (Vioxx), and are characterized as non-steroidal anti-inflammatory drugs (NSAIDs). However, their use is associated with an increased risk of ulcerations, heart attacks, and strokes [6,7]. The inhibition of lipoxygenase is an alternative for the treatment of inflammatory processes, and it can also be a combination strategy in conjunction with COX-2 inhibitors [8–11]. Zileuton (Zyflo®), which is available in the USA for the treatment of asthma, is currently the only 5-LOX inhibitor marketed for therapy in humans. Nevertheless, zileuton shows low potential in other diseases, such as allergic rhinitis, rheumatoid arthritis, and inflammatory bowel disease, revealing a great need for novel inhibitors of 5-LOX [12].

As has been extensively reviewed, biodiversity provides exclusive chemical scaffolds for drug discovery [13,14]. Chalcones are known for their various biological activities, including antioxidant, anticancer, antimicrobial, anti-inflammatory, and antiprotozoal activities [15–20]. Prenylated chalcones are found as natural products, such as the derricidin and isocordoin used in this work as an inspiration, and they and others could be retrieved from the natural product database of Brazilian biodiversity (Nuclei of Bioassays, Ecophysiology and Biosynthesis of Natural Products Database, NuBBE$_{DB}$) [21,22]. In the present work, we report the screening for LOX-1 inhibitors and the identification of a chalcone scaffold with prenyl groups as competitive inhibitors. The use of a polarography assay was applied in order to further explore the anti-inflammatory potential of the most potent compounds. Together with the in vitro experiments, molecular modeling studies were also conducted in order to reveal the molecular features related to the binding of this inhibitor to the LOX binding site.

2. Results and Discussion

The present work is part of a screening campaign to identify biologically-active natural products from Brazilian biodiversity to be used as models for further optimization in a drug discovery pipeline. A combination of in vitro assays, enzyme kinetics, and molecular modeling studies were used to evaluate the molecular reasons for the LOX inhibition activity of the most promising compound.

For the in vitro screening to identify the inhibitors of the enzymatic activity of LOX-1, a spectrophotometric assay was employed in order to measure the conjugated diene formed by linoleic acid oxidation. Soybean LOX-1, which converts linoleic acid to 13-hydroperoxylinoleic acid, is inhibited by NSAIDs in a qualitatively similar way to that of the rat mast cell line 5-LOX, and it may be used as a reliable screening target for this activity [23]. The screening was performed with compounds from Brazilian biodiversity (NuBBE$_{DB}$) [15,16] and synthetic derivatives previously synthesized by using natural products as models. In this context, a variety of natural products were evaluated, such as triterpenes, chalcones, casearins, alkaloids, guanidines, and flavonoids. Among all of the screened compounds, prenyl chalcones were identified as LOX-1 inhibitors.

Hydroxyl chalcones (**1–3**) were previously obtained by Claisen–Schmidt condensation [24] followed by prenylation to obtain chalcones **4–6**. Prenyl chalcones and their precursor hydroxyl chalcones were evaluated in order to verify the relevance of the prenyl group for LOX-1 inhibition. Prenyl units are frequently found in natural products, and are associated with an increase in hydrophobicity and facilitated interaction with cell membranes. LOX-1 inhibition by (*E*)-2-*O*-farnesyl chalcone (**6**, IC$_{50}$ = 5.7 μM) was twofold more potent than 3-*O*-geranyl chalcone (**5**, IC$_{50}$ = 11.8 μM), and >10-fold more potent than 2-*O*-prenyl chalcone (**4**, IC$_{50}$ > 50.0 μM). Chalcones containing hydroxyl groups did not show significant inhibitory effects on LOX-1, regardless of the position of these groups (Table 1). Compound **6** was almost twofold more potent than the positive control, curcumin (IC$_{50}$ = 10.1 μM), and twofold less potent than the positive control, nordihydroguaiaretic acid (NDGA) (IC$_{50}$ = 2.7 μM). The graph demonstrating the inhibitory activity of the compounds on LOX-1 at different concentrations is available online in the Supplementary Materials (Figure S1).

The results indicate that a larger extension of the prenyl chains is directly related to an enhanced inhibition of LOX-1. The extension of this chain results in higher lipophilicity and an increase in the capacity to perform hydrophobic interactions in the binding site. The binding site of LOX shows that it is primarily composed of hydrophobic amino acids, as will be discussed further in this paper within the molecular docking evaluation, specifying its preference for hydrophobic ligands, such as **6** and **5**. Lipophilicity is an important physicochemical property which is often related to the bioactivity of the compounds. The tendency observed herein is similar to that previously reported for curcumin analog LOX inhibitors [25]. The log *P* values were calculated based on the consensus method provided by the SwissADME web tool [26,27]. The increase in the molecular hydrophobicity of prenyl chalcones (log *P*) was correlated with the more potent inhibitory activities of the compounds, as shown in Table 1. A higher hydrophobicity was also previously correlated with a more potent antileishmanial activity for these compounds [24].

The catalytic activity of LOX-1 comprises the conversion of its substrate to polyunsaturated molecules by the consumption of oxygen atoms, which is a limiting rate of that mechanism. A polarography-based assay was used to evaluate the rate of the oxygen consumption during LOX-1 inhibition, during which oxygen consumption is associated with the formation of a hydroperoxyl group [28]. In order to evaluate the decrease in oxygen consumption upon LOX-1 inhibition by the prenyl chalcone **6**, an O_2-selective electrode coupled to an oxygen monitoring probe was employed. A delay of the oxygen consumption was detected in the presence of prenyl chalcone **6** at 10 μM, to a similar extent to that of the positive control NDGA, which suggests a decline in LOX-1 catalysis and a decrease in the detectable polyunsaturated product (Figure 1).

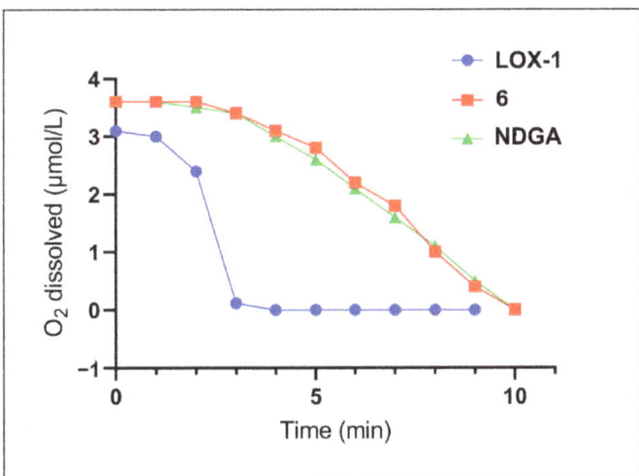

Figure 1. The activity of LOX-1 (1 μM) was monitored by O_2 consumption. LOX-1 inhibition led to a lower rate of oxygen consumption, which was observed in the presence of (*E*)-2-*O*-farnesyl chalcone (**6**) and nordihydroguaiaretic acid (NDGA), both at 10 μM.

Enzyme kinetics studies were carried out in order to determine the mechanism of inhibition of the most potent compound **6**. Primarily, the Michaelis–Menten constant (K_M) and maximum velocity values (V_{max}) were determined for the enzymatic activity without an inhibitor (K_M = 45.8 μM and V_{max} = 10 μmol/min). Increasing the concentrations of **6** (5–10 μM) allowed us to confirm the competitive mechanism against LOX-1 inhibition, with a K_i = 7 μM.

Table 1. IC$_{50}$ values for the inhibition of lipoxygenase-1.

Compound	Structure	IC$_{50}$ (μM) [a]	Calculated log P [b]
1	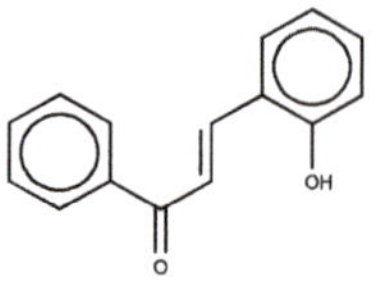	>50	3.09
2	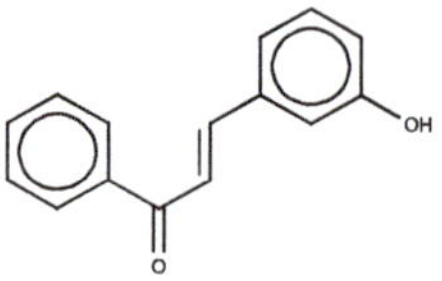	>50	2.87
3	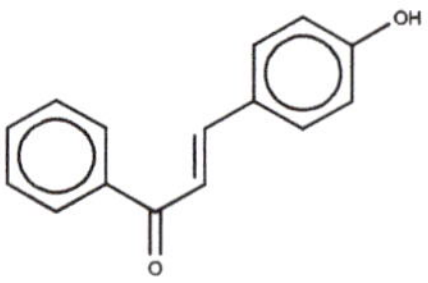	>50	2.87
4	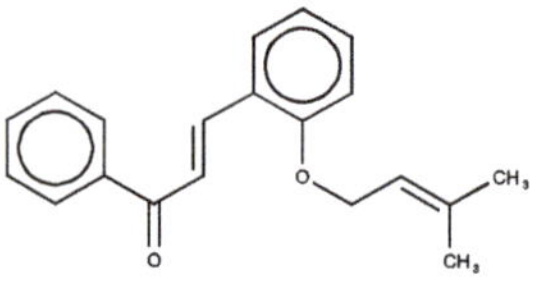	>50	4.52
5	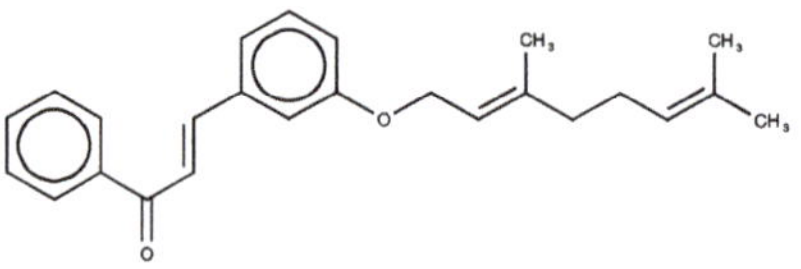	11.8 ± 0.9	5.94
6	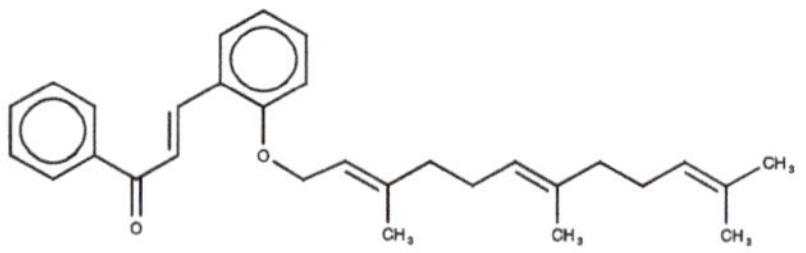	5.7 ± 0.8	7.46
Curcumin	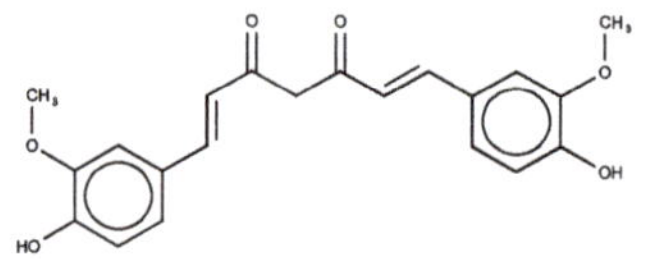	10.1 ± 0.5	
Nordihydroguaiaretic acid (NDGA)	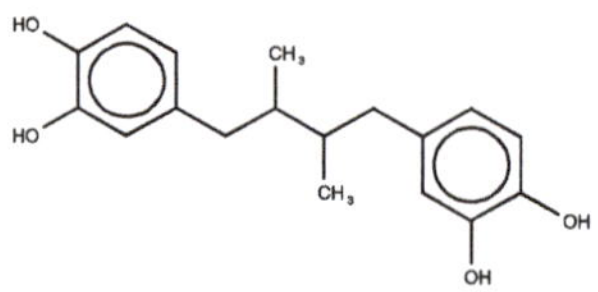	2.7 ± 0.5	

[a] The values represent the mean and standard deviation of three independent experiments. [b] Calculated by the consensus method provided by the SwissADME web tool [26,27].

Molecular docking studies were performed in order to assess the molecular connections related to the experimental results obtained. In general, the binding site of eukaryotic LOXs is conservative, allowing us to use LOX-5 as structural model to determine ligand interactions [29,30]. The binding site of LOX-5 is primarily composed (approximately 50%) of hydrophobic amino acids (Trp147, Phe169, Phe177, Leu179, Phe229, Leu368, Ile406, Ala410, Phe555, Ala603, Leu607, Phe610, Val671, and Ala672), clarifying the increased potency for the hydrophobic compounds **6** and **5**. The increased potency of **6** compared to **5** is justified by the enhanced number of hydrophobic interactions due to the increased length of the prenyl chain, as shown in Figure 2. The docking simulations showed that **5** (Figure 2A), a chalcone with two prenyl groups at position 3 of ring B, interacts with the LOX binding cavity, where the chalcone moiety interacts with His367, Phe555, Tyr558, Leu607, and Phe610, while the prenyl chain is stabilized by the nonpolar residues Phe177, Ala410, Gln413 and Lys409. Compound **6** (Figure 2B) adopts a similar conformation to that of **5** in the LOX binding cavity. The same residues—His367, Phe555, Tyr558, Leu607, and Phe610—interact with the chalcone moiety; an additional hydrogen bond, formed between the chalcone carbonyl oxygen and the Asn554 side-chain nitrogen atom, contributes to the stabilization of the compound. Compound **6** presents a longer and more hydrophobic substituent in ring B at position 2, which is composed of three prenyl groups. Consequently, a higher number of protein-ligand hydrophobic interactions take place. In this case, the hydrophobic chain is stabilized by the nonpolar residues Trp147, Leu368, Ile406, Ala410, and Ala672, in addition to the basic amino acids Arg411, His372, and His550. The molecular docking study corroborated the previous experimental results by revealing an increased number of intermolecular interactions of **6** ($\Delta G_{\text{binding}}$ = −41.84) and the binding site compared to that of compound **5** ($\Delta G_{\text{binding}}$ = −38.15).

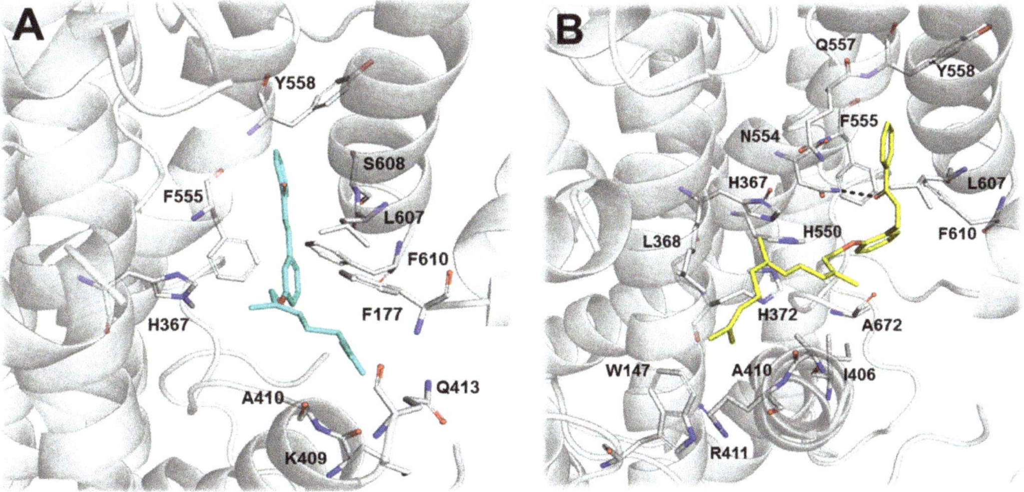

Figure 2. 3D representation of the top-scored docking poses of **5** in cyan (**A**) and **6** in yellow (**B**), rendered as stick models in the human 5-LOX binding site (PDB ID: 3V99). The relevant binding site residues are highlighted and represented by the respective letter code: A (Ala) = alanine; F (Phe) = phenylalanine; H (His) = histidine; I (Ile) = isoleucine; K (Lys) = lysine; L (Leu) = leucine; N (Asn) = asparagine; Q (Gln) = glutamine; R (Arg) = arginine; S (Ser) = serine; Y (Tyr) = tyrosine; W (Trp) = tryptophan.

Previous studies performed computational approaches to screen chalcone and flavone derivatives for their potential as inhibitors of 5-LOX, and showed the lipoxygenase activity inhibition. This study was important to identify prenyl chalcones as inhibitors of LOX. The mode of binding for chalcones was assessed in order to verify the importance of the prenyl moieties for binding. The mode of binding was similar to other chalcones previously reported in the literature [31].

3. Materials and Methods

3.1. Synthesis of Compounds

Compound **1–3** and prenylchalcones **4–6** were synthesized as described previously [24]. The NMR data for compound **3** are shown below:

(*E*)-4-hydroxychalcone (16). Yellow solid, 33% yield. 1H NMR (500 MHz, DMSO-d6), δH 7.62 (d; J = 7.5 Hz, H-2/H-6), 6.82 (d; J = 8.5 Hz, H-3/H-5), 8.08 (dd; J = 1.5, 7.5 Hz, H-2′/H-6′), 7.53 (dd; J = 7.5, 7.5 Hz, H-3′/H-5′), 7.69 (m; H-4′/H-α/H-β). 13C NMR (500 MHz, DMSO-d6), δC 125.7 (C-1), 128.3 (C-2/C-6), 115.8 (C-3/C-5), 160.1 (C-4), 137.9 (C-1′), 130.9 (C-2′/C-6′), 128.6 (C-3′/C-5′), 132.7 (C-4′), 118.4 (C-α), 144.5 (C-β), 189.0 (C-β′).

3.2. LOX Bioassay

The ability of chalcones **1–6** to act as LOX-1 inhibitors was evaluated according to a method from Ha et al. (2004) [28]. Initially, 2.910 mL Tris-HCl buffer (0.1 mM (pH 8.0), Sigma-Aldrich, Brazil) was added to a quartz cuvette, followed by the addition of 25 µL LOX-1 (final concentration 1 µM, Sigma-Aldrich, Brazil) and 10 µL of the compounds tested (the final concentration ranged from 1 to 20 µM). The solution was incubated for 5 min at 25 °C. The reaction was started with the addition of linoleic acid (55 µL, 1 mM, Sigma-Aldrich, Brazil), and the LOX-1 activity was monitored at λ = 221 nm, pH 8.0, and 25 °C by an absorbance spectrophotometer in UV-Vis (Perkin Elmer Lambda 22, Waltham, MA, USA). For the determination of K_M, varying substrate concentrations were used (linoleic acid, 13.7–91.7 µM) and 1 µM LOX-1. Nordihydroguaiaretic acid (NDGA, CAS Number: 500-38-9, Sigma-Aldrich, St. Louis, MO, USA ≥97%) was used as a positive control. The spectrophotometric measurements at λ = 221 nm are associated with the conjugated double bonds formed from the linoleic acid substrate.

3.3. Oxygen Consumption by the Catalytic Activity of LOX

The oxygen consumption was monitored using an O_2-selective electrode coupled to a YSI 5300 Oxygen Monitor (Yellow Springs, OH, USA) at 25 °C. The reactions were initiated by adding O_2 (100 µM), followed by the addition of LOX (20 nM) in 0.1 M Phosphate buffered saline (PBS), with pH 7.4 at 25 °C and the compounds being tested [32].

3.4. Docking Simulation

The 3D chemical structure of each compound was generated using the standard tools available in the molecular modeling software package Sybyl 8.0 [33]. The conformational energy was minimized using the Tripos force field and Powell's method. The partial atomic charges were calculated using the Gasteiger–Hückel method, which is available in Sybyl 8.0, considering the molecules to be in an implicit aqueous environment (a dielectric constant of 80.0). The simulations were carried out using GOLD [34], a software package based on a genetic algorithm to explore the ligand conformational space. The protein structure of 5-LOX was PDB ID: 3V99, available at the Protein Data Bank (PDB). The protein was prepared for the docking studies by adding hydrogen atoms, removing the water, and cocrystallizing the inhibitors. The enzyme–inhibitor interactions within a radius of 12 Å centered on the PHE177 CZ atom were evaluated. The binding site was defined according to co-crystalized ligand arachidonic acid. The docking poses were obtained by applying the ChemScore fitness function. LIGPLOT [35] was used to identify the contacts among the inhibitors and LOX.

3.5. Theoretical Calculation of log P

The molecular hydrophobicity of the chalcone derivatives was calculated based on their log P values (partitioning coefficient in n-octanol/water) based on the consensus method (arithmetic mean of the values predicted by the following proposed methods: XLOGP3, WLOGP, MLOGP, SILICOS-IT, and iLOGP), and was performed using the SwissADME web tool [26,27].

4. Conclusions

This paper contributes to the field of inflammatory diseases with information on the molecular features which are most important for the development of LOX inhibitors. The LOX inhibition strategy is an important advancement to overcome the unwanted effects of anti-inflammatories that are selective for COX-2 inhibition. The research on LOX inhibitors published herein is important to address the treatment of inflammatory diseases for which the current drugs lack efficacy. The initial approach employing an in vitro screening was effective in identifying potent LOX-1 inhibitors, namely, (*E*)-2-*O*-farnesyl chalcone (**6**, IC_{50} = 5.7 μM) and 3-*O*-geranyl chalcone (**5**, IC_{50} = 11.8 μM). Next, the enzyme kinetics study allowed for the determination of the mode of action of the competitive inhibitor **6**, providing information on the binding site to be explored in the computational studies. The molecular reasons for the binding of compounds **5** and **6** to lipoxygenase were analyzed by computational molecular docking studies, highlighting the contribution of the prenyl moieties to the binding.

Finally, it is worthy to note that compound **6** was not more effective than NDGA, a well-established control for lipoxygenase inhibition. However, some advances can be expected, as chalcone derivatives showed lower cell toxicity [36]. On the other hand, the usefulness of NDGA is limited due to its toxicity. For example, its long-term use has been correlated with liver damage and kidney dysfunction [37].

Supplementary Materials: Supplementary Materials are available online.

Author Contributions: Conceptualization, M.L.Z., V.F.X. and V.S.B.; Formal analysis, M.L.Z., I.P., L.A.D. and M.V.; Investigation, M.L.Z., I.P., L.A.D., R.S.C., L.C.P. and C.M.Q.G.d.F.; Methodology, M.L.Z., I.P., L.A.D., V.F.X. and L.O.R.; Project administration, M.L.Z., V.F.X., A.D.A. and V.S.B.; Resources, V.F.X., A.D.A. and V.S.B.; Supervision, V.F.X., L.O.R., A.D.A. and V.S.B.; Writing—original draft, M.L.Z., I.P. and L.A.D.; Writing—review and editing, M.V., L.C.P., V.F.X., L.O.R., A.D.A. and V.S.B. All authors have read and agreed to the published version of the manuscript.

Funding: This work was supported by the Fundação de Amparo à Pesquisa do Estado de São Paulo (FAPESP), grants #2010/01506-7 and #2013/07600-3, #2019/18445-5. M.L.Z., R.S.C., and M.V. acknowledge the FAPESP for scholarships #2011/03017-6, #2009/15457-0, and #2019/05967-3. The authors acknowledge the Araucaria Foundation (FA), CNPq (Project #429073/2016-0), INCT-CNPq (Project #2014/465637-0), and INCT-FAPESP (Project #2014/50926-0) agencies for their financial support.

Institutional Review Board Statement: Not applicable.

Informed Consent Statement: Not applicable.

Data Availability Statement: Not applicable.

Conflicts of Interest: The authors declare that they have no conflicts of interest.

Sample Availability: Samples of the compounds are available from the authors.

References

1. Werz, O. 5-lipoxygenase: Cellular biology and molecular pharmacology. *Curr. Drug Targets Inflamm. Allergy* **2002**, *1*, 23–44. [CrossRef]
2. Steele, V.E.; Holmes, C.A.; Hawk, E.T.; Kopelovich, L.; Lubet, R.A.; Crowell, J.A.; Sigman, C.C.; Kelloff, G.J. Lipoxygenase inhibitors as potential cancer chemopreventives. *Cancer Epidemiol. Biomarkers Prev.* **1999**, *8*, 467–483.
3. Grechkin, A. Recent developments in biochemistry of the plant lipoxygenase pathway. *Prog. Lipid Res.* **1998**, *37*, 317–352. [CrossRef]
4. Brash, A.R. Lipoxygenases: Occurrence, functions, catalysis, and acquisition of substrate. *J. Biol. Chem.* **1999**, *274*, 23679–23682. [CrossRef] [PubMed]

5. Huang, K.; Masuda, A.; Chen, G.; Bushra, S.; Kamon, M.; Araki, T.; Kinoshita, M.; Ohkawara, B.; Ito, M.; Ohno, K. Inhibition of cyclooxygenase-1 by nonsteroidal anti-inflammatory drugs demethylates MeR2 enhancer and promotes Mbnl1 transcription in myogenic cells. *Sci. Rep.* **2020**, *10*, 2558. [CrossRef] [PubMed]

6. Allison, M.C.; Howatson, A.G.; Torrance, C.J.; Lee, F.D.; Russel, R.I.N. Gastrointestinal damage associated with the use of nonsteroidal anti-inflammatory drugs. *N. Engl. J. Med.* **1992**, *20*, 749–754. [CrossRef] [PubMed]

7. Sibbald, B. Rofecoxib (Vioxx) voluntarity withdrawn from market. *CMAJ* **2004**, *171*, 1027–1028. [CrossRef]

8. Doiron, J.L.; Boudreau, H.; Picot, N.; Villebonet, B.; Surette, M.E.; Touaibia, M. Synthesis and 5-lipoxygenase inhibitory activity of new cinnamoyl and caffeoyl clusters. *Bioorg. Med. Chem. Lett.* **2009**, *19*, 1118–1121. [CrossRef]

9. Kurihara, H.; Kagawa, Y.; Konno, R.; Kim, S.M.; Takahashi, K. Lipoxygenase inhibitors derived from marine macroalgae. *Bioorg. Med. Chem. Lett.* **2014**, *24*, 1383–1385. [CrossRef]

10. Martel-Pelletier, J.; Lajeunesse, D.; Reboul, P.; Pelletier, J.P. Therapeutic role of dual inhibitors of 5-LOX and COX, selective and non-selective non-steroidal anti-inflammatory drugs. *Ann. Rheum. Dis.* **2003**, *62*, 501–509. [CrossRef]

11. Jaismy, J.P.; Manju, S.L.; Ethiraj, K.R.; Elias, G. Safer anti-inflammatory therapy through dual COX-2/5-LOX inhibitors: A structure-based approach. *Eur. J. Pharm. Sci.* **2018**, *121*, 356–381.

12. Carter, G.W.; Young, P.R.; Albert, D.H.; Bouska, J.; Dyer, R.; Bell, R.L.; Summers, J.B.; Brooks, D.W. 5-Lipoxygenase inhibitory activity of zileuton. *J. Pharmacol. Exp. Ther.* **1991**, *256*, 929–937.

13. Newman, D.J.; Cragg, G.M. Natural Products as Sources of New Drugs from 1981 to 2014. *J. Nat. Prod.* **2016**, *79*, 629–661. [CrossRef]

14. De Luca, V.; Salim, V.; Atsumi, S.M.; Yu, F. Mining the biodiversity of plants: A revolution in the making. *Science* **2012**, *206*, 1658–1661. [CrossRef]

15. Hsieh, H.K.; Lee, T.H.; Wang, J.P.; Wang, J.J.; Lin, C.N. Synthesis and anti-inflammatory effect of chalcones and related compounds. *Pharm. Res.* **1998**, *15*, 39–46. [CrossRef] [PubMed]

16. Zeraik, M.L.; Ximenes, V.F.; Regasini, L.O.; Dutra, L.A.; Silva, D.H.S.; Fonseca, L.M.; Coelho, D.; Machado, S.A.; Bolzani, V.S. 4′-Aminochalcones as novel inhibitors of the chlorinating activity of myeloperoxidase. *Curr. Med. Chem.* **2012**, *19*, 5405–5413. [CrossRef] [PubMed]

17. Babu, M.A.; Shakya, N.; Prathipati, P.; Kaskhedikar, S.G.; Saxena, A.K. Development of 3D-QSAR models for 5-lipoxygenase antagonists: Chalcones. *Bioorg. Med. Chem.* **2002**, *10*, 4035–4041. [CrossRef]

18. Irfan, R.; Mousavi, S.; Alazmi, M.; Saleem, R.S.Z. A Comprehensive Review of Aminochalcones. *Molecules* **2020**, *25*, 5381. [CrossRef] [PubMed]

19. Reddy, N.P.; Aparoy, P.; Reddy, T.C.; Achari, C.; Sridhar, P.R.; Reddanna, P. Design, synthesis, and biological evaluation of prenylated chalcones as 5-LOX inhibitors. *Bioorg. Med. Chem.* **2010**, *18*, 5807–5815. [CrossRef]

20. Detsi, A.; Majdalani, M.; Kontogiorgis, C.A.; Hadjipavlou-Litina, D.; Kefalas, P. Natural and synthetic 2′-hydroxy-chalcones and aurones: Synthesis, characterization and evaluation of the antioxidant and soybean lipoxygenase inhibitory activity. *Bioorg. Med. Chem.* **2009**, *17*, 8073–8085. [CrossRef]

21. Valli, M.; dos Santos, R.N.; Figueira, L.D.; Nakajima, C.H.; Castro-Gamboa, I.; Andricopulo, A.D.; Bolzani, V.S. Development of a natural products database from the biodiversity of Brazil. *J. Nat. Prod.* **2013**, *76*, 439–444. [CrossRef]

22. Pilon, A.C.; Valli, M.; Dametto, A.C.; Pinto, M.E.F.; Freire, R.T.; Castro-Gamboa, I.; Andricopulo, A.D.; Bolzani, V.S. NuBBE$_{DB}$: An updated database to uncover chemical and biological information from Brazilian biodiversity. *Sci. Rep.* **2017**, *7*, 7215. [CrossRef] [PubMed]

23. Taraporewala, I.B.; Kauffman, J.M. Synthesis and structure-activity relationships of anti-inflammatory 9,10-dihydro-9-oxo-2-acridine-alkanoic acids and 4-(2-carboxyphenyl)aminobenzenealkanoic acids. *J. Pharm. Sci.* **1990**, *79*, 173–178. [CrossRef]

24. Passalacqua, T.G.; Dutra, L.A.; Almeida, L.; Velásquez, A.M.; Torres, F.A.; Yamasaki, P.R.; Santos, M.B.; Regasini, L.O.; Michels, P.A.; Bolzani, V.S.; et al. Synthesis and evaluation of novel prenylated chalcone derivatives as anti-leishmanial and anti-trypanosomal compounds. *Bioorg. Med. Chem. Lett.* **2015**, *25*, 3342–3345. [CrossRef] [PubMed]

25. Katsori, A.M.; Chatzopoulou, M.; Dimas, K.; Kontogiorgis, C.; Patsilinakos, A.; Trangas, T.; Hadjipavlou-Litina, D. Curcumin analogues as possible anti-proliferative & anti-inflammatory agents. *Eur. J. Med. Chem.* **2011**, *46*, 2722–2735. [PubMed]

26. Daina, A.; Michielin, O.; Zoete, V. SwissADME: A free web tool to evaluate pharmacokinetics, drug-likeness and medicinal chemistry friendliness of small molecules. *Sci. Rep.* **2017**, *7*, 42717. [CrossRef] [PubMed]

27. SwissADME. Available online: http://www.swissadme.ch/ (accessed on 25 November 2020).

28. Ha, T.J.; Nihei, K.; Kubo, I. Lipoxygenase inhibitory activity of octyl gallate. *J. Agric. Food Chem.* **2004**, *52*, 3177–3181. [CrossRef]

29. Steczko, J.; Donoho, G.P.; Clemens, J.C.; Dixon, J.E.; Axelrod, B. Conserved histidine residues in soybean lipoxygenase: Functional consequences of their replacement. *Biochemistry* **1992**, *31*, 4053–4057. [CrossRef] [PubMed]

30. Boyington, J.C.; Gaffney, B.J.; Amzel, L.M. The three-dimensional structure of an arachidonic acid 15-lipoxygenase. *Science* **1993**, *260*, 1482–1486. [CrossRef]

31. Amin, S.N.M.; Idris, M.H.M.; Selvaraj, M.; Amin, S.N.M.; Jamari, H.; Kek, T.L.; Salleh, M.Z. Virtual screening, ADME study, and molecular dynamic simulation of chalcone and flavone derivatives as 5-Lipoxygenase (5-LO) inhibitor. *Mol. Simul.* **2020**, *46*, 487–496. [CrossRef]

32. Clapp, C.H.; Grandizio, A.M.; Yang, Y.; Kagey, M.; Turner, D.; Bicker, A.; Muskardin, D. Irreversible inactivation of soybean lipoxygenase-1 by hydrophobic thiols. *Biochemistry* **2002**, *24*, 11504–11511. [CrossRef] [PubMed]

33. Certara. *Sybyl*; Version 8.0; Certara: Headquarters Princeton, NJ, USA, 2012.

34. CCDC. *GOLD*; Version 1.10; CCDC: Hong Kong, China, 2020.
35. Wallace, A.C.; Laskowski, R.A.; Thornton, J.M. LIGPLOT: A program to generate schematic diagrams of protein-ligand interactions. *Prot. Eng.* **1995**, *8*, 127–134. [CrossRef] [PubMed]
36. Santos, M.B.; Marques, B.C.; Ayusso, G.M. Chalcones and their B-aryl analogues as myeloperoxidase inhibitors: In silico, in vitro and ex vivo investigations. *Bioorg. Chem.* **2021**, *110*, 104773. [CrossRef] [PubMed]
37. John, G.S.M.; Takeuchi, S.; Venkatraman, G.; Rayala, S.K. Nordihydroguaiaretic Acid in Therapeutics: Beneficial to Toxicity Profiles and the Search for its Analogs. *Curr. Cancer Drug Targets* **2020**, *20*, 86–103.

Communication

Phloridzin Acts as an Inhibitor of Protein-Tyrosine Phosphatase MEG2 Relevant to Insulin Resistance

Sun-Young Yoon [1,2,†], Jae Sik Yu [1,†], Ji Young Hwang [1], Hae Min So [1], Seung Oh Seo [1], Jung Kyu Kim [3], Tae Su Jang [4], Sang J. Chung [1,*] and Ki Hyun Kim [1,*]

[1]   School of Pharmacy, Sungkyunkwan University, Suwon 16419, Korea; dalae1104@gmail.com (S.-Y.Y.); jsyu@bu.edu (J.S.Y.); smailemaster@naver.com (J.Y.H.); haemi9312@gmail.com (H.M.S.); wjstkz@naver.com (S.O.S.)

[2]   Department of Cosmetic Science, Kwangju Women's University, Gwangju 62396, Korea

[3]   School of Chemical Engineering, Sungkyunkwan University, Suwon 16419, Korea; legkim@skku.edu

[4]   Department of Medicine, Dankook University, Cheonan, Chungnam 31116, Korea; jangts@dankook.ac.kr

[*]   Correspondence: sjchung@skku.edu (S.J.C.); khkim83@skku.edu (K.H.K.); Tel.: +82-31-290-7703 (S.J.C.); +82-31-290-7700 (K.H.K.)

[†]   These authors contributed equally to this work.

Abstract: Inhibition of the megakaryocyte protein tyrosine phosphatase 2 (PTP-MEG2, also named PTPN9) activity has been shown to be a potential therapeutic strategy for the treatment of type 2 diabetes. Previously, we reported that PTP-MEG2 knockdown enhances adenosine monophosphate activated protein kinase (AMPK) phosphorylation, suggesting that PTP-MEG2 may be a potential antidiabetic target. In this study, we found that phloridzin, isolated from *Ulmus davidiana* var. *japonica*, inhibits the catalytic activity of PTP-MEG2 (half-inhibitory concentration, $IC_{50} = 32 \pm 1.06$ μM) in vitro, indicating that it could be a potential antidiabetic drug candidate. Importantly, phloridzin stimulated glucose uptake by differentiated 3T3-L1 adipocytes and C2C12 muscle cells compared to that by the control cells. Moreover, phloridzin led to the enhanced phosphorylation of AMPK and Akt relevant to increased insulin sensitivity. Importantly, phloridzin attenuated palmitate-induced insulin resistance in C2C12 muscle cells. We also found that phloridzin did not accelerate adipocyte differentiation, suggesting that phloridzin improves insulin sensitivity without significant lipid accumulation. Taken together, our results demonstrate that phloridzin, an inhibitor of PTP-MEG2, stimulates glucose uptake through the activation of both AMPK and Akt signaling pathways. These results strongly suggest that phloridzin could be used as a potential therapeutic candidate for the treatment of type 2 diabetes.

Keywords: protein tyrosine phosphatases (PTPs); PTP-MEG2; phloridzin; type 2 diabetes; glucose-uptake

Citation: Yoon, S.-Y.; Yu, J.S.; Hwang, J.Y.; So, H.M.; Seo, S.O.; Kim, J.K.; Jang, T.S.; Chung, S.J.; Kim, K.H. Phloridzin Acts as an Inhibitor of Protein-Tyrosine Phosphatase MEG2 Relevant to Insulin Resistance. *Molecules* **2021**, *26*, 1612. https://doi.org/10.3390/molecules26061612

Academic Editor: Young Hae Choi

Received: 5 January 2021
Accepted: 7 March 2021
Published: 14 March 2021

1. Introduction

Protein tyrosine phosphatases (PTPs) constitute a diverse family of enzymes and control the cellular protein tyrosine phosphorylation levels associated with cell growth, metabolism, and apoptosis [1]. Loss of the regulation in protein tyrosine phosphorylation has been implicated in many diseases, including cancer, diabetes, and autoimmune disorders, suggesting that PTPs may act as potential drug targets [1,2]. Some PTPs, such as PTPN1, PTPN9, PTPN11, PTPRF, PTPRS, and dual specificity phosphatase 9 (DUSP-9), lead to type 2 diabetes associated insulin resistance through antagonism of insulin action [3,4]. Since the activation of adenosine monophosphate -activated protein kinase (AMPK) stimulates glucose uptake by promoting GLUT4 translocation, antidiabetic effects are associated with AMPK phosphorylation, suggesting that AMPK is a target for the treatment of type 2 diabetes [5–8]. We have previously shown that knockdown of the megakaryocyte protein tyrosine phosphatase 2 (PTP-MEG2, also named PTPN9) enhances AMPK phosphorylation in 3T3-L1 preadipocytes, suggesting that PTP-MEG2 can be a potential antidiabetic drug

159

target [9]. In addition, it has been reported that PTP-MEG2 is an antagonist of liver insulin signaling and that depletion of PTP-MEG2 in the liver of diabetic (db/db) mice leads to insulin sensitization and normalization of hyperglycemia [10]. Moreover, PTP-MEG2 inhibitor has shown to enhance Akt phosphorylation in primary hepatocytes and improve insulin resistance and glucose homeostasis in diet-induced obese C57BL/6 mice, suggesting that the inhibition of PTP-MEG2 activity may be a potential therapeutic approach in treating type 2 diabetes [1]. Based on previous research, identification of the PTP-MEG2 inhibitors may be an effective strategy for treating type 2 diabetes. In this study, phloridzin was identified as a slow binding inhibitor of PTP-MEG2, suggesting that it could potentially be used as an antidiabetic drug.

Phloridzin, phloretin 2′-*O*-glucoside (sometimes also referred to as phlorizin), is a flavonoid belonging to the chemical class of dihydrochalcones and is associated with flavonoid biosynthesis [11]. Phloridzin is one of the major polyphenols found in apple (*Malus* sp.), and it is mainly present in the leaves and bark of apple trees. Low amounts of phloridzin exist in apple fruits as well. A few other species belonging to Rosaceae and Ericaceae also possess this compound, but only in low amounts [11]. Phloridzin has been widely used as an alternative medicine and possesses various pharmacological traits: scavenging effect for superoxide anions [12], inhibitory effect on lipid peroxidation [13], melanogenesis stimulatory effect through the inhibition of protein kinase C activity [14], and proliferative effect on human CD49f$^+$/CD29$^+$ keratinocytes [15]. In particular, the effects of phloridzin on glucose uptake and diabetes have been intensively investigated [11,16–18]. Phloridzin lowered the blood glucose levels and reversed *Sglt1* expression in streptozotocin-induced diabetic mice [17], and was able to inhibit the sodium-dependent glucose-linked transporter, SGLT [18]. Due to its potential applicability as a pharmacological agent, phloridzin has recently received significant attention from many research groups. However, the direct target for the antidiabetic function of phloridzin remains unknown. In this study, phytochemical analysis of the root barks of *Ulmus davidiana* var. *japonica* led to the isolation of phloridzin. We observed that phloridzin inhibited the enzymatic activity of PTP-MEG2 in vitro, indicating that phloridzin targets PTP-MEG2. We subsequently examined the effects of phloridzin in insulin-responsive cell lines such as mouse 3T3-L1 adipocytes and C2C12 muscle cells. Herein, we describe the isolation of phloridzin and its potential as an inhibitor of PTP-MEG2 relevant to insulin resistance.

2. Results

2.1. Isolation and Identification of Phloridzin

The crude extract of the root barks of *U. davidiana* var. *japonica* was sequentially solvent partitioned between water and organic solvents, including hexane, dichloromethane, ethyl acetate (EtOAc), and *n*-butanol of increasing polarity, yielding four fractions. Phytochemical analysis of the EtOAc-soluble fraction was performed using repeated column chromatography and semi-preparative HPLC separation, resulting in the isolation of phloridzin. The structure of phloridzin (Figure 1) was determined by comparing the ^{1}H and ^{13}C-NMR spectra with those previously reported in the literature [11,14,17], and by LC/MS analysis.

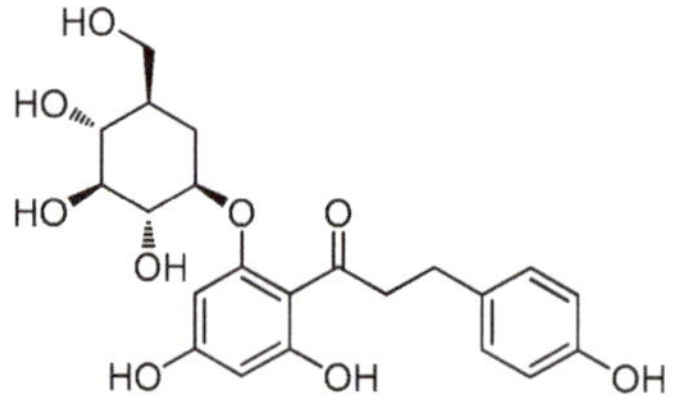

Figure 1. The chemical structure of phloridzin.

2.2. Suppression of PTP-MEG2 Enhanced AMPK Phosphorylation

To identify the antidiabetic target of phloridzin, we performed PTP-MEG2 knockdown in 3T3-L1 preadipocytes using PTP-MEG2 siRNA for 72 h. Scrambled siRNA was used as a negative control. At the same time, we also incubated cells with 10 μM phloridzin for 72 h. Efficient suppression of PTP-MEG2 was confirmed by quantitative RT-PCR or Western blotting (Figure 2A,B). Consistent with our previous findings [9], we found that PTP-MEG2 knockdown significantly increased AMPK phosphorylation, suggesting that PTP-MEG2 can be a potential antidiabetic drug target (Figure 2B,C). In addition, AMPK phosphorylation was also enhanced in 10 μM phloridzin-treated cells, as compared to the control (Figure 2B,C). Furthermore, concurrent treatment with PTP-MEG2 siRNA and 10 μM phloridzin increased AMPK phosphorylation less than that of PTP-MEG2 siRNA knockdown alone, indicating that phloridzin targets PTP-MEG2 (Figure 2B,C).

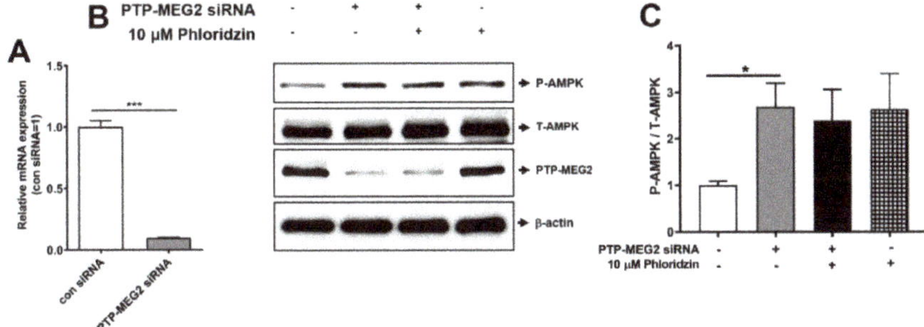

Figure 2. Suppression of PTP-MEG2 enhanced AMPK phosphorylation. (**A,B**) 3T3-L1 preadipocytes were transfected with 30 nM PTP-MEG2 siRNA or scrambled siRNA as a control (con). After 72 h, cells were lysed and analyzed by quantitative real-time PCR (**A**) or Western blotting (**B**). (**C**) Quantification of phospho-AMPK/total-AMPK using the ATTO image analysis software (CS analyzer 4). All experiments were independently repeated at 3 times. Results are presented as mean $\pm$ standard error of the mean (SEM). *** $p < 0.001$, * $p < 0.05$ compared to the control siRNA condition (con siRNA).

2.3. Phloridzin Inhibited the Catalytic Activity of PTP-MEG2 In Vitro

Next, PTP-MEG2 was overexpressed and purified using a cobalt affinity resin (Figure 3A). To measure its catalytic activity, kinetic constants of PTP-MEG2 were determined using DiFMUP as a fluorogenic PTP substrate (Table 1). Phloridzin showed better inhibition of DiFMUP hydrolysis by PTP-MEG2 than other PTPs such as PTPN1, PTPRS, PTPRF, and DUSP9 (Table 2). The progress curves of DiFMUP hydrolysis by PTP-MEG2 were linear in the absence of an inhibitor but hyperbolic progress curves were observed in the presence of phloridzin, indicating that phloridzin acted as a slow-binding inhibitor of PTP-MEG2 (Figure 3B). To estimate its half-inhibitory concentration (IC$_{50}$), PTP-MEG2 was pre-incubated with phloridzin for 20 min and then a DiFMUP solution was added to it. The IC$_{50}$ value of phloridzin against PTP-MEG2 was determined to be 32 $\pm$ 1.06 μM (Figure 3C). Since the Hill coefficient (*n*) obtained based on the slope of the Hill plot represents the degree of cooperativity in ligand–protein interactions (*n* > 1, positively cooperative binding; *n* = 1, noncooperative binding; *n* < 1, negative cooperative binding) [19], we estimated the extent of cooperativity between phloridzin and PTP-MEG2. We found that *n* was estimated to be 2.5 for PTP-MEG2, indicating positive cooperation in the binding of phloridzin to PTP-MEG2 (Figure 3D). Moreover, surface plasmon resonance (SPR) analysis was performed to measure the molecular interaction between phloridzin and PTP-MEG2. To this end, the association ($k = 48.4$ M^{-1}s^{-1}) and dissociation ($k = 2.26 \times 10^{-4}$ s^{-1}) constants were estimated. SPR analysis revealed the specific interaction of phloridzin with PTP-MEG2, which had an equilibrium constant ($K = k/k$) value of 4.66 μM (Figure 3E). In addition, we measured the enzymatic activity of purified PTP-MEG2 by antidiabetic agents such as rosiglitazone and metformin. We found that both rosiglitazone and metformin did not inhibit the catalytic activity of PTP-MEG2 in vitro, indicating that inhibition of catalytic activity of PTP-MEG2 was phloridzin-specific among glucose-lowering agents (Supplementary Materials Figure S1).

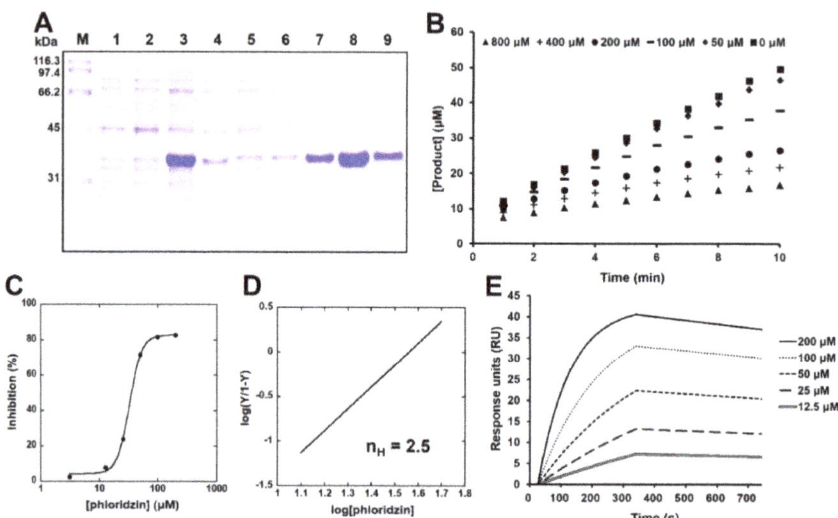

Figure 3. PTP-MEG2 inhibition by phloridzin. (**A**) Polyacrylamide gel electrophoresis (PAGE) analysis of PTP-MEG2 (molecular weight: 37.8 kDa). M: molecular weight marker, lane 1: total sample before induction, lane 2: supernatant before induction, lane 3: total sample after induction, lane 4: supernatant after induction, lane 5: sample passed through the column after sonication, lane 6: sample after washing the column using lysis buffer, lane 7: sample after washing the column using 10 mM imidazole, lanes 8–9: protein eluted using 100 mM imidazole. (**B**) Progress curves showing inhibition of PTP-MEG2 by phloridzin at 800 μM (▲), 400 μM (+), 200 μM (●), 100 μM (▬), 50 μM (♦), and 0 μM (■). (**C**) IC_{50} values estimated by fitting the sigmoid plot for % inhibition versus various concentrations of phloridzin (200, 100, 50, 25, 12.5, 3.125, and 0 μM). (**D**) Hill coefficient (nH) was calculated from the slopes of the Hill plots created according to the Hill equation. (**E**) SPR analysis was performed in running buffer (5% DMSO in PBS) using a Reichert SR7500DC system. Purified PTP-MEG2 was injected for 3 min at a flow rate of 20 μL/min. Interaction analyses were performed by injection of increasing concentrations (12.5 to 200 μM) of phloridzin at a flow rate of 30 μL/min. Association and dissociation signals were monitored for 300 and 420 s, respectively. All experiments were independently repeated at 3 times.

Table 1. Kinetic constants for 6,8-difluoro-4-methylumbelliferyl phosphate (DiFMUP)hydrolysis by PTP-MEG2. Kinetic constants of PTP-MEG2 for measuring enzyme activity were determined using DiFMUP as a fluorogenic PTP substrate.

	[E] (pM)	K (μM)	V (μM min^{-1})	k (min^{-1})	k/K (μM^{-1} min^{-1})
PTP-MEG2	62.5	163.1	2.166	3.47×10^4	2.12×10^2

Table 2. Inhibition of PTPs by phloridzin treatment. PTPs were added to solutions containing 200 μM phloridzin in reaction buffer with DiFMUP ($2 \times K$).

PTPs	PTPN1	PTPRS	PTP-MEG2	PTPRF	DUSP9
Inhibition (%)	39 ± 3	18 ± 1	81 ± 1	11 ± 2	13 ± 2

2.4. Glucose Uptake Is Increased Following Phloridzin Treatment

Previously, PTP-MEG2 inhibitors have been shown to increase insulin signaling and improve glucose homeostasis in diet-induced obese mice [1]. Since glucose uptake by insulin target tissues, such as skeletal muscle, liver, and fat, plays an important role in maintaining glucose homeostasis [6], we examined the effects of phloridzin in 3T3-L1 adipocytes and C2C12 muscle cells. To assess the appropriate concentrations of phloridzin for use in cell-based studies, its cytotoxicity was studied using an EZ-Cytox assay kit. Incubation of 3T3-L1 adipocytes and C2C12 muscle cells with 1 or 10 μM phloridzin for 48 h had no effects on cell viability, indicating that these concentrations were suitable for cell treatment (Figure 4A,B).

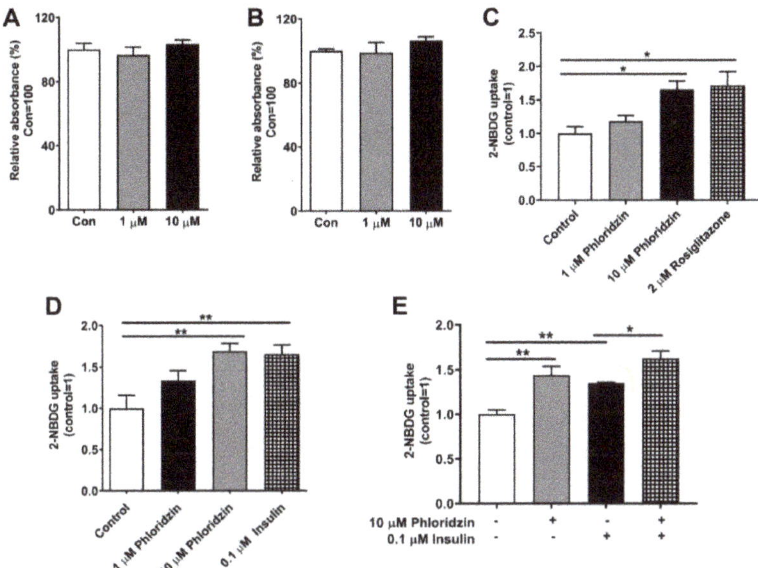

Figure 4. Phloridzin increases glucose uptake. (**A,B**) 3T3-L1 adipocytes (**A**) and C2C12 muscle cells (**B**) were incubated with the indicated concentrations of phloridzin for 2 days and cell viability was assessed by the EZ-Cytox assay kit. (**C–E**) Differentiated 3T3-L1 preadipocytes (**C**) and C2C12 muscle cells (**D,E**) were incubated for 1 h or 72 h (**E**) with 1 or 10 μM phloridzin, 0.1 μM insulin, or 2 μM rosiglitazone, and then treated with the fluorescent glucose indicator, 2-NBDG, for 30 min. For the detection of 2-NBDG, cells were washed with PBS and the fluorescence intensity was evaluated using a fluorescence microplate reader. All experiments were independently repeated at 3 times. Results are expressed as mean ± SEM. Data were analyzed using a two-tailed unpaired *t*-test. ** $p < 0.01$, * $p < 0.05$ compared to the control group.

We next elucidated the effects of phloridzin on the glucose uptake of differentiated 3T3-L1 adipocytes and C2C12 muscle cells using the fluorescent glucose probe, 2-(N-(7-nitrobenz-2-oxa-1, 3-diazol-4-yl) amino)-2-deoxyglucose (2-NBDG). Cells were incubated with 1 or 10 μM phloridzin, 0.1 μM insulin (positive control), or 2 μM rosiglitazone (an antidiabetic drug used as the positive control) for 1 h. After changing the culture medium, the cells were treated with 2-NBDG for 30 min. After washing the cells with PBS, the fluorescence intensity (excitation/emission = 485/535 nm) was measured. We found that the phloridzin-treated cells had significantly enhanced fluorescence intensity as compared to the control, suggesting that phloridzin stimulated glucose uptake by differentiated 3T3L1 adipocytes as well as C2C12 muscle cells (Figure 4C,D). In addition, C2C12 muscle cells were cultured with 10 μM phloridzin or 0.1 μM insulin for 72 h and glucose uptake was assessed. We found that incubation with 10 μM phloridzin enhanced glucose uptake. Furthermore, concurrent treatment with 10 μM phloridzin and 0.1 μM insulin increased glucose uptake, as compared to cells treated with 0.1 μM insulin alone (Figure 4E). In addition, treatment with 10 μM phloridzin for 72 h induced the translocation of glucose transporter type 4 (GLUT4) in C2C12 muscle cells, similar to 0.1 μM insulin-treated effects

(Supplementary Materials Figure S2). Treatment with 2 μM rosiglitazone or 0.1 μM insulin also increased the cellular glucose uptake, indicating that the fluorescent probe worked properly in our cell systems (Figure 4C–E).

2.5. Phloridzin Enhanced the Phosphorylation of AMPK and Akt

We next investigated the cellular mechanisms associated with increased glucose uptake. AMPK controls glucose homeostasis by regulating cellular metabolism of peripheral tissues, such as skeletal muscle, liver, and adipose tissues. Activation of AMPK leads to the stimulation of glucose uptake and lipid oxidation [7,8]. Akt signaling is associated with insulin-stimulated glucose uptake and Akt phosphorylation promotes GLUT-4 translocation and glucose uptake [20,21]. In this study, mature 3T3-L1 adipocytes were incubated with 1 or 10 μM phloridzin for 72 h and Western blotting was performed. We found that treatment with 10 μM phloridzin markedly induced AMPK phosphorylation compared to the control (Figure 5A,B).

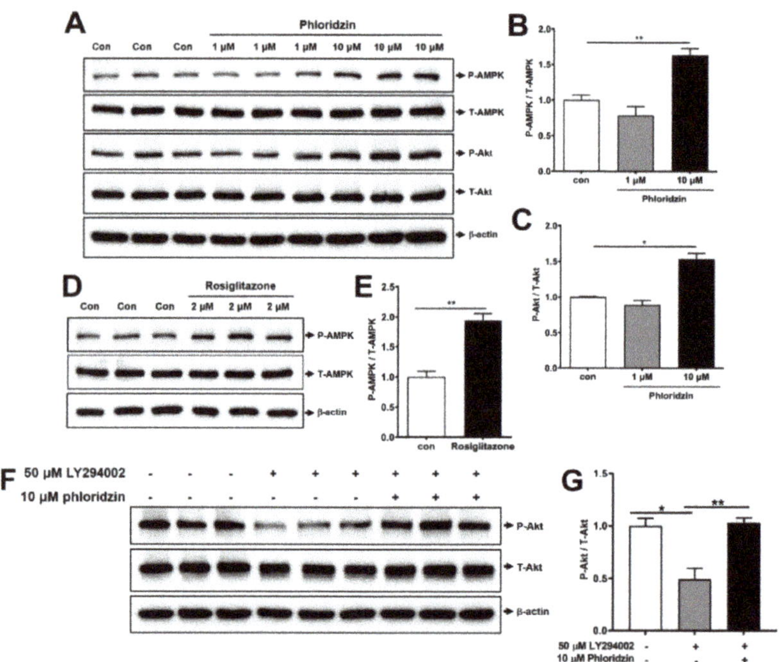

Figure 5. Phloridzin enhances the phosphorylation of AMPK and Akt. (**A,D**) Mature adipocytes were incubated with 1 or 10 μM phloridzin or 2 μM rosiglitazone. After 72 h, the cells were lysed, and Western blotting was performed using antibodies against phosphorylated AMPK, total AMPK, phosphorylated Akt, total Akt, and beta-actin. (**B,C,E,G**) Quantification of phospho-AMPK/total-AMPK (**B,E**) and phospho-Akt/total-Akt (**C,G**) using the ATTO image analysis software (CS analyzer 4). (**F**) 50 μM LY294002 or 10 μM phloridzin were treated in mature 3T3-L1 adipocytes. After 24 h, cells were lysed and analyzed by Western blotting. All experiments were independently repeated at 3 times. Results are presented as mean ± SEM. Data were analyzed using two-tailed unpaired *t*-test. ** $p < 0.01$, * $p < 0.05$ compared to the control group.

In agreement with this result, incubation with 10 μM phloridzin significantly enhanced Akt phosphorylation compared to the control (Figure 5A,C). Rosiglitazone also remarkably increased AMPK phosphorylation, indicating that the Western blotting experiments were properly performed (Figure 5D,E). These results indicate that phloridzin increases glucose uptake through activation of both AMPK and Akt signaling pathways.

2.6. Phloridzin Restored LY294002-Blocked Akt Phosphorylation

We next examined the role of phloridzin in the PI3K/Akt pathway. Since LY294002, a chemical inhibitor of phosphatidylinositol-3-kinase (PI3K), reduces Akt phosphorylation, it is commonly used to study the role of the PI3K/Akt pathway [22]. In our study, we incubated mature 3T3-L1 adipocytes with 50 μM LY294002 for 24 h. Treatment with 50 μM LY294002 markedly decreased Akt phosphorylation, while 10 μM phloridzin restored LY294002-reduced Akt phosphorylation (Figure 5F,G). Our study of the PI3K inhibitor revealed that phloridzin significantly stimulated the downstream Akt signaling pathway.

2.7. Phloridzin Ameliorated Palmitate-Induced Insulin Resistance in C2C12 Muscle Cells

Next, we investigated whether phloridzin might attenuate insulin resistance in C2C12 muscle cells. Palmitate, a saturated fatty acid, has been shown to decrease insulin-stimulated glucose uptake and reduce the insulin-stimulated phosphorylation of Akt, suggesting that palmitate causes insulin resistance in C2C12 muscle cells [23,24]. When we incubated C2C12 muscle cells with 500 μM palmitate for 48 h, palmitate-treated cells decreased insulin-stimulated glucose uptake, indicating that palmitate contributes to insulin resistance (Figure 6A).

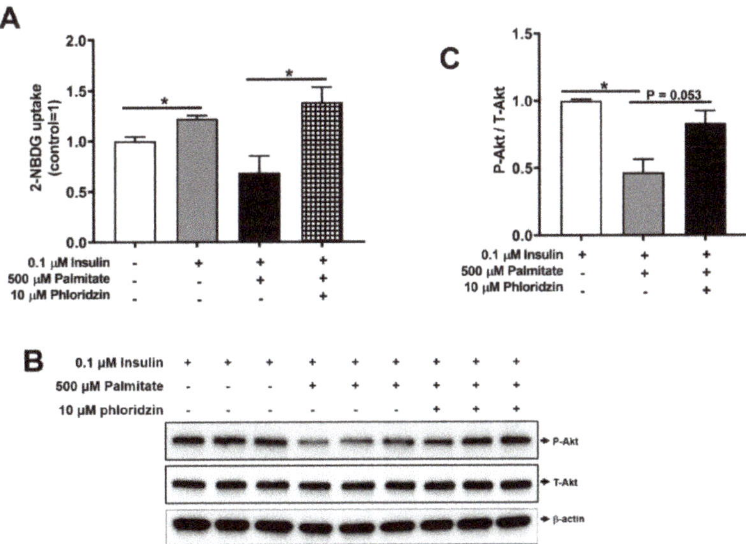

Figure 6. Phloridzin attenuates palmitate-induced insulin resistance in C2C12 muscle cells. (**A**,**B**) Differentiated C2C12 muscle cells were treated with 10 μM phloridzin or 500 μM palmitate for 48 h and then incubated with and without 0.1 μM insulin for 30 min. Cells were analyzed by glucose uptake (**A**) or Western blotting (**B**). (**C**) Quantification of phospho-Akt/total-Akt using the ATTO image analysis software. All experiments were independently repeated at 3 times. Results are expressed as mean ± SEM. Data were analyzed using two-tailed unpaired *t*-test. * *p* < 0.05 compared to the control group.

Importantly, 10 μM phloridzin attenuated palmitate-induced insulin resistance by markedly increasing glucose uptake (Figure 6A). In addition, palmitate inhibited the insulin-stimulated phosphorylation of Akt in C2C12 muscle cells using Western blotting (Figure 6B,C). Interestingly, incubation with 10 μM phloridzin ameliorated palmitate-induced insulin resistance by enhancing Akt phosphorylation (Figure 6B,C). These results indicate that phloridzin attenuates palmitate-induced insulin resistance in C2C12 muscle cells.

2.8. Phloridzin Did Not Enhance Lipid Accumulation

We next elucidated the effect of phloridzin on lipid accumulation. 3T3-L1 preadipocytes were differentiated in the presence of phloridzin or rosiglitazone. On day 6 of differentiation, the degree of adipocyte differentiation was evaluated by Oil Red O staining. In addition, we performed quantification of lipid accumulation. To this end, Oil Red O dye was eluted by incubation with isopropanol and the absorbance was measured using a microplate reader. Adipogenesis is associated with CCAAT/enhancer-binding protein α (C/EBPα) and peroxisome proliferator-activated receptor-γ (PPARγ) [25]. Rosiglitazone, a PPARγ agonist, is an insulin-sensitizing agent with side effects such as rodents and human weight gain [26,27]. In line with this, we found that rosiglitazone significantly stimulated the formation of lipid droplets compared to the control (Figure 7A,B).

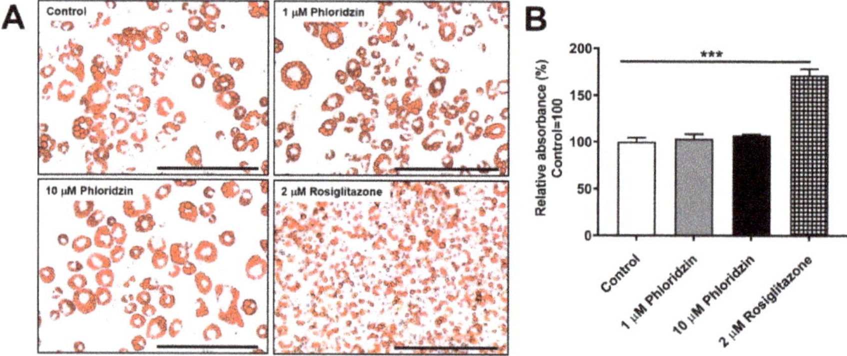

Figure 7. Phloridzin does not accelerate lipid accumulation. (**A**) 3T3-L1 preadipocytes were differentiated into mature adipocytes in the presence of phloridzin or rosiglitazone. The extent of lipid droplets was monitored by Oil Red O staining on day 6 of differentiation. Images of fat accumulation were acquired using an EVOS FL Imaging System. (**B**) For quantification of lipid accumulation, Oil Red O dye was eluted by incubation with isopropanol and the absorbance was measured at 490 nm using a microplate reader. All experiments were independently repeated at 3 times. Results are presented as mean ± SEM. Data were analyzed using two-tailed unpaired *t*-test. *** $p < 0.001$ compared to the control group. Scale bar: 200 µm.

On the contrary, phloridzin treatment did not accelerate adipocyte differentiation compared to the control, indicating that phloridzin does not enhance fat accumulation (Figure 7A,B). These results indicate that phloridzin may act as a glucose lowering agent through the PPARγ-independent signaling pathway, suggesting that phloridzin stimulates glucose uptake without significant lipid accumulation.

3. Discussion

Since PTPs are associated with diverse diseases, including cancer, diabetes, and autoimmune dysfunctions [2–4], they are suggested as next-generation drug targets. Some reports have shown that treatment with PTP-MEG2 inhibitors results in improved insulin action in both in vitro and in diet-induced obese mice [1,28], and a decrease in PTP-MEG2 in hepatocytes increases insulin sensitivity [10]. In line with our previous findings [9], we also showed that PTP-MEG2 knockdown increased AMPK phosphorylation, suggesting that PTP-MEG2 may be a potential antidiabetic drug target. Therefore, the use of PTP-MEG2 inhibitors is considered an effective strategy for the treatment of type 2 diabetes. In this study, phloridzin was identified as an inhibitor of PTP-MEG2, and, therefore, a potential anti-diabetes candidate.

Phloridzin is one of the dihydrochalcones usually found in apples and contributes to the inhibition of oxidative stress [11,17]. Some reports have shown that phloretin and phloridzin improve insulin sensitivity and enhance glucose uptake by inhibiting Cdk5 activation and corresponding PPARγ phosphorylation in differentiated 3T3L1 cells [29], and that phloretin exerts beneficial effects on the skeletal system in estrogen-deficient mice [30]. On the other hand, long-term intake of products containing high doses of phloridzin has been reported to adversely affect the musculoskeletal system in type 2 diabetic rat models [31]. Sodium–glucose co-transporter-2 (SGLT2) plays an important role in reabsorption of glucose in the kidney, and inhibition of SGLT2 has been reported as a new therapeutic strategy for diabetes [32]. Phloridzin has been shown to lack specificity for SGLT2, and phloridzin has been reported previously to improve hyperglycemia through the reduction of the blood glucose level and reverse the streptozotocin-mediated induction of SGLT1 rather than SGLT2 in the small intestine in streptozotocin-induced diabetic mice [17,32–34]. In addition, it has been reported that phloridzin has the proliferative effects on CD49f$^+$/CD29$^+$ keratinocytes via the extracellular signal-regulated kinase (ERK)-dependent mammalian target of rapamycin (mTOR) pathway [15]. However, the direct target for the antidiabetic function of phloridzin is unknown. Here, we found that phloridzin inhibited the enzymatic activity of PTP-MEG2 in vitro, indicating that phloridzin targets PTP-MEG2. Moreover, in the present study, we demonstrated that phloridzin enhances glucose uptake by differentiated 3T3-L1 adipocytes and C2C12 muscle cells through the activation of both AMPK and Akt signaling pathways.

Insulin binds to the insulin receptor on the surface of insulin-responsive tissues such as liver, adipose tissue, and skeletal muscle to induce membrane translocation of GLUT4 [35]. The activation of AMPK stimulates glucose uptake by promoting GLUT4 translocation, and reduction of GLUT4 is associated with cellular insulin resistance in diabetic animal models [5,8,36]. We also found that phloridzin induced translocation of GLUT4 in C2C12 muscle cells, suggesting that phloridzin stimulates glucose uptake by promoting GLUT4 translocation.

Insulin resistance is a major feature of type 2 diabetes and refers to insulin dysfunction in peripheral tissues, such as skeletal muscles, liver, and adipose tissues [37]. Palmitate, a saturated fatty acid, has been found to induce insulin resistance through downregulation of the insulin-stimulated phosphorylation of Akt [23,24]. Some reports have indicated that *cis*-9, *trans*-11-conjugated linoleic acid attenuates palmitate-induced insulin resistance by increasing glucose and fatty acid consumption [38], and that palmitate-induced insulin resistance was blocked by salsalate and adiponectin through inhibition of selenoprotein *p* in HepG2 cells [24]. In line with this, we also found that phloridzin ameliorated palmitate-induced insulin resistance by significantly enhancing Akt phosphorylation.

Rosiglitazone, a member of the thiazolidinedione class, is commonly used as an antidiabetic drug. Rosiglitazone has been shown to enhance glucose uptake into muscle and adipose tissue via activation of AMPK [26,39]. Rosiglitazone has adverse effects, one of which is weight gain, and has been shown to cause fat accumulation through activation of PPARγ [26,27,40]. Consistent with these findings, we found that rosiglitazone induced lipid accumulation. In contrast, phloridzin does not enhance fat accumulation compared to the control, suggesting that phloridzin improves insulin sensitivity without significant lipid accumulation.

4. Materials and Methods

4.1. Chemicals

Phloridzin (purity ≥98%, by LC/MS) was obtained from the root barks of *U. davidiana* var. *japonica* in the continuing systematic projects to isolate natural products from medicinal plants in our group. The root barks of *U. davidiana* var. *japonica* were collected from Wonhwasan-ro, Jecheon-si, Korea, in 2016, and authenticated by one of the authors (K. H. Kim). A voucher specimen (SKKU-NR 0401) of the plant was deposited at the herbarium of the School of Pharmacy, Sungkyunkwan University, Suwon, Korea. Briefly, extracts of dried root barks of *U. davidiana* var. *japonica* were prepared in 50% aqueous EtOH for 2 days at 70 °C and filtered. The filtrate was concentrated under a reduced pressure to obtain the EtOH extract, which was successively solvent-partitioned using hexane, dichloromethane, ethyl acetate (EtOAc), and *n*-butanol. The EtOAc-soluble fraction was subjected to chromatography in a Diaion HP-20 column in a gradient solvent system of MeOH (100% H$_2$O, 20% MeOH, 40% MeOH, 60% MeOH, 80% MeOH, and 100% MeOH) to yield six fractions (E0, E2, E4, E6, E8, and E10). Fraction E8 was separated by preparative reversed-phase HPLC using a gradient solvent system of MeOH-H$_2$O (45–100% MeOH) to yield six fractions (E8A–E8F). Fraction E8D was separated by preparative reversed-phase HPLC using a gradient solvent system of MeCN-H$_2$O (25–40% MeCN) to further yield six fractions (E8D1–E8D6). Phloridzin was purified from fraction E8D3 using semi-preparative HPLC using a solvent system of MeOH-H$_2$O (formic acid 0.1% (*v*/*v*)) (46% MeOH).

Phloridzin: Colorless gum; ^{1}H NMR (400 MHz, CD$_3$OD): δ 2.90 (2H, t, *J* = 7.5 Hz, H-9), 3.41 (1H, m, H-5″), 3.45 (1H, m, H-2″), 3.46 (1H, m, H-3″), 3.47 (1H, m, H-4″), 3.48 (2H, m, H-8), 3.73 (1H, dd, *J* = 5.0 and 12.0 Hz, H-6″), 3.93 (1H, d, *J* = 12.0 Hz, H-6″), 5.06 (1H, d, *J* = 7.0 Hz, H-1″), 5.98 (1H, d, *J* = 2.0 Hz, H-5), 6.20 (1H, d, *J* = 2.0 Hz, H-3), 6.70 (2H, d, *J* = 8.5 Hz, H-2′ and H-6′), 7.08 (2H, d, *J* = 8.5 Hz, H-3′ and H-5′); ^{13}C NMR (100 MHz, CD$_3$OD): δ 30.8 (t, C-9), 46.9 (t, C-8), 62.2 (t, C-6″), 71.3 (d, C-4″), 74.6 (d, C-2″), 78.1 (d, C-5″), 78.3 (d, C-3″), 95.2 (d, C-5), 98.3 (d, C-3), 102.1 (d, C-1″), 106.8 (s, C-1), 115.8 (d, C-2′ and C-6′), 130.2 (d, C-3′ and C-5′), 133.6 (s, C-10), 156.3 (s, C-4′), 162.2 (s, C-4), 165.6 (s, C-6), 167.5 (s, C-2), 206.5 (s, C-7); Electrospray ionization mass spectrometry (ESIMS) (positive-ion mode) *m/z*: 457.1 [M + Na]$^+$.

4.2. Cell Culture

C2C12 muscle cells and 3T3-L1 preadipocytes obtained from the American Type Culture Collection (ATCC, Manassas, VA, USA) were cultured as previously described [5,25]. 3T3-L1 preadipocytes were grown in high glucose Dulbecco's modified Eagle's medium (DMEM; Welgene, Seoul, Republic of Korea) containing 10% bovine calf serum (BCS; Thermo Fisher Scientific, Massachusetts, USA) and an antibiotic-antimycotic solution (Welgene). C2C12 muscle cells were cultured in DMEM supplemented with 15% fetal bovine serum (FBS; Welgene) and the antibiotic-antimycotic solution. Cells were incubated at 37 °C in a 5% CO$_2$ incubator.

4.3. Cell Differentiation

The methods used for differentiating 3T3-L1 preadipocytes and C2C12 cells were as described previously [5,25]. Once 3T3-L1 preadipocytes reached 100% confluency, they were cultured in DMEM supplemented with 10% FBS, antibiotic-antimycotic solution, 0.5 mM isobutylmethylxanthine (IBMX; Merck KGaA, Darmstadt, Germany), 1 μM dexamethasone (Sigma-Aldrich, Saint Louis, MO, USA), and 5 μg/mL insulin (Merck KGaA, Darmstadt, Germany) for 2 days. Cells were then maintained in DMEM supplemented with 10% FBS, antibiotic-antimycotic solution, and 5 μg/mL insulin for 2 days. Next, cells were incubated for 4 days in DMEM containing 10% FBS and antibiotic-antimycotic solution. When C2C12 muscle cells reached 100% confluency, they were differentiated into myotubes in DMEM supplemented with 2% horse serum (Thermo Fisher Scientific), antibiotic-antimycotic solution, and 5 μM insulin for 4 days. In all the experiments, the culture medium was replenished daily.

4.4. Glucose Uptake Assay

The methods used for evaluating glucose-uptake in C2C12 muscle cells and 3T3-L1 preadipocytes were as described previously [5]. Differentiated cells were grown in low-glucose DMEM (Gibco BRL) for 4 h and then incubated with phloridzin or rosiglitazone (an antidiabetic drug used as a positive control [41]) in glucose-depleted DMEM (Gibco BRL) for 1 h. The cells were treated with 5 μM of the fluorescent glucose probe 2-[N-(7-nitrobenz-2-oxa-1,3-diazol-4-yl)amino]-2-deoxyglucose (2-NBDG; Thermo Fisher Scientific) for 30 min. After washing the cells with phosphate-buffered saline (PBS, Welgene), the cell pellets were resuspended in PBS and passed through a 40 μm cell strainer. The fluorescence intensity (excitation/emission = 485/535 nm) of the cells was measured using a fluorescence microplate reader (Victor^{TM} X4, PerkinElmer, Waltham, MA, USA).

4.5. Overexpression and Purification of Recombinant PTPs

The methods used for the overexpression and purification of PTPs were as described previously [21]. The N-terminal His6-tagged human PTP-MEG2 and PTPN1 were transformed into *Escherichia coli* (*E. coli*) Rosetta (DE3) (Merck Millipore, Darmstadt, Germany). The human genes of PTPRS, PTPNF, and DUSP9 with both an N-terminal maltose-binding protein (MBP) and a C-terminal His6-tag were transformed into *E. coli* Rosetta (DE3). For the expression of the recombinant PTPs, cells were induced for 16 h at 18 °C using 0.1 mM isopropyl β-D-1-thiogalactopyranoside (IPTG). Cells were then harvested by centrifugation (3570× *g* for 15 min at 4 °C), washed with buffer A consisting of 50 mM Tris pH 7.5, 500 mM NaCl, 5% glycerol, 0.025% 2-mercaptoethanol, and 1 mM phenyl-methylsulfonyl fluoride (PMSF), and then lysed in buffer A by ultrasonication. After centrifugation (29,820× *g* at 4 °C for 30 min), the supernatant was incubated with a cobalt affinity resin (TALON®, Takara Korea, Seoul, Korea) on a shaker at 4 °C for 1 h. The resin was then washed with buffer A containing 10 mM imidazole. PTPs were eluted with buffer A supplemented with 100 mM imidazole and stored at −80 °C until further use.

4.6. Measurement of Enzymatic Activity and Half-Inhibitory Concentration (IC$_{50}$) Value

The enzymatic activity of purified PTP-MEG2 was assayed using 6,8-difluoro-4-methylumbelliferyl phosphate (DiFMUP) (100 μM), a commonly used PTP substrate, as previously described [42]. To determine the K values, 62.5 pM of PTP-MEG2 was added to the reaction buffer (20 mM tris pH 6.0, 150 mM NaCl, 2.5 mM dithiothreitol (DTT), 0.01% Triton X-100) containing DiFMUP (800, 400, 200, 100, 50, 25, 12.5, or 6.25 μM) in 96-well plates (to a final volume of 100 μL). Fluorescence intensities were measured continuously for 10 min (excitation/emission = 355/460 nm) using a Victor^{TM} X4 multi-label plate reader (Perkin Elmer, Norwalk, CT, USA), and K values were determined using Lineweaver-Burk plots. To identify the mode of inhibition of phloridzin, phloridzin at concentrations of 800, 400, 200, 100, 50, and 0 μM was added to DiFMUP ($2 \times K$; 326.2 μM) in the reaction buffer. After addition of 62.5 pM of PTP-MEG2, progress curves were plotted against the product concentration. To estimate IC$_{50}$ values, various concentrations of phloridzin (200, 100, 50, 25, 12.5, 3.125, and 0 μM) were pre-incubated with 62.5 pM of PTP-MEG2 for 20 min, following which, the solution containing 326.2 μM DiFMUP was added to these mixtures. IC$_{50}$ values were determined by fitting the sigmoid plots for the % inhibition versus various concentrations of phloridzin (KaleidaGraph, Synergy Software, Reading, Pennsylvania, USA) used. The Hill coefficient (n), which measures the degree of cooperativity between phloridzin and PTP-MEG2, is defined as the slope of the Hill plot created using the Hill equation [22].

4.7. Surface Plasmon Resonance

Surface plasmon resonance (SPR) analysis was performed in running buffer (5% dimethyl sulfoxide, DMSO in PBS) using a Reichert SR7500DC system (Reichert Technologies, Depew, NY, USA). For immobilization of PTP-MEG2, carboxymethyl dextran hydrogel (CMDH) surface sensor chips were activated with a mixture of 0.1 M EDC (N-ethyl-N'-(3-dimethylaminopropyl)carbodiimide) and 0.05 M NHS (N-hydroxysuccinimide) for 7 min and purified PTP-MEG2 (32 μg/mL 10 mM sodium acetate, pH 6.0) was injected for 3 min at a flow rate of 20 μL/min. Deactivation of the surface was performed using 1 M ethanolamine (pH 8.5) for 8 min. Interaction analyses were performed by injection of increasing concentrations (12.5 to 200 μM) of Phloridzin at a flow rate of 30 μL/min. Association and dissociation signals were monitored for 300 and 420 s, respectively. For the analysis, background signal recorded in reference channel was subtracted and then binding signals were fitted. Data was analyzed using Scrubber2 software (BioLogic Software, Campbell, Australia).

4.8. Western Blotting

Proteins were extracted using the lysis buffer containing 25 mM hydroxyethylpiperazine ethane sulfonic acid (HEPES), 150 mM NaCl, 1% Triton X-100, 10% glycerol, 5 mM EDTA, 10 mM NaF, 2 mM Na VO, and a protease inhibitor cocktail (Roche Korea, Seoul, Republic of Korea). Proteins were separated by electrophoresis using a 10% sodium dodecyl sulfate-polyacrylamide gel and transferred to a polyvinylidene fluoride membrane (Merck KGaA, Darmstadt, Germany) using a wet transfer system. Membranes were incubated overnight at 4 °C with the following primary antibodies: anti-total AMPK, anti-phosphorylated AMPK, anti-total Akt, anti-phosphorylated Akt (Cell Signaling Technology, Beverly, MA, USA), PTP-MEG2 (Santa Cruz Biotechnology, Dallas, TX, USA), and anti-beta-actin (AbFrontier, Seoul, Republic of Korea). Membranes were then probed with an anti-rabbit-IgG-horseradish peroxidase-conjugated secondary antibody (Santa Cruz Biotechnology). Antibody–antigen complexes were detected using enhanced chemiluminescence (ECL) reagents (GE Healthcare Korea, Songdo, Republic of Korea).

4.9. RNA Interference

Knockdown of PTP-MEG2 in 3T3-L1 preadipocytes was performed using small interfering RNAs (siRNA, Bioneer, Daejeon, Korea). Scramble siRNA was used as the negative control (Bioneer). Transfections were performed using Dharmafect (Dharmacon, GE Healthcare Korea, Songdo, Korea) according to the manufacturer's instructions.

4.10. Quantitative Real-Time Polymerase Chain Reaction (qRT-PCR)

Total RNA was isolated from 3T3-L1 preadipocytes using NucleoSpin RNA Plus (Macherey-Nagel, Duren, Germany) and treated with DNase (Macherey-Nagel) to remove genomic DNA. The total RNA (1 µg) was used to synthesize cDNA with the High-Capacity Reverse Transcription kit (Applied Biosystems, Foster City, CA, USA). PCR was performed on a CFX Connect Real-Time PCR Detection System (Bio-Rad, Hercules, CA, USA) using SsoAdvanced Universal SYBR green super-mix (Bio-Rad) according to the manufacturer's instructions. Gene expression levels of PTP-MEG2 were normalized to levels of the control gene, glyceraldehyde 3-phosphate dehydrogenase (GAPDH). Primer information is provided in Supplementary Materials Table S1.

4.11. Oil Red O Staining

Lipid droplets of 3T3-L1 adipocytes were assessed by Oil Red O staining, as described previously [25]. On day 6 after differentiation, cells were washed with PBS, and fixed with 4% paraformaldehyde for 15 min. The cells were then washed with PBS and stained with filtered 0.3% Oil Red O solution in isopropanol for 30 min. After washing the cells with PBS, images of fat accumulation were acquired using an EVOS FL Imaging System (Thermo Fisher Scientific Korea Ltd., Seoul, Korea). For quantification of lipid accumulation, Oil Red O dye was eluted by incubation with isopropanol and the absorbance was measured at 490 nm using a microplate reader (Victor^{TM} X4).

4.12. Statistical Analysis

Statistical significance ($p < 0.05$) was determined using a two-tailed unpaired *t*-test (GraphPad Software, San Diego, CA, USA). All experiments were independently repeated at 3 times. ATTO image analysis software (CS analyzer 4) was used for protein quantification. Results are presented as mean $\pm$ standard error of the mean (SEM). *** $p < 0.001$, ** $p < 0.01$, * $p < 0.05$ compared to the control group.

5. Conclusions

We previously reported that PTP-MEG2 knockdown enhances AMPK phosphorylation, suggesting that PTP-MEG2 may be a potential antidiabetic target [9]. Here, phloridzin was obtained from the root barks of *U. davidiana* var. *japonica* as part of our ongoing research projects to discover bioactive natural products [43–49], and for the first time, we found that phloridzin inhibited the catalytic activity of PTP-MEG2 in vitro, indicating that phloridzin targets PTP-MEG2. Through our cell-based studies, we demonstrated that phloridzin stimulated glucose uptake by differentiated 3T3-L1 adipocytes and C2C12 muscle cells through the activation of both AMPK and Akt signaling pathways. Importantly, phloridzin ameliorated palmitate-induced insulin resistance in C2C12 muscle cells. Moreover, we found that phloridzin did not promote fat accumulation, suggesting that phloridzin improves insulin sensitivity without significant lipid accumulation. Taken together, our results suggest that phloridzin, an inhibitor of PTP-MEG2, could be a potential therapeutic candidate for the treatment of type 2 diabetes.

Supplementary Materials: Figure S1: Rosiglitazone and metformin do not inhibit the catalytic activity of PTP-MEG2, Figure S2: GLUT4 translocation was increased by 10 µM phloridzin. Table S1: Primer sequences used to target mouse genes, Table S2: Kinetic constants for DiFMUP hydrolysis by PTPs.

Author Contributions: Conceptualization, S.-Y.Y., S.J.C. and K.H.K.; Formal Analysis, S.-Y.Y., J.S.Y., J.K.K. and T.S.J.; Investigation, S.-Y.Y., J.S.Y., H.M.S., S.O.S. and J.Y.H.; Resources, J.S.Y. and H.M.S.; Writing—Original Draft Preparation, S.-Y.Y., J.S.Y., S.J.C. and K.H.K.; Writing—Review and Editing, S.-Y.Y., S.J.C. and K.H.K.; Project Administration, S.J.C. and K.H.K.; Funding Acquisition, S.J.C. and K.H.K. All authors have read and agreed to the published version of the manuscript.

Funding: This research was supported by the Bio & Medical Technology Development Program of the National Research Foundation of Korea (NRF) funded by the Korean government (MSIT) (NRF-2012M3A9C4048775 and NRF-2017M3A9C8031995). This work was also supported by the National Research Foundation of Korea (NRF) grant funded by the Korea government (MSIT) (2018R1A2B2006879). This work was supported by a grant from the KRIBB Research Initiative Program.

Institutional Review Board Statement: Not applicable.

Informed Consent Statement: Not applicable.

Data Availability Statement: Data is contained within the article or Supplementary Material.

Acknowledgments: The authors thank Jin Soo Kim and Jae Kwan Kim for their help in measuring enzymatic activity of PTPs for inhibitors.

Conflicts of Interest: The authors declare no conflict of interest.

References

1. Zhang, S.; Liu, S.; Tao, R.; Wei, D.; Chen, L.; Shen, W.; Yu, Z.H.; Wang, L.; Jones, D.R.; Dong, X.C.; et al. A highly selective and potent PTP-MEG2 inhibitor with therapeutic potential for type 2 diabetes. *J. Am. Chem. Soc.* **2012**, *134*, 18116–18124. [CrossRef]
2. Gurzov, E.N.; Stanley, W.J.; Brodnicki, T.C.; Thomas, H.E. Protein tyrosine phosphatases: Molecular switches in metabolism and diabetes. *Trends Endocrinol. Metab.* **2015**, *26*, 30–39. [CrossRef] [PubMed]
3. Hendriks, W.J.; Pulido, R. Protein tyrosine phosphatase variants in human hereditary disorders and disease susceptibilities. *Biochim. Biophys. Acta* **2013**, *1832*, 1673–1696. [CrossRef] [PubMed]
4. He, R.J.; Yu, Z.H.; Zhang, R.Y.; Zhang, Z.Y. Protein tyrosine phosphatases as potential therapeutic targets. *Acta Pharmacol. Sin.* **2014**, *35*, 1227–1246. [CrossRef] [PubMed]
5. Kim, M.S.; Hur, H.J.; Kwon, D.Y.; Hwang, J.T. Tangeretin stimulates glucose uptake via regulation of AMPK signaling pathways in C2C12 myotubes and improves glucose tolerance in high-fat diet-induced obese mice. *Mol. Cell Endocrinol.* **2012**, *358*, 127–134. [CrossRef] [PubMed]
6. Balasubramanian, R.; Robaye, B.; Boeynaems, J.M.; Jacobson, K.A. Enhancement of glucose uptake in mouse skeletal muscle cells and adipocytes by P2Y6 receptor agonists. *PLoS ONE* **2014**, *9*, e116203. [CrossRef] [PubMed]
7. Zhou, G.; Sebhat, I.K.; Zhang, B.B. AMPK activators–potential therapeutics for metabolic and other diseases. *Acta Physiol.* **2009**, *196*, 175–190. [CrossRef] [PubMed]
8. Long, Y.C.; Zierath, J.R. AMP-activated protein kinase signaling in metabolic regulation. *J. Clin. Investig.* **2006**, *116*, 1776–1783. [CrossRef] [PubMed]
9. Yoon, S.Y.; Lee, J.H.; Kwon, S.J.; Kang, H.J.; Chung, S.J. Ginkgolic acid as a dual-targeting inhibitor for protein tyrosine phosphatases relevant to insulin resistance. *Bioorg. Chem.* **2018**, *81*, 264–269. [CrossRef]
10. Cho, C.Y.; Koo, S.H.; Wang, Y.; Callaway, S.; Hedrick, S.; Mak, P.A.; Orth, A.P.; Peters, E.C.; Saez, E.; Montminy, M.; et al. Identification of the tyrosine phosphatase PTP-MEG2 as an antagonist of hepatic insulin signaling. *Cell Metab.* **2006**, *3*, 367–378. [CrossRef] [PubMed]

11. Gosch, C.; Halbwirth, H.; Stich, K. Phloridzin: Biosynthesis, distribution and physiological relevance in plants. *Phytochemistry* **2010**, *71*, 838–843. [CrossRef]
12. Ridgway, T.; O'Reilly, J.; West, G.; Tucker, G.; Wiseman, H. Antioxidant action of novel derivatives of the apple-derived flavonoid phloridzin compared to oestrogen: Relevance to potential cardioprotective action. *Biochem. Soc. Trans.* **1997**, *25*, 106s. [CrossRef] [PubMed]
13. Robak, J.; Gryglewski, R.J. Flavonoids are scavengers of superoxide anions. *Biochem. Pharmacol.* **1988**, *37*, 837–841. [CrossRef]
14. Shoji, T.; Kobori, M.; Shinmoto, H.; Tanabe, M.; Tsushida, T. Progressive effects of phloridzin on melanogenesis in B16 mouse melanoma cells. *Biosci. Biotechnol. Biochem.* **1997**, *61*, 1963–1967. [CrossRef]
15. Lee, J.; Jung, E.; Kim, Y.S.; Park, D.; Toyama, K.; Date, A.; Lee, J. Phloridzin isolated from Acanthopanax senticosus promotes proliferation of alpha6 integrin (CD 49f) and beta1 integrin (CD29) enriched for a primary keratinocyte population through the ERK-mediated mTOR pathway. *Arch. Dermatol. Res.* **2013**, *305*, 747–754. [CrossRef] [PubMed]
16. Ehrenkranz, J.R.; Lewis, N.G.; Kahn, C.R.; Roth, J. Phlorizin: A review. *Diabetes Metab. Res. Rev.* **2005**, *21*, 31–38. [CrossRef] [PubMed]
17. Masumoto, S.; Akimoto, Y.; Oike, H.; Kobori, M. Dietary phloridzin reduces blood glucose levels and reverses Sglt1 expression in the small intestine in streptozotocin-induced diabetic mice. *J. Agric. Food Chem.* **2009**, *57*, 4651–4656. [CrossRef] [PubMed]
18. Panayotova-Heiermann, M.; Loo, D.D.; Wright, E.M. Kinetics of steady-state currents and charge movements associated with the rat Na^+/glucose cotransporter. *J. Biol. Chem.* **1995**, *270*, 27099–27105. [CrossRef] [PubMed]
19. Stefan, M.I.; Le Novere, N. Cooperative binding. *PLoS Comput. Biol.* **2013**, *9*, e1003106. [CrossRef] [PubMed]
20. Mackenzie, R.W.; Elliott, B.T. Akt/PKB activation and insulin signaling: A novel insulin signaling pathway in the treatment of type 2 diabetes. *Diabetes Metab. Syndr. Obes.* **2014**, *7*, 55–64. [CrossRef] [PubMed]
21. Sharma, N.; Arias, E.B.; Bhat, A.D.; Sequea, D.A.; Ho, S.; Croff, K.K.; Sajan, M.P.; Farese, R.V.; Cartee, G.D. Mechanisms for increased insulin-stimulated Akt phosphorylation and glucose uptake in fast- and slow-twitch skeletal muscles of calorie-restricted rats. *Am. J. Physiol. Endocrinol. Metab.* **2011**, *300*, E966–E978. [CrossRef] [PubMed]
22. Jiang, H.; Fan, D.; Zhou, G.; Li, X.; Deng, H. Phosphatidylinositol 3-kinase inhibitor(LY294002) induces apoptosis of human nasopharyngeal carcinoma in vitro and in vivo. *J. Exp. Clin. Cancer Res.* **2010**, *29*, 34. [CrossRef]
23. Feng, X.T.; Wang, T.Z.; Leng, J.; Chen, Y.; Liu, J.B.; Liu, Y.; Wang, W.J. Palmitate contributes to insulin resistance through downregulation of the Src-mediated phosphorylation of Akt in C2C12 myotubes. *Biosci. Biotechnol. Biochem.* **2012**, *76*, 1356–1361. [CrossRef] [PubMed]
24. Jung, T.W.; Choi, H.Y.; Lee, S.Y.; Hong, H.C.; Yang, S.J.; Yoo, H.J.; Youn, B.S.; Baik, S.H.; Choi, K.M. Salsalate and Adiponectin Improve Palmitate-Induced Insulin Resistance via Inhibition of Selenoprotein P through the AMPK-FOXO1alpha Pathway. *PLoS ONE* **2013**, *8*, e66529. [CrossRef] [PubMed]
25. Cho, Y.L.; Min, J.K.; Roh, K.M.; Kim, W.K.; Han, B.S.; Bae, K.H.; Lee, S.C.; Chung, S.J.; Kang, H.J. Phosphoprotein phosphatase 1CB (PPP1CB), a novel adipogenic activator, promotes 3T3-L1 adipogenesis. *Biochem. Biophys. Res. Commun.* **2015**, *467*, 211–217. [CrossRef] [PubMed]
26. Fryer, L.G.; Parbu-Patel, A.; Carling, D. The Anti-diabetic drugs rosiglitazone and metformin stimulate AMP-activated protein kinase through distinct signaling pathways. *J. Biol. Chem.* **2002**, *277*, 25226–25232. [CrossRef] [PubMed]
27. Kim, J.; Kwak, H.J.; Cha, J.Y.; Jeong, Y.S.; Rhee, S.D.; Cheon, H.G. The role of prolyl hydroxylase domain protein (PHD) during rosiglitazone-induced adipocyte differentiation. *J. Biol. Chem.* **2014**, *289*, 2755–2764. [CrossRef] [PubMed]
28. Wang, M.; Li, X.; Dong, L.; Chen, X.; Xu, W.; Wang, R. Virtual screening, optimization, and identification of a novel specific PTP-MEG2 Inhibitor with potential therapy for T2DM. *Oncotarget* **2016**, *7*, 50828–50834. [CrossRef] [PubMed]
29. Kumar, S.; Sinha, K.; Sharma, R.; Purohit, R.; Padwad, Y. Phloretin and phloridzin improve insulin sensitivity and enhance glucose uptake by subverting PPARγ/Cdk5 interaction in differentiated adipocytes. *Exp. Cell Res.* **2019**, *383*, 111480. [CrossRef] [PubMed]
30. Lee, E.J.; Kim, J.L.; Kim, Y.H.; Kang, M.K.; Gong, J.H.; Kang, Y.H. Phloretin promotes osteoclast apoptosis in murine macrophages and inhibits estrogen deficiency-induced osteoporosis in mice. *Phytomedicine* **2014**, *21*, 1208–1215. [CrossRef]
31. Londzin, P.; Siudak, S.; Cegiela, U.; Pytlik, M.; Janas, A.; Waligora, A.; Folwarczna, J. Phloridzin, an Apple Polyphenol, Exerted Unfavorable Effects on Bone and Muscle in an Experimental Model of Type 2 Diabetes in Rats. *Nutrients* **2018**, *10*, 1701. [CrossRef]
32. Chao, E.C.; Henry, R.R. SGLT2 inhibition–a novel strategy for diabetes treatment. *Nat. Rev. Drug Discov.* **2010**, *9*, 551–559. [CrossRef] [PubMed]
33. Rieg, T.; Vallon, V. Development of SGLT1 and SGLT2 inhibitors. *Diabetologia* **2018**, *61*, 2079–2086. [CrossRef] [PubMed]
34. Zhao, H.; Yakar, S.; Gavrilova, O.; Sun, H.; Zhang, Y.; Kim, H.; Setser, J.; Jou, W.; LeRoith, D. Phloridzin improves hyperglycemia but not hepatic insulin resistance in a transgenic mouse model of type 2 diabetes. *Diabetes* **2004**, *53*, 2901–2909. [CrossRef] [PubMed]
35. Brewer, P.D.; Habtemichael, E.N.; Romenskaia, I.; Mastick, C.C.; Coster, A.C. Insulin-regulated Glut4 translocation: Membrane protein trafficking with six distinctive steps. *J. Biol. Chem.* **2014**, *289*, 17280–17298. [CrossRef]
36. Yoon, S.Y.; Kang, H.J.; Ahn, D.; Hwang, J.Y.; Kwon, S.J.; Chung, S.J. Identification of chebulinic acid as a dual targeting inhibitor of protein tyrosine phosphatases relevant to insulin resistance. *Bioorg. Chem.* **2019**, *90*, 103087. [CrossRef] [PubMed]
37. Goldstein, B.J. Insulin resistance as the core defect in type 2 diabetes mellitus. *Am. J. Cardiol.* **2002**, *90*, 3g–10g. [CrossRef]

38. Qin, H.; Liu, Y.; Lu, N.; Li, Y.; Sun, C.H. cis-9,trans-11-Conjugated linoleic acid activates AMP-activated protein kinase in attenuation of insulin resistance in C2C12 myotubes. *J. Agric. Food Chem.* **2009**, *57*, 4452–4458. [CrossRef] [PubMed]
39. Ye, J.M.; Dzamko, N.; Hoy, A.J.; Iglesias, M.A.; Kemp, B.; Kraegen, E. Rosiglitazone treatment enhances acute AMP-activated protein kinase-mediated muscle and adipose tissue glucose uptake in high-fat-fed rats. *Diabetes* **2006**, *55*, 2797–2804. [CrossRef] [PubMed]
40. Minge, C.E.; Bennett, B.D.; Norman, R.J.; Robker, R.L. Peroxisome proliferator-activated receptor-gamma agonist rosiglitazone reverses the adverse effects of diet-induced obesity on oocyte quality. *Endocrinology* **2008**, *149*, 2646–2656. [CrossRef] [PubMed]
41. Gerstein, H.C.; Yusuf, S.; Bosch, J.; Pogue, J.; Sheridan, P.; Dinccag, N.; Hanefeld, M.; Hoogwerf, B.; Laakso, M.; Mohan, V.; et al. Effect of rosiglitazone on the frequency of diabetes in patients with impaired glucose tolerance or impaired fasting glucose: A randomised controlled trial. *Lancet* **2006**, *368*, 1096–1105. [PubMed]
42. Lee, S.Y.; Kim, W.; Lee, Y.G.; Kang, H.J.; Lee, S.H.; Park, S.Y.; Min, J.K.; Lee, S.R.; Chung, S.J. Identification of sennoside A as a novel inhibitor of the slingshot (SSH) family proteins related to cancer metastasis. *Pharmacol. Res.* **2017**, *119*, 422–430. [CrossRef] [PubMed]
43. Lee, S.; Lee, D.; Ryoo, R.; Kim, J.C.; Park, H.B.; Kang, K.S.; Kim, K.H. Calvatianone, a Sterol Possessing a 6/5/6/5-Fused Ring System with a Contracted Tetrahydrofuran B-Ring, from the Fruiting Bodies of *Calvatia nipponica*. *J. Nat. Prod.* **2020**, *83*, 2737–2742. [CrossRef] [PubMed]
44. Lee, S.R.; Kang, H.S.; Yoo, M.J.; Yi, S.A.; Beemelmanns, C.; Lee, J.C.; Kim, K.H. Anti-adipogenic Pregnane Steroid from a *Hydractinia*-associated Fungus, *Cladosporium sphaerospermum* SW67. *Nat. Prod. Sci.* **2020**, *26*, 230–235.
45. Lee, S.; Ryoo, R.; Choi, J.H.; Kim, J.H.; Kim, S.H.; Kim, K.H. Trichothecene and tremulane sesquiterpenes from a hallucinogenic mushroom *Gymnopilus junonius* and their cytotoxicity. *Arch. Pharm. Res.* **2020**, *43*, 214–223. [CrossRef] [PubMed]
46. Trinh, T.A.; Park, E.J.; Lee, D.; Song, J.H.; Lee, H.L.; Kim, K.H.; Kim, Y.; Jung, K.; Kang, K.S.; Yoo, J.E. Estrogenic Activity of Sanguiin H-6 through Activation of Estrogen Receptor α Coactivator-binding Site. *Nat. Prod. Sci.* **2019**, *25*, 28–33. [CrossRef]
47. Ha, J.W.; Kim, J.; Kim, H.; Jang, W.; Kim, K.H. Mushrooms: An Important Source of Natural Bioactive Compounds. *Nat. Prod. Sci.* **2020**, *26*, 118–131.
48. Yu, J.S.; Li, C.; Kwon, M.; Oh, T.; Lee, T.H.; Kim, D.H.; Ahn, J.S.; Ko, S.K.; Kim, C.S.; Cao, S.; et al. a unique rearranged benzoquinone-chromanone from the hawaiian volcanic soil-associated fungal strain *Penicillium herquei* FT729. *Bioorg. Chem.* **2020**, *105*, 104397. [CrossRef]
49. Yu, J.S.; Park, M.; Pang, C.; Rashan, L.; Jung, W.H.; Kim, K.H. Antifungal Phenols from *Woodfordia uniflora* Collected in Oman. *J. Nat. Prod.* **2020**, *83*, 2261–2268. [CrossRef]

Article

Microextraction of *Reseda luteola*-Dyed Wool and Qualitative Analysis of Its Flavones by UHPLC-UV, NMR and MS

Elbert van der Klift [1], Alexandre Villela [1], Goverdina C. H. Derksen [2], Peter P. Lankhorst [3] and Teris A. van Beek [1,*

[1] Laboratory of Organic Chemistry, Wageningen University, Stippeneng 4,
     6708 WE Wageningen, The Netherlands; ejcvanderklift@gmail.com (E.v.d.K.);
     alexandre.villela@naturalproductschemistry.com (A.V.)
[2] Research Group Biobased Products, Avans University of Applied Sciences, P.O. Box 90.116,
     4800 RA Breda, The Netherlands; gch.derksen@avans.nl
[3] DSM, Biotechnology Center, A. Fleminglaan 1, 2613 AX Delft, The Netherlands; peter.lankhorst@planet.nl
*   Correspondence: teris.vanbeek@wur.nl; Tel.: +31-317-482376

Abstract: Detailed knowledge on natural dyes is important for agronomy and quality control as well as the fastness, stability, and analysis of dyed textiles. Weld (*Reseda luteola* L.), which is a source of flavone-based yellow dye, is the focus of this study. One aim was to reduce the required amount of dyed textile to ≤50 µg for a successful chromatographic analysis. The second aim was to unambiguously confirm the identity of all weld flavones. By carrying out the extraction of 50 µg dyed wool with 25 µL of solvent and analysis by reversed-phase UHPLC at 345 nm, reproducible chromatographic fingerprints could be obtained with good signal to noise ratios. Ten baseline separated peaks with relative areas ≥1% were separated in 6 min. Through repeated polyamide column chromatography and prepHPLC, the compounds corresponding with the fingerprint peaks were purified from dried weld. Each was unequivocally identified, including the position and configuration of attached sugars, by means of 1D and 2D NMR and high-resolution MS. Apigenin-4′-*O*-glucoside and luteolin-4′-*O*-glucoside were additionally identified as two trace flavones co-eluting with other flavone glucosides, the former for the first time in weld. The microextraction might be extended to other used dye plants, thus reducing the required amount of precious historical textiles.

Keywords: weld; *Reseda luteola* L.; natural yellow dye; µ-analysis; micro-extraction of wool; UHPLC; fingerprinting; flavones; NMR; structure elucidation

Citation: van der Klift, E.; Villela, A.; Derksen, G.C.H.; Lankhorst, P.P.; van Beek, T.A. Microextraction of *Reseda luteola*-Dyed Wool and Qualitative Analysis of Its Flavones by UHPLC-UV, NMR and MS. *Molecules* **2021**, *26*, 3787. https://doi.org/10.3390/molecules26133787

Academic Editors: Young Hae Choi, Young Pyo Jang, Yuntao Dai and Luis Francisco Salomé-Abarca

Received: 18 May 2021
Accepted: 11 June 2021
Published: 22 June 2021

Publisher's Note: MDPI stays neutral with regard to jurisdictional claims in published maps and institutional affiliations.

1. Introduction

Since the stone age, mankind has dyed his clothes for decoration. Later also rugs and tapestries were dyed. For this purpose, minerals, insects, and especially plant extracts have been used. The most used dye plants in Europe throughout the ages include madder (*Rubia tinctorum* L.), weld (*Reseda luteola* L.) and woad (*Isatis tinctoria* L.) [1]. During the second half of the 19th century, natural dyes were replaced by cheaper, synthetic dyes. However, in recent times for various reasons there is a renewed interest in natural dyes. To facilitate the re-introduction of natural dyes, over the last 25 years many papers have appeared, for instance on screening methods for agronomic studies of dye plants [2,3], the fastness of dyed textiles [4], the colorimetric properties of dye plants and their individual constituents [5], mild extraction methods for textiles [6–8], degradation of dyes [9–11], safety of natural dyes [12], the development of new production routes [13], and analysis of dyed textiles [1,14]. Concerning the latter topic, natural dye analysis is of great importance for historians and conservators. Detailed knowledge about natural dyes used in old garments and tapestries can provide answers concerning the plant(s) used for dyeing, dyeing practices, age, origin, original colour, and trade routes. It can also help to shed light on previous restorations and help future restorations [1].

As historic textiles are precious, non-destructive analysis is preferred. Fibre optics reflectance spectroscopy (FORS) for the detection of dye molecules in the visible wavelength spectrum is such a technique. Unfortunately, the discriminatory power is limited and moreover the method is not suitable for distinguishing between different yellow dyes [15–17]. Surface-enhanced Raman scattering (SERS) has been used as a micro invasive technique [18,19]. SERS has also been coupled on-line with HPLC for greater resolution [20]. This methodology is better at distinguishing between yellow dyes. Jurasekova et al. have shown that different ratios of the flavones luteolin and apigenin generate different spectra [18]. A disadvantage of this method is that the textile has to be coated with silver nanoparticles.

Still the most powerful—albeit destructive—technique to obtain highly detailed knowledge on which dye plants have been used is HPLC after the extraction of one or more dyed threads. Each plant has its own unique blend of major and minor dye molecules and this fingerprint is thus specific even if some compounds are partially the same. Analyses can become more complex when significant decomposition has occurred. A prerequisite for a successful analysis is the intact isolation of all dyes. Until fairly recently strong mineral acids were employed to release dyes [9,21,22], e.g., from metals used for mordanting. Unfortunately, this leads to the hydrolysis of glycosidic dyes, such as those present in weld, and then no representative fingerprint is obtained. However, much progress has been made in developing milder yet efficient extraction methods [6]. Formic acid or EDTA instead of HCl provide higher recovery and glycosidic dyes remain intact. Moreover, other acids, such as oxalic acid, citric acid, or trifluoroacetic acid, have been proposed [23,24].

Detection after the HPLC separation can be by UV-Vis or UV-Vis followed by MS. The amount of material typically taken for HPLC analysis varies from 0.1-3.9 mg with 0.2 mg as average [1,6,7,9,21–29]. Obviously, the less material is needed, the better. On the other hand, analyte quantities are minute. Rosenberg estimated that typically less than 1 mg of dye is present per gram of fabric [7] and this estimate is in good correspondence with actual measurements on pure compounds [5]. This translates to 50 ng per 50 µg of wool. Even with 100% extraction efficiency—Willemen et al. have shown it might actually be <50% [5]—this equals, with a final sample volume of 125 µL and a 10 µL injection, about 4 ng on column. For a 1% component this amounts to only 40 pg. Thus, concentrations should be as high as possible. Earlier we have shown that by using short UHPLC columns for dye substances significant time reduction and sensitivity gains are possible [30]. Since then, UHPLC has been used for dye separations by several other groups [27,31,32]. Thus, the first aim was to investigate if it was possible by combining miniaturisation of the extraction step and using a modern UHPLC for the separation and detection to obtain clear chromatograms while using significantly less than 0.1 mg of textile. Modern UV detectors have become significantly more sensitive with femtogram (fg) limits of detection (LOD) [33]. Weld-dyed wool was taken as test material.

Upon going through assigned weld chromatograms in the literature, different assignments were found, and not always all stereochemical features of the glycosides were clear. Although ESI-MS is a powerful tool, it is less suited to assign where sugars are attached to the aglycone or to determine the configuration of sugars (α or β). Thus, several partial identifications of weld flavone mono and diglucosides have been reported after LC-MS studies [25,26,28]. The most detailed studies so far are those by Moiteiro et al., Marques et al., Mantzouris et al. and Otłowska et al. [24,26,28,34]. In contrast, the study by Burger is confusing with the major peaks reported as unknown [35]. Further, it is unclear where exactly luteolin-4′-*O*-glucoside elutes. According to Otlowska et al., it elutes just prior to luteolin while Moiteiro et al. place it before apigenin-7-*O*-glucoside [28,34]. Later, the group of Gaspar switched the assignment of the luteolin-3′ and 4′-*O*-glucosides [36]. The study by Serrano et al. is confusing as luteolin-3′,7-di-*O*-glucoside **3**, apigenin-7-*O*-glucoside **5**, and apigenin **9** are each assigned to two different peaks in the chromatogram, which is plainly impossible [32]. Moreover, monoglucosides elute before diglucosides [32], which is highly unlikely in a reversed-phase system. To arrive at an unambiguously assigned weld fingerprint these knowledge gaps have to be filled. Thus, the second aim was to carry out a preparative extraction of weld plant material to enable the isolation and purification of all flavonoids corresponding with peaks having a normalised peak area $\geq$1% in the weld fingerprint chromatogram. This was to be followed by the unequivocal identification of each isolated compound by means of 1-dim and 2-dim NMR and HRMS. Results on the micro-extraction work and the structure elucidation of all weld flavones are reported below.

2. Results

2.1. UHPLC Separation

As a first step, a UHPLC method for obtaining a fast chromatographically well-resolved fingerprint chromatogram of weld was developed. Starting point was the chromatogram obtained earlier with a 50 mm × 3.0 mm UHPLC column in combination with a standard HPLC pump [37]. To obtain more resolution of the three minor flavones eluting in between lut-7-*O*-glu and lut (see Figure 1 for structures), a 100 × 2.1 mm 1.8 µm UHPLC column was used instead. To obtain less band broadening, higher reproducibility and higher sensitivity, the column was mounted in a modern UHPLC set-up. The A (40 mM formic acid buffer pH = 3 with 0.04 mM EDTA) and B (acetonitrile) solvents remained the same as before [37] but gradient and flow were adapted because of the increased length and decreased diameter of the new column. With a gradient from 15% to 45% of acetonitrile in 8 min, the chromatogram shown in Figure 2 was obtained for the weld extract showing baseline resolution of the ten peaks with $\geq$1% relative abundance (based on normalized peak areas at UV 350 nm). Weld extracts appear to be remarkably constant in their composition as this chromatogram shows much correspondence with LC-UV weld chromatograms published by other groups [3,23,25,26,28]. The profiles published by Moiteiro et al. and Gaspar et al. somewhat deviate, but this could be due to a different selectivity of the used C18 stationary phase [34,36]. A simple quadrupole mass spectrometer operating in negative mode was connected in series and provided useful initial data about the MW of the different peaks, especially in extracted ion chromatography (EIC) mode.

luteolin-7,4'-di-*O*-β-D-glucopyranoside **2** (lut-7,4'-di-*O*-glu): R_1 = glu, R_2 = H, R_3 = glu
luteolin-7,3'-di-*O*-β-D-glucopyranoside **3** (lut-7,3'-di-*O*-glu): R_1 = glu, R_2 = glu, R_3 = H
luteolin-7-*O*-β-D-glucopyranoside **4** (lut-7-*O*-glu): R_1 = glu, R_2 = H, R_3 = H
luteolin-4'-*O*-β-D-glucopyranoside **tr2** (lut-4'-*O*-glu): R_1 = H, R_2 = H, R_3 = glu
luteolin-3'-*O*-β-D-glucopyranoside **7** (lut-3'-*O*-glu): R_1 = H, R_2 = glu, R_3 = H
luteolin **8** (lut): R_1 = H, R_2 = H, R_3 = H

apigenin-6,8-di-*C*-β-D-glucopyranoside **1** (api-6,8-di-*C*-glu): R_1 = glu, R_2 = H, R_3 = glu, R_4 = H

Figure 1. *Cont.*

apigenin-7-*O*-β-D-glucopyranoside **5** (api-7-*O*-glu), R₁ = H, R₂ = glu, R₃ = H, R₄ = H

apigenin-4′-*O*-β-D-glucopyranoside **tr1** (api-4′-*O*-glu): R₁ = H, R₂ = H, R₃ = H, R₄ = glu

apigenin **9** (api): R₁ = H, R₂ = H, R₃ = H, R₄ = H

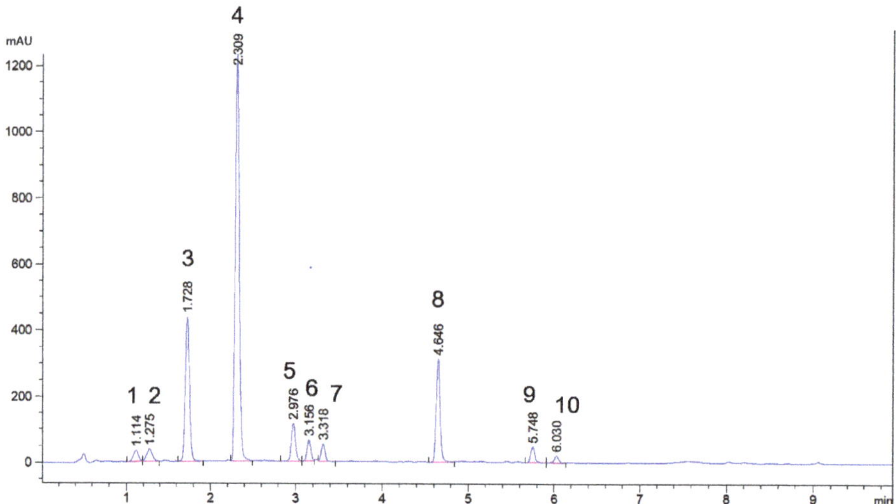

chrysoeriol-7-*O*-β-D-glucopyranoside **6** (chry-7-*O*-glu): R₁ = glu

chrysoeriol **10** (chry): R₁ = H

Figure 1. Structures of all flavones identified in weld (*R. luteola*). Numbers in bold, refer to peak numbers in Figure 2; **tr1** & **tr2** = two trace flavones not visible in Figure 2; glu = β-D-glucopyranoside.

Figure 2. Reversed phase UHPLC profile of an extract of weld; UV detection at 345 nm. Peaks of the ten main flavones are numbered and correspond with the bold numbers in Figure 1.

2.2. Microextraction Method for Weld-Dyed Wool

As a next step, the extraction procedure (300 μL of solvent in a 2 mL vial) used earlier by Villela et al. for weld-dyed wool was downscaled to enable the analysis of 50 μg of sample [30]. Initially, we attempted to use a procedure used for the extraction of ng amounts of sex pheromone [38]. Said procedure used 5 μL of solvent in a shortened melting point tube to extract an SPME needle. Although successful analyses of weld-dyed wool were obtained, in practice it proved extremely difficult to transfer <100 μg of wool threads into a melting point tube. Additionally, due to the small volume available for injection, the standard UHPLC autosampler could not be used and hardware adaptations to enable manual injection were required. For the same reason, filtration posed a problem. Because of these limitations, this approach was abandoned and 250 μL commercially available inserts for autosampler vials were tried instead.

A minute amount of yellow lab-dyed wool was taken by means of chirurgic scissors and transferred into an empty insert previously weighed on a μg balance (readable up to

0.1 µg) (Figure 3a). After weighing again, 0.5 µL of extraction solvent (methanol-water-formic acid (80:15:5, *v/v/v*) was added per µg of wool. After a 30 min extraction step at 60 °C, the extract was diluted five times with water and filtered through a 5-µm zero-dead volume needle filter into a clean empty insert (Figure 3b). This dilution step enabled the filtration step. A filtration step is advisable to avoid clogging of the UHPLC column with wool fibres. The dilution step did not affect the sensitivity as the higher water content of the sample allowed a five times bigger injection volume. Ten µL were injected into the UHPLC by means of the standard autosampler.

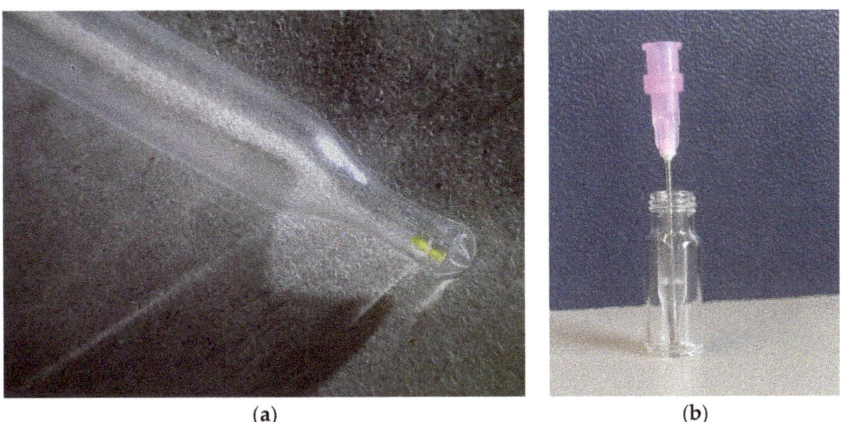

(a)   (b)

Figure 3. (**a**) bottom part of 250 µL insert for 1.8 mL HPLC autosampler vial with <0.1 mg weld-dyed wool sample at the very bottom; (**b**) autosampler vial with insert containing ~100 µL of extracted wool sample after filtration through a zero dead volume 5 µm syringe filter. The purple needle filter can still be seen.

In Figure 4a,b, the chromatograms of 56.6 µg and 49.2 µg of weld-dyed wool respectively are depicted. The fingerprints are qualitatively identical to the one of weld extract in Figure 2. Quantitatively the flavone diglucosides (peaks **1**, **2**, and **3**) are less prominent in the extracted wool sample as compared to the original weld extract (Figure 2). This can be due to either the dyeing or the extraction. Willemen et al. noted that flavone diglucosides are less well bound during the dyeing process than the monoglucosides and aglycones [5]. The total sample volumes (~125 µL) allowed for duplicate injections into the UHPLC. As expected, intraday retention times were constant within 0.003 min and variation of relative peak areas was ≤1% (Supplementary Figures S68 and S69). In both chromatograms, the 10 peaks constituting the fingerprint are clearly resolved and well above background noise. Although peak **1** appears small, it can still be integrated. To determine whether the proposed technique is capable of differentiating wool samples dyed with another natural yellow dye, onion-dyed wool was extracted in duplicate. This also served to see if even smaller samples can be processed and successfully analysed. In Figure 4c, a chromatogram of 28.3 µg onion-dyed wool is shown. From the UHPLC profile, it can be concluded that the microextraction methodology also works for other yellow flavonoid-based dyes and that wool quantities <50 µg can be processed. Onion contains flavonoids of the flavonol type, with quercetin-3,4′-di-*O*-glucoside, quercetin-4′-*O*-glucoside, and quercetin as major compounds [39].

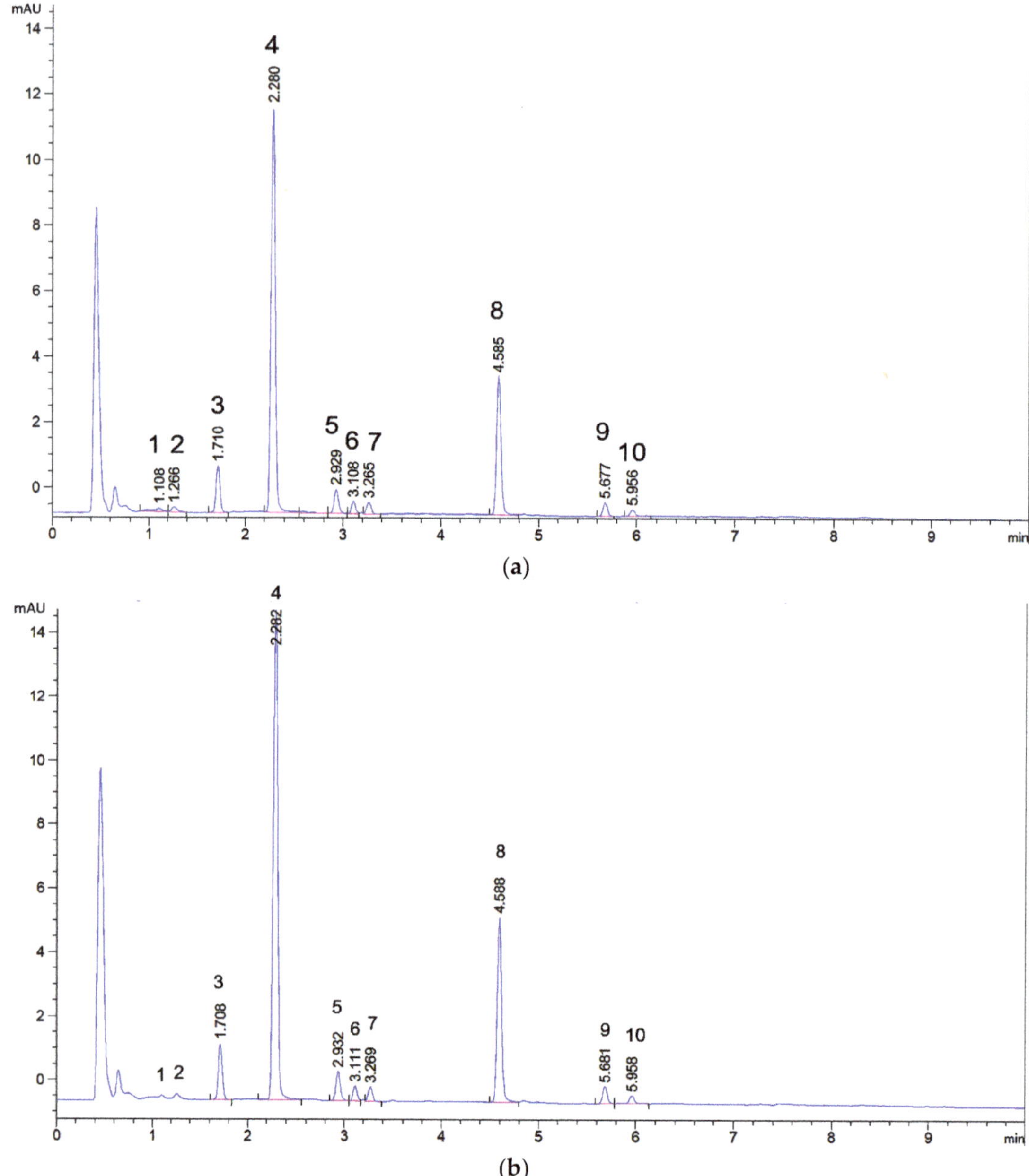

Figure 4. *Cont.*

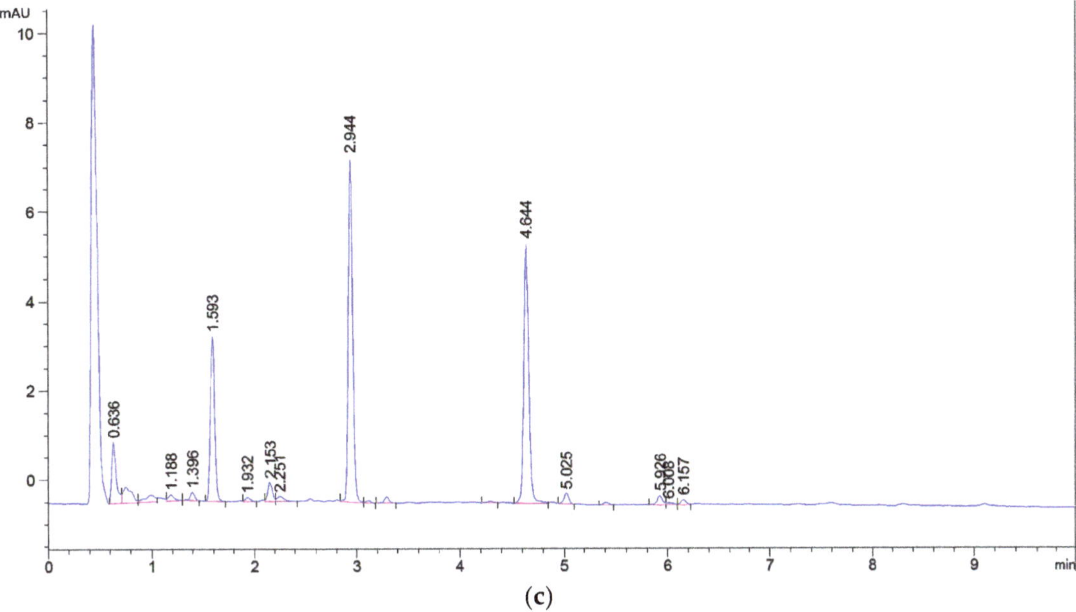

Figure 4. UHPLC profiles of extracted wool samples (**a**). extract of 56.6 μg weld-dyed wool. (**b**). extract of 49.2 μg weld-dyed wool. Peak numbers refer to Figure 1. (**c**). extract of 28.3 μg of onion-dyed wool.

2.3. Preparative Isolation of Major and Minor Flavones of Weld

As the literature on the identity of weld flavones was not 100% clear, next it was decided to isolate at least a few mg of each flavone corresponding with peaks **1–10** in Figures 2 and 4a. This should suffice to record 1-dim and 2-dim NMR spectra as well as high-resolution mass spectra (HRMS). This was carried out by extracting 200 g of weld followed by repeated column chromatography on polyamide and preparative RP-HPLC. In addition to the successful isolation of the ten flavones corresponding to peaks **1–10**, two additional trace flavones (<1% normalised peak area), named **tr1** & **tr2** (Figure 1) were isolated. So, in total, 12 flavones were isolated from weld in pure form, five of which elute between 2.8 and 3.4 min. Due to their low concentration and overlap, the two trace flavones are not visible in the UHPLC fingerprint, but their NMR spectra were recorded nonetheless.

2.4. Identification of Isolated Flavones by NMR and HRMS

The 12 compounds isolated were expected to be flavones based on the extraction method, retention times and especially their distinctive on-line UV spectra (see for example Figures S73–S75). Their 1-dimensional and 2-dimensional 600 or 700 MHz NMR spectra were recorded in DMSO-d_6 and fully assigned. All spectra can be found in the supplementary materials (Supplementary Figures S1–S67). In Table 1, ^{1}H NMR shifts and coupling constants are presented together with ^{13}C NMR shifts. Particular attention was given to the point of attachment of sugar units and the configuration of the sugar units. When available, retention times were compared with those of authentic standards. In all cases, high-resolution mass spectra (HRMS) were also recorded to confirm the NMR assignments. In the following three paragraphs, starting with aglycones and ending with the diglucosides, results are discussed and also compared with published NMR spectra.

Table 1. ^{1}H-NMR and ^{13}C-NMR data of weld flavones; numbering of compounds as in Figure 1 (DMSO-d_6, 300 K).

Position	lut (8) ^{1}H δ (ppm), Mult. [b]; J (Hz)	lut (8) ^{13}C δ (ppm)	lut-7-O-glu (4) ^{1}H δ (ppm), Mult.; J (Hz)	lut-7-O-glu (4) ^{13}C δ (ppm)	lut-3′-O-glu (7) ^{1}H δ (ppm), Mult.; J (Hz)	lut-3′-O-glu (7) ^{13}C δ (ppm)	lut-4′-O-glu (tr2) ^{1}H δ (ppm), Mult.; J (Hz)	lut-4′-O-glu (tr2) ^{13}C δ (ppm)	lut-7,3′-di-O-glu (3) [a] ^{1}H δ (ppm), Mult.; J (Hz)	lut-7,3′-di-O-glu (3) [a] ^{13}C δ (ppm)	lut-7,4′-di-O-glu (2) [a] ^{1}H δ (ppm), Mult.; J (Hz)	lut-7,4′-di-O-glu (2) [a] ^{13}C δ (ppm)
2	-	164.9	-	165.3	-	164.4	-	162.8	-	164.8	-	164.7
3	6.70	103.2	6.80	103.9	6.78	103.2	6.64	104.0	6.95	104.0	6.94	104.8
4	-	182.5	-	182.6	-	182.4	-	n.d.	-	182.9	-	183.0 **
5	-	162.3	-	162.0	-	162.3	-	n.d.	-	161.7	-	162.0
4a	-	104.4	-	106.1	-	104.3	-	101.9	-	106.3	-	106.3
6	6.22, d; 1.2	99.6	6.49, d; 1.9	100.2	6.20, d; 1.5	99.8	5.92	101.4	6.48	100.2	6.49	100.4
7	-	165.5	-	163.7	-	166.2	-	n.d.	-	163.9	-	164.0
8	6.48, d; 1.2	94.6	6.84, d; 1.9	95.4	6.55, d; 1.5	94.9	6.22	95.8	6.94	95.3	6.91	95.5
8a	-	158.4	-	157.7	-	158.4	-	n.d.	-	158.0	-	157.9
1′	-	122.1	-	122.1	-	121.1	-	126.2	-	123.2	-	125.2
2′	7.44, d; 2.1	113.9	7.47, d; 2.2	114.2	7.80, d; 1.7	115.5	7.46	114.4	7.90	115.3	7.60, d; 2.0	114.7
3′	-	146.8	-	146.6	-	147.0	-	148.7	-	146.7	-	148.4
4′	-	150.9	-	150.9	-	153.5	-	149.5	-	151.9	-	149.9
5′	6.92, d; 8.3	116.6	6.95, d; 8.4	116.7	6.96, d; 8.5	117.6	7.24, d; 8.4	116.7	7.02, d; 8.3	117.1	7.28, d; 8.6	116.5
6′	7.46, dd; 8.3, 2.1	119.6	7.50, dd; 8.4, 2.2	119.9	7.66, dd; 1.7, 8.5	122.8	7.43, d; 8.4	118.0	7.73, d; 8.3	122.7	7.57, dd; 8.6, 2.0	118.9
1″			5.13, d; 7.6	100.6	4.88, d; 7.1	103.1	4.87, d; 7.3	102.2	5.11, d; 7.2	100.4	5.12, d; 7.7	100.5
2″			3.31, pt; 7.6, 8.9	73.8	3.36, n.r.	74.0	3.36, n.r.	74.0	3.32, n.r.	73.7	3.37, n.r.	73.9
3″			3.35, pt; 8.9, 8.9	77.1	3.36, n.r.	76.8	3.35, n.r.	76.6	3.37, n.r.	76.7	3.37, n.r.	76.5
4″			3.23, pt; 8.9, 8.9	70.2	3.21, pt; 8.8, 8.8	70.6	3.22, pt; 8.7, 8.7	70.4	3.26, n.r.	70.1	3.23, n.r.	70.4
5″			3.50, n.r.	77.9	3.48, m	78.1	3.42, n.r.	77.9	3.49, n.r.	77.7	3.51, n.r.	77.8
6″			3.76/3.53, d; 11.6/n.r.	61.3	3.83/3.56, d; 11.5/dd; 11.5, 6.4	61.7	3.77/3.53, d; 11.2/dd; 11.2, 5.7	61.4	3.75/3.56, d; 9.4 */d; 9.4 *	61.1	3.76/3.53, n.r./n.r.	61.3
1‴									4.92, d; 7.1	102.6	4.90, d; 7.1	102.1
2‴									3.39, n.r.	74.1	3.31, n.r.	73.8
3‴									3.34, n.r.	77.1	3.36, n.r.	77.1
4‴									3.18, n.r.	70.9	3.23, n.r.	70.4
5‴									3.54, n.r.	78.1	3.43, n.r.	78.0
6‴									3.83/3.53, d 11.6 */d; 11.6 *	61.8	3.76/3.53, n.r./n.r.	61.3

Table 1. *Cont.*

Position	api (9) 1H δ (ppm), Mult.; J (Hz)	api (9) 13C δ (ppm)	api-7-O-glu (5) 1H δ (ppm), Mult.; J (Hz)	api-7-O-glu (5) 13C δ (ppm)	api-4'-O-glu (tr1) 1H δ (ppm), Mult.; J (Hz)	api-4'-O-glu (tr1) 13C δ (ppm)	api-6,8-di-C-glu (1) [c] 1H δ (ppm), Mult.; J (Hz)	api-6,8-di-C-glu (1) [c] 13C δ (ppm)	chry (10) 1H δ (ppm), Mult.; J (Hz)	chry (10) 13C δ (ppm)	chry-7-O-glu (6) 1H δ (ppm), Mult.; J (Hz)	chry-7-O-glu (6) 13C δ (ppm)
2	-	164.3	-	165.4	-	162.6	-	166.9	-	163.8	-	n.d.
3	6.75	103.1	6.84	102.7	6.74	104.0	6.64	104.4	6.80	103.3	6.75	101.0
4	-	182.2	-	182.6	-	n.d.	-	n.d.	-	n.d.	-	n.d.
5	-	162.2	-	161.8	-	n.d.	-	163.4	-	n.d.	-	161.9
4a	-	103.7	-	106.2	-	102.1	-	106.2	-	n.d.	-	105.9
6	6.15	100.1	6.47, d; 1.4	100.1	5.94	101.2	-	108.9	6.06	100.3	6.43	99.7
7	-	167.2	-	163.8	-	n.d.	-	n.d.	-	n.d.	-	163.3
8	6.43	94.9	6.86, d; 1.4	95.3	6.22	95.6	-	106.9	6.35	95.1	6.83	95.1
8a	-	158.3	-	157.7	-	n.d.	-	157.9	-	n.d.	-	157.5
1'	-	121.7	-	117.5	-	125.4	-	124.0	-	121.9	-	n.d.
2'	7.93, d; 8.8	128.8	7.95, d; 8.4	129.2	8.01, d; 8.6	128.3	7.97, d; 8.5	130.6	7.54	110.4	7.42	110.0
3'	6.96, d; 8.8	116.7	6.89, d; 8.4	117.2	7.21, d; 8.6	117.1	6.91, d; 8.5	117.4	-	149.0	-	n.d.
4'	-	162.6	-	165.4	-	160.8	-	163.4	-	n.d.	-	150.3
5'	equivalent to 3'	eq. 2'	eq. 3'		eq. 3'		eq. 3'		6.94, d; 8.0	116.4	6.66	117.8
6'			eq. 2'		eq. 2'		eq. 2'		7.55, d; 8.0	120.8	7.55	122.5
OCH₃									3.92	56.5	3.85	56.3
1''			5.11, d; 7.7	100.4	5.06, d; 7.4	100.5	4.92, d; 9.7	76.4			5.08, d; 7.6	100.7
2''			3.30, pt; 8.5, 8.5	73.7	3.32, n.r.	73.9	n.a., n.r.	n.a.			3.29, pt; 8.1, 8.1	73.9
3''			3.35, pt; 8.9, 8.9 *	77.0	3.34, n.r.	77.2	n.a., n.r.	n.a.			3.35, pt; 8.9, 8.9	77.1
4''			3.22, pt; 9, 9	70.2	3.22, pt; 9, 9	70.2	n.a., n.r.	n.a.			3.22, pt; 9.1, 9.1	70.3
5''			3.49, n.r.	77.8	3.43, n.r.	77.7	n.a., n.r.	n.a.			3.48, n.r.	77.8
6''			3.76/3.53, d; 11.4/n.r.	61.2	3.74/3.53, d; 11.5/n.r.	61.1	n.a./n.a., n.r./n.r.	n.a.			3.76/3.51, d; 11.1/n.r.	61.3
1'''							4.90, d; 9.7	75.4				
2'''							n.a., n.r.	n.a.				
3'''							n.a., n.r.	n.a.				
4'''							n.a., n.r.	n.a.				
5'''							n.a., n.r.	n.a.				
6'''							n.a./n.a., n.r./n.r.	n.a.				

DMSO: 2.55 ppm/40.45 ppm; pt = pseudo t; n.d. = not detected; n.r. = not resolved; n.a. = not assigned. Error of J-values may be 0.5 Hz, or a bit more; * = severe line-broadening, not accurate; ** = cross-peak signal with very low intensity at position of H-3. [a] = signals of 2''–6'' and 2'''–6''' were assigned tentatively, by analogy with a mono-substituted lut. Their assignment to a specific glucose moiety was not attempted, except for 1'' and 1'''. Assignment of '' and ''' moieties is arbitrary; [b] = 1H signals: only multiplicities other than s are mentioned; [c] = DMSO-d_6–MeOH-d_4 1:3 (v/v) at 300 K, with J-values taken from spectrum in DMSO-d_6 at 320 K.

2.4.1. Flavone Aglycones

The literature indicates that the three aglycones present in weld are luteolin, apigenin and chrysoeriol [40]. Based on their lower polarity and correspondence with a published fingerprint [40], the three last eluting peaks were thus expected to be luteolin (lut, peak **8** in Figure 2)—the main aglycone of weld—apigenin (api, **9**) and chrysoeriol (chry, **10**). The structures of all flavones are depicted in Figure 1. For lut and api, the peak assignment could be confirmed by authentic references. Their identity was further fully confirmed by comparison with ^{1}H and ^{13}C NMR data published in literature [41–44] and full interpretation of their HMBC spectra (see Supplementary Figures S3, S7 and S12). The NMR data of chry were in good agreement with the literature except for the assignment of H-6 and H-8 by Kim [44,45]. Assignment based solely on 1-dim NMR data are less reliable than those based on both 1-dim and 2-dim NMR data as in this study. This has been shown for other flavones too [42]. The position of the methyl group in chry was unequivocally assigned to be 3′ based on its ROESY spectrum (Supplementary Figure S13). The HRMS data (Supplementary Table S1) further substantiated the NMR data and assignments. The NMR data of the three aglycones were useful for the interpretation of the NMR spectra of the monoglucosides and diglucosides discussed in the next two paragraphs.

2.4.2. Flavone Monoglucosides

Based on the literature, the following flavone monoglucosides were expected: lut-7-*O*-glucoside, lut-3′-*O*-glucoside, lut-4′-*O*-glucoside, api-7-*O*-glucoside and a chry-*O*-glucoside with the main component being lut-7-*O*-glu (peak **4** in Figure 2). Again, all NMR data can be found in the Supplementary Materials. Based on their intermediate polarity, the monoglucosides (peaks **4–7**, Figure 2) elute in between the diglucosides (peaks **1–3**) and aglycones (peaks **8–10**). Their identification is discussed below, peak by peak in order of increasing retention times.

The major compound (peak **4**) should be lut-7-*O*-glucoside. This was confirmed by comparison of its retention time with a reference and HRMS. Additionally, there was a good correspondence with published NMR data except for the assignments of H-3, H-6 and H-8 by Chung, which seem erroneous [41,43,46,47]. This confirmed the hexose to be glucose. The HMBC spectrum further substantiated the attachment of the glucose unit to O-7. As $J_{1''2''}$ is 7.6 Hz, lut-7-*O*-glu isolated from weld is lut-7-*O*-β-D-glucopyranoside **4**. This exactly matches an earlier assignment [48].

The HRMS and ^{1}H NMR spectrum of the flavone corresponding with peak **5** suggested the presence of both apigenin and a hexose unit. The downfield shift of H-6 and H-8 indicated that the sugar unit was probably attached to O-7. The C-7/H-1″ cross-peak in the HMBC spectrum confirmed this. Its ^{13}C NMR spectrum was virtually identical to the one reported by Oyama and Kondo for api-7-*O*-β-glu [49]. This, and a $J_{1''2''}$ of 7.7 Hz, proved that peak **5** corresponded with api-7-*O*-β-D-glucopyranoside **5**.

The HRMS and ^{1}H NMR of the flavone corresponding with peak **6** pointed in the direction of chrysoeriol as aglycone and a hexose. Again, the downfield shift of H-6 and H-8 made it clear that the sugar was attached to O-7. This was confirmed by a C-7/H-1″ cross-peak in the HMBC spectrum. The location of the methoxy group at C-3′ as well as that of the sugar moiety to C-7 were confirmed based on the ROESY spectrum. The ^{13}C NMR spectrum gave a good match with the one reported by Harput et al. [47] for chry-7-*O*-glucopyranoside. This, and a $J_{1''2''}$ of 7.6 Hz, confirmed the main compound of the sample to be chry-7-*O*-D-glucopyranoside **6**.

The HRMS as well as the ^{1}H NMR data of the trace flavone **tr1** were indicative of an apigenin-hexose. Relative to the ^{1}H NMR of apigenin itself, H-3′ and H-5′ had shifted downfield suggesting the hexose to be attached to O-4′. The C-4′ and H-1″ cross-peak in the HMBC confirmed this. The spectra matched those reported by Oyama and Kondo [49] and Ding et al. [50] for api-4′-*O*-β-D-glucopyranoside well. This, and a $J_{1''2''}$ of 7.4 Hz, led to the conclusion that this minor compound is api-4′-*O*-β-D-glucopyranoside **tr1**. This compound has not yet been reported to occur in *R. luteola*.

The mass of the second trace flavone **tr2** was that of a luteolin-hexose. Based on the identification of lut-7-glu (peak **4**) and lut-3′-glu (peak **7**, vide infra) in combination with the identification of api-4′-glu, lut-4′-*O*-glu appeared a plausible candidate. The 0.3 ppm downfield shift of H-5′ supported this hypothesis. Furthermore, C-4′ can be discriminated from C-3′ by long range couplings in the HMBC of C-4′ to both H-2′and H-6′. The anomeric proton couples with C-4′ and therefore this compound was confirmed as lut-4′-*O*-glu. A good fit with the ^{13}C spectrum published by Lee et al. [51] for lut-4′-*O*-β-D-glucopyranoside and a $J_{1''2''}$ of 7.3 Hz, led to the identification of this second minor compound as lut-4′-*O*-β-D-glucopyranoside **tr2**.

The HRMS indicated that peak **7** corresponded with a luteolin-hexose. In the ^{1}H NMR H-3, 6 and 8 had the same chemical shifts as in luteolin **8** itself but H-2′ and 6′ had shifted suggesting attachment of the sugar to O-3′. This was proven by a C-3′ and H-1″ cross-peak in the HMBC spectrum. The ^{1}H data matched published data [43]. The ^{13}C data of lut-3′-*O*-glu differed less than 1 ppm (sugar moiety C-atoms) and less than 3 ppm (other C-atoms) from those reported by Markham et al. [41]. This confirmed the identity of the main compound of the sample, including the nature of the hexose. As L-glucose does not occur in nature, the sugar moiety of the flavonoid is D-glucose. Furthermore, as the magnitude of the coupling constant H-1″/H-2″ is in the range 7–8 Hz, the configuration of the anomeric C-atom could be ascertained as β, and the cyclic form of the glucose as pyranosidic [43]. Thus, lut-3′-*O*-glu isolated from weld is lut-3′-*O*-β-D-glucopyranoside **7**. Whereas this configuration of the anomeric C-atom is the same as that of an earlier assignment [48], the glucose is in the pyranose form, and not in the furanose form as claimed by Batirov et al. [48].

In the past, also HPLC has been used by us to separate the weld flavones [37]. With a run time of one hour, lut-4′-*O*-glu **tr2** can be observed as a minor peak eluting just after chry-7-*O*-glu **6** while api-4′-*O*-glu **tr1** co-elutes with both with chry-7-*O*-glu **6** and lut-4′-*O*-glu **tr2** (Supplementary Figure S70). According to the Extracted Ion Chromatograms (EICs), the UHPLC system has a slightly different selectivity, which could be due to a different C18 phase, different gradient, or the use of acetonitrile instead of methanol. In the UHPLC system the trace flavones api-4′-*O*-glu **tr1** and lut-4′-*O*-glu **tr2** partially coelute with api-7-*O*-glu **5**. For EICs see Figures S76–S83.

2.4.3. Flavone Diglucosides

Based on the literature, the following diglucosides were expected: lut-7,3′-di-*O*-glucoside **3**, an isomer thereof (peak **2**), and api-6,8-di-*C*-glucoside (peak **1**). The identification is discussed in order of decreasing retention times (Figure 2).

The mass of the main diglucoside (peak **3**) was in agreement with lut-7,3′-di-*O*-glucoside, one of the main flavones of weld. In the ^{1}H NMR spectrum H-2′, H6′ and H-5′ had near identical shifts as the same protons in lut-3′-*O*-glu **7** while the shifts of H-3, H-6, H-8 corresponded well with those of the same protons in lut-7-*O*-glucoside **4**. The attachment of hexoses to O-7 and O-3′ was further confirmed by cross-peaks in the HMBC between C-7 and H-1″, and C-3′ and H-1‴. The ^{13}C NMR was in excellent agreement with the one reported by Markham et al. [41] for lut-7,3′-di-*O*-glu. This confirmed both hexoses to be glucose. As $J_{1''2''}$ is 7.2 Hz and $J_{1'''2'''}$ is 7.1 Hz, the compound isolated from weld is lut-7,3′-di-*O*-β-D-glucopyranoside **3**. The spectral identification was confirmed by chromatographic evidence: an authentic reference had the same retention time.

The mass of the compound corresponding with peak **2** was the same as that of the major diglucoside lut-7,3′-di-*O*-glu (peak **3**). In the ^{1}H NMR spectrum H-2′, H6′ and H-5′ had near identical shifts as the same protons in lut-4′-*O*-glu while the shifts of H-3, H-6, H-8 corresponded well with those of the same protons in lut-7-*O*-glucoside **4**. This pointed in the direction of lut-7,4′-di-*O*-glu. Cross-peaks in the HMBC spectrum between C-7 and H-1″, and C-4′ and H-1‴ proved that the sugar moieties were indeed attached to O-7, and O-4′. The NMR data of the compound are consistent with the aglycone being lut. As there are no ^{1}H- or ^{13}C-NMR data reported for lut-7,4′-di-*O*-glu in the literature, the ^{13}C shifts of its sugar moieties were compared with those obtained for lut-7-*O*-glu and lut-4′-*O*-glu (Table 1). The differences between them were <1.0 ppm, confirming both hexoses to be glucose. Similarly, there were excellent matches with the shifts of C-2 to C-8a of lut-7-*O*-glu **4** and of C-1′ to C-6′ of lut-4′-*O*-glu **tr2** (Table 1). As $J_{1''2''}$ is 7.7 Hz and $J_{1'''2'''}$ is 7.1 Hz, the compound is lut-7,4′-di-*O*-β-D-glucopyranoside **2**.

The HRMS corresponding with peak **1** gave as elemental composition $C_{27}H_{30}O_{15}$ (Supplementary Table S1), which is the composition of the expected api-6,8-di-*C*-glucoside **1**. The absence of H-6 and H-8 signals in the ^{1}H and HSQC spectra suggested that the sugar moieties were indeed attached to C-6 and C-8. This was supported by cross-peaks in the HMBC spectrum between C-5 and H-1″, C-6 and H-1″, C-8a and H-1‴, and C-8 and H-1‴. The remaining ^{1}H NMR data were consistent with the aglycone being api and the presence of two hexoses. The ^{13}C shifts of the anomeric carbons further supported the assignment as their signals (~75 ppm) appeared more upfield than those of the *O*-glycosides (~100 ppm), see Table 1. Additionally, there was an excellent fit with the ^{13}C NMR spectrum reported by Sato et al. [52] for api-6,8-di-*C*-β-D-glucopyranoside in DMSO-d_6 at 90 °C. The H-1″ and H-1‴ shifts of 4.9 ppm, and the $J_{1''2''}$ and $J_{1'''2'''}$ of 10 Hz were consistent with 6- and 8-*C*-β-D-glucosyl moieties [43]. The combined data show this flavone to be api-6,8-di-*C*-β-D-glucopyranoside **1** (synonym: vicenin 2).

3. Discussion

Starting from only 50 μg of weld-dyed wool and using a simple extraction procedure, high quality chromatographic fingerprints consisting of ten peaks can be obtained (Figure 4a,b), whereby "high quality" is defined in terms of overall baseline stability, chromatographic resolution, peak shape and signal-to-noise ratio. Due to the mild extraction conditions, the flavone aglycones and glucosides originally present in a weld extract are all detectable in the wool extract. This makes the fingerprint representative for weld. For instance, distinction from another yellow dye proved easy (Figure 4c). After membrane filtration to remove extracted wool threads, the final volume available for injection suffices for duplicate injections. As the injections are highly reproducible, one could also resort to single injections. This could in principle increase the sensitivity further, either by injecting a larger volume or by decreasing the initial extraction volume. However, the current sensitivity is sufficient as exemplified by an estimated signal-to-noise ratio of at least 100 for a 1% component like chrysoeriol **10**. Actually, the limiting factor appears to be chemical noise and baseline stability rather than electronic noise. Telling UV spectra could be obtained as evidenced by the those of luteolin **8**, apigenin **9** and even chrysoeriol **10**, which has a peak height <0.25 mAU (Supplementary Figures S73–S75). Starting with smaller quantities of wool would be possible too, but amounts smaller than 25 μg are in practice difficult to handle. If needed, another possibility for increasing the sensitivity would be to mount a 60 mm flow cell, which has a five times lower LOD [33]. The current surplus sensitivity could facilitate the investigation of historic textiles, which possibly contain lower concentrations of dye components due to decomposition. The method might also be used for following artificial ageing of dyed textiles as relative changes in flavone concentrations can be easily monitored even if decomposition products do not absorb at 345 nm.

The separation was carried out by UHPLC with diode-array detection (DAD). This allowed for a fast (6 min), efficient (baseline separation of all 10 flavones), selective and sensitive detection. Mass spectrometric detection in series with the DAD was attempted, but for this particular analysis, the simple quadrupole MS was less sensitive than a modern DAD. ESI-MS in negative mode could be used for the original weld extract but not for extracts of 50 μg of dyed wool as these contain 100 times less material. A more expensive MS will be able to provide the required sensitivity. Although on-line MS was still useful for peak assignments, for dye analysis mass spectrometry is perhaps slightly less selective than UV-Vis, as it may detect also non-dye compounds and is poorly suitable for distinguishing between different isomers such luteolin 3′- and 4′-glucosides. Thus, when only MS [27] is used for dye analyses instead of the combination DAD-MS, important information is lost.

To distinguish between isomers, NMR is a much more powerful method. A disadvantage of NMR is that it requires many preliminary purification steps. In this study 1-dim and 2-dim NMR successfully allowed the unequivocal identification of the ten flavones corresponding with the ten fingerprint peaks. Additionally, two trace flavones were identified. All of the twelve flavones had been reported for weld earlier with the exception of apigenin-4′-*O*-β-D-glucopyranoside **tr1**. The analysis of weld by Marques et al. is correct [26] although they did not detect either lut-3′-*O*-glu or lut-4′-*O*-glu. The correction by Gaspar et al. [36] of their earlier assignment [34] of these two compounds is justified, indeed the 3′-isomer elutes later than 4′-isomer. In the chromatographic system used by Gaspar et al., these two isomers are much better separated than in our system. Thus, some controversy with regard to the elution order of the different flavones was solved and this may assist future analyses of weld-dyed textiles. Future work should focus on expansion of the micro-extraction and UHPLC approach to other dye plants like madder.

4. Materials and Methods

4.1. Chemicals and Materials

Luteolin-7-monoglucoside and luteolin-7,3′-diglucoside (analytical control grade) were from Extrasynthèse. Luteolin was from Extrasynthèse or Sigma. Ammonium formate was from Aldrich or Fluka and EDTA tetrasodium salt dihydrate was from Aldrich. Aluminum chloride and formic acid were from Acros. The purity of formic acid used in extractions was 98%+ and for wool treatment 99%. Methanol (LC-MS grade) was from Prolabo and ultrapure water was prepared with an EasyPure system of Barnstead/Thermoline. Other chemicals were of analytical grade. Wool was weighed on a Mettler UM3 μg balance. *Reseda luteola* L. plants were grown in the province of Groningen, The Netherlands.

4.2. Wool Dyeing

Wool was dyed by students with weld as described by Villela et al. [30]. Five by five cm pieces of Kova medium weight wool sateen white (with a sateen finish on one side; natural cream colour; fabric is ready for dyeing and/or printing; approximately 290 g/m^2; from Whaleys, Bradford) were pre-mordanted with aluminum potassium sulphate dodecahydrate (alum). Reseda extract was prepared by sonicating 1.5 g of dried and ground weld with 30 mL of 96% alcohol-water 3:1 (*v*/*v*) for 10 min in an ultrasonic bath for the filtration and removal of alcohol. The resulting extract was diluted with 60 mL of water, heated to 80 °C, after which 4 pieces of pre-soaked, pre-mordanted wool were added and stirred during 15 min. Afterwards the pieces were rinsed with hot and cold water and dried. Wool was dyed in 2011 and stored in the dark at room temperature until extracted in 2019. Wool threads were removed just prior to the analysis. Onion-dyed wool: alum pre-mordanted wool was dyed during educational activities with the outer (paper-like) scales of onions in the same way as weld with the exception of starting from dry onion extract. The onion extract–wool ratio was approximately 1:100. Photos of both weld-dyed and onion-dyed wool can be found in Supplementary Figures S71 and S72 respectively.

4.3. Wool Micro-Extraction

Approximately 50 μg of weld-dyed wool was taken by means of chirurgic scissors and transferred into an empty 250 μL insert for a standard 1.8 mL autosampler vial previously weighed on a μg balance (readable up to 0.1 μg) (Figure 3a). After weighing again to determine the amount of wool taken, 0.5 μL of extraction solvent (methanol-water-formic acid (80:15:5, *v*/*v*/*v*) was added per μg of wool. After firmly closing the cap with an inert seal to avoid any evaporation of solvent during the extraction, a 30 min extraction step at 60 °C followed by placing the vial in a water bath. Next, the extract was diluted five times with ultrapure water and filtered through a 5 μm zero-dead volume needle filter (Sol-M blunt needle fill needle, W/5 micron filter, 110022F, Sol-Millennium) into a clean and empty 250 μL insert (Figure 3B). This dilution step enabled the filtration step. A filtration step is advisable to avoid clogging of the UHPLC column. Ten μL were injected into the UHPLC by means of the standard autosampler. The photo of dyed wool in the insert (Figure 3a) was taken by means of a digital microscope (DNT, The Netherlands).

4.4. Preparative Isolation and Purification of Weld Flavones

Aerial parts of dried *Reseda luteola* L. (cultivar code B) plants were used. The plants were grown, harvested and dried as described by Villela [37]. This plant material was ground with a cutting mill (Retsch GmbH, type SM1, Haan/Germany) equipped with a 0.5 mm sieve.

Compounds of low polarity were removed from 200 g of the ground plant material by stirring overnight with 2 L of petroleum ether 40°/60°-MTBE (1:1). After removal of solvents, the compounds of interest were extracted from the defatted plant material by stirring overnight with 2 L of 96% alcohol-water (8:2). The next morning, the solvent was heated till near boiling after which the plant material was removed by vacuum filtration through a glass filter (por 40). The plant material was extracted anew with 0.5 L of fresh alcohol-water (8:2) and filtered through the same glass filter. The two extracts were combined after which all solvent was removed with a rotary evaporator. The hydroalcoholic extract was fractionated via repeated polyamide column chromatography using different water–acetone gradients as eluent. Final separation and purification took place by preparative RP-HPLC (Shimadzu prepHPLC, Altima C18 column, 250 × 22 mm, 5 μm, A eluent: 2.1 L of water with 12.9 mL of formic acid and 5.05 g of ammonium bicarbonate, B eluent: methanol, various gradients were used, DAD monitoring wavelength 350 nm). Eluents were removed by rotary evaporation and freeze drying. The progress of the purification was monitored by TLC, RP-HPLC–UV, and ^{1}H-NMR.

4.5. UHPLC

An Agilent 1290 Infinity system equipped with binary pumps (G4220A), autosampler (G4226A), thermostatted (25 °C) column compartment (G1316C) and DAD (G4212A) in combination with a 60008 10 mm flow cell was used. The column was an Agilent Zorbax Eclipse Plus C18, 2.1 × 100 mm 1.8 μm). Mobile phase A: 40 mM formic acid buffer pH = 3 in water with 0.04 mM EDTA; mobile phase B: acetonitrile. Flow rate: 0.45 mL/min, initial pressure 511 bar. Linear gradient: 0 min 15% B; 8 min 45% B; Monitoring wavelength: 345 nm. Injection volume: 10 μL. For some analysis of the extract, the effluent of the DAD was connected to an Agilent Mass Spectrometer Detector (G6125B). The MSD was operated in negative mode with a m/z 150–800 scan range. Data were viewed as TIC and EIC plots.

4.6. NMR Spectroscopy

All 1-dim and 2-dim NMR spectra were recorded either on a Bruker Avance III 600 MHz NMR spectrometer or a Bruker Avance III 700 MHz NMR. Both were equipped with a 5 mm cryoprobe. Spectra were recorded in either 5 or 3 mm NMR tubes. The temperature of the sample was kept at 300 K. HSQC and HMBC spectra were recorded with standard pulse sequences. ROESY spectra were recorded on a Bruker Avance III 700 MHz NMR spectrometer with a standard pulse sequence and 225 μs mixing time. The ^{1}H spectra of api-7-*O*-glu **5**, api-4'-*O*-glu **tr1** and chry-7-*O*-glu **6** were acquired with the standard Bruker pulse sequence zgcppr for suppression of the water signal. ^{1}H spectra of api-6,8-di-*C*-glu **1** were recorded in DMSO-d_6 in a 3 mm probe at 300 and 320 K prior to dilution of the sample with 2 volumes MeOH-d_4 and transfer of the solution to a 5 mm tube; the ^{1}H spectrum and 2D spectra of api-6,8-di-*C*-glu **1** in DMSO-d_6—MeOH-d4 (1:2) were recorded at 300 K.

Between 0.7 and 2.7 mg of each of the 12 flavones was dissolved in 0.6 mL (5 mm probe) or 0.2 mL (3 mm probe) of DMSO-d_6. Data interpretation: C–H pairs were assigned in the HSQC spectrum and assignments were confirmed using the HMBC spectrum. ^{13}C δ of methine C-atoms reported in Table 1 were obtained from the HSQC experiment, and those of the quaternary C-atoms from the HMBC experiment. Chemical shifts were referenced against the peak of DMSO (δ (^{1}H) = 2.55 ppm and δ (^{13}C) = 40.45 ppm).

4.7. High-Resolution Mass Spectrometry

The twelve fractions containing the purified flavones were analysed by high-resolution mass spectrometry. The solutions were directly infused at a flow rate of 3 µL/min via a syringe pump into a Thermo Fisher Exactive Orbitrap mass spectrometer. The analyses were carried by electrospray, in negative ionisation mode. Settings were: 250–2000 m/z (scan range), ultrahigh resolution, 3.50 kV (spray voltage), 400 °C (capillary temperature), −25.00 V (capillary voltage). Measurements were carried out continuously during 0.6–1.7 min, and the data displayed in Supplementary Table S1 were obtained from 31–70 scans averaged spectra.

Supplementary Materials: The following are available online. Figures S1–S67: 1-dim and 2-dim NMR spectra of all 12 flavones; Figures S68–S69: UHPLC profiles of duplicate injections of extracted wool samples; Figure S70: HPLC profile of weld extract; Figures S71 and S72: photos of weld and onion dyed wool; Figures S73–S75: on-line UV spectra of luteolin, apigenin and chrysoeriol; Figures S76–S84: EICs and TIC of HPLC-DAD-MS analysis of weld extract; Table S1: High-resolution mass spectrometric data of all 12 flavones.

Author Contributions: Conceptualization, T.A.v.B. and A.V.; validation, E.v.d.K. and A.V.; investigation, E.v.d.K., A.V., T.A.v.B. and P.P.L.; writing—original draft preparation, T.A.v.B. and A.V.; writing—review and editing, T.A.v.B., A.V., G.C.H.D. and P.P.L.; supervision, T.A.v.B. and P.P.L.; funding acquisition, G.C.H.D., A.V., P.P.L. and T.A.v.B. All authors have read and agreed to the published version of the manuscript.

Funding: This research was partially funded by CAPES.

Institutional Review Board Statement: Not applicable.

Informed Consent Statement: Not applicable.

Data Availability Statement: Additional data can be found in the Supplementary Materials.

Acknowledgments: The authors are grateful to Wanda Olsder and Rubia Natural Colours for supplying the plant material, as well as Aniek Jongerius and Frank Claassen (Laboratory of Organic Chemistry) for analysing extracts of weld, mass spectrometric measurements and technical support. Financial support from CAPES—Brasília/Brazil (Ph.D. scholarship of Alexandre Villela), Wageningen University, Rubia Natural Colours, and DSM (NMR measurements) is gratefully acknowledged.

Conflicts of Interest: The authors declare no conflict of interest.

Sample Availability: Samples of the compounds are not available from the authors.

References

1. Ferreira, E.S.B.; Hulme, A.N.; McNab, H.; Quye, A. The natural constituents of historical textile dyes. *Chem. Soc. Rev.* **2004**, *33*, 329–336. [CrossRef]
2. Derksen, G.C.H.; van Beek, T.A.; de Groot, Æ.; Capelle, A. A high-performance liquid chromatographic method for the analysis of anthraquinone glycosides and aglycones in madder root (*Rubia tinctorum*). *J. Chromatogr. A* **1998**, *816A*, 277–281. [CrossRef]
3. Cristea, D.; Bareau, I.; Vilarem, G. Identification and quantitative HPLC analysis of the main flavonoids present in weld (*Reseda luteola* L.). *Dyes Pigm.* **2003**, *57*, 267–272. [CrossRef]
4. Zarkogianni, M.; Mikropoulou, E.; Varella, E.; Tsatsaroni, E. Colour and fastness of natural dyes: Revival of traditional dyeing techniques. *Color. Technol.* **2010**, *127*, 18–27. [CrossRef]
5. Willemen, H.; van den Meijdenberg, G.J.P.; van Beek, T.A.; Derksen, G.C.H. Comparison of madder (*Rubia tinctorum* L.) and weld (*Reseda luteola* L.) total extracts and their individual dye compounds with regard to their dyeing behaviour, colour, and stability towards light. *Color. Technol.* **2019**, *135*, 40–47. [CrossRef]
6. Zhang, X.; Laursen, R.A. Development of mild extraction methods for the analysis of natural dyes in textiles of historical interest using LC-diode array detector-MS. *Anal. Chem.* **2005**, *77*, 2022–2025. [CrossRef] [PubMed]
7. Rosenberg, E. Characterisation of historical organic dyestuffs by liquid chromatography-mass spectrometry. *Anal. Bioanal. Chem.* **2008**, *391*, 33–57. [CrossRef]
8. Ford, L.; Henderson, R.L.; Rayner, C.M.; Blackburn, R.S. Mild extraction methods using aqueous glucose solution for the analysis of natural dyes in textile artefacts dyed with Dyer's madder (*Rubia tinctorum* L.). *J. Chromatogr. A* **2017**, *1487*, 36–46. [CrossRef]
9. Degano, I.; Biesaga, M.; Colombini, M.P.; Trojanowicz, M. Historical and archaeological textiles: An insight on degradation products of wool and silk yarns. *J. Chromatogr. A* **2011**, *1218*, 5837–5847. [CrossRef]

10. Villela, A.; van Vuuren, M.S.A.; Willemen, H.M.; Derksen, G.C.H.; van Beek, T.A. Photo-stability of a flavonoid dye in presence of aluminium ions. *Dyes Pigm.* **2019**, *162*, 222–231. [CrossRef]

11. Ahn, C.; Zeng, X.; Li, L.; Obendorf, S.K. Thermal degradation of natural dyes and their analysis using HPLC-DAD-MS. *Fash. Text.* **2014**, *1*, 22. [CrossRef]

12. Derksen, G.C.H.; van Holthoon, F.L.; Willemen, H.M.; Krul, C.A.M.; Franssen, M.C.R.; van Beek, T.A. Development of a process for obtaining non-mutagenic madder root (*Rubia tinctorum*) extract for textile dyeing. *Ind. Crops Prod.* **2021**, *164*, 113344. [CrossRef]

13. Taghizadeh Borujeni, R.; Akbari, A.; Gharehbaii, A.; Yunessnia lehi, A. Extraction and preparation of dye powders from *Reseda luteola* L. using membrane processes and its dyeing properties. *Environ. Technol. Innov.* **2021**, *21*, 101249. [CrossRef]

14. Beldean-Galea, M.S.; Copaciu, F.-M.; Coman, M.-V. Chromatographic analysis of textile dyes. *J. AOAC Int.* **2018**, *101*, 1353–1370. [CrossRef]

15. Gulmini, M.; Idone, A.; Diana, E.; Gastaldi, D.; Vaudan, D.; Aceto, M. Identification of dyestuffs in historical textiles: Strong and weak points of a non-invasive approach. *Dyes Pigm.* **2013**, *98*, 136–145. [CrossRef]

16. De Ferri, L.; Tripodi, R.; Martignon, A.; Ferrari, E.S.; Lagrutta-Diaz, A.C.; Vallotto, D.; Pojana, G. Non-invasive study of natural dyes on historical textiles from the collection of Michelangelo Guggenheim. *Spectrochim. Acta A Mol. Biomol. Spectrosc.* **2018**, *204*, 548–567. [CrossRef] [PubMed]

17. Tamburini, D.; Dyer, J. Fibre optic reflectance spectroscopy and multispectral imaging for the non-invasive investigation of Asian colourants in Chinese textiles from Dunhuang (7th–10th century AD). *Dyes Pigm.* **2019**, *162*, 494–511. [CrossRef]

18. Jurasekova, Z.; Domingo, C.; Garcia-Ramos, J.V.; Sanchez-Cortes, S. In situ detection of flavonoids in weld-dyed wool and silk textiles by surface-enhanced Raman scattering. *J. Raman Spectrosc.* **2008**, *39*, 1309–1312. [CrossRef]

19. Zaffino, C.; Bruni, S.; Guglielmi, V.; de Luca, E. Fourier-transform surface-enhanced Raman spectroscopy (FT-SERS) applied to the identification of natural dyes in textile fibers: An extractionless approach to the analysis. *J. Raman Spectrosc.* **2014**, *45*, 211–218. [CrossRef]

20. Zaffino, C.; Bedini, G.D.; Mazzola, G.; Guglielmi, V.; Bruni, S. Online coupling of high-performance liquid chromatography with surface-enhanced Raman spectroscopy for the identification of historical dyes. *J. Raman Spectrosc.* **2016**, *47*, 607–615. [CrossRef]

21. Surowiec, I.; Quye, A.; Trojanowicz, M. Liquid chromatography determination of natural dyes in extracts from historical Scottish textiles excavated from peat bogs. *J. Chromatogr. A* **2006**, *1112*, 209–217. [CrossRef]

22. Karadag, R.; Torgan, E.; Taskopru, T.; Yildiz, Y. Characterization of dyestuffs and metals from selected 16-17th-century Ottoman silk brocades by RP-HPLC-DAD and FESEM-EDX. *J. Liq. Chromatogr. Rel. Technol.* **2015**, *38*, 591–599. [CrossRef]

23. Valianou, L.; Karapanagiotis, I.; Chryssoulakis, Y. Comparison of extraction methods for the analysis of natural dyes in historical textiles by high-performance liquid chromatography. *Anal. Bioanal. Chem.* **2009**, *395*, 2175–2189. [CrossRef] [PubMed]

24. Mantzouris, D.; Karapanagiotis, I.; Valianou, L.; Panayiotou, C. HPLC-DAD-MS analysis of dyes identified in textiles from Mount Athos. *Anal. Bioanal. Chem.* **2011**, *399*, 3065–3079. [CrossRef]

25. Zhang, X.; Laursen, R. Application of LC-MS to the analysis of dyes in objects of historical interest. *Int. J. Mass Spectrom.* **2009**, *284*, 108–114. [CrossRef]

26. Marques, R.; Sousa, M.M.; Oliveira, M.C.; Melo, M.J. Characterization of weld (*Reseda luteola* L.) and spurge flax (*Daphne gnidium* L.) by high-performance liquid chromatography-diode array detection-mass spectrometry in Arraiolos historical textiles. *J. Chromatogr. A* **2009**, *1216*, 1395–1402. [CrossRef] [PubMed]

27. Tamburini, D. Investigating Asian colourants in Chinese textiles from Dunhuang (7th–10th century AD) by high performance liquid chromatography tandem mass spectrometry—Towards the creation of a mass spectra database. *Dyes Pigm.* **2019**, *163*, 454–474. [CrossRef]

28. Otłowska, O.; Ślebioda, M.; Kot-Wasik, A.; Karczewski, J.; Śliwka-Kaszyńska, M. Chromatographic and spectroscopic identification and recognition of natural dyes, uncommon dyestuff components, and mordants: Case study of a 16th century carpet with chintamani motifs. *Molecules* **2018**, *23*, 339. [CrossRef] [PubMed]

29. Liu, J.; Li, W.; Kang, X.; Zhao, F.; He, M.; She, Y.; Zhou, Y. Profiling by HPLC-DAD-MSD reveals a 2500-year history of the use of natural dyes in Northwest China. *Dyes Pigm.* **2021**, *187*, 109143. [CrossRef]

30. Villela, A.; Derksen, G.C.H.; van Beek, T.A. Analysis of a natural yellow dye: An experiment for analytical organic chemistry. *J. Chem. Educ.* **2014**, *91*, 566–569. [CrossRef]

31. Han, J.; Wanrooij, J.; van Bommel, M.; Quye, A. Characterisation of chemical components for identifying historical Chinese textile dyes by ultra high performance liquid chromatography-photodiode array-electrospray ionisation mass spectrometer. *J. Chromatogr. A* **2017**, *1479*, 87–96. [CrossRef] [PubMed]

32. Serrano, A.; van Bommel, M.; Hallett, J. Evaluation between ultrahigh pressure liquid chromatography and high-performance liquid chromatography analytical methods for characterizing natural dyestuffs. *J. Chromatogr. A* **2013**, *1318*, 102–111. [CrossRef]

33. Huesgen, A.G. Comparison of Sensitivity and Linearity of the Agilent 1200 Infinity Series High Dynamic Range Diode Array Detector Solution and the Conventional Agilent 1290 Infinity DAD Technical Overview. 2021. Available online: https://www.agilent.com/cs/library/technicaloverviews/Public/5991-3877EN.pdf (accessed on 8 May 2021).

34. Moiteiro, C.; Gaspar, H.; Rodrigues, A.I.; Lopes, J.F.; Carnide, V. HPLC quantification of dye flavonoids in *Reseda luteola* L. from Portugal. *J. Sep. Sci.* **2008**, *31*, 3683–3687. [CrossRef]

35. Burger, P.; Monchot, A.; Bagarri, O.; Chiffolleau, P.; Azoulay, S.; Fernandez, X.; Michel, T. Whitening agents from *Reseda luteola* L. and their chemical characterization using combination of CPC, UPLC-HRMS and NMR. *Cosmetics* **2017**, *4*, 51. [CrossRef]

36. Gaspar, H.; Moiteiro, C.; Turkman, A.; Coutinho, J.; Carnide, V. Influence of soil fertility on dye flavonoids production in weld (*Reseda luteola* L.) accessions from Portugal. *J. Sep. Sci.* **2009**, *32*, 4234–4240. [CrossRef]
37. Villela, A.; van der Klift, E.J.C.; Mattheussens, E.S.G.M.; Derksen, G.C.H.; Zuilhof, H.; van Beek, T.A. Fast chromatographic separation for the quantitation of the main flavone dyes in *Reseda luteola* (weld). *J. Chromatogr. A* **2011**, *1218*, 8544–8550. [CrossRef] [PubMed]
38. Van Beek, T.A.; Silva, I.M.M.S.; Posthumus, M.A.; Melo, R. Partial elucidation of Trichogramma putative sex pheromone at trace levels by solid-phase microextraction and gas chromatography-mass spectrometry studies. *J. Chromatogr. A* **2005**, *1067*, 311–321. [CrossRef]
39. Lee, J.; Mitchell, A.E. Quercetin and isorhamnetin glycosides in onion (*Allium cepa* L.): Varietal comparison, physical distribution, coproduct evaluation, and long-term storage stability. *J. Agric. Food Chem.* **2011**, *59*, 857–863. [CrossRef]
40. Peggie, D.A.; Hulme, A.N.; McNab, H.; Quye, A. Towards the identification of characteristic minor components from textiles dyed with weld (*Reseda luteola* L.) and those dyed with Mexican cochineal (*Dactylopius coccus* Costa). *Microchim. Acta* **2008**, *162*, 371–380. [CrossRef]
41. Markham, K.R.; Ternai, B.; Stanley, R.; Geigner, H.; Mabry, T.J. Carbon-13 NMR studies of flavonoids-III Naturally occurring flavonoid glycosides and their acylated derivatives. *Tetrahedron* **1978**, *34*, 1389–1397. [CrossRef]
42. Van Loo, P.; de Bruyn, A.; Budesinsky, M. Reinvestigation of the structural assignment of signals in the ^{1}H and ^{13}C NMR spectra of the flavone apigenin. *Magn. Reson. Chem.* **1986**, *24*, 879–882. [CrossRef]
43. Markham, K.R.; Geigner, H. ^{1}H Nuclear magnetic resonance spectroscopy of flavonoids and their glycosides in hexadeuterodimethylsulfoxide. In *The Flavonoids: Advances in Research since 1986*, 1st ed.; Harborne, J.B., Ed.; Chapman & Hall: London, UK, 1994; pp. 441–497.
44. Kim, J.H.; Cho, Y.H.; Park, S.M.; Lee, K.E.; Lee, J.J.; Lee, B.C.; Pyo, H.B.; Song, K.S.; Park, H.D.; Yun, Y.P. Antioxidants and inhibitor of matrix metalloproteinase-1 expression from leaves of *Zostera marina* L. *Arch. Pharm. Res.* **2004**, *27*, 177–183. [CrossRef] [PubMed]
45. Park, Y.; Moon, B.-H.; Yang, H.; Lee, Y.; Lee, E.; Lim, Y. Complete assignments of NMR data of 13 hydroxymethoxyflavones. *Magn. Reson. Chem.* **2007**, *45*, 1072–1075. [CrossRef] [PubMed]
46. Chung, H.S. Inhibition of monoamine oxidase by a flavone and its glycoside from *Ixeris dentata* Nakai. *Nutraceut. Food* **2003**, *8*, 141–144.
47. Harput, Ü.Ş.; Çaliş, İ.; Saracoğlu, İ.; Dönmez, A.A.; Nagatsu, A. Secondary metabolites from *Phlomis syriaca* and their antioxidant activities. *Turk. J. Chem.* **2006**, *30*, 383–390.
48. Batirov, E.K.; Tadzhibaev, M.M.; Malikov, V.M. Flavonoids of *Reseda luteola*. *Khim. Prir. Soedin.* **1979**, *15*, 728–729.
49. Oyama, K.-i.; Kondo, T. Total synthesis of apigenin 7,4′-di-*O*-β-glucopyranoside, a component of blue flower pigment of *Salvia patens*, and seven chiral analogs. *Tetrahedron* **2004**, *60*, 2025–2034. [CrossRef]
50. Ding, H.-Y.; Chen, Y.-Y.; Chang, W.-L.; Lin, H.-C. Flavonoids from the flowers of *Pueraria lobata*. *J. Chin. Chem. Soc.* **2004**, *51*, 1425–1428. [CrossRef]
51. Lee, M.H.; Son, Y.K.; Han, Y.N. Tissue factor inhibitory flavonoids from the fruits of *Chaenomeles sinensis*. *Arch. Pharm. Res.* **2002**, *25*, 842–850. [CrossRef] [PubMed]
52. Sato, S.; Akiya, T.; Nishizawa, H.; Suzuki, T. Total synthesis of three naturally occurring 6,8-di-C-glycosylflavonoids: Phloretin, naringenin, and apigenin bis-C-β-D-glucosides. *Carbohydr. Res.* **2006**, *341*, 964–970. [CrossRef]

Article

The Three Pillars of Natural Product Dereplication. Alkaloids from the Bulbs of *Urceolina peruviana* (C. Presl) J.F. Macbr. as a Preliminary Test Case

Mariacaterina Lianza [1], Ritchy Leroy [2], Carine Machado Rodrigues [2], Nicolas Borie [2], Charlotte Sayagh [2], Simon Remy [2], Stefan Kuhn [3], Jean-Hugues Renault [2] and Jean-Marc Nuzillard [2,*]

[1] Department of Pharmacy and Biotechnology, University of Bologna, 40126 Bologna, Italy; mariacaterina.lianz3@unibo.it
[2] Université de Reims Champagne Ardenne, CNRS, ICMR UMR 7312, 51097 Reims, France; ritchy.leroy@univ-reims.fr (R.L.); carine.machado@univ-reims.fr (C.M.R.); nicolas.borie@univ-reims.fr (N.B.); charlotte.sayagh@univ-reims.fr (C.S.); simon.remy@univ-reims.fr (S.R.); jh.renault@univ-reims.fr (J.-H.R.)
[3] School of Computer Science and Informatics, De Montfort University, Leicester LE1 9BH, UK; stefan.kuhn@dmu.ac.uk
* Correspondence: jm.nuzillard@univ-reims.fr; Tel.: +33-32-691-8210

Abstract: The role and importance of the identification of natural products are discussed in the perspective of the study of secondary metabolites. The rapid identification of already reported compounds, or structural dereplication, is recognized as a key element in natural product chemistry. The biological taxonomy of metabolite producing organisms, the knowledge of metabolite molecular structures, and the availability of metabolite spectroscopic signatures are considered as the three pillars of structural dereplication. The role and the construction of databases is illustrated by references to the KNApSAcK, UNPD, CSEARCH, and COCONUT databases, and by the importance of calculated taxonomic and spectroscopic data as substitutes for missing or lost original ones. Two NMR-based tools, the PNMRNP database that derives from UNPD, and KnapsackSearch, a database generator that provides taxonomically focused libraries of compounds, are proposed to the community of natural product chemists. The study of the alkaloids from *Urceolina peruviana*, a plant from the Andes used in traditional medicine for antibacterial and anticancer actions, has given the opportunity to test different approaches to dereplication, favoring the use of publicly available data sources.

Keywords: natural products; dereplication; databases; spectroscopy; taxonomy; molecular structures

Citation: Lianza, M.; Leroy, R.; Machado Rodrigues, C.; Borie, N.; Sayagh, C.; Remy, S.; Kuhn, S.; Renault, J.-H.; Nuzillard, J.-M. The Three Pillars of Natural Product Dereplication. Alkaloids from the Bulbs of *Urceolina peruviana* (C. Presl) J.F. Macbr. as a Preliminary Test Case. *Molecules* **2021**, *26*, 637. https://doi.org/10.3390/molecules26030637

Academic Editors: Young Hae Choi, Young Pyo Jang, Yuntao Dai and Luis Francisco Salomé-Abarca

Received: 17 December 2020
Accepted: 22 January 2021
Published: 26 January 2021

Publisher's Note: MDPI stays neutral with regard to jurisdictional claims in published maps and institutional affiliations.

1. Introduction

1.1. General Considerations

Organic natural products are produced by living organisms to ensure their own basic functional requirements through primary metabolism and to fine-tune the relationships with their surrounding through specialized or secondary metabolism. The term "natural product" (NP) generally refers to an organic specialized metabolite. NP biosynthesis is controlled by the genes and therefore depends on organism species. The biological evolution led to the preservation of some NPs across related species while others were left over. A set of species may consequently share a set of identical specialized metabolites. The taxonomic classification of species relied on phenotype comparison at the time biologists would not even dream to have access to the genome of living organisms. The identification of preserved NP structures or structural features could then assist the classification task through chemotaxonomic studies, as NP structures are part of the phenotype.

The investigation of NPs is not only bound to taxonomic studies but is motivated by the uses human beings make of them. NPs with therapeutic, organoleptic, psychoactive, poisonous, tinctorial (non-exhaustive list) properties were generally not produced to be used by humans for the purposes organisms produce them. Most of NP properties are related to their interaction with other biologically produced systems, referred to as NP targets. An NP would thus more likely interact with a target of therapeutic interest (an enzyme to be inhibited, for example) than with a randomly chosen molecule drawn from the chemical space of organic molecules, due to the co-evolution of all living species over hundreds of millions of years.

The understanding of the interaction between an NP and a target is a challenging task and is often a necessary step for the design of chemical compounds with enhanced properties [1]. This step requires a precise knowledge of the structure of the NP (and of its target) at the atomic level, a concern that converges with the one of chemotaxonomy.

Finding the structure of a compound that is already known should be, at least seemingly, much easier than the one of an unknown compound. The tentative identification of known compounds is one of the aspects of what is covered by the term "dereplication", because earlier efforts for purification and/or structure determination have not to be replicated [2]. Undertaking dereplication in first place makes sense because an organism for which nothing is known about its chemistry may share compounds with an already studied organism with close taxonomic relationship for the reason invoked in the first paragraph. Compounds that resist dereplication may be false (known) unknowns when the employed dereplication tools fail or true (unknown) unknowns [3], for which isolation and structure elucidation tools have to be deployed [4]. The determination of the molecular structure of NPs by dereplication constitutes an important part of this article.

Dereplication is a matter of collective memory by essence. This raises the questions of what information has to be preserved and of how to do it. Proving that two substances are identical at the atomic level is currently achieved by physico-chemical methods. The data produced by the analytical instruments and the related conditions in which they are obtained, namely the meta-data, are of prime importance. By language abuse, the analytic data and their associated meta-data will be referred to here as "spectroscopic data". If the molecular structure of compound A is known and if compound B is proved to be identical to compound A by spectroscopic data comparison, then the structure of compound B can be asserted as being also the one compound A, without having to interpret the data obtained from compound B, hence providing the expected time and effort gain.

Obviously, structures must be preserved along with associated spectroscopic data as the end of the currently described dereplication process is the labeling of a sample with the structure of a compound (compound naming will be discussed hereafter). A theoretical dereplication strategy would be to preserve the structure and spectroscopic data of all, probably less than 400,000, known NPs to date, and to compare the data from a presumably known compound with all the preserved data. If ever possible, this approach would be highly inefficient and it would be more efficient to limit the comparison work to the compounds from organisms that are taxonomically close to the one from which the currently considered NP comes from, in the way used by NP chemists during the pre-computer age [5]. This means that a link should be preserved between a NP structure and the taxonomy of the organism(s) it originates from. It clearly appears at this point that structure description, spectroscopy and taxonomy constitutes the three pillars of dereplication. Selected aspects of each of them are detailed hereafter.

1.2. The Three Pilars of Dereplication

1.2.1. Molecular Structures

The structure of purely organic compounds, excluding organometallic species, is remarkably well described by mathematical graphs, with atoms as nodes and bonds as edges. The idea of matching a single compound with a single structure is valid at least when a compound cannot be described by more than one tautomeric form. While InChI [6] and SMILES [7] linear notations retain all the necessary structural features of a compound, including chirality, the text-based MOL format (and the derived SDF format) is widely used as it includes atom coordinates necessary either for 2D depictions or for 3D viewing [8]. The three representation modes evoked here may coexist in order to avoid conversion operations, even though a computer tool can facilitate them [9]. Structures may be surrounded by various calculated properties (molecular formula, molecular mass, chemical classification, topological descriptors, for example) coded as tag-value pairs. The compound name may be also considered as calculated property. Aspirin is indeed acetylsalicylic acid (another compound name for the same substance) for which IUPAC [10] indicates it is 2-acetoxybenzoic acid in English but "acide 2-acétoxybenzoïque" in French and "2-(Acetyloxy)benzoesäure" in German, thus precluding any kind of simple character-by-character name comparison. Considering the name as a molecular property, a compound is better referenced by a list of synonyms rather than by a single name. A structure that is proposed to be a new one because dereplication did not prove it was already known must be searched for in the literature, a task that is simplified by looking, if possible, in comprehensive structure collections such as the one provided by CAS [11] or PubChem [12].

1.2.2. Spectroscopy

Spectroscopy is considered here in the broadest sense, namely as any physico-chemical methods of characterization. This includes the methods that truly rely on the interaction between electromagnetic waves and matter (UV-visible, IR, Raman, NMR, vibrational and electronic dichroism spectroscopies, optical rotation measurements) but also mass spectrometry (MS), fusion and boiling temperature measurements, and others. Even a single optical rotation value has to be associated to meta-data, such as the nature of the used solvent, the sample concentration, the temperature, and possibly the model of the measurement device. Reporting in NMR spectroscopy is a much more complex task as it must encompass the conditions of raw data acquisition, the nature of the processing operations that lead to spectra and the feature recognition processes that produce "reduced data" (a list of chemical shift values correspond to the position of spectral peaks, for example) and ultimately contributes to molecular structure proposals [13]. The diversity of spectrometer manufacturers, each proposing its own file formats, clearly precludes the easy comparison of spectroscopic data, even though a universal, text-based format named JCAMP [14] is supported by the IUPAC but could be possibly superseded in a near future by the ADF format of the Allotrope foundation [15]. The result of spectroscopic analysis is only meaningful if the link between data and compound structure is preserved, possibly leading to spectra interpretation, thus making possible to associate a particular spectral feature (the mass of a molecular fragment in MS) and a structural feature (a fragment of a molecular structure). There is no easy way to access the spectroscopic data of known natural products [16]. Most of visible efforts in this direction were devoted to the characterization of primary metabolites in the perspective of metabolomic studies [17,18].

A set of NMR and of MS^n spectra constitute a better way to identify a known compound than a fusion temperature and an optical rotation value, even though the two latter may suffice to rule out an incorrect structure hypothesis. Dereplication by MS-based methods has earned a high level of interest with the advent of MS^2-driven molecular network analysis [19,20]. Alternatively, a molecular (elemental) formula deduced from high-resolution MS (HRMS) associated to 1D and 2D NMR spectra may suffice to identify a known compound with a high level of confidence if reference data are available. A workaround to the lack of experimental spectroscopic data can be found, more or less accurately depending on the analytical technique, by means of computerized prediction tools. Dereplication of NPs based on ^{13}C-NMR predicted data has been reported and discussed [21,22]. Such predictions may be carried out by various software, including proprietary or free methods, available on local computers or through web interfaces, with possible automated use or not, and with performances that can be difficult to evaluate. CNMR Predictor, NMRPREDICT, ChemDraw, nmrshiftdb2 are such software, among which nmrshiftdb2 may be used for free in an automated may on a local computer while CNMR Predictor is a commercial product renowned for its accuracy.

1.2.3. Taxonomy

Taxonomy of living beings is a science in permanent evolution, where the findings of molecular biology separate species that were assumed to be close parents according to phenotype similarity and possibly finds similarities where none was apparent, while taxa names might evolve during time. Tools such as "NCBI Taxonomy Browser" [23] and "Tree of Life" [24] are of great help to navigate through taxonomic information and to locate species belonging to a given taxon. Answering the question of which species produce a given compound and of which compounds are known to be produced by an organism of a given species is possible by means of the Dictionary of Natural Products (DNP) with limitations inherent to a commercial product.

1.2.4. Databases

The results of the chemical study of living beings are diluted among a profusion of specialized scientific journals. The construction of a collective memory about NPs is not a spontaneous process, so that the initial dilution of results has to be counterbalanced by efforts for the re-concentration of the knowledge at some well-defined places named databases. Databases that link structural, spectroscopic and taxonomic knowledge should constitute the basis of well-managed NP chemistry. A must-read article recently published focuses on where to find data about NPs in 2020 [25]. Open databases for NP research containing structural and spectroscopic data were reported earlier in [26]. The grouping of structural, taxonomic, and experimental spectroscopic data of natural products was undertaken in the '90s in the framework of the SISTEMAT project [27]. The data and software resulting from this visionary undertaking are unfortunately not accessible to the general public [28]. Other databases dedicated to the study of NP chemistry always miss some aspect. Biological activity studies are purposely left aside in this article as they do not constitute an entry point for dereplication. As well, bibliographic databases are not considered here, as the creation of NP databases from primary literature is not discussed, even though everyone understands that this is an important aspect of NP research.

2. Results

This article reports the availability of a computer software for the creation of taxonomy-focused NP databases named KnapsackSearch and of a database named PNMRNP, exemplified by the study of alkaloids from *Urceolina peruviana*.

2.1. KnapsakSearch

The KNApSAcK website exposes multiple databases searchable by organism name, metabolite name, and other commonly used compound identifiers [29]. Searching KNApSAcK for a given genus name displays a series of lines, each one showing a compound identifier, the related CAS identifier, metabolite name, molecular formula, molecular weight, and the name of the species in which it was reported. A genus name refers to a set of species names and each species is related to a set of compounds, each one, as discussed earlier, being possibly present in organisms from different species. Directly querying KNApSAcK for family names of organisms fails. About 54,000 compounds are referenced in KNApSAcK, thus giving access to an incomplete but still non-negligible part of the chemical space of NPs.

When starting the chemical study of an organism, the search for taxonomy-related ones may not be limited to those of the same species and may extend to the entire family, to parts of it, or to super-sets of it. The goal of the KnapsackSearch (KS) project is to process a list of user-defined genera to produce a list of chemical compounds that are related to one or more of these genera. The result is obtained as an SDF file, so that each compound is associated to a 2D molecular structure with chiral center flags and to a taxonomic and a spectroscopic description. The SDF chemistry file format is not a database format by itself but is sufficiently widespread to be read by most of chemistry software and computational toolkits. The source code of KS is made of Python scripts that rely on the RDKit library of functions for cheminformatics [30]. The freely available EdiSDF software is useful for the viewing of 2D structures and related tag-value pairs [31].

The workflow of KS (Figure S1) starts with the collection of all pairs made of a compound identifier and a binomial name obtained as replies to queries for the queried genus names. Each compound identifier (C_ID) is associated to a list of organism binomial names. As an example, Figure S2a shows the beginning of the list of compounds (columns 1–5) and organisms (column 6) related to genus *Galanthus*. C_IDs are then used as keys for compound search. Figure S2b shows the result of such a query for galanthamine, C_ID C00001570. The resulting in data aggregates containing a compound name, a molecular formula, a molecular weight, a CAS number (if any), an InChI string, the InChIKey hashed form of the InChI, and a SMILES string. The latter is decoded to produce atom and bond lists reshaped as a 2D MOL block. A compound is validated at this stage if the InChI calculated from the MOL block is identical to the one given by KNApSAcK. Molecular formula calculated from the MOL blocks are also compared to the KNApSAcK ones because the latter always lack the electric charge indication if there is one and because they may correspond to [M+H]$^+$ ion formula; in these cases, the retained molecular formula are deduced from the ones of [M+H]. All InChiKeys are recalculated as it may happen that compounds with different C_IDs yield identical InChIKeys. Such compounds are withdrawn from the regular compound list and processed separately to produce compound aggregates in which the alternative attributes, such as C_IDs and names, are joined together. Each compound is then associated to the taxonomic information retrieved during the first stage of the data collection process and to the ^{13}C-NMR chemical shifts as predicted by the nmrshiftdb2 software. The data record of galanthamine, as displayed by the EdiSDF software, is presented in Figure S2c.

The source code of KS is freely available [32] and a few KS-generated SDF files are given with the corresponding lists of organism genera related to a taxonomic family. The ^{13}C-NMR chemical shifts included in KS files may be reformatted to be imported by NMR spectroscopy software by ACD/Labs and to facilitate compound selection according to chemical shift values, thus allowing the user to benefit from the easy prediction of ^{13}C-NMR chemical values on a massive scale (massive meaning without one-by-one manual operation on structure records) by nmrshiftdb2 and from the friendly graphical interface of software from ACD/Labs. The future of the web-based approach to family-focused NP databases in KS obviously relies on the continuation of the KNApSAcK web service [33]. Database, service, and software discontinuations obviously constitute serious threats, whatever the considered domain of scientific activity.

2.2. Predicted NMR Data for Natural Products (PNMRNP)

This section reports the transformation of a discontinued NP database, the Universal Natural Product Database (UNPD), into a "two-pillar" NP database, PNMRNP, in which biological taxonomy data is missing. Chemical classification is tentatively proposed as a remedy to this lack. The initial data used in this process is a set of Comma Separated Value (CSV) files from UNPD, provided as a part of the In Silico Data Base (ISDB) dedicated to MS-based dereplication [34]. Most of the data transformations were carried out using RDKit and locally developed Python scripts.

The Comma Separated Values (CSV) files from UNPD contain SMILES and InChI character strings as structure descriptors of NPs (213,210 compounds). Attempts to decode the SMILES chains led to the detection of a non-negligible amount of badly formed chains, so that only InChIs were considered for 2D structure generation with retained chirality information. A set of 43 compounds was discarded, containing duplicate or organometallic or inorganic compounds.

Decoding an InChI is achieved through the dedicated software library linked to RDKit and may result into unexpected results. For example, aliphatic amides were reconstructed from their InChI as their iminol tautomer, which is correct because all tautomers of a given molecule share the same InChI. Transforming aliphatic iminols into alphatic amides was undertaken using a chemical transformation rule coded as a reaction "SMILES arbitrary target specification" also known as reaction SMARTS or SMIRKS [35]. A set of such rules was applied to fix unlikely tautomeric forms. This step would have benefited from the application of a recent molecule standardization software related to RDKit [36].

RDKit does not handle the axial chirality of substituted allenes or spirans, possibly resulting in incorrect structures upon InChI decoding. Structures of compounds for which the InChI to structure conversion and back-conversion to InChI (the so-called "round-trip") fails to be consistent were tentatively obtained by means of the ChemDraw software driven by a python for win32 script. An identifier resolution using the Chemical Identifier Resolver (CIR) from the US National Health Institute (NIH) is attempted in case of persisting failure [37]. After final checking of round-trip consistency, the nmrshiftdb2-predicted ^{13}C-NMR chemical shift lists [38] were appended to compound data, resulting in a SDF file containing 211,280 records.

Even though the initial CSV files assigned a chemical name to some of the compounds in UNPD, an alternative naming procedure was carried out. The PubChem website offers a file that relates InChI and PubChem Compound Identifier (CID) and another one that relates CID and synonym lists. A set of synonyms was associated by this means to the PNMRNP compounds that are named in PubChem.

The assignment of chemical classification data to NPs in PNMRNP does not replace genuine but unavailable biological taxonomic data but may assist NP chemists to reduce the size of the chemical space to investigate when facing a dereplication problem. The link between biological and chemical taxonomy was already exploited in SISTEMAT [39]. The production of chemical classification data constitutes a remedy to the absence of a way to associate easily and at no cost a set of living organism names to an NP identifier. Chemical classification in itself is a fuzzy concept. Discussing about the definition of an alkaloid may result in an answer such as "an alkaloid is like my wife. I can recognize her when I see her, but I can't define her" [40]. Two independent classification systems are available in PNMRNP, one (CL1) is the result of a locally developed attempt that is not comprehensive but that may meet some needs while the other one (CL2) relies on the well-established ClassyFire software [41].

Chemical classes in CL1 are defined according to the presence of specific substructures (subgraphs or molecular graphs) and are identified using SMILES that are interpreted as SMARTS [42]. Chemical classification is organized in PNMRNP with four levels, so that menthol is reported to be a secondary metabolite, a terpene, a monoterpene, and a menthane compound. More precisely substructures are identified as deriving from primary metabolites (identifier: 01) such as amino-acids, sugars, or lipids and are otherwise classified as being specifically related to secondary metabolites (identifier: 02). Terpenes (02-02) include monoterpenes (02-02-01) that share the menthane skeleton (02-02-01-001). Sugar containing compounds (01-01) were identified through a set of 1296 SMILES chains covering open chain and cyclic sugars with possible features such as deoxy- and amino-substitution (63 classes of sugars, overall). Hexopyranoses (01-01-14), with their five asymmetric carbons, thus featuring alpha- and beta-anomeric forms, are identified by a set of 32 SMILES chains to which a generic one without chirality indicator is added. The rather ubiquitous α- and β-D-glucopyranose molecular sub-units are identified as 01-01-14-005 and -006, respectively. The idea of searching for sugars in NPs was put into practice recently in the framework of the COCONUT NP database development and the possible *in silico* deglycosylation [43]. The CL1 data items in PNMRNP include the lists of atoms concerned by each detected substructure. A part of the classification was inspired by "Pharmacognosy", a book by J. Bruneton [44], and another part from the skeleton library included in the resource files of the LSD software, a library itself borrowed from the SISTEMAT knowledge base [45]. The catalog of SMILES that resulted from the CL1 effort toward a chemical classification of NPs is available as a supplementary information file in Excel format.

Chemical taxonomy in the second classification system (CL2) in PNMRNP results from replies to queries sent to the web interface of ClassyFire. This system deals with chemistry as a whole, distinguishing between organic and inorganic compounds at the first level, named "Kingdom" by reference to the classification of living beings. The overall hierarchy of chemical classes covers up to eleven levels. The recently reported classification tool named NPClassifier specifically targets NPs [46]. Classification CL2 was introduced with version 2 of PNMRNP [47]. The link between biological and chemical classifications is highlighted by considering that a molecule can be recognized by ClassyFire as a *Strychnos* alkaloid (*i.e.*, from a plant of the *Strychnos* genus) on a sole structural basis, without any reference to its source, possibly natural or synthetic. The natural origin of so-called "organic" compounds has become difficult to ascertain without resorting to proprietary databases, so that a NP-likeness score, a calculated molecular property, is invoked in order to evaluate to which extent a natural product is natural [48]. This approach fits with the current belief according to which a human being is better known by the algorithm of a popular social network than by her- or himself.

2.3. CSEARCH

The web interface of CSEARCH was also considered for NP structure dereplication besides of KS and PNMRNP. The CSEARCH web server accepts requests made of a list of ^{13}C-NMR chemical shifts, at best with each value associated to a multiplicity indication (number of attached hydrogen atoms, as deduced from DEPT or multiplicity-edited 2D HSQC spectra) and returns within a few minutes a list of structures sorted in the decreasing order of likelihood, proposed from a database containing several tens of millions of compounds and their predicted chemical shift values [49]. This database mostly contains structures of synthesized molecules and has no built-in concept of NP, resulting in hard to exploit results if the query is not accurate enough but may also give the solution of the submitted problem ranked in the first places, if not in first place.

2.4. Databases and Dereplication

To sum up briefly, KnapsachSearch may be considered as a part of a "two-pillars and half" approach to dereplication, while a "true three pillars" would have been achieved if spectroscopic data were of experimental origin instead of being predicted. PNMRNP can be qualified as "two-pillars" with its predicted spectroscopic data (a half-pillar) and biological taxonomy replaced by chemical taxonomy (a second half-pillar). The "one-pillar and half" NMRPREDICT/CSEARCH approach, dealing with structures and predicted ^{13}C-NMR spectroscopy only, should be considered before any other one, if pertinent. A tentatively exhaustive (and even more than that) source of NP data, COCONUT, collects structures from various sources to propose a publicly available document-oriented database of about 400,000 compounds, some of them being clearly not so natural. COCONUT version 1 was a "one-pillar" database, devoid of spectroscopic and taxonomic data but was recently supplemented with chemical classification (a half-pillar) and could be possibly supplemented in the future with predicted spectroscopic data (another half-pillar) to provide a useful "two-pillar" tool for NP structural dereplication.

2.5. Application to the Alkaloids of Urceolina peruviana (Amaryllidaceae)

Urceolina peruviana (C. Presl) J.F. Macbr., also known as *Stenomesson miniatum* (Herb.) Ravenna, is a bulbous perennial plant, which grows wild in the Andean regions of Peru and Bolivia (Figure 1). It has a scape up to 40 cm long, an umbrella of six or more red or orange tubular flowers, blooms in the spring or summer, the leaves are narrow, long until 28 cm.

Figure 1. *Urceolina peruviana.*

There is scarce information on this species of Amaryllidaceae in the scientific literature. The only article about the alkaloid composition of its bulbs was written in 1957 by Boit and Döpke, who reported the identification of three alkaloids (tazettine, haemanthamine and lycorine) and two others that could be traced back to nerinine and albomaculine [50]. Girault, in his book "Kallawaya, guérisseurs itinérants des Andes: recherches sur les pratiques médicinales et magiques", on a survey carried out in the Andes on the uses of medicinal plants by the indigenous South Americans, mentions *Urceolina peruviana* whose fresh bulbs were mixed with pork or llama fat and used in the form of ointment to treat tumors and abscesses [51]. Amaryllidaceae alkaloids constitute a set of about 600 compounds, some of them, such as galanthamine, having been intensively studied for their therapeutic action [52]. The present article illustrates the use of the aforementioned NMR-based dereplication tools by the study of *U. peruviana* and on its alkaloids.

The freeze-dried bulbs of *U. peruviana* were ground before being subjected to extraction. Extract 1 (11 mg) resulted from a non-selective solid-liquid extraction of a single bulb by methanol followed by acid-base liquid-liquid extractions for basic compound isolation. Extract 2 (20 mg) was obtained by lixiviation of alkalinized powder from a single bulb by EtOAc followed by acid-base liquid-liquid extractions according to patent [53]. The method used for the preparation of extract 2 was also applied on 270 g of dry bulb powder to yield 2.742 g of extract 3. A comparison of 1D ^{1}H- and ^{13}C-NMR spectra of extracts 1, 2, and 3 is provided in Figure S3.

Crude extract 3 was fractionated by Centrifugal Partition Chromatography (CPC) in the so-called "pH-zone refining" development mode, which is particularly adapted to the preparative scale fractionation and purification of H$^+$ ion exchanging compounds, without resorting to a solid-state chromatographic support. Emergence order of analytes from the CPC column depends on their acidity constant (K$_a$) and on the distribution constant (K$_D$) of their neutral form between the two liquid chromatographic phases. The chromatogram looks like trapezoidal blocks of analytes separated by steep boundaries, the so-called shock layers and forms an isotachic train of analytes [54]. The fractionation process led to 13 fractions, hereafter named A1 to A13, among which A4, A7, A9, and A11 were each found to contain a highly major compound. Purity and content of fractions A3 and A5 were very similar to the one of A4. Fraction A1 had a very low mass and a high complexity and was therefore not studied further. Fractions A2, A6, A8, A10, A12 and A13 are "intermediates" and concentrate minor compounds between the shock layers of the trapezoidal zones corresponding to the emergence of the major compounds of the injected sample.

The LC-HRMS analysis of a crude alkaloid extract 2 of *U. peruviana* monitored by UV absorbance at 287 nm showed 4 major peaks, to which molecular formula were assigned through accurate mass analysis of the [M+H]$^+$ ion: $C_{16}H_{17}NO_3$ (peak 3), $C_{17}H_{19}NO_4$ (peak 4), $C_{18}H_{21}NO_5$ (peak 2), and $C_{19}H_{23}NO_5$ (peak 6) as indicated in Figure 2. Two other minor peaks, one visible in the ion-current chromatogram and the other one in the UV chromatogram were also considered for further analysis, associated to molecular formula $C_{19}H_{25}NO_5$ (peak 1) and $C_{18}H_{21}NO_4$ (peak 5), respectively. The LC-HRMS analysis of crude extract 3 results in the same list of formulas but with $C_{18}H_{21}NO_4$ replaced with $C_{18}H_{19}NO_4$ and with $C_{18}H_{18}N_2O_4$ and $C_{19}H_{26}N_2O_5$ as supplementary proposals; the two latter suggest the presence of compounds containing two nitrogen atoms, a feature that is not common among Amaryllidaceae alkaloids and the pertinence of which was not ascertained. The ^{1}H, ^{13}C, ^{1}H-^{1}H COSY, ^{1}H-^{1}H ROESY, ^{1}H-^{13}C multiplicity-edited HSQC, and ^{1}H-^{13}C HMBC NMR spectra of most of fractions from extract 3 were recorded. A ^{1}H-^{15}N HMBC spectrum of extract 3 was also recorded, also offering a rapid and rough estimate of extract complexity by inspection of the projection of this 2D spectrum on the ^{15}N chemical shift axis (Figure 3).

Database creation was undertaken prior to and during the course of *U. peruviana* compound identification. The search by means of KS for the compounds reported in KNAp-SAcK and related to 67 genera from the Amaryllidaceae family resulted in 249 structures, among which 209 contained at least one nitrogen atom and were thus considered as possible alkaloids. These structures were imported by ACD/Labs "C+H NMR Predictors and Database" software as a new database and semi-automatically supplemented with ACD/Labs-predicted ^{1}H- and ^{13}C-NMR data by means of the protocol reported in Figure S4 to produce database DB1. The same set of 209 records, each including nmrshiftdb2-predicted ^{13}C-NMR data, was imported by the same ACD/Labs software after appropriate reformatting of the writing of chemical shift values to yield database DB2. Six small databases containing 2 to 15 records where derived from DB2 by selecting the molecules according to the molecular formula obtained by LC-HRMS analysis of extract 3, after having verified that no compound in DB2 contains two nitrogen atoms. Database DB3 was created by the same process as DB2 but starting from the 211,280 records of PNMRNP. The latter has also been filtered to retain compounds that include one of the eight substructures that are commonly found in Amaryllidaceae alkaloids [55] (Figure S5) to give DB3′, with 635 structures. A collection of 693 compounds was created from COCONUTv1 and named DB4, retaining the compounds that contain one of the eight Amaryllidaceae substructures after an initial step that selected 109,638 compounds with more than 12 carbon atoms and with one or two nitrogen atoms. DB4 was supplemented with ^{13}C-NMR chemical shifts from nmrshiftdb2 and formatted as an ACD/Labs database.

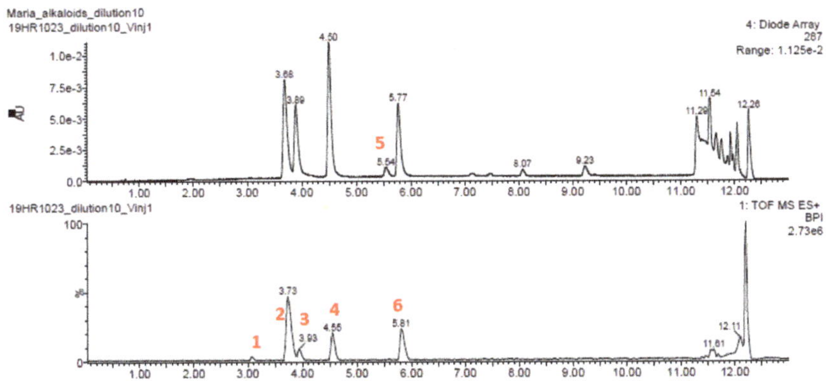

Figure 2. LC-HRMS ESI$^+$ analysis of extract 2, UV detection (top) and ion current intensity (bottom). HRMS data are compatible with [M+H]$^+$ ions of formula $[C_{19}H_{26}NO_5]^+$ (peak 1), $[C_{18}H_{22}NO_5]^+$ (peak 2), $[C_{16}H_{18}NO_3]^+$ (peak 3), $[C_{17}H_{20}NO_4]^+$ (peak 4), $[C_{18}H_{22}NO_4]^+$ (peak 5), $[C_{19}H_{24}NO_5]^+$ (peak 6).

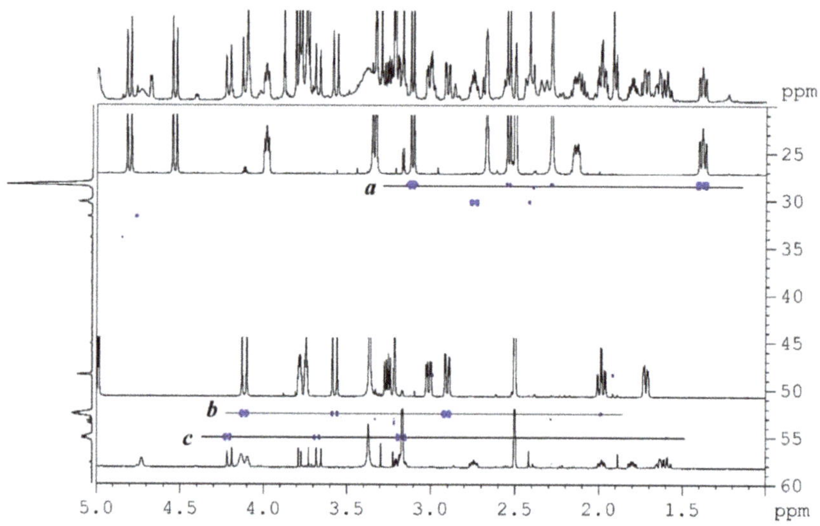

Figure 3. The ^{1}H-^{15}N HMBC spectrum of extract 3. The projection on the ^{15}N chemical shift axis provides of rough estimation of extract complexity. Traces *a*, *b*, and *c* are the ^{1}H-NMR spectra of tazettine, haemanthamine, and crinine recorded from fractions A4, A9, and A11, respectively.

2.5.1. Fraction A4, Major Compound

This compound is also the major compound in fractions A3 and A5. The CSEARCH algorithm succeeded to retrieve tazettine **1** (structure in Figure 4) as a likely compound from the list of the 18 ^{13}C-NMR chemical shifts and multiplicities from fraction A4. The molecular formula was constrained to include only C, H, N, and O atoms with a molecular mass comprised between 250 u and 400 u. Only a single chemical shift value was considered slightly unsatisfactory with δ_C 29.6 predicted by CSEARCH at position 4 and δ_C 25.9 observed (full atom numbering is reported in Figure S5). The analysis of the NMR spectra led to the identification of an aromatic ring substituted by a methylenedioxo bridge, a N-Me group, an ether O-Me group, and a hemiacetal group. The list of the $C_{18}H_{21}NO_5$ Amaryllidaceae alkaloids in the KNApSAcK database contained two compounds among 12 that shared these NMR-derived structural features. The HMBC correlations of the ^{1}H-NMR signal of the OH group lead to retain only the planar structure proposed for compound A4. None of the five $C_{18}H_{21}NO_4$ Amaryllidaceae compound structures present in the KNApSAcK database satisfied the NMR-derived constraints. The proposed planar structure is the one of tazettine and criwelline, which are epimers at position 3 [56]. CSEARCH ranked the 6-OMe criwelline in second position. Tazettine was retained as the structure of the major compound in fraction A4 after the analysis of the ROESY spectrum and the measurement and ^{1}H-^{1}H coupling constants. Its molecular formula relates it to peak 2 in the chromatograms in Figure 2.

2.5.2. Fraction A7, Major Compound

The CSEARCH algorithm failed to retrieve a likely structure from the list of the 19 chemical shifts drawn from the ^{13}C-NMR spectrum of fraction A7. The molecular formula was constrained to include only C, H, N, and O atoms with a molecular mass comprised between 300 u and 400 u. Only two $C_{19}H_{23}NO_5$ molecular structures of compounds from Amaryllidaceae were found in the KNApSAcK (DB2) database, among which only one contained three methoxy groups bound to an aromatic ring. This structural constraint was derived from the presence of three methyl signals in the ^{1}H-NMR spectrum that correlate in the HMBC spectrum with signals of aromatic carbons. This planar structure was confirmed by all available NMR data. None of the two $C_{19}H_{25}NO_5$ Amaryllidaceae compound structures present in the KNApSAcK database satisfied the NMR-derived constraints. The retained structure was indeed present in the solutions proposed by CSEARCH, but with a poor ranking, due to the low-quality matching between the experimental (δ_C 161.4, 110.9, and 155.0) and the predicted (δ_C 166.9, 103.1, and 156.6) chemical shifts for carbons at positions 6, 6a, and 7, respectively. Prediction by nmrshiftdb2 gave values of δ_C 169.8, 108.9, and 161.2 while CNMR Predictor (ACD Labs) gives δ_C 162.1, 111.3, and 157.1 at the same positions. The proposed structure is the one of albomaculine **2** (structure in Figure 4). Its molecular formula relates it to peak 2 in the chromatograms in Figure 2.

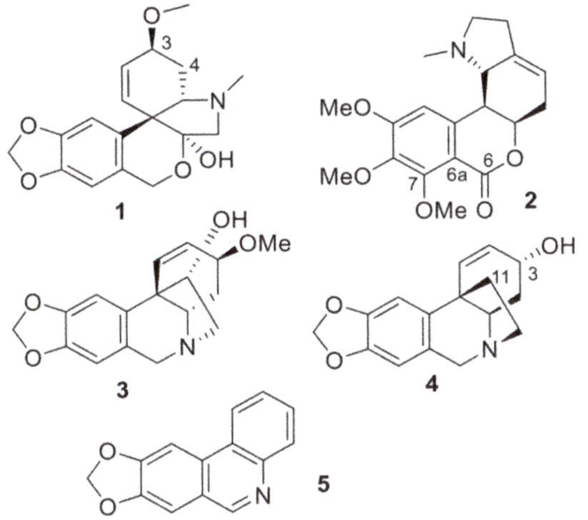

Figure 4. Structure of tazettine **1**, albomaculine **2**, haemanthamine **3**, crinine **4**, and trisphaeridine **5**.

2.5.3. Fraction A9, Major Compound

The list of 17 ^{13}C-NMR chemical shifts and associated multiplicities was submitted to a spectral similarity search through the CSEARCH web interface. The molecular formula of candidate structures was constrained to include only C, H, N, and O atoms, accounting for a molecular mass comprised between 250 u and 350 u. A structure without chirality information was given as best solution, with a mean deviation of δ_C 1 between experimental and CSEARCH-proposed chemical shift values. KNApSAcK was also considered for the identification of the major compound in fraction A9 as a possible alternative to CSEARCH. KNApSAcK (DB2) contains 11 molecules from Amaryllidaceae with molecular formula $C_{17}H_{19}NO_4$, the only one found by LC-MS of the total alkaloid extract accounting for 17 ^{13}C resonances. From NMR data, compound A9 contains an aromatic ring with a methylenedioxo substituent and hydrogens in para position, a carbon-carbon double bond between two CH groups, and a methoxy group attached to an aliphatic carbon. The only two compounds that fit with these constraints are crinamine and haemanthamine, who present the same planar formula as the one proposed by CSEARCH. This planar structure was confirmed by the analysis of all available NMR data. The analysis of the ROESY spectrum and the ^{1}H-^{1}H coupling constants led to the identification of haemanthamine **3** (structure in Figure 4). Its molecular formula relates it to peak 4 in the chromatograms in Figure 2.

2.5.4. Fraction A11, Major Compound

The ^{13}C-NMR spectrum of fraction A11 shows 16 peaks from a major compound whose positions were used as search keys in the CSEARCH data base. The molecular formula was constrained to include only C, H, N, and O atoms with a molecular mass comprised between 250 u and 400 u. The most likely proposed structure was the one of crinine **4**, $C_{16}H_{17}NO_3$ (structure in Figure 4). Only a single chemical shift value was considered slightly unsatisfactory (δ_C 40.0 predicted by CSEARCH, δ_C 44.2 experimental, at position 11). The KNApSAcK database of Amaryllidaceae compounds contains four compounds for this molecular formula, and only three that contain four aromatic or olefinic methine groups: crinine, vittatine, and epivittatine which only differ by the absolute configuration of asymmetric centers. More precisely, crinine and vittatine are two enantiomers, for which unambiguous identification would rely on chiroptical methods. The same situation holds for epi-crinine and epi-vittatine, epimers of the former at position 3. The identification the correct epimer was obtained by the detailed analysis of J-coupling values supported by the 2D ROESY spectrum. A comparison of the ^{13}C-NMR chemical shift values in A11 with those published for synthetic crinine and epi-crinine supports our conclusion [57]. NP identification up to the absolute configuration by optical rotation measurement is possible for pure or highly major compounds but is not possible for minor compounds in fractions without isolation. Its molecular formula relates it to peak 3 in the chromatograms in Figure 2.

2.5.5. Fraction A2, a Minor Compound

Fraction A2 contains a major compound, tazettine **1**, which is also the very major compound in fractions A3–A5, and many minor compounds. The ^{1}H-NMR spectrum of fraction A2 shows an isolated singlet at δ_H 9.16 that was used as an entry point for compound identification. This highly deshielded proton is directly bound to a methine carbon at δ_C 151.83 according to HSQC data and is surrounded by carbons at δ_C 100.36 (CH), 105.40 (CH), 122.82 (C), 124.03 (C), 129.6 (C), and 143.74 (C) according to HMBC data. Querying for δ_H 9.16 ± 0.2 in DB1 (the only one among our DBs with predicted ^{1}H-NMR data) resulted in three candidate structures: angustine, vittacarboline, and trispheridine (or trisphaeridine). Searching then for δ_C 100.36, 105.4, 122.82, 124.03, 129.6, 143.74, and 151.83 with a 5 ppm tolerance resulted reduced the list of candidates to trispheridine **5** only (structure in Figure 4). Using DB2 and DB3 avoided to rely on proprietary NMR chemical shift prediction. Querying DB2 for the same list of seven ^{13}C-NMR chemical shift with a 2 ppm tolerance yielded deoxylycobetaine chloride, trispheridine, and vasconine as proposals. A reduced tolerance of 1 ppm resulted in trispheridine only, thus also proving the good quality of the prediction by nmrshiftdb2 for this compound. Querying DB3 for the same seven chemical shift values with a tolerance of 2 ppm resulted in 628 compounds among which 124 contain at least 12 carbon atoms and 1 or 2 nitrogen atoms, using C(12-100) H(1-100) N(1-2) O(1-100) as molecular formula filter. Trispheridine is present in this compound list but reducing the number of hits would require supplementary constraints, thus demonstrating the usefulness of taxonomy-based filtering for dereplication. The presence of trispheridine in fraction A2 and its NMR spectra assignment was confirmed by further studies. Searching in DB3′ or in DB4 for trispheridine cannot be successful because its structure does not fit with any of those used in the definition of what an Amaryllidaceae alkaloid should be, even though this compound is present in the PNMRNP and COCONUTv1 database. The ClassyFire algorithm itself does not consider trisphaeridine as an alkaloid but NPClassifier identifies it as an Amarylidaceae (*sic*) alkaloid. Its NP-likeness is −0.08, a value that would make it slightly closer to a non-NP (lowest value is −5) than to an NP (highest value is +5). Exploring the philosophical implications of this observation is left as an exercise to the reader.

2.5.6. Database Searches

The structural identification of compounds A4, A7, A9 and A11 reported hereabove was carried out using lists of ^{13}C-NMR chemical shifts that were unambiguously drawn from spectra due to high sample purity (Table S1). After this study, a question arose about the possible results of an identification process solely relying on these lists, without any other NMR information source, only taking into account the possible molecular formula derived from LC-MS data acquired on crude extract 3. The chemical shift lists were used as search keys in DB1 (209 structures from KNApSAcK), DB2 (209 structures from KNAp-

SAcK), DB3 (211,280 structures from full PNMRNP), DB3' (635 structures from PNMRNP filtered for Amaryllidaceae-type alkaloids), and DB4 (693 structures from COCONUTv1 filtered for Amaryllidaceae-type alkaloids) with predicted chemical shifts by ACD/Labs software in DB1 and predicted by nmrshiftdb2 in all other DBs. All DBs were formatted for being read by the ACD/Labs DB software so that the same search tool can be used for compound identification. The poor prediction of a single chemical shift in the targetted compound may result in a global failure of the search, to which it can be remedied either by decreasing the number of experimental chemical shifts to be taken into account or by increasing the allowed chemical shift deviation. Table S2 shows the influence of these parameters on the number and nature of solutions, it illustrates the difficulty of identifying pure compounds without ambiguity solely on the basis of lists of ^{13}C-NMR chemical shift values and molecular formula.

3. Materials

3.1. Chemicals

Acetonitrile (CH_3CN), methanol (MeOH), methyl-tert-butyl ether (MtBE), chloroform ($CHCl_3$), triethylamine (Et_3N), and sulfuric acid (H_2SO_4) were purchased from Carlo Erba Reactifs SDS (Val de Reuil, France). Hexadeuterated dimethylsulfoxide (DMSO-d_6) was purchased from Eurisotop (Saclay, France). Deionized water was used to prepare aqueous solutions.

3.2. NMR

NMR analyses were performed in DMSO-d_6 at 298 K on an Avance AVIII-600 spectrometer (Bruker, Karlsruhe, Germany) equipped with a cryoprobe optimized for ^{1}H detection and fitted with cooled ^{1}H, ^{13}C and ^{2}H coils and preamplifiers. TopSpin 3.2 (Bruker, Karlsruhe, Germany) was used for data acquisition using standard microprograms. Data processing relied on TopSpin 4.0. The central resonance of DMSO-d_6 (septet) was set at δ_C 39.8 for ^{13}C-NMR spectrum referencing. The central resonance of residual DMSO-d_5 (quintet) was set at δ_H 2.5 for ^{1}H-NMR spectrum referencing.

3.3. UPLC-HRMS

Ultra Performance Liquid Chromatography coupled to Mass Spectrometry (UPLC-MS) analyses were performed with an Acquity UPLC H-Class (Waters, Manchester, UK) system coupled to a Synapt G2-Si (Waters) equipped with an electrospray (ESI) ion source. Chromatographic separation was achieved on a Uptisphere Strategy C18-HQ column (150 $\times$ 2.1 mm, 2.2 µm; Interchim, Montluçon, France). A gradient elution mode was used with solvent A (ammonium acetate 1%, pH 6.6) and solvent B (CH_3CN) at flow rate of 0.4 mL min^{-1}. Starting from 10%B, the gradient was linearly increased to 20%B in 6 min, then to 25%B in other 6 min, after 0.2 min the percentage of B was increased to 100% keeping it constant for 1 min. Finally, the gradient returned in the initial conditions in 0.2 min, maintaining it constant for 2 min for equilibration. The injection volume was 1 µL, the column temperature was regulated at 30 °C. All samples were solubilized in methanol and analyzed at concentration of 200 ppm. MS data acquisition parameters were: capillary voltage 3 kV, desolvation temperature 450 °C, desolvation gas flow 950 L/h, source temperature 120 °C, cone voltage 40 V, cone gas flow 50 L/h and scanning range of *m/z* 50–2000.

3.4. CPC

Centrifugal Partition Chromatography (CPC) fractionations were carried out using a lab-scale FCPE300®column of 303 mL inner volume (Kromaton Technology, Angers, France). The column was composed of 7 circular partition disks, each engraved with 33 twin-cells of 1.0 mL. The liquid phases were pumped by a preparative 1800 V7115 pump (Knauer, Berlin, Germany). Fractions of 20 mL were collected by a Labocol Vario 4000 (Labomatic Instruments, Allschwil, Switzerland). MtBE, CH_3CN and H_2O were equilibrated according the proportion 5:2:3 (*v/v*) and the two phases were separated. The lower aqueous phase was used as stationary phase and acidified with H_2SO_4 10 mM (retainer). The upper organic phase was alkalinized with Et_3N 8 mM (displacer) and used as mobile phase. The column was filled with the stationary phase at 300 rpm column rotation speed and 50 mL/min and then the mobile phase was pumped at 1200 rpm and 20 mL/min for hydrodynamic column equilibration. 1 g of extract was solubilized in 10 mL of retainer phase (acidified aqueous phase) and 5 mL of neutral organic phase. After sample loading through a 6-port Rheodyne valve (UpChurch Scientific, Oak Harbor, WA, USA) equipped with a 20 mL sample loop, the mobile phase was pumped into the column in ascending mode at flow-rate of 20 mL/min and 1200 rpm. The fractions were collected from the basic organic mobile phase and pooled according to TLC offline analysis to give 13 fractions namely A1–A13. TLC analysis was achieved on Merck TLC Silica gel 60 F254 plates, using $CHCl_3$/ MeOH (8.5/1.5) as eluent. All experiments were conducted at room temperature $(20 \pm 2\ °C)$.

3.5. Plant Material

Fresh bulbs of *U. peruviana* (1090.3 g) were purchased at the horticultural nursery Quatro Estaciones (Cochabamba, Bolivia) in August 2019. Some bulbs were grown, and the plants were identified by Dr. Umberto Mossetti, a voucher specimen (BOLO0602041) was deposited in the Herbarium of University of Bologna. The bulbs were stored in a cold room at 5 °C until the use, then they were freeze-dried and crushed, resulting in 220 g of plant material.

4. Conclusions

The rapid identification of natural products, either pure or in mixtures, depends on the availability of databases that connect together molecular structures, taxonomic information, and spectroscopic data, which constitute the three pillars of dereplication. We propose to the scientific community two easily findable NMR-based tools, the PNMRNP database that derives from earlier works on MS^2 spectra prediction, and KnapsackSearch, a database generator that provides focused libraries of compounds whose content is oriented by biological taxonomy. These tools were involved in the study of an iberoamerican plant, *Urceolina peruviana*, in a way that relies strongly on ^{13}C-NMR spectroscopy but also on other 1D and 2D NMR techniques as well as on preparative fractionation methods particularly suitable for alkaloid purification and on liquid chromatography coupled to high-resolution MS. The identification of five known compounds by these means is reported. The fully unambiguous characterization of a compound within a mixture may be reached only after purification and an exhaustive analytical study. However, the rapid and context adapted structure analysis is feasible by means of an approach that relies on computer databases and that adequately contributes to the study of complex natural substances.

Supplementary Materials: The following are available online, Figures S1 to S5 and Tables S1 to S2: titles as indicated in Table of Contents.

Author Contributions: Conceptualization, J.-M.N. and M.L.; methodology, J.-M.N., M.L. and J.-H.R.; software, J.-M.N., S.K. and R.L.; investigation, M.L., C.M.R., C.S., N.B. and S.R.; data curation, R.L., M.L. and J.-M.N.; writing—original draft preparation, J.-M.N.; writing—review and editing, C.M.R., C.S., N.B., S.R., S.K., J.-H.R.; project administration, J.-H.R. and J.-M.N. All authors have read and agreed to the published version of the manuscript.

Funding: This research received no external funding.

Institutional Review Board Statement: Not applicable.

Informed Consent Statement: Not applicable.

Data Availability Statement: NMR time-domain and spectral data files are temporarily available (1.6 Gbytes) from http://eos.univ-reims.fr/LSD/OpenData_Molecules_2021_Nuzillard.zip and permanently available from https://doi.org/10.5281/zenodo.4309769 so that all acquisition and processing parameters may be freely consulted. These archives also contain the database files or ways to access them and files that were intermediately created for the composition of this article.

Acknowledgments: Financial support to the PlAneT CPER project was provided by CNRS, Conseil Régional Champagne Ardenne, Conseil Général de la Marne, Ministry of Higher Education and Research (MESR), and EU-programme FEDER.

Sample Availability: Samples of compounds are not available from the authors.

Conflicts of Interest: The authors declare no conflict of interest.

References

1. Chen, Y.; Kirchmair, J. Cheminformatics in natural product based drug discovery. *Mol. Inform.* **2020**, *39*, 2000171. [CrossRef]
2. Hubert, J.; Nuzillard, J.-M.; Renault, J.-H. Dereplication strategies in natural product research: How many tools and methodologies behind the same concept? *Phytochem. Rev.* **2017**, *16*, 55–95. [CrossRef]
3. Logan, D.C. Known knowns, known unknowns, unknown unknowns and the propagation of scientific enquiry. *J. Exp. Bot.* **2009**, *60*, 712–714. [CrossRef]
4. Bakiri, A.B.; Plainchont, B.; Emerenciano, V.d.P.; Reynaud, R.; Hubert, J.; Renault, J.-H.; Nuzillard, J.-M. Computer-aided dereplication and structure elucidation of natural products at the university of Reims. *Mol. Inform.* **2017**, *36*, 1700027. [CrossRef]
5. Rutz, A.; Dounoue-Kubo, M.; Ollivier, S.; Bisson, J.; Bagheri, M.; Saesong, T.; Ebrahimi, S.N.; Ingkaninan, K.; Wolfender, J.-L.; Allard, P.-M. Taxonomically informed scoring enhances confidence in natural products annotation. *Front. Plant Sci.* **2019**, *10*, 1329. [CrossRef]
6. Heller, S.R.; McNaught, A.; Pletnev, I.; Stein, S.; Tchekhovskoi, D. Inchi, the IUPAC international chemical identifier. *J. Cheminform.* **2015**, *7*, 23. [CrossRef]
7. Weininger, D. SMILES, a chemical language and information system, 1. Introduction to methodology and encoding rules. *J. Chem. Inf. Model.* **1988**, *28*, 31–36. [CrossRef]
8. Dalby, A.; Nourse, J.G.; Hounshell, W.D.; Gushurst, A.K.I.; Grier, D.L.; Leland, B.A.; Laufer, J. Description of several chemical structure file formats used by computer programs developed at Molecular Design Limited. *J. Chem. Inf. Model.* **1992**, *32*, 244–255. [CrossRef]
9. O'Boyle, N.M.; Banck, M.; James, C.A.; Morley, C.; Vandermeersch, T.; Hutchison, G.R. Open Babel: An open chemical toolbox. *J. Cheminform.* **2011**, *3*, 33. [CrossRef]
10. Blue Book—IUPAC. International Union of Pure and Applied Chemistry. Available online: https://iupac.org/what-we-do/books/bluebook/ (accessed on 13 December 2020).
11. Empowering Innovation & Scientific Discoveries. CAS. Available online: https://www.cas.org (accessed on 13 December 2020).
12. Kim, S.; Chen, J.; Cheng, T.; Gindulyte, A.; He, J.; He, S.; Li, Q.; Shoemaker, B.A.; Thiessen, P.A.; Yu, B.; et al. PubChem 2019 update: Improved access to chemical data. *Nucleic Acids Res.* **2018**, *47*, D1102–D1109. [CrossRef]
13. Pupier, M.; Nuzillard, J.-M.; Wist, J.; Schlörer, N.E.; Kuhn, S.; Erdelyi, M.; Steinbeck, C.; Williams, A.J.; Butts, C.; Claridge, T.D.W.; et al. NMReDATA, a standard to report the NMR assignment and parameters of organic compounds. *Magn. Reson. Chem.* **2018**, *56*, 703–715. [CrossRef]
14. IUPAC CPEP Subcommittee on Electronic Data Standards. Available online: http://www.jcamp-dx.org/protocols.html (accessed on 13 December 2020).
15. Solution. Allotrope Foundation. Available online: https://www.allotrope.org/solution (accessed on 13 December 2020).
16. Wolfender, J.-L.; Nuzillard, J.-M.; van der Hooft, J.J.J.; Renault, J.-H.; Bertrand, S. Accelerating metabolite identification in natural product research: Toward an ideal combination of liquid chromatography–high-resolution tandem mass spectrometry and NMR profiling, in silico databases, and chemometrics. *Anal. Chem.* **2019**, *91*, 704–742. [CrossRef]
17. Wishart, D.S.; Feunang, Y.D.; Marcu, A.; Guo, A.X.; Liang, K.; Vázquez-Fresno, R.; Sajed, T.; Johnson, D.; Li, C.; Karu, N.; et al. HMDB 4.0: The human metabolome database for 2018. *Nucleic Acids Res.* **2018**, *46*, D608–D617. [CrossRef]
18. Ulrich, E.L.; Akutsu, H.; Doreleijers, J.F.; Harano, Y.; Ioannidis, Y.E.; Lin, J.; Livny, M.; Mading, S.; Maziuk, D.; Miller, Z.; et al. BioMagResBank. *Nucleic Acids Res.* **2008**, *36*, D402–D408. [CrossRef]
19. Wang, M.; Carver, J.J.; Phelan, V.V.; Sanchez, L.M.; Garg, N.; Peng, Y.; Nguyen, D.D.; Watrous, J.; Kapono, C.A.; Luzzatto-Knaan, T.; et al. Sharing and community curation of mass spectrometry data with global natural products social molecular networking. *Nat. Biotechnol.* **2016**, *34*, 828–837. [CrossRef]
20. Kuhn, S.; Colreavy-Donnelly, S.; Santana de Souza, J.; Borges, R.M. An integrated approach for mixture analysis using MS and NMR techniques. *Faraday Discuss.* **2019**, *218*, 339–353. [CrossRef]
21. Hubert, J.; Nuzillard, J.-M.; Purson, S.; Hamzaoui, M.; Borie, N.; Reynaud, R.; Renault, J.-H. Identification of natural metabolites in mixture: A pattern recognition strategy based on ^{13}C NMR. *Anal. Chem.* **2014**, *86*, 2955–2962. [CrossRef]
22. Bruguière, A.; Derbré, S.; Coste, C.; Le Bot, M.; Siegler, B.; Leong, S.T.; Sulaiman, S.N.; Awang, K.; Richomme, P. ^{13}C-NMR dereplication of *Garcinia* extracts: Predicted chemical shifts as reliable databases. *Fitoterapia* **2018**, *131*, 59–64. [CrossRef]
23. Taxonomy Browser (Root). Available online: https://www.ncbi.nlm.nih.gov/Taxonomy/Browser/wwwtax.cgi (accessed on 13 December 2020).

24. Tree of Life Web Project. Available online: http://tolweb.org/tree/ (accessed on 13 December 2020).
25. Sorokina, M.; Steinbeck, C. Review on natural products databases: Where to find data in 2020. *J. Cheminform.* **2020**, *12*, 20. [CrossRef]
26. Johnson, S.R.; Lange, B.M. Open-access metabolomics databases for natural product research: Present capabilities and future potential. *Front. Bioeng. Biotechnol.* **2015**, *3*, 22. [CrossRef]
27. Fromanteau, D.L.G.; Gastmans, J.P.; Vestri, S.A.; Emerenciano, V.d.P.; Borges, J.H.G. A constraints generator in structural determination by microcomputer. *Comput. Chem.* **1993**, *17*, 369–378. [CrossRef]
28. Scotti, M.T.; Herrera-Acevedo, C.; Oliveira, T.B.; Costa, R.P.O.; Santos, S.Y.K.O.; Rodrigues, R.P.; Scotti, L.; Da-Costa, F.B. SistematX, an online web-based cheminformatics tool for data management of secondary metabolites. *Molecules* **2018**, *23*, 103. [CrossRef]
29. Afendi, F.M.; Okada, T.; Yamazaki, M.; Hirai-Morita, A.; Nakamura, Y.; Nakamura, K.; Ikeda, S.; Takahashi, H.; Altaf-Ul-Amin, M.; Darusman, L.K.; et al. KNApSAcK family databases: Integrated metabolite-plant species databases for multifaceted plant research. *Plant Cell Physiol.* **2012**, *53*, e1. [CrossRef]
30. RDKit. Available online: https://www.rdkit.org/ (accessed on 13 December 2020).
31. ISIDA/EdiSDF—Laboratoire de Chemoinformatique. Available online: http://infochim.u-strasbg.fr/spip.php?article83 (accessed on 13 December 2020).
32. nuzillard/KnapsackSearck: Automated Data Search in the KNApSAcK Database. Available online: https://github.com/nuzillard/KnapsackSearch (accessed on 13 December 2020).
33. KNApSAcK Core System. Available online: http://www.knapsackfamily.com/knapsack_core/top.php (accessed on 13 December 2020).
34. Allard, P.-M.; Péresse, T.; Bisson, J.; Gindro, K.; Marcourt, L.; Pham, V.C.; Roussi, F.; Litaudon, M.; Wolfender, J.-L. Integration of molecular networking and in silico MS/MS fragmentation for natural products dereplication. *Anal. Chem.* **2016**, *88*, 3317–3323. [CrossRef]
35. Daylight Theory: SMIRKS—A Reaction Transform Language. Available online: https://www.daylight.com/dayhtml/doc/theory/theory.smirks.html (accessed on 13 December 2020).
36. Bento, A.P.; Hersey, A.; Félix, E.; Landrum, G.; Gaulton, A.; Atkinson, F.; Bellis, L.J.; De Veij, M.; Leach, A.R. An open source chemical structure curation pipeline using RDKit. *J. Cheminform.* **2020**, *12*, 51. [CrossRef]
37. CIRpy—CIRpy 1.0.2 Documentation. Available online: https://cirpy.readthedocs.io/en/latest/ (accessed on 13 December 2020).
38. Steinbeck, C.; Kuhn, S. NMRShiftDB—Compound identification and structure elucidation support through a free community-built web database. *Phytochemistry* **2004**, *65*, 2711–2717. [CrossRef]
39. Vestri Alvarenga, S.A.; Gastmans, J.-P.; do Vale Rodrigues, G.; Moreno, P.R.H.; Emerenciano, V.d.P. A computer-assisted approach for chemotaxonomic studies—Diterpenes in Lamiaceae. *Phytochemistry* **2001**, *56*, 583–595. [CrossRef]
40. Pelletier, S.W. The nature and definition of an alkaloid. In *Alkaloids: Chemical and Biological Perspectives*; Pelletier, S.W., Ed.; Wiley: New York, NY, USA, 1983; Volume 1, pp. 1–31.
41. Djoumbou Feunang, Y.; Eisner, R.; Knox, C.; Chepelev, L.; Hastings, J.; Owen, G.; Fahy, E.; Steinbeck, C.; Subramanian, S.; Bolton, E.; et al. ClassyFire: Automated chemical classification with a comprehensive, computable taxonomy. *J. Cheminform.* **2016**, *8*, 61. [CrossRef]
42. Daylight Theory: SMARTS—A Language for Describing Molecular Patterns. Available online: https://www.daylight.com/dayhtml/doc/theory/theory.smarts.html (accessed on 13 December 2020).
43. Schaub, J.; Zielesny, A.; Steinbeck, C.; Sokorina, M. Too sweet: Cheminformatics for deglycosylation in natural products. *J. Cheminform.* **2020**, *12*, 67. [CrossRef]
44. Bruneton, J. *Pharmacognosie*, 5th ed.; Lavoisier collection Tec&Doc: Cachan, France, 2016.
45. Nuzillard, J.-M.; Emerenciano, V.d.P. Automatic structure elucidation through data base search and 2D NMR spectral analysis. *Nat. Prod. Commun.* **2006**, *1*, 57–64.
46. Kim, H.W.; Wang, M.; Leber, C.A.; Nothias, L.-F.; Reher, R.; Kang, K.B.; van der Hooft, J.J.J.; Dorrestein, P.C.; Gerwick, W.H.; Cottrell, G.W. NPClassifier: A deep neural network-based structural classification tool for natural products. *ChemRxiv. Prepr.* **2020**. [CrossRef]
47. Predicted Carbon-13 NMR Data of Natural Products (PNMRNP). Zenodo. Available online: https://zenodo.org/record/3825257 (accessed on 13 December 2020).
48. Sorokina, M.; Steinbeck, C. NaPLeS: A natural products likeness scorer—Web application and database. *J. Cheminform.* **2019**, *11*, 55. [CrossRef]
49. Robien, W. Computer-assisted peer reviewing of spectral data: The CSEARCH protocol. *Monatsh. Chem.* **2019**, *150*, 927–932. [CrossRef]
50. Boit, H.-G.; Döpke, W. Alkaloide aus Urceolina-, Hymenocallis-, Elisena-, Calostemma- und Hippeastrum-Arten. *Chem. Ber.* **1957**, *90*, 1827–1830. [CrossRef]
51. Giraud, L. *Kallawaya: Guérisseurs Itinérants des Andes. Recherche sur les Pratiques Médicinales et Magiques*; IRD Editions: Montpellier, France, 1984.
52. Berkov, S.; Osorio, E.; Viladomat, F.; Bastida, J. Chapter Two—Chemodiversity, chemotaxonomy and chemoecology of Amaryllidaceae alkaloids. In *The Alkaloids: Chemistry and Biology*; Knölker, H.-J., Ed.; Academic Press: Cambridge, MA, USA, 2020; Volume 83, pp. 113–185. [CrossRef]
53. Renault, J.-H.; Nuzillard, J.-M.; Maciuk, A.; Zèches-Hanrot, M. Use of Centrifugal Partition Chromatography for Purifying. Galanthamine. Patent WO 2006/064105, 22 June 2006.

54. Renault, J.-H.; Le Crouerour, G.; Thépenier, P.; Nuzillard, J.-M.; Zèches-Hanrot, M.; Le Men-Olivier, L. Isolation of indole alkaloids from *Catharanthus roseus* by centrifugal partition chromatography in the pH-zone refining mode. *J. Chromatogr. A* **1999**, *849*, 421–431. [CrossRef]
55. Martin, S.F. Chapter 3—The Amaryllidaceae alkaloids. In *The Alkaloids: Chemistry and Pharmacology*; Brossi, A., Ed.; Academic Press: Cambridge, MA, USA, 1987; Volume 30, pp. 251–376. [CrossRef]
56. Haughwitz, R.D.; Jeffs, P.W.; Wenkert, E. 358. Proton magnetic resonance studies of some Amaryllidaceae alkaloids of the 5,10*b*-ethanophenanthridine series and of criwelline and tazettine. *J. Chem. Soc.* **1965**, 2001–2009. [CrossRef]
57. Das, M.K.; Kumar, N.; Bisai, A. Catalytic asymmetric total syntheses of naturally occurring Amaryllidaceae alkaloids, (−)-crinine, (−)-*epi*-crinine, (−)-oxocrinine, (+)-*epi*-elwesine, (+)-vittatine, and (+)-*epi*-vittatine. *Org. Lett.* **2018**, *20*, 4421–4424. [CrossRef]

Article

Dereplication of Natural Extracts Diluted in Glycerin: Physical Suppression of Glycerin by Centrifugal Partition Chromatography Combined with Presaturation of Solvent Signals in ^{13}C-Nuclear Magnetic Resonance Spectroscopy

Marine Canton [1,2], Jane Hubert [2], Stéphane Poigny [1], Richard Roe [1], Yves Brunel [1], Jean-Marc Nuzillard [2,*] and Jean-Hugues Renault [2,*]

[1] Laboratoires Pierre Fabre Dermo-Cosmétique, 3 avenue Hubert Curien, BP 13562, CEDEX 1, 31035 Toulouse, France
[2] Université de Reims Champagne Ardenne, CNRS, ICMR UMR 7312, CEDEX 2, 51097 Reims, France
* Correspondence: jm.nuzillard@univ-reims.fr (J.-M.N.); jh.renault@univ-reims.fr (J.-H.R.)

Academic Editors: Young Hae Choi, Young Pyo Jang, Yuntao Dai and Luis Francisco Salomé-Abarca
Received: 12 October 2020; Accepted: 28 October 2020; Published: 31 October 2020

Abstract: For scientific, regulatory, and safety reasons, the chemical profile knowledge of natural extracts incorporated in commercial cosmetic formulations is of primary importance. Many extracts are produced or stabilized in glycerin, a practice which hampers their characterization. This article proposes a new methodology for the quick identification of metabolites present in natural extracts when diluted in glycerin. As an extension of a ^{13}C nuclear magnetic resonance (NMR) based dereplication process, two complementary approaches are presented for the chemical profiling of natural extracts diluted in glycerin: A physical suppression by centrifugal partition chromatography (CPC) with the appropriate biphasic solvent system EtOAc/CH$_3$CN/water 3:3:4 (*v/v/v*) for the crude extract fractionation, and a spectroscopic suppression by presaturation of ^{13}C-NMR signals of glycerin applied to glycerin containing fractions. This innovative workflow was applied to a model mixture containing 23 natural metabolites. Dereplication by ^{13}C-NMR was applied either on the dry model mixture or after dilution at 5% in glycerin, for comparison, resulting in the detection of 20 out of 23 compounds in the two model mixtures. Subsequently, a natural extract of *Cedrus atlantica* diluted in glycerin was characterized and resulted in the identification of 12 metabolites. The first annotations by ^{13}C-NMR were confirmed by two-dimensional NMR and completed by LC-MS analyses for the annotation of five additional minor compounds. These results demonstrate that the application of physical suppression by CPC and presaturation of ^{13}C-NMR solvent signals highly facilitates the quick chemical profiling of natural extracts diluted in glycerin.

Keywords: natural extract; glycerin; chemical profiling; centrifugal partition chromatography; ^{13}C nuclear magnetic resonance; cosmetic industry

---

1. Introduction

Natural extracts are commonly used all over the world as ingredients in cosmetic formulations for their antioxidant activity, their scent, their antimicrobial properties, and as natural dyes or hydration agents [1–3]. Over the last few years, regulation has become more demanding, as evidenced by the European cosmetics regulation in force since 2014 (no 1223/2009), which gives a prominent role to consumer safety [4,5]. This regulation includes a list of prohibited and restricted compounds. Moreover, it requires the suppliers to establish a product information file (PIF) for each cosmetic product. The PIF summarizes data including a description of the cosmetic product, the proof of the claimed effect, a safety assessment, and product stability data [6]. Consequently, the composition of the natural extracts contained in the cosmetic formulations must be known to prevent toxicity related risks.

The most common way to characterize an extract requires the isolation of its constituents [7]. Although being exhaustive, this process is very time consuming and usually leads to the purification of known compounds. Quicker methods more compatible with industry constraints must be found. To respond to this need, dereplication strategies mainly based on liquid chromatography hyphenated to mass spectrometry (LC-MS), liquid chromatography hyphenated to ultraviolet spectroscopy with diode array detection (LC-UV-DAD), thin layer chromatography (TLC), and/or nuclear magnetic resonance (NMR) data combined with the use of in silico or experimental databases were developed in order to avoid the purification of all components of a mixture [8–10]. The term dereplication has been introduced in the 1990s as a method for the quick identification of known compounds in a sample [11,12].

The dereplication process developed by Hubert et al. relies on ^{13}C-NMR chemical shift comparisons for compound identification [10]. It starts with the extract fractionation by centrifugal partition chromatography (CPC) followed by ^{13}C-NMR analysis of all fractions. After automatic collection and pooling of ^{13}C-NMR signals, hierarchical clustering analysis (HCA) highlights clusters of ^{13}C-NMR chemical shifts in which all elements share similar chromatographic behaviors. Finally, the clusters are assigned to chemical structures using a database that contains the chemical structures and calculated ^{13}C-NMR chemical shifts of known natural metabolites of low molecular weight [10]. The structures that best match the experimental data are then confirmed by two-dimensional (2D) NMR and, if required, by complementary MS analysis. According to the Chemical Analysis Working Group of the Metabolomics Standard Initiative, this compound annotation process has a confidence level of 2 because it relies on spectral similarity [13,14]. After complementary techniques confidently defining 2D structures like 2D NMR, the confidence level increases to 1. The efficiency of this ^{13}C-NMR based workflow has been largely proven by the identification of major compounds in numerous dry extracts [15–18].

Nowadays, many natural extracts of industrial interest are produced or stabilized in matrices or support solvents for easy handling and preservation [19–21]. Glycerin, also called glycerol, is the most used carrier solvent in marketed cosmetic ingredients, mainly due to its non-toxicity as well as its humectant and bacteriostatic properties [22]. In addition, it can be claimed as a natural ingredient when produced as a by-product of the synthesis of biodiesel [23]. A high concentration of glycerin in an extract interferes unfavorably with the dereplication procedure briefly described hereabove. The high boiling point of glycerin (BP = 286 °C at 105 Pa, with decomposition) and its very low vapor pressure (P_{vap} = 0,02 Pa at 25 °C) makes it impossible to evaporate without metabolite degradation [24].

The aim of this study was to develop an improved method for the quick identification of the major compounds contained in a glycerinated natural extract. Two strategies were jointly elaborated, including a physical suppression of glycerin by CPC and a spectroscopic suppression of the ^{13}C-NMR resonances of glycerin. For the physical suppression of glycerin, an appropriate CPC biphasic system able to separate glycerin from other metabolites was investigated. Then, spectroscopic suppression by presaturation in ^{13}C-NMR using a previously described method [25] was applied to fractions which still contained glycerin. As a proof of concept, the ^{13}C-NMR dereplication methodology for glycerinated extracts was first applied to a model mixture of 23 analytical standards of diverse polarities. A comparison was performed between the metabolites identified in the dry model mixture and diluted at 5% in glycerin. Secondly, the method was tested on a genuine extract of *Cedrus atlantica* bark diluted in glycerin.

2. Results and Discussion

2.1. General Methodology for ^{13}C-NMR Dereplication of Metabolite Mixtures Diluted in Glycerin

The aim of the present work was to develop a ^{13}C-NMR dereplication process dedicated to natural extracts diluted in glycerin. As described by the flowchart in Figure 1, the approach consists of the combination of two glycerin removal methods: A physical suppression by CPC fractionation followed by a spectroscopic suppression of glycerin signals in ^{13}C-NMR of glycerin-containing fractions.

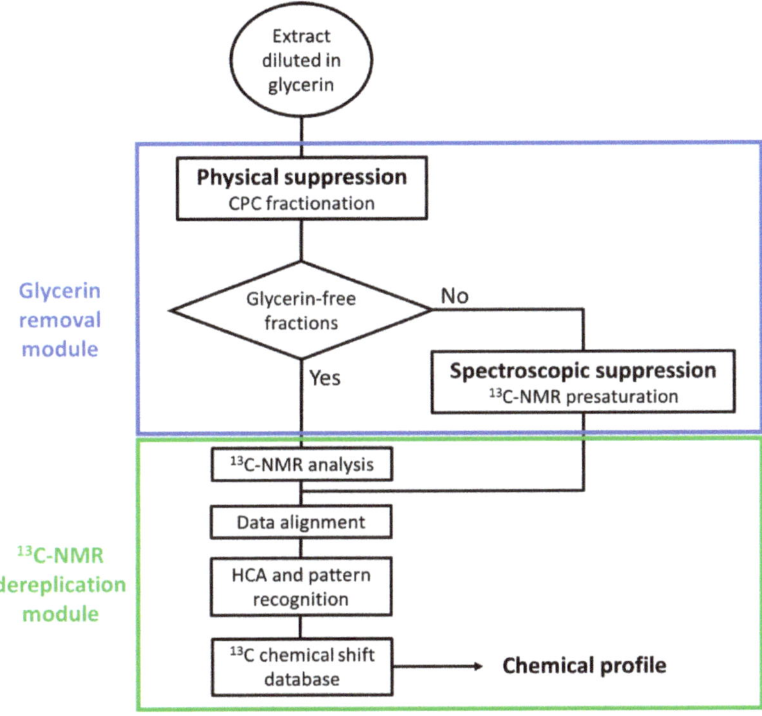

Figure 1. Global workflow for the dereplication of natural extracts diluted in glycerin. CPC: Centrifugal partition chromatography; HCA: Hierarchical clustering analysis.

CPC is a widely used tool for natural compound separation [26,27]. It is a solid-support-free chromatographic technique that uses the two liquid phases of a biphasic solvent system as stationary and mobile phases. The choice of the biphasic solvent system is of prime importance in this context. A high proportion of glycerin in a high volume of injected sample may disrupt the biphasic nature of the solvent system, resulting in separation failure [28]. Indeed, the two phases of the biphasic system must remain stable in the presence of a large amount of glycerin and, at the same time, must allow a gradual elution of metabolites during the fractionation process. To obtain a glycerin extract, the plant is usually treated with water or ethanol and, after filtration and evaporation of the ethanol when present, glycerin is added. In other cases, the plant is directly extracted in glycerin and water, and then the extract is filtered. Whatever the method used, the resulting extract composition is usually complex and contains metabolites covering a wide range of polarities. Our approach was to retain the glycerin in the stationary phase while eluting the compounds of interest.

At the end of this first step, glycerin was expected to be contained in only a few CPC fractions. These fractions may contain sugars for which separation is not easily achieved. While the dry fractions can be directly analyzed by ^{13}C-NMR, those containing glycerin in large proportions cannot. However, the global chemical profile of the natural extract requires the characterization of these fractions. The direct analysis by ^{13}C-NMR of fractions containing predominantly glycerin can lead to the production of decoupling artifacts linked to the very high intensity of the two glycerin NMR signals [25]. A specific ^{13}C signal suppression sequence was used to remove these artifacts and reduce the intensity of glycerin signals. A presaturation technique reported in [25] achieved a 97% signal intensity decrease, thus providing an efficient elimination of decoupling artifacts. Moreover, the chemical shift band for which a 50% signal intensity decrease occurred was sharp—about 0.1 ppm wide —avoiding information loss related to signal suppression.

The ^{13}C-NMR spectra of glycerin-free and glycerin-containing fractions were subjected to the usual dereplication methodology: Automatic collection and alignment of ^{13}C peaks, HCA and identification of metabolites present in a database by ^{13}C chemical shift fingerprint comparison. The feasibility of this strategy was demonstrated on a model mixture containing 23 analytical standards from different chemical families as well as by the characterization of a genuine glycerin extract of *Cedrus atlantica* Carrière.

2.2. Proof of Concept on a Model Mixture

To simulate the chemical diversity of natural extracts obtained with non-selective solvents such as water, alcohol, or glycerin, 23 analytical standards of various phytochemical classes have been selected. Sugars, organic acids, fatty acids, phenolic acids, hydroxycinnamic acids, stilbenes, tannins, flavonoids, alkaloids, saponins, and amino acids were part of the model mixture. In the case of flavonoids and hydroxycinnamic acids, the relatively lipophilic aglycone form (quercetin and ferulic acid), as well as the glycosylated form (rutin and chlorogenic acid) were included. For comparison purposes, two model mixtures were prepared: One in dry form and one at 5% wt. in glycerin/water (1:1 *w/w*).

The first step of our method was to fractionate the model mixtures by CPC with an appropriate solvent system. The biphasic system ethyl acetate (EtOAc)/acetonitrile (CH_3CN)/water 3:3:4 (*v/v/v*) was investigated because of the combination of a medium polar binary system (EtOAc/water) with CH_3CN as a bridging solvent was promising for our study. This type of system has already been referenced for the fractionation of natural polar extracts [29–31]. The influence of glycerin on the biphasic system EtOAc/CH_3CN/water 3:3:4 (*v/v/v*) was evaluated according to the method developed by Marchal et al. [32], which relies on the drawing of a pseudo ternary diagram in which the mobile phase, the stationary phase, and glycerin intervene at the apexes. Glycerin was successively added to mobile and stationary phases present in predefined ratios: 0/1, 1/3, 1/1, 3/1, and 1/0 (*w/w*). Then, the solvent mixture was shaken, and the visual inspection for homogeneity of one or two phases was carried out. If the mixture becomes monophasic, the corresponding composition (*w/w/w*) of this mixture in terms of upper phase, mobile phase, and glycerin was reported on the phase diagram in its orthogonal representation, and then the binodal curve was drawn by connecting the different demixing points.

The biphasic nature of the EtOAc/CH_3CN/water 3:3:4 (*v/v/v*) system was very little modified by the addition of glycerin (maximum tested mass of glycerin was 90% wt. without forming a single phase), which promises high stability of this solvent system (Figure 2a). Conversely, the same experiment carried out on *n*-butanol (*n*-BuOH)/acetic acid/water 4:1:5 (*v/v/v*), one of the most polar CPC systems, resulted in single-phase solvent mixtures. The maximum demixing point was reached at 47% wt. of glycerin added in the 1:1 (*w/w*) lower/upper phase mixture. The pseudo ternary system of glycerin in *n*-BuOH/acetic acid/water 4:1:5 (*v/v/v*) is presented in Figure 2b as an example of a CPC solvent system which is not suitable for glycerin-containing samples.

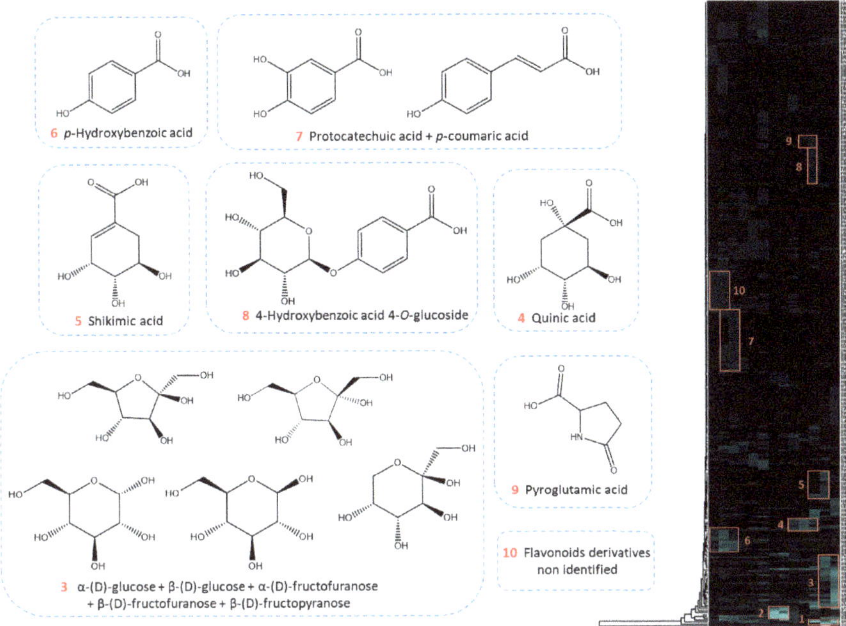

Figure 2. Ternary diagram of upper phase, lower phase, and glycerin (*w/w/w*) for the solvent systems: (**a**) ethyl acetate (EtOAc)/acetonitrile (CH$_3$CN)/water 3:3:4 (*v/v/v*) and (**b**) *n*-butanol (n-BuOH)/acetic acid/water 4:1:5 (*v/v/v*) as an example of appropriate (**a**) and inappropriate (**b**) solvent systems for our physical glycerin suppression strategy. Demixing points are symbolized by triangles. In (**a**), no demixing point was observed until the addition of 90% wt. of glycerin (maximum points studied symbolized by squares). Experiments were performed at 20 °C.

The selection of the biphasic solvent system EtOAc/CH$_3$CN/water 3:3:4 (*v/v/v*) as a solvent system compatible with the presence of high amounts of glycerin was followed by an evaluation of its suitability for natural product fractionation. The partition coefficient (K$_D$) indicates the distribution of a compound in the upper and the lower phases of a biphasic system. For a crude plant extract containing compounds covering a wide range of polarities, the ideal CPC system for our study should cover a large range of K$_D$ values with a significantly different one for glycerin. Moreover, the partition coefficients K$_D$ of the compounds of interest should be comprised between 0.2 and 5 to retain the highest selectivity while maintaining short CPC run times [33]. The partition coefficients of glycerin and ten metabolites from various chemical families and with diverse polarities contained in the model extract were measured in the EtOAc/CH$_3$CN/water 3:3:4 (*v/v/v*) solvent system: Caffeine, chlorogenic acid, ferulic acid, glycyrrhizin, linoleic acid, polydatin, quercetin, rutin, succinic acid, and vanillin. The shake flask method [34] was used to determine each K$_D$ individually. The construction of response-concentration curves was performed for each of the ten compounds, ensuring that each analytical measurement was carried out within the linearity range (Table S1). The measured partition coefficients of the ten standards, shown in Table 1, ranged from 0.02 for the low polarity linoleic acid to 5.28 for rutin, a medium-polar compound. This result showed a large K$_D$ diversity, thus predicting a successful separation of metabolites by CPC elution. Moreover, we measured the K$_D$ of glycerin using the shake flask method and an analysis by gas chromatography with flame ionization detection GC/FID (data not shown). The glycerin did not distribute at all in the system and remained exclusively present in the lower phase. CPC in the ascending mode was thus selected for the elution of standard compounds in order to maintain glycerin in the column. The system EtOAc/CH$_3$CN/water 3:3:4 (*v/v/v*) seemed appropriate for the resolution of our initial problem: It is stable in the presence of glycerin and provides a good selectivity over a wide range of analyte polarities. The obtained result provides a way to physically separate glycerin from the other compounds and thus obtain, after CPC fractionation, glycerin-free fractions from the mobile phase and glycerin-containing fractions from the stationary phase, as collected by column extrusion (i.e., back-flushing of the column stationary phase by fresh stationary phase).

Table 1. Analytical standards selected for the measurement of their distribution coefficients (K_D) individually and as a mixture in the biphasic system EtOAc/CH$_3$CN/water 3:3:4 (*v/v/v*). Their thin-layer chromatography (TLC) retardation factor (Rf) with EtOAc/toluene/acetic acid/formic acid 6:4:1:1 (*v/v/v/v*) as elution solvents, retention time and scan mode chosen are listed. Each LC-MS analysis was performed in triplicate. ESI: Electrospray ion.

Compound	Chemical Family	Formula	Rf	Retention Time (min)	Exact Mass (g/mol)	Scan Mode	Individual K_D by LC-MS	K_D in Mixture
Caffeine	Alkaloid	$C_8H_{10}N_4O_2$	0.39	3.63	194.0804	ESI$^+$	1.13 ± 0.09	1.03 ± 0.01
Chlorogenic acid	Hydroxycinnamic acid	$C_{16}H_{18}O_9$	0.11	3.69	354.0951	ESI$^-$	1.11 ± 0.09	1.23 ± 0.01
Ferulic acid	Hydroxycinnamic acid	$C_{10}H_{10}O_4$	0.66	5.42	194.0579	ESI$^-$	0.14 ± 0.01	0.17 ± 0.01
Glycyrrhizin	Saponin	$C_{42}H_{62}O_{16}$	0.04	10.15	822.4038	ESI$^-$	2.09 ± 0.13	0.78 ± 0.02
Linoleic acid	Fatty acid	$C_{18}H_{32}O_2$	0.81	14.67	280.2402	ESI$^-$	0.02 ± 0.01	0.01 ± 0.01
Polydatin	Stilbene	$C_{20}H_{22}O_8$	0.21	5.55	390.1315	ESI$^-$	0.62 ± 0.02	0.72 ± 0.01
Quercetin	Flavonoid	$C_{15}H_{10}O_7$	0.63	7.81	302.0427	ESI$^+$	0.02 ± 0.01	0.05 ± 0.01
Rutin	Flavonoid	$C_{27}H_{30}O_{16}$	0.04	5.61	610.1534	ESI$^+$	5.28 ± 0.34	2.52 ± 0.14
Succinic acid	Organic acid	$C_4H_6O_4$	-	0.80	118.0266	ESI$^-$	1.18 ± 0.09	1.24 ± 0.14
Vanillin	Other phenolic	$C_8H_8O_3$	0.66	4.55	152.0473	ESI$^+$	0.08 ± 0.01	0.17 ± 0.01

The CPC fractionation of the two model mixtures containing 23 analytical standards in a dry form or diluted at 5% wt. in glycerin/water 1:1 (*w/w*) was performed. The injected sample represented 2 g of the dry model mixture, which corresponded to 40 g of the glycerin-containing model mixture. The injected sample volume was approximately 35 mL or 12% of the column capacity, which is still acceptable and sufficient for the compounds to be partitioned in the remaining part of the column. No disturbance of the solvent system was observed in the fractionation of the glycerin-containing mixture. As a result, 13 simplified fractions were obtained in order of decreasing polarity of their constituents. Sample recoveries were 92% and 110% wt. for the dry and glycerinated model mixtures, respectively. The hygroscopic character of glycerin explains the recovery of more than 100%. The details of the CPC fractionation mass balances are presented in Tables S2 and S3. Regarding the elution fractions (F_{01}–F_{09} in each experiment), the recovered masses were almost similar in both experiments, 311 mg and 320 mg for dry and glycerin extracts, respectively. Fractions F_{10} to F_{13} correspond to the column extrusion step (see experimental part). The summary HPTLC chromatograms allowed a visual analysis of the two CPC fractionations in Figure 3.

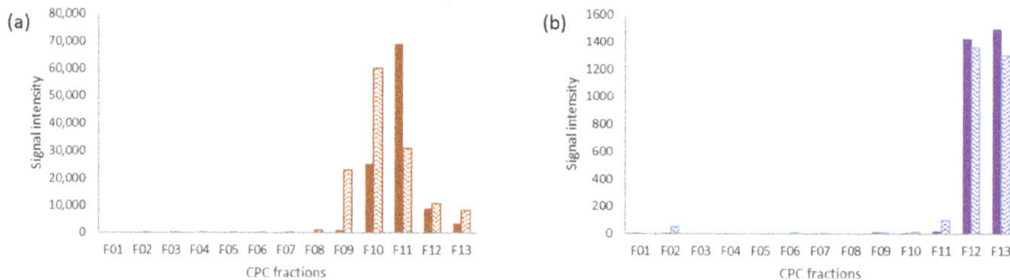

Figure 3. Recapitulative high performance (HP)TLC plates of the CPC fractionations. (**a**,**b**) refer to glycerin model mixture; (**c**,**d**) refer to dry model mixture. Vanillin/sulphuric acid reagent was used for plates (**a**,**c**) while (**b**,**d**) were revealed by the Neu reagent.

According to the high-performance (HP)TLC plates of the analytical standards available in Figure S1, several standards have been assigned: 1: Linoleic acid; 2: Quercetin; 3: Ferulic acid, vanillin; 4: Gallic acid; 5: Polydatin; 6: Glycyrrhizin; 7: Chlorogenic acid; 8: Rutin; 9: Glycerin; 10: D-fructose, D-glucose, and sucrose.

As shown in Figure 3a, the blue polar spot of glycerin was only present in fractions F_{11}, F_{12}, and to a lesser extend in F_{13}, confirming the non-partitioning of glycerin in the CPC biphasic system. Moreover, the presence of glycerin was easily deduced because the drying of these fractions left a liquid residue. These three fractions corresponded to the extrusion of the CPC column. Glycerin was maintained in the stationary phase, while many analytical standards could be eluted during the CPC fractionation.

As expected from partition coefficient measurements, the analytical standards were relatively well separated, which was highly encouraging for future dereplication works. Fractionation of the dry sample led to a separation almost identical to that obtained with the glycerinated mixture. Some analytical standards were eluted in exactly the same fractions, while the elution of others was slightly modified. For example, gallic acid and polydatin (blue spot at a retardation factor (Rf) of 0.54 and 0.21 in Figure 3b,d, respectively) were both predominantly present in fractions F_{03} and F_{04}. Chlorogenic acid (Rf 0.11 blue spot in Figure 3b,d) was eluted in fractions F_{05}–F_{07} and F_{05}–F_{08} for glycerinated and dry mixture fractionation, respectively. Sugars (browns spot at Rf 0.03 in Figure 3a,c), as very polar compounds, were present in the three last fractions in both experiments. Some differences appeared for quercetin (yellow spot at Rf 0.63 in Figure 3b,d), which eluted more slowly in the fractionation of the glycerinated mixture than for the dry mixture. These differences can be explained by the interface and/or flow pattern modifications due to the larger volume of glycerin-rich injected samples.

Glycerin-free fractions were directly characterized by ^{13}C-NMR spectroscopy. The three glycerin-containing fractions F_{11}, F_{12}, and F_{13} were analyzed by ^{13}C-NMR with presaturation of glycerin signals. 200 mg of fractions were mixed with 600 µL of DMSO-d_6. Such a quantity was necessary for metabolite detection. Figure 4 presents the results of the presaturation of the two signals of glycerin in fraction F_{11}.

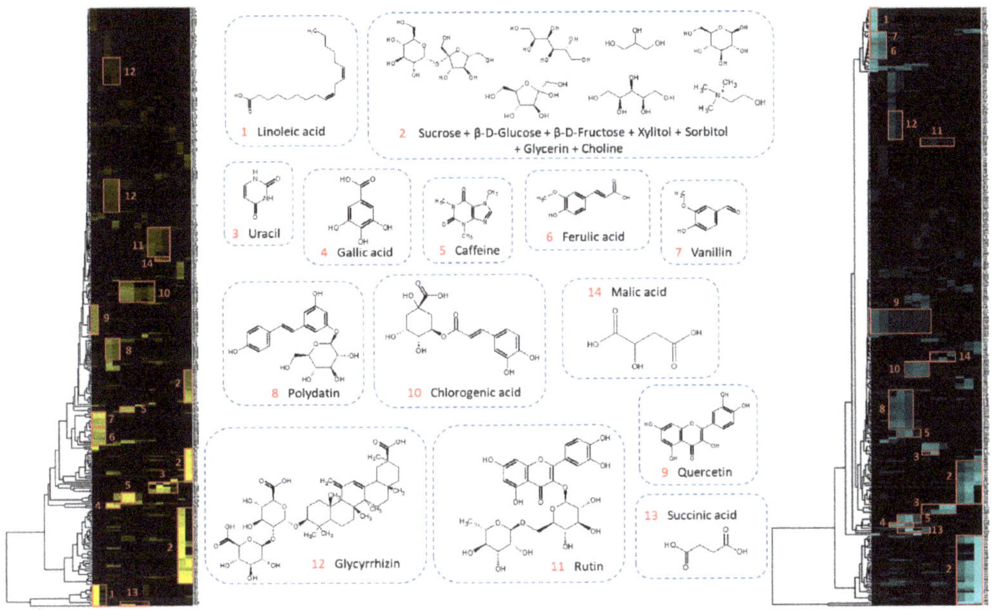

Figure 4. (a) ^{13}C-NMR spectrum of glycerin-containing fraction F_{11}. * refers to decoupling artifacts. (b) Analysis of the same sample as in (a) with multiple presaturation of glycerin signals. The frame inserts show spectra overviews drawn at the same vertical scale, without peak clipping. Both acquisitions required the recording of 1024 scans and eight dummy scans.

Without presaturation (Figure 4a), the resonance peaks of DMSO-d_6 were much smaller than those of glycerin. After presaturation (Figure 4b), glycerin signals were significantly reduced, thus demonstrating the efficiency of the glycerin resonance peak elimination. Quantitatively, there was a 97% decrease in the signal at 63.7 ppm and a decrease of 94% for the signal at 73.1 ppm. Furthermore, decoupling artifact signals were present in the spectrum without presaturation (Figure 4a). Some of these were easily recognizable because they were out of phase, but others were easy to confuse with genuine signals in the spectrum. These decoupling artifacts could interfere with the dereplication process by presuming signals where there were none. Presaturation of glycerin signals also led to the removal of decoupling artifacts, as expected.

The alignment of the ^{13}C-NMR data of the 13 fractions could then be carried out. The 2D heat map visualization and the HCA were performed. Figure 5 shows the heat maps obtained for the glycerinated model mixture after HCA on the 13 fractions with a classical ^{13}C-NMR analysis (Figure 5a,b) and with glycerin signals presaturation on fractions F_{11}, F_{12}, and F_{13} (Figure 5c).

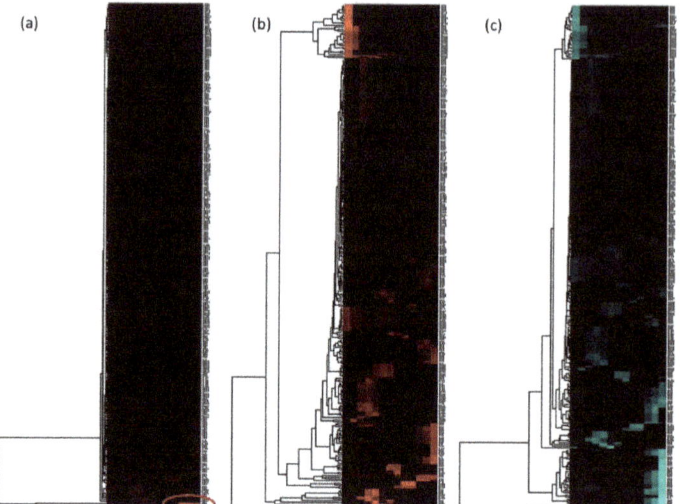

Figure 5. Comparison of heat maps for glycerin model extract: (**a**) without presaturation of glycerin signals. The red circle corresponds to the intense cluster of glycerin (δ 63.7, 73.1); (**b**) similar to (**a**) with manual suppression in the text file of the two bins 63.7 and 73.1 ppm corresponding to the glycerin signals; (**c**) with presaturation of glycerin signals during ^{13}C-NMR analysis of F_{11}, F_{12}, and F_{13}.

Figure 5a seemed to have mainly one cluster, as indicated by the large correlation branch of the HCA made on chemical shift lines. The manual suppression of this intense cluster (δ 63.7 and 73.1) in the text file before visualization in 2D heat map improved chemical shift clusters observation (Figure 5b). The presaturation of glycerin signals in the fractions F_{11}, F_{12}, and F_{13} further improved the heat map with well-defined chemical shift clusters (Figure 5c) and fewer glycerin artifacts. In addition to the decoupling artifacts, other parasitic signals related to the truncation of the intense glycerin signals were observed, as highlighted in Figure S2. The presaturation allowed a strong decrease in the intensity of the glycerin signals and the strong reduction of all the artifacts related to glycerin.

The chemical profiling of model mixtures was performed by database search (see experimental part). Heat maps and annotations of chemical shift clusters are given in Figure 6. The same 20 analytical standards were characterized in the dry and the glycerinated mixtures.

The most apolar compounds, linoleic acid, ferulic acid, vanillin, and quercetin, were the first to elute (clusters 1, 6, 7, and 9, respectively). This result was expected given their K_D, which was less than or equal to 0.1 in this solvent system. Polydatin (cluster 8), which has a K_D around 0.7, was eluted from fraction three in both experiments. Caffeine, chlorogenic acid, and succinic acid (clusters 5, 10, 13 respectively), which have K_D between 1 and 1.5, were then eluted. Rutin, with the highest K_D (5.3), emerged from the column at the end of the process. Only the elution behavior of glycyrrhizin was not in agreement with its measured K_D, which could be due to its strong amphiphilic character. The simultaneous measurement of the K_D of the ten metabolites in the model mixture was undertaken (Table 1). Among the ten compounds, glycyrrhizin and rutin presented a strongly changed K_D value, depending on measurement condition, either as a pure compound or as a mixture component. The K_D value of rutin was high individually and in a mixture, which explained its late elution. However, for glycyrrhizin, the K_D measured when pure was greater than 2 whereas it was less than 1 when in a mixture. This result explained the behavior of glycyrrhizin, which eluted starting from the third fraction.

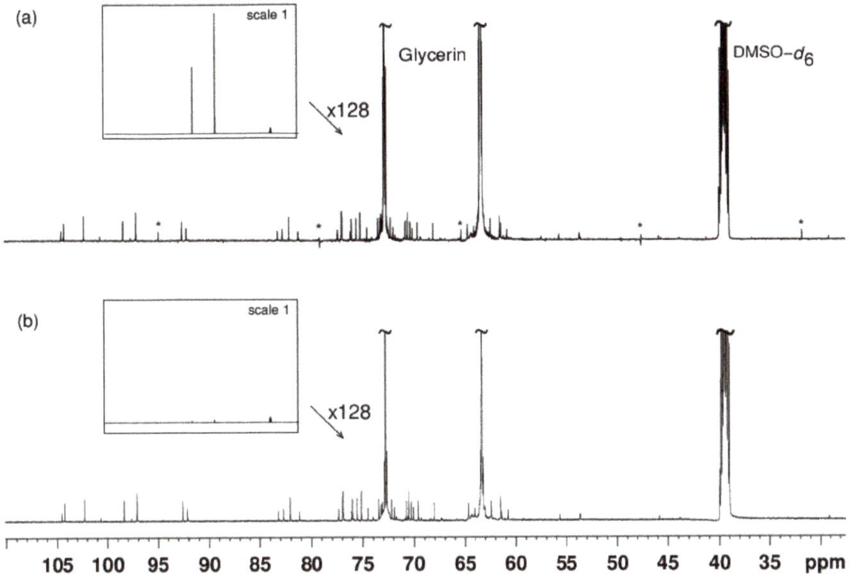

Figure 6. ^{13}C-NMR chemical shift clusters obtained by applying HCA on dry model extract (left) and diluted in glycerin (right).

Among the 23 metabolites constituting the model mixture, three metabolites could not be annotated: Tannic acid, citric acid, and proline. With hindsight, tannic acid was found not to be an appropriate candidate for this study. Commercial tannic acid was usually a gallotanins mixture which tended to undergo autodegradation [35,36]. We thus chose to exclude it from the discussions. Regarding proline and citric acid, two hypotheses were put forward. A long lasting elution of a compound could generate a very low concentration in each of the fractions and then a non-observation of the signals in ^{13}C-NMR. A second explanation, the most likely, was that the clusters of the two missing compounds were broken up. Indeed, when a compound has one or more chemical displacements very close to another compound, the usual δ tolerance for cluster detection is unusable. For example, several chemical shifts of proline can be in the same 0.2 ppm bin as polydatin (δ 62.1), glycyrrhizin (δ 46.8), or caffeine (δ 30.0). Additional experiments were performed on each CPC fraction by LC-MS in order to observe the elution of the analytical standards. Figure 7 shows the elution profile of the two missing compounds: Citric acid and proline. The graphical visualization of their elution showed a similar elution behavior in dry and glycerin extracts fractionation. Moreover, their elution was limited to a few fractions. The hypothesis of a broken cluster on the heat map was therefore preferred.

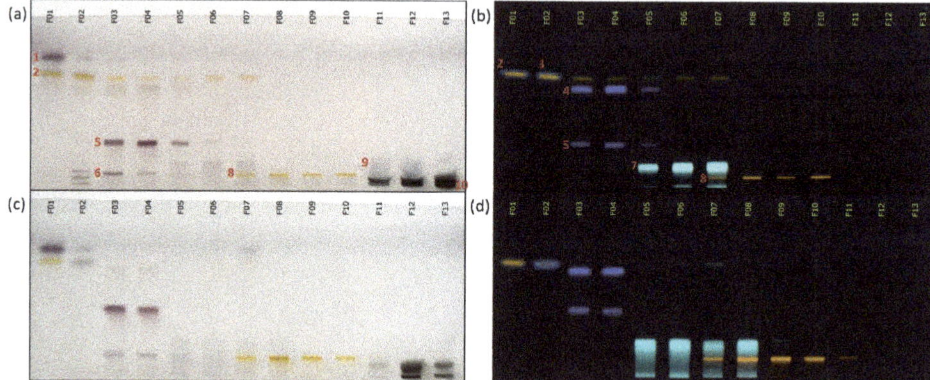

Figure 7. LC-MS elution profiles of (**a**) citric acid and (**b**) proline in dry and glycerin model mixture CPC fractionations. The full boxes correspond to the area of the signal in the fractionation of the dry extract and the hatched boxes to the fractionation of the glycerin extract.

These results confirmed that despite the large amount injected and the high proportion of glycerin in the extract, it was possible to characterize the compounds contained in a model mixture diluted in glycerin with the method described hereafter: A physical suppression of glycerin by CPC fractionation with the EtOAc/CH$_3$CN/water 3:3:4 (*v/v/v*) solvent system in ascending mode followed by a spectroscopic suppression of glycerin signals for fractions containing glycerin in large proportion. In addition, the proof of concept of this method for the quick chemical profiling of glycerin extracts was promising since the results of the dereplication of the dry and of the glycerinated form of the same mixture were identical. It was then necessary to demonstrate the robustness of this method by the analysis of a genuine glycerinated extract.

2.3. Dereplication of a Genuine Natural Extract Diluted in Glycerin

The selected extract was an experimental aqueous extract of *Cedrus atlantica* Carrière bark diluted in glycerin. Due to its origin, this aqueous extract contained a complex mixture of metabolites covering a wide polarity range. The major compounds were identified by the glycerin-insensitive new dereplication method illustrated by the flow chart in Figure 1.

CPC fractionation was carried out as described for the model mixture. The injection of 17.5 g of glycerinated extract was equivalent to the injection of 2.5 g of dry extract as the solution contains 14.5% wt. of dry natural extract. The fractionation process resulted in 13 fractions. Eleven of them could be evaporated to dryness whilst the last two fractions contained glycerin. The recovery of the sample was 99%. The mass of each CPC fraction is available in Table S3. The first nine fractions and the last two extruded fractions represented 1% wt. and 98.5% wt. of the recovered material, respectively. As shown on the HPTLC profiles (Figure S3), the extruded fractions contained glycerin and sugars predominantly.

Spectroscopic glycerin suppression was therefore undertaken by ^{13}C-NMR presaturation on F$_{12}$ and F$_{13}$ in order to decrease glycerin signal intensity and to remove possible decoupling artifacts. ^{13}C-NMR spectra were then processed with an alignment of NMR data in regular chemical shift windows of 0.3 ppm width. Indeed, even though glycerin was removed by spectroscopy, significant matrix effects appeared in the last two fractions. Table 2 reports these matrix effects on the chemical shifts at the anomeric positions of the five sugars present in fractions F$_{12}$ and F$_{13}$.

Table 2. Matrix effects of chemical shifts of sugars anomeric position.

Compound	δ in F_{12}	δ in F_{13}	$\Delta\delta$
α-(D)-glucose	92.94	92.71	0.23
β-(D)-glucose	97.51	97.35	0.16
α-(D)-fructofuranose	104.89	104.65	0.24
β-(D)-fructopyranose	102.65	102.47	0.18
β-(D)-fructopyranose	98.79	98.59	0.20

Chemical shift bins, of widths generally set at 0.2 ppm to be selective enough while including a possible chemical shift deviation, therefore had to be increased to 0.3 ppm. HCA was then performed. The resulting heat map containing the correlated ^{13}C-NMR signal revealed ten major chemical shift groups (Figure 8). The assignment of the clusters led to the identification of 12 metabolites: *p*-hydroxybenzoic acid, protocatechuic acid, quinic acid, *p*-coumaric acid, shikimic acid, 4-hydroxybenzoic acid 4-*O*-glucoside, pyroglutamic acid, α-(D)-glucose, β-(D)-glucose, α-(D)-fructofuranose, β-(D)-fructopyranose, β-(D)-fructofuranose. Clusters one and two referred to residual signals from glycerin and a glycerin derivative identified as glycerin acetate.

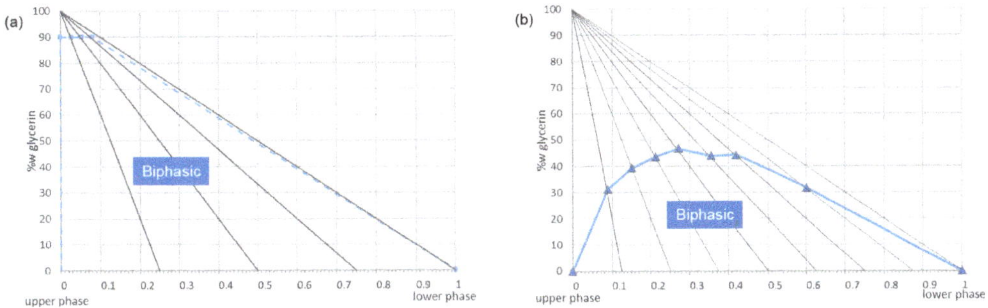

Figure 8. ^{13}C-NMR chemical shift clusters obtained by applying HCA on CPC fractions of glycerinated extract of *C. atlantica*.

The structures of all the compounds annotated on the heat map were confirmed by checking for all carbon atoms of the neighboring correlations in HSQC, HMBC, and COSY spectra. Other metabolites reported in the genus *Cedrus* [37] were finally searched by complementary LC-MS analyses, leading to the annotation of catechin, taxifolin, and astragalin in the less polar fractions and syningetin-3-*O*-glucoside and kaempferol-3-*O*-rutinoside in the medium polar fractions. The work presents a ^{13}C-NMR dereplication process permitting the characterization of major compounds of a natural extract diluted in glycerin. LC-MS, as a complementary technique with very high sensitivity, allows annotating other minor compounds referenced in the literature.

3. Materials and Methods

3.1. Chemical and Natural Extract

Ethyl acetate, acetonitrile, *n*-butanol, and methanol (MeOH) were purchased from Carlo Erba (Val de Reuil, France). Acetic acid was purchased from VWR (Radnor, PA, USA). Two model mixtures were prepared with 23 analytical standards for a total mass of 2 g: 500 mg of sucrose, 500 mg of D-glucose, 500 mg of D-fructose, and 25 mg of each of the following 20 compounds: Caffeine, chlorogenic acid, choline, citric acid, ferulic acid, gallic acid, glycerin, glycyrrhizin, linoleic acid, malic acid, polydatin, L-proline, quercetin, rutin, D-sorbitol, succinic acid, tannic acid, uracil, vanillin, and xylitol. The purity of each analytical standard was at least 95%. Their suppliers are listed in

Table S4. While one mixture was left as the dry model sample, the second was diluted at 5% wt. in glycerin and water in a ratio of 1:1 (*w/w*), giving a model mixture diluted in glycerin (glycerinated model mixture, hereafter) of 40 g. The genuine natural extract was a glycerinated extract of *Cedrus atlantica* Carrière. It was produced by maceration of *C. atlantica* bark in water, followed by clarification, concentration, and glycerin addition, resulting in a 14.5% wt. *C. atlantica* extract in 75.5% wt. glycerin and 10% wt. water.

3.2. Distribution Coefficients Determination

The distribution coefficients of glycerin and of ten analytical standards (caffeine, chlorogenic acid, ferulic acid, glycyrrhizin, linoleic acid, polydatin, quercetin, rutin, succinic acid, and vanillin) were measured individually or as a mixture in the EtOAc/CH$_3$CN/water 3:3:4 (*v/v/v*) biphasic system. The shake flask method was used to determine each K_D [34]. The temperature of the study was 20 °C. After solvent system equilibration, a defined volume of upper and lower phase (5 mL for glycerin study, 2 mL for analytical standards study) were transferred into a vial and vortexed for 5 s in the presence of the compound (11 mg for glycerin, 3 mg for the other analytical standards). Each vial was then placed for 15 min in an ultrasonic bath and left to settle (centrifugation was required in the case of glycyrrhizin to break the emulsion and separate the two phases). 500 μL of the lower phase and 500 μL of the upper phase were separately transferred into two clean vials. The solvents were evaporated in a vacuum oven at 40 °C. Finally, the lower and upper phase residues were diluted in 1 mL of MeOH for LC-MS analysis. The distribution coefficient K_D is defined as the ratio of the concentration of the solute in the upper phase (C_{up}, organic phase) to its concentration in the lower phase (C_{down}, aqueous phase) at thermodynamic equilibrium.

$$K_D = \frac{C_{up}}{C_{down}} \tag{1}$$

For K_D measurement in the mixture, a total of 240 mg of the 23 analytical standards in the same proportion as in the model sample were diluted in 160 mL of each lower and upper phase of the solvent system. After ultrasonic agitation for 5 min, the two phases were separated, evaporated under vacuum, and the residues were diluted in 3 mL MeOH for LC-MS analyses.

Measurements were performed in triplicate. Means of the LC-MS peak areas and standard deviations were calculated to give a distribution coefficient value (Table 1).

3.3. LC-MS Analyses

LC-MS analyses were performed with an Acquity UPLC H-Class (Waters, Manchester, UK) system coupled to a Synapt G2-Si (Waters) equipped with an electrospray (ESI) ion source. Chromatographic separation was achieved on a Kinetex C18 column (150 × 2.1 mm, 2.6 μm; Phenomenex, Torrance, CA, USA). The column temperature was regulated at 30 °C. Compounds were eluted with a gradient of water (Eluent A) and CH$_3$CN (Eluent B) with 0.1% formic acid in each mobile phase. The mobile phase flow rate was maintained at 0.6 mL/min and the gradient was designed as follows: t = 0 min, 5% B; t = 1 min, 5% B; t = 9 min, 40% B; t = 15 min, 100% B; t = 17 min, 100% B; t = 17.1 min, and 5% B until 19 min. All injection volumes were 0.5 μL. MS analyses were performed in both positive (ESI$^+$) and negative (ESI$^-$) ion modes. For each compound, the most suitable mode was chosen. Electrospray interface fitted to the following parameters: Capillary voltage 3 kV and 2 kV for ESI$^+$ and ESI$^-$, respectively; desolvation temperature 450 °C; desolvation gas flow 950 L/h; source temperature 120 °C; cone voltage 40 V; cone gas flow 50 L/h and scanning range of m/z 50–2000 both in the positive and negative ionization modes.

3.4. Centrifugal Partition Chromatography

CPC experiments were carried out with a lab-scale FCPE300® column of 303.5 mL capacity (Kromaton Technology, Angers, France). The column was composed of seven circular partition disks. Each disk is engraved with 33 twin-cells. The liquid phases were pumped by a preparative 1800 V7115 pump (Knauer, Berlin, Germany). All experiments were performed in the same conditions. The biphasic system EtOAc/CH$_3$CN/water 3:3:4 (*v/v/v*) was prepared in a separatory funnel. After decantation, the phases were separated. CPC fractionation was achieved in the ascending mode: The lower phase was used as the stationary phase and the upper one as the mobile phase. The CPC column was filled at a flow rate of 100 mL/min and at a rotation speed of 500 rpm. Then the rotation speed was increased to 1200 rpm. For the dry extract, 2 g of dry extract was dissolved in 12 mL of stationary phase and 6 mL of mobile phase and injected through a 20 mL injection loop. The glycerinated mixtures were directly injected through a 35 mL injection loop (40 g of the model glycerinated mixture and 17.5 g of the *C. atlantica* extract, respectively). The mobile phase was pumped progressively from 0 to 20 mL/min in 5 min and then maintained at 20 mL/min for 60 min. The most hydrophilic compounds retained inside the column were finally extruded by pumping fresh stationary phase for 25 min, maintaining the flow rate at 20 mL/min. Fractions of 20 mL were collected by a Labocol Vario 4000 (Labomatic Instruments, Allschwil, Switzerland). All collected fractions were spotted on Merck TLC plates coated with silica gel 60 F$_{254}$ and developed with EtOAc/toluene/acetic acid/formic acid (6:4:1:1 (*v/v/v/v*) for the model mixtures and 7:3:1:1 (*v/v/v/v*) for the genuine extract). After inspection at 254 nm, 366 nm, and under visible light after chemical revelation with acidified vanillin solution, fractions were pooled according to their composition. The pooling after fractionation of the two model mixtures was identical and resulted in 13 elution fractions. The genuine natural extract was also fractionated in 13 fractions. All fractions were dried under vacuum, except several extruded fractions still containing glycerin.

3.5. Nuclear Magnetic Resonance

NMR analyses were performed at 298 K on an Avance AVIII-600 spectrometer (Bruker, Karlsruhe, Germany) equipped with a cryoprobe optimized for ^{1}H detection and with cooled ^{1}H, ^{13}C and ^{2}H coils and preamplifiers. Dry fractions were dissolved in 600 µL of DMSO-d_6 and analyzed with the Uniform Driven Equilibrium Fourier Transform (UDEFT) sequence with an acquisition time of 0.36 s, and a relaxation delay of 3 s. 1024 scans were recorded on samples containing more than 20 mg, 2048 scans on those weighing between 10 mg and 20 mg and 4096 scans for fractions weighing less than 10 mg in order to obtain comparable signal-to-noise ratios. The receiver gain was set to the highest possible value. The spectra were manually phased and baseline-corrected using the TOPSPIN v4.0.5 software (Bruker). The central resonance of DMSO-d_6 was set at 39.8 ppm for spectrum referencing. Additional homonuclear correlation spectroscopy (COSY), heteronuclear single-quantum coherence (HSQC), and heteronuclear multiple bond correlation (HMBC) spectra were also recorded on dry fractions. For fractions in which glycerin was still present in a large proportion, the NMR sample was prepared with 200 mg of the fraction in 600 µL of DMSO-d_6. The presaturation sequence described in Canton et al. [25] was used to decrease the two dominating ^{13}C-NMR signals of glycerin and facilitate the identification of metabolites eluted in these fractions. The presaturation field of intensity $\Omega_1/2\pi = 11.7$ Hz was focused on the two resonances of glycerin applied during 3 s before the 90° pulse in a modified Bruker *zgpg* pulse sequence.

3.6. Data Processing

All ^{13}C-NMR signals were automatically collected on each spectrum. The resulting peak lists were stored in a text file from which a locally developed algorithm written in Python language aligns the NMR peak positions in regularly spaced chemical shift windows ($\Delta\delta$ = 0.2 ppm for the model studies and 0.3 ppm for the genuine extract study). The table of peak intensity values according to chemical shift bin and fraction indexes was imported into PermutMatrix software (version 1.9.3, LIRMM, France) for hierarchical clustering analysis. The resulting ^{13}C-NMR chemical shift clusters were visualized as dendrograms on a 2D heat map. In order to identify compounds, the elements of each chemical shift cluster were used as a search key in a local database built using the ACD/C+H NMR Predictors and DB software (Advanced Chemistry Development, Inc., ACD/Labs, Toronto, ON, Canada). This database contains more than 3000 compounds to date and associates structures to the predicted NMR chemical shifts of the proton and carbon atoms, as calculated by the ACD/Labs predictor. A literature survey was carried out on the genus *Cedrus*, resulting in 40 metabolites stored in the database [37–40]. A tolerated ^{13}C-NMR chemical shift difference between the predicted database spectrum and the real spectrum was established at 2 ppm. Finally, each proposition given by the database was confirmed by the interpretation of 1D and 2D NMR data (^{1}H NMR, HSQC, HMBC, and COSY) and LC-MS analyses.

3.7. High-Performance Thin-Layer Chromatography

HPTLC analyses were performed on Merck HPTLC plates 10 × 20 coated with silica gel 60 F_{254} with an automatic TLC sampler (ATS 4) and an automatic development chamber (ADC 2) (CAMAG, Muttenz, Switzerland). 5 µg of pure analytical standards, 12 µg of dry CPC fractions or 80 µg of glycerin-containing CPC fractions were applied with a band length of 8 mm. The developing solvents were: EtOAc/toluene/acetic acid/formic acid 6:4:1:1 (*v/v/v/v*) for the analysis of standards and of the CPC fractions obtained from the model mixtures while the same developing solvents in 7:3:1:1 (*v/v/v/v*) proportions were used for the *C. atlantica* extract. First, the plates were visualized at 254 nm and 366 nm. Then, the plates were heated at 100 °C with a TLC plate heater 3 (CAMAG) for 3 min and immersed in a solution of Neu reagent (1.25 g of diphenylborinic acid aminoethylester dissolved in 250 mL of EtOAc) and recorded at 366 nm and in white light. Finally, the plates were immersed in an acidified vanillin solution (375 mg of vanillin, 3.75 mL of 96% H_2SO_4 diluted in 125 mL of ethanol) and heated at 100 °C for 5 min. An HPTLC chromatogram of the 23 analytical standards, available in Figure S1, was performed to calculate their retardation factor. HPTLC analyses were carried out at the end of each CPC fractionation.

4. Conclusions

The combination of physical suppression by CPC fractionation and spectroscopic suppression by presaturation in ^{13}C-NMR were reported for the dereplication of natural extract metabolites diluted in glycerin. The biphasic solvent system EtOAc/CH$_3$CN/water 3:3:4 (*v/v/v*) was selected for its stability in the presence of glycerin and its ability to separate compounds in a wide polarity range. The proof of principle of this approach was successfully achieved on a model mixture of 23 analytical standards. The 20 compounds identified in the dry model mixture were also identified in the glycerinated one. The method applied on a bark extract of *Cedrus atlantica* Carrière diluted in glycerin resulted in the identification of 12 metabolites of different chemical classes. This approach is, therefore, efficient for the fast chemical profiling of natural extracts diluted in glycerin. In the future, the characterization of major metabolites in extracts diluted in other solvents, notably glycols, produced for the cosmetics market, will be studied.

Supplementary Materials: The following are available online, Figure S1: HPTLC plates of the 23 analytical standards, Figure S2: Emphasis artifacts on the ^{13}C-NMR spectra of glycerin fractions by comparison between a conventional ^{13}C-NMR analysis and an analysis with presaturation of glycerin signals, Figure S3: HPTLC plates summarizing the CPC fractionation of *C. atlantica* glycerin extract, Table S1: Calibration curves for the 10 analytical standards for which K_D has been measured, Table S2: Summary of the grouping for CPC fractionations of the two model extracts, Table S3: Summary of the grouping for CPC fractionation of *C. atlantica* diluted in glycerin/water, Table S4: Summary of the 23 analytical compounds used as model extracts.

Author Contributions: M.C. has performed experiments, data treatment and wrote the original draft; J.H. and J.-H.R. supervised the methodology of physical suppression of glycerin by CPC; J.-M.N. supervised the methodology spectroscopic suppression of glycerin signals by ^{13}C-NMR; S.P. contributed to obtaining the genuine extract of *C. atlantica*; S.P., R.R., and Y.B. reviewed the text and the figures of the paper as well as J.H., J.-H.R., and J.-M.N. J.-H.R. and J.-M.N. are academic PhD thesis supervisors of M.C.; S.P. and R.R. are industrial PhD thesis supervisors of MC. All authors have read and agreed to the published version of the manuscript.

Funding: Financial support was provided by the laboratories Pierre Fabre Dermo-cosmétique and the Association Nationale de la Recherche et de la Technologie through the CIFRE grant no. 2017/1032. CNRS Conseil Régional Champagne Ardenne, Conseil Général de la Marne, Ministry of Higher Education and Research (MESR) and EU-programme FEDER also provided financial support to the PlAneT Contrat de Plan Etat-Région (CPER) project.

Acknowledgments: We thank GREENTECH S.A. for providing an experimental glycerin extract of *Cedrus atlantica* used for the example in this study. We also thank our colleagues from the University of Reims Champagne Ardenne and from the laboratories Pierre Fabre Dermo-cosmétique who greatly assisted the research.

Conflicts of Interest: The authors declare no conflict of interest.

References

1. Duke, J.A. *Handbook of Medicinal Herbs*, 2nd ed.; CRC Press: Boca Raton, FL, USA, 2002.
2. Kusumawati, I.; Indrayanto, G. Natural Antioxidants in Cosmetics. *Stud. Nat. Prod. Chem.* **2013**, *15*, 485–505. [CrossRef]
3. Hughes, K.; Ho, R.; Butaud, J.-F.; Filaire, E.; Ranouille, E.; Berthon, J.-Y.; Raharivelomanana, P. A selection of eleven plants used as traditional Polynesian cosmetics and their development potential as anti-aging ingredients, hair growth promoters and whitening products. *J. Ethnopharmacol.* **2019**, *245*, 112159. [CrossRef] [PubMed]
4. The European Parliament and the Council of the European Union. Regulation (EC) No 1223/2009 of the European Parliament and of the Council of 30 November 2009 on cosmetic products. *Off. J. Eur. Union L* **2009**, *342*, 59–209.
5. European Commission. *The SCCS Notes of Guidance for the Testing of Cosmetic Ingredients and Their Safety Evaluation 9th Revision; Adopted at the 11th Plenary Meeting of SCCS*; European Commission: Luxembourg, 2016.
6. Fernandez, X.; Michel, T.; Kerdudo, A. Analyse des principes actifs et substances réglementées en cosmétique. Technique de l'ingénieur. 2015; Ref J3300 V1. 1–30. (In French)
7. Demirci, F. Natural products isolation. In *Methods in Biotechnology*, 2nd ed.; Sarker, S.D., Latif, Z., Gray, A.I., Eds.; Humana Press: Totowa, NJ, USA, 2005.
8. Hubert, J.; Nuzillard, J.-M.; Renault, J.-H. Dereplication strategies in natural product research: How many tools and methodologies behind the same concept? *Phytochem. Rev.* **2015**, *16*, 55–95. [CrossRef]
9. Gaudêncio, S.P.; Pereira, F. Dereplication: Racing to speed up the natural products discovery process. *Nat. Prod. Rep.* **2015**, *32*, 779–810. [CrossRef] [PubMed]
10. Hubert, J.; Nuzillard, J.-M.; Purson, S.; Hamzaoui, M.; Borie, N.; Reynaud, R.; Renault, J.-H. Identification of Natural Metabolites in Mixture: A Pattern Recognition Strategy Based on ^{13}C NMR. *Anal. Chem.* **2014**, *86*, 2955–2962. [CrossRef]
11. Beutler, J.A.; Alvarado, A.B.; Schaufelberger, D.E.; Andrews, P.; McCloud, T.G. Dereplication of Phorbol Bioactives: *Lyngbya majuscula* and *Croton cuneatus*. *J. Nat. Prod.* **1990**, *53*, 867–874. [CrossRef]
12. Corley, D.G.; Durley, R.C. Strategies for Database Dereplication of Natural Products. *J. Nat. Prod.* **1994**, *57*, 1484–1490. [CrossRef]
13. Sumner, L.W.; Amberg, A.; Barrett, D.; Beale, M.H.; Beger, R.; Daykin, C.A.; Fan, T.W.-M.; Fiehn, O.; Goodacre, R.; Griffin, J.L.; et al. Proposed minimum reporting standards for chemical analysis. *Metabolomics* **2007**, *3*, 211–221. [CrossRef]
14. Blaženović, I.; Kind, T.; Ji, J.; Fiehn, O. Software Tools and Approaches for Compound Identification of LC-MS/MS Data in Metabolomics. *Metabolites* **2018**, *8*, 31. [CrossRef]
15. Angelis, A.; Hubert, J.; Aligiannis, N.; Michalea, R.; Abedini, A.; Nuzillard, J.-M.; Gangloff, S.C.; Skaltsounis, A.-L.; Renault, J.-H. Bio-Guided Isolation of Methanol-Soluble Metabolites of Common Spruce (*Picea abies*) Bark by-Products and Investigation of Their Dermo-Cosmetic Properties. *Molecules* **2016**, *21*, 1586. [CrossRef]
16. Hubert, J.; Angelis, A.; Aligiannis, N.; Rosalia, M.; Abedini, A.; Bakiri, A.; Reynaud, R.; Nuzillard, J.-M.; Gangloff, S.C.; Skaltsounis, L.A.; et al. In Vitro Dermo-Cosmetic Evaluation of Bark Extracts from Common Temperate Trees. *Planta Medica* **2016**, *82*, 1351–1358. [CrossRef]
17. Tisserant, L.-P.; Hubert, J.; Lequart, M.; Borie, N.; Maurin, N.; Pilard, S.; Jeandet, P.; Aziz, A.; Renault, J.-H.; Nuzillard, J.-M.; et al. ^{13}C NMR and LC-MS Profiling of Stilbenes from Elicited Grapevine Hairy Root Cultures. *J. Nat. Prod.* **2016**, *79*, 2846–2855. [CrossRef] [PubMed]
18. Nivelle, L.; Martiny, L.; Courot, E.; Jeandet, P.; Tarpin, M.; Renault, J.-H.; Renault, J.-H.; Clément, C.; Martiny, L.; Delmas, D.; et al. Anti-Cancer Activity of Resveratrol and Derivatives Produced by Grapevine Cell Suspensions in a 14 L Stirred Bioreactor. *Molecules* **2017**, *22*, 474. [CrossRef] [PubMed]

19. Herman, A. Antimicrobial Ingredients as Preservative Booster and Components of Self-Preserving Cosmetic Products. *Curr. Microbiol.* **2018**, *76*, 744–754. [CrossRef] [PubMed]

20. Jeong, K.M.; Oh, J.Y.; Zhao, J.; Jin, Y.; Yoo, D.E.; Han, S.Y.; Lee, J. Multi-functioning deep eutectic solvents as extraction and storage media for bioactive natural products that are readily applicable to cosmetic products. *J. Clean. Prod.* **2017**, *151*, 87–95. [CrossRef]

21. Shehata, E.; Grigorakis, S.; Loupassaki, S.; Makris, D.P. Extraction optimisation using water/glycerol for the efficient recovery of polyphenolic antioxidants from two *Artemisia* species. *Sep. Purif. Technol.* **2015**, *149*, 462–469. [CrossRef]

22. Kerdudo, A.; Fontaine-Vive, F.; Dingas, A.; Faure, C.; Fernandez, X. Optimization of cosmetic preservation: Water activity reduction. *Int. J. Cosmet. Sci.* **2014**, *37*, 31–40. [CrossRef]

23. Chemat, F.; Vian, M.A.; Cravotto, G. Green Extraction of Natural Products: Concept and Principles. *Int. J. Mol. Sci.* **2012**, *13*, 8615–8627. [CrossRef]

24. Ardi, M.; Aroua, M.; Hashim, N.A. Progress, prospect and challenges in glycerol purification process: A review. *Renew. Sustain. Energy Rev.* **2015**, *42*, 1164–1173. [CrossRef]

25. Canton, M.; Roe, R.; Poigny, S.; Renault, J.-H.; Nuzillard, J.-M. Multiple solvent signal presaturation and decoupling artifact removal in $^{13}C\{^1H\}$ nuclear magnetic resonance. *Magn. Reson.* **2020**, *1*, 155–164. [CrossRef]

26. Friesen, J.B.; McAlpine, J.B.; Chen, S.-N.; Pauli, G.F. Countercurrent Separation of Natural Products: An Update. *J. Nat. Prod.* **2015**, *78*, 1765–1796. [CrossRef] [PubMed]

27. Ingkaninan, K.; Hazekamp, A.; Hoek, A.C.; Balconi, S.; Verpoorte, R. Application of centrifugal partition chromatography in a general separation and dereplication procedure for plant extracts. *J. Liq. Chromatogr. Relat. Technol.* **2000**, *23*, 2195–2208. [CrossRef]

28. Marston, A.; Hostettmann, K. Counter-current chromatography as a preparative tool—applications and perspectives. *J. Chromatogr. A* **1994**, *658*, 315–341. [CrossRef]

29. Hubert, J.; Chollet, S.; Purson, S.; Reynaud, R.; Harakat, D.; Martinez, A.; Nuzillard, J.-M.; Renault, J.-H. Exploiting the Complementarity between Dereplication and Computer-Assisted Structure Elucidation for the Chemical Profiling of Natural Cosmetic Ingredients: *Tephrosia purpurea* as a Case Study. *J. Nat. Prod.* **2015**, *78*, 1609–1617. [CrossRef] [PubMed]

30. Lehbili, M.; Magid, A.A.; Hubert, J.; Kabouche, A.; Voutquenne-Nazabadioko, L.; Renault, J.-H.; Nuzillard, J.-M.; Morjani, H.; Abedini, A.; Gangloff, S.C.; et al. Two new bis-iridoids isolated from *Scabiosa stellata* and their antibacterial, antioxidant, anti-tyrosinase and cytotoxic activities. *Fitoterapia* **2018**, *125*, 41–48. [CrossRef]

31. Schmitt, M.; Magid, A.A.; Hubert, J.; Etique, N.; Duca, L.; Voutquenne-Nazabadioko, L. Bio-guided isolation of new phenolic compounds from *Hippocrepis emerus* flowers and investigation of their antioxidant, tyrosinase and elastase inhibitory activities. *Phytochem. Lett.* **2020**, *35*, 28–36. [CrossRef]

32. Marchal, L.; Intes, O.; Foucault, A.; Legrand, J.; Nuzillard, J.-M.; Renault, J.-H. Rational improvement of centrifugal partition chromatographic settings for the production of 5-n-alkylresorcinols from wheat bran lipid extract I. Flooding conditions—Optimizing the injection step. *J. Chromatogr. A* **2003**, *1*, 51–62. [CrossRef]

33. Renault, J.-H.; Nuzillard, J.-M.; Intes, O.; Maciuk, A. Chapter 3: Solvent systems. In *Comprehensive Analytical Chromatography: Countercurrent Chromatography*; Elsevier BV: Amsterdam, The Netherlands, 2002; Volume 38, pp. 49–83.

34. Berthod, A.; Carda-Broch, S. Determination of liquid–liquid partition coefficients by separation methods. *J. Chromatogr. A* **2004**, *1037*, 3–14. [CrossRef]

35. Yang, L.; Yin, P.; Ho, C.-T.; YuJun, L.; Sun, L.; Liu, Y. Effects of thermal treatments on 10 major phenolics and their antioxidant contributions in *Acer truncatum* leaves and flowers. *R. Soc. Open Sci.* **2018**, *5*, 180364. [CrossRef]

36. Rodríguez, H.; Rivas, B.D.L.; Gómez-Cordovés, C.; Muñoz, R. Degradation of tannic acid by cell-free extracts of *Lactobacillus plantarum*. *Food Chem.* **2008**, *107*, 664–670. [CrossRef]

37. Douros, A.; Hadjipavlou-Litina, D.; Nikolaou, K.; Skaltsa, H. The Occurrence of Flavonoids and Related Compounds in *Cedrus brevifolia* A. Henry ex Elwes & A. Henry Needles. Inhibitory Potencies on Lipoxygenase, Linoleic Acid Lipid Peroxidation and Antioxidant Activity. *Plants* **2017**, *7*, 1. [CrossRef]

38. Dakir, M.; El Hanbali, F.; Mellouki, F.; Akssira, M.; Benharref, A.; Del Moral, J.F.Q.; Barrero, A.F.; Del Moral, J.F.Q. Antibacterial diterpenoids from *Cedrus atlantica*. *Nat. Prod. Res.* **2005**, *19*, 719–722. [CrossRef] [PubMed]

39. Barrero, A.F.; Quilezdelmoral, J.; Herrador, M.M.; Arteaga, J.F.; Akssira, M.; Benharref, A.; Dakir, M. Abietane diterpenes from the cones of *Cedrus atlantica*. *Phytochemistry* **2005**, *66*, 105–111. [CrossRef] [PubMed]
40. Nam, A.-M.; Paoli, M.; Castola, V.; Casanova, J.; Bighelli, A. Identification and quantitative determination of lignans in *Cedrus atlantica* resins using ^{13}C NMR spectroscopy. *Nat. Prod. Commun.* **2011**, *6*. [CrossRef]

Sample Availability: Samples of the compounds are not available from the authors.

Publisher's Note: MDPI stays neutral with regard to jurisdictional claims in published maps and institutional affiliations.

Article

Facile and Rapid Isolation of Oxypeucedanin Hydrate and Byakangelicin from *Angelica dahurica* by Using [Bmim]Tf₂N Ionic Liquid

Alice Nguvoko Kiyonga, Gyeongmin Hong, Hyun Su Kim, Young-Ger Suh and Kiwon Jung *

Institute of Pharmaceutical Sciences, College of Pharmacy, CHA University, Sungnam 13844, Korea; gabriella@chauniv.ac.kr (A.N.K.); gmini93@chauniv.ac.kr (G.H.); khs8812@snu.ac.kr (H.S.K.); ygsuh@cha.ac.kr (Y.-G.S.)
* Correspondence: pharmj@cha.ac.kr; Tel.: +82-31-8817173

Abstract: Ionic liquids (ILs) have sparked much interest as alternative solvents for plant materials as they provide distinctive properties. Therefore, in this study, the capacity of ILs to extract oxypeucedanin hydrate and byakangelicin from the roots of *Angelica dahurica* (*A. dahurica*) was investigated. The back-extraction method was examined to recover target components from the IL solution as well. Herein, [Bmim]Tf₂N demonstrated outstanding performance for extracting oxypeucedanin hydrate and byakangelicin. Moreover, factors including solvent/solid ratio, extraction temperature and time were investigated and optimized using a statistical approach. Under optimum extraction conditions (solvent/solid ratio 8:1, temperature 60 °C and time 180 min), the yields of oxypeucedanin hydrate and byakangelicin were 98.06% and 99.52%, respectively. In addition, 0.01 N HCl showed the most significant ability to back-extract target components from the [Bmim]Tf₂N solution. The total content of both oxypeucedanin hydrate (36.99%) and byakangelicin (45.12%) in the final product exceeded 80%. Based on the data, the proposed approach demonstrated satisfactory extraction ability, recovery and enrichment of target compounds in record time. Therefore, the developed approach is assumed essential to considerably reduce drawbacks encountered during the separation of oxypeucedanin hydrate and byakangelicin from the roots of *A. dahurica*.

Keywords: ionic liquids; oxypeucedanin hydrate; byakangelicin; *A. dahurica*; back-extraction; enrichment

Citation: Kiyonga, A.N.; Hong, G.; Kim, H.S.; Suh, Y.-G.; Jung, K. Facile and Rapid Isolation of Oxypeucedanin Hydrate and Byakangelicin from *Angelica dahurica* by Using [Bmim]Tf₂N Ionic Liquid. *Molecules* **2021**, *26*, 830. https://doi.org/10.3390/molecules26040830

Academic Editors: Young Hae Choi, Young Pyo Jang, Yuntao Dai and Luis Francisco Salomé-Abarca

Received: 11 January 2021
Accepted: 2 February 2021
Published: 5 February 2021

Publisher's Note: MDPI stays neutral with regard to jurisdictional claims in published maps and institutional affiliations.

1. Introduction

Angelica dahurica (Fisch. ex Hoffm.) Benth. et Hook. f. (*A. dahurica*) is a folk medicinal plant for which the roots have been used over the years as a remedy for cold, fever, headache, nasal congestion, toothache, asthma, stomachache and dysmenorrhea in East Asian countries (Korea, China, Russia, Taiwan, etc.) [1–4]. Several previous studies have also reported that the roots of *A. dahurica* exhibit various pharmacological functions, including antibacterial, anti-asthmatic, hypotensive, anti-inflammatory, antioxidation, anti-cancer, and anti-Alzheimer effects [5–11]. Pharmacological functions of *A. dahurica* are generally attributed to coumarins (oxypeucedanin, bergapten, imperatorin, cnidilin, isoimperatorin, xanthotoxol, oxypeucedanin hydrate, byakangelicin, etc.), which are major constituents of this herb [12–14]. Oxypeucedanin hydrate (Figure 1a) has inhibitory effects on tyrosinase activity [15] and exhibits anti-inflammatory and antioxidant activities [16]. Byakangelicin (Figure 1b) has been reported to modulate the brain distribution of various bioactive compounds by improving accumulation and consequently boosting their curative effectiveness [17]. Byakangelicin shows effect against carbon tetrachloride-induced liver injury and fibrosis in mice as well [18].

231

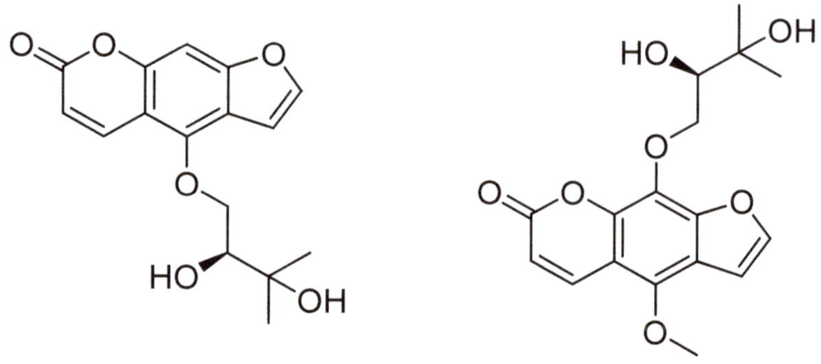

(a)

(b)

Figure 1. Chemical structures of (**a**) Oxypeucedanin hydrate, (**b**) Byakangelicin.

Ionic liquids (ILs) are a broad class of molten salts entirely composed of ions, and they possess a melting point below 100 °C [19,20]. ILs possess distinctive properties such as great selectivity and electric conductivity, negligible vapor pressure and low volatility [21], poor flammability [22], strong thermal stability [23], and solvation power of organic [24] and inorganic [25] compounds. ILs are also referred to as "designer" solvents because their physicochemical characteristics can be tuned by altering their cations or anions [26]. The remarkable properties of ILs have engendered enormous interest among scientists in diverse fields, enabling them to be considered as green media and a promising substitute for volatile organic solvents (VOCs). Therefore, ILs have been applied in diverse fields such as biochemistry [27], electrochemistry [28], analytical chemistry [29], pharmaceutics [30], medicine [31], and so forth. In analytical chemistry, ILs have been applied for the extraction and separation of several bioactive components from herbal products, namely, phenolics [32], alkaloids [33,34], flavonoids [35,36], ginsenosides [37], coumarins [38], terpenoids [39,40], and so forth [41,42].

Back-extraction, also called liquid-liquid extraction, is a purification process in which a compound or a mixture of compounds transfer from one phase to another based on their differential solubility. In this process, two essentially immiscible or partially soluble solvents are utilized to allow compound transfer and separation. The back-extraction process is routinely employed by chemists due to numerous advantages, such as simplicity, low cost, ease of scale up and rapidity [43]. In their study, Absalan and co-workers successfully employed [BMIm]PF$_6$ for extracting 3-indole-butyric acid from its aqueous concentrate and effectively back-extracted the compound from the IL phase [44]. Moreover, Larriba and co-workers were able to back-extract tyrosol from the ILs using 0.1M NaOH aqueous solution as well [45].

The separation of active components in general and particularly of minor compounds from plant raw materials is exhausting and tedious. In that regard, this study aimed to develop an effective and relevant approach for the extraction and enrichment of minor coumarins, oxypeucedanin hydrate and byakangelicin, from the roots of *A dahurica* using ILs as the extraction solvent and the back-extraction method for their recovery. The ultimate purpose was to overcome drawbacks of traditional separation techniques. Conventional separation methods are exhausting and time-consuming because they involve repetitive chromatographic steps. However, in this study, IL was effectively utilized as an extraction solvent while the back-extraction method was satisfactorily employed for the enrichment of target molecules from the IL solutions. To improve the rates of extraction and recovery of two compounds, several parameters (solvent/solid ratio, extraction temperature and extraction time) were studied and optimized by response surface methodology (RSM). As far as we know, this is the first time a procedure using ILs and back-extraction technique has been consistently achieved for the separation and enrichment of oxypeucedanin hydrate and byakangelicin from the roots of *A. dahurica*. This study also confirmed the ability of ILs as separation media for active substances of plant origin.

2. Results and Discussion

2.1. Screening of ILs

Two imidazolium-based hydrophobic ILs, [Bmim]PF$_6$ and [Bmim]Tf$_2$N, were evaluated and compared for their ability to extract oxypeucedanin hydrate and byakangelicin from the roots of *A. dahurica*. These two ILs were selected in this study due to their easy availability in the market and moderate cost compared to commonly employed room temperature ILs. For the extraction experiments, powdered samples (one gram each) were extracted with IL at a 6:1 solvent/solid ratio and 40 °C extraction temperature for 120 min. The agitation speed was 500 rpm. Note that samples were soaked for 30 min prior to their extraction. The results revealed that [Bmim]Tf$_2$N demonstrated the highest extraction performance for oxypeucedanin hydrate and byakangelicin, 71.64% (872.996 µg)) and 71.38% (1074.78 µg)), respectively. Extraction yields of oxypeucedanin hydrate and byakangelicin obtained from (Bmim)PF$_6$ were, respectively, 56.48% (688.22 µg)) and 57.11% (859.98 µg)) (Figure 2). Based on the principle "like dissolves like", it was assumed that target components were extracted with excellent yields in [Bmim]Tf$_2$N because they possess similar polarity with [Bmim]Tf$_2$N than with [Bmim]PF$_6$. On the other hand, the difference in viscosity between [Bmim]Tf$_2$N and [Bmim]PF$_6$ [46,47] was assumed to have impacted the extraction abilities of these ILs. Scientists have suggested that an increase in viscosity hinders the imbibement of the solvent into the plant tissues, which therefore has a negative impact on the mass transfer of compounds [48,49]. In this regard, the closeness of [Bmim]Tf$_2$N polarity with that of the target components and its low viscosity compared to [Bmim]PF$_6$ were presumed as the driving forces leading to the improvement of oxypeucedanin hydrate and byakangelicin extraction yields. Because [Bmim]Tf$_2$N showed the highest extraction efficiency for both target components, it was selected for further experiments.

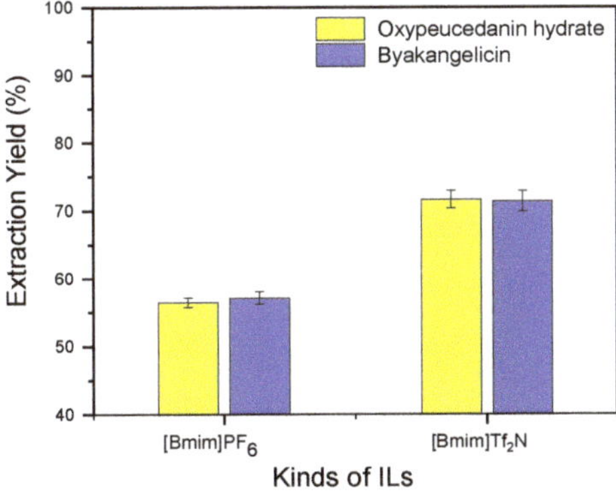

Figure 2. Effect of ionic liquids (ILs) on the extraction of oxypeucedanin hydrate and byakangelicin.

2.2. Single Factor Experiments

2.2.1. Effect of the Solvent/Solid Ratio on the Extraction of Oxypeucedanin Hydrate and Byakangelicin

The extraction process was carried out at 2:1, 4:1, 6:1, 8:1 and 10:1 solvent/solid ratios to examine their effect on the extraction of oxypeucedanin hydrate and byakangelicin. This was in the aim of improving the extraction of both compounds while preventing incomplete extraction and waste of the solvent. Herein, the extraction temperature and time were set as 40 °C and 120 min, respectively, and the agitation speed was 500 rpm. Note that the samples were soaked for 30 min prior to their extraction. From the results, the extraction of oxypeucedanin hydrate and byakangelicin increased with the increment of the solvent/solid ratio. In addition, the most significant extraction of approximately 81% was achieved when the ratio was 10:1 (Figure 3).

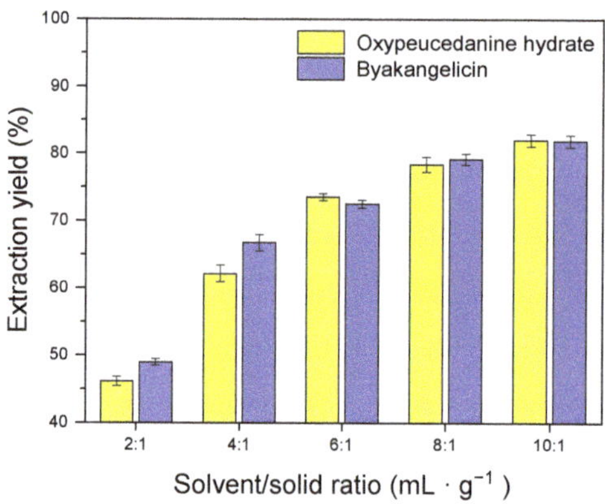

Figure 3. Effect of solvent/solid ratio on the extraction of oxypeucedanin hydrate and byakangelicin.

2.2.2. Effect of Temperature on the Extraction of Oxypeucedanin Hydrate and Byakangelicin

Temperature is one of several factors investigated during the extraction process as it is known for either positively (by improving the extraction) or negatively (by causing the degradation of heat-sensible components) impacting the extraction of active components. In this regard, experiments were performed to investigate the impact of various extraction temperatures (20 °C, 40 °C, 50 °C, 60 °C, and 70 °C) on the extraction of oxypeucedanin hydrate and byakangelicin. The solvent/solid ratio was maintained at 10:1, the extraction time at 120 min and the agitation speed at 500 rpm. Note that the samples were soaked for 30 min prior to their extraction. As shown in Figure 4, the extraction of oxypeucedanin hydrate and byakangelicin greatly improved when the temperature increased from 20 °C to 60 °C. At 60 °C, the extraction yields of oxypeucedanin hydrate and byakangelicin were 92.67% and 94.16%, respectively. Thus, 60 °C was selected for further experiments.

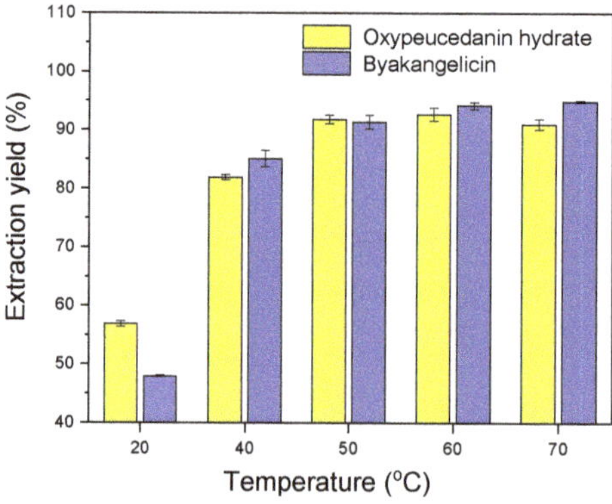

Figure 4. Effect of temperature on the extraction of oxypeucedanin hydrate and byakangelicin.

2.2.3. Effect of Time on the Extraction of Oxypeucedanin Hydrate and Byakangelicin

The effect of time on the extraction of oxypeucedanin hydrate and byakangelicin was studied for improving the extraction of both compounds while preventing incomplete extraction and time-waste. The time parameter was varied as follows: 30, 60, 120, 180 and

240 min. Other conditions were: extraction temperature 60 °C, solvent/solid ratio 10:1 and agitation speed 500 rpm. Note that the samples were soaked for 30 min prior to their extraction. As depicted in Figure 5, the most notable extraction of two compounds was observed when the extraction time was 180 min. At this time, the extraction efficiency of oxypeucedanin hydrate and byakangelicin reached 94.64% and 95.29%, respectively.

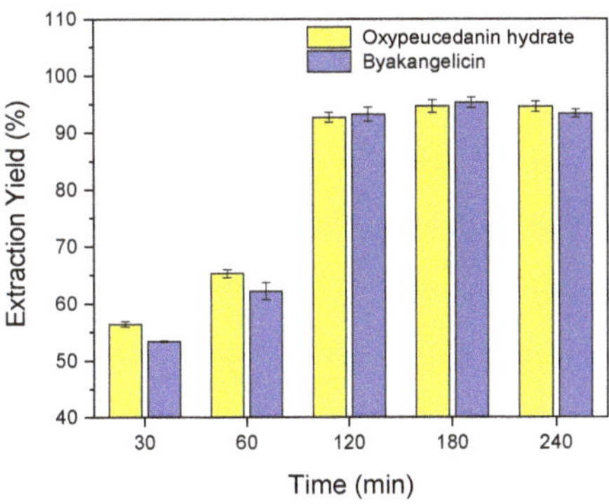

Figure 5. Effect of time on the extraction of oxypeucedanin hydrate and byakangelicin.

2.3. Statistical Optimization of the Extraction Conditions of Oxypeucedanin Hydrate and Byakangelicin

For the statistical optimization of the extraction of oxypeucedanin hydrate and byakangelicin from the roots of *A. dahurica*, three variables including solvent/solid ratio, extraction temperature, and extraction time were optimized using RSM. The conditions for extraction variables were established based on data obtained in Section 2.2 and are illustrated in Tables 1 and 2 along with corresponding experimental data. X_1 is a code for the extraction time variable, X_2 is the extraction temperature variable and X_3 is the solvent/solid ratio variable.

Table 1. Three level factorials for optimization of oxypeucedanin hydrate and byakangelicin extraction.

Levels	X_1 Extraction Time (min)	X_2 Extraction Temperature (°C)	X_3 Solvent/Solid Ratio (mL·g^{-1})
Low level (−1)	60	40	6:1
Average (0)	120	50	9:1
High level (1)	180	60	12:1

The analysis of variance for response surface quadratic models is summarized in Table 3. The *p*-values ($p < 0.05$) and the regression coefficient ($R^2 > 0.95$) were used to assess the significance of the model. The regression coefficients ($R^2 = 0.989$ for oxypeucedanin hydrate and $R^2 = 0.987$ for byakangelicin) indicated that only 1.0% of the variations were not explained by the model and that the model had high reliability (Table 3).

Table 2. Box–Behnken design and response values for optimization of oxypeucedanin hydrate and byakangelicin extraction.

Run	X_1 Extraction Time (min)	X_2 Extraction Temperature (°C)	X_3 Solvent/Solid Ratio (mL·g^{-1})	Extraction Yield of Oxypeucedanin Hydrate (%)	Extraction Yield of Byakangelicin (%)
1	120	50	9:1	87.55	88.54
2	120	50	9:1	88.01	89.07
3	180	60	9:1	100.20	101.19
4	180	50	6:1	90.74	89.56
5	120	60	12:1	97.88	99.06
6	60	50	6:1	77.21	77.94
7	120	50	9:1	86.36	90.87
8	120	60	6:1	88.64	85.59
9	60	50	12:1	85.82	88.09
10	180	40	9:1	91.86	92.22
11	60	60	9:1	82.10	89.66
12	180	50	12:1	100.23	100.65
13	120	40	6:1	79.74	77.89
14	60	40	9:1	74.35	76.43
15	120	40	12:1	90.41	90.09

Table 3. Variance analysis for the quadratic response surface regression model.

	Degrees of Freedom	Sum of Squares	Mean Square	F-Value	p-Value	R^2	R^2 Adjusted
Extraction yield of oxypeucedanin hydrate (%)							
Model	9	828.83	92.09	49.90	0.00		
X_1	1	504.87	504.87	147.58	0.00		
X_2	1	131.72	131.725	71.37	0.00		
X_3	1	180.57	180.57	97.83	0.00		
X_1^2	1	1.16	0.67	0.36	0.57		
X_2^2	1	0.05	0.22	0.12	0.74	0.989	0.969
X_3^2	1	9.66	9.66	5.23	0.07		
$X_1 X_2$	1	0.09	0.09	0.05	0.83		
$X_1 X_3$	1	0.19	0.19	0.10	0.76		
$X_2 X_3$	1	0.51	0.51	0.28	0.62		
Residual	5	9.23	1.85				
Lack of fit	3	7.78	2.59	3.58	0.23		
Pure error	2	1.45	0.724				
Total	14	838.06					
Extraction yield of byakangelicin (%)							
Model	9	807.01	89.67	42.82	0.00		
X_1	1	331.42	331.42	158.25	0.00		
X_2	1	188.95	188.95	90.22	0.00		
X_3	1	275.08	275.08	131.35	0.00		
X_1	1	2.01	1.50	0.72	0.43		
X_2^2	1	0.12	0.26	0.12	0.74		
X_3^2	1	4.28	4.28	2.04	0.21	0.987	0.964
$X_1 X_2$	1	4.53	4.53	2.16	0.20		
$X_1 X_3$	1	0.22	0.22	0.11	0.76		
$X_2 X_3$	1	0.40	0.40	0.19	0.68		
Residual	5	10.47	2.09				
Lack of fit	3	7.48	2.49	1.67	0.40		
Pure error	2	2.99	1.49				
Total	14	817.48					

As seen in Table 3, sole linear term coefficients X_1, X_2, and X_3 were significant with p-values inferior to 0.05. This confirmed that a linear model was adequate to represent our model. The "lack of fit"for *p*-values ($p = 0.23$ for oxypeucedanin hydrate and $p = 0.40$ for byakangelicin) and the "lack of fit" for F-values (F = 3.58 for oxypeucedanin hydrate and F = 1.67 for byakangelicin) indicated that the model was significant and suitable for obtaining optimum extraction of oxypeucedanin hydrate and byakangelicin (Table 3). The second-order differential equations for obtaining oxypeucedanin hydrate and byakangelicin are illustrated below.

$$\text{Yield of oxypeucedanin hydrate} = 87.3076 + 7.944X_1 + 4.0578X_2 + 4.7509X_3 - 0.4253X_1{}^2 + 0.2437X_2{}^2 + 1.6173X_3{}^2 + 0.1489X_1X_2 + 0.2199X_1X_3 - 0.3581\,X_2X_3 \tag{1}$$

$$\text{Yield of byakangelicin} = 89.4976 + 6.4364X_1 + 4.8599X_2 + 5.8638X_3 + 0.6384X_1{}^2 - 0.2631X_2{}^2 - 1.0769X_3{}^2 - 1.0640X_1X_2 + 0.2363X_1X_3 - 0.3168X_2X_3 \tag{2}$$

In Figure 6, the response surface plots indicating the interaction between parameters and their effects on the extraction of oxypeucedanin hydrate and byakangelicin are displayed. The solvent/solid ratio of 8:1 at an extraction temperature of 60 °C for 180 min were the optimum conditions proposed by RSM to improve the extraction yields of oxypeucedanin hydrate and byakangelicin up to 97.98% and 98.01%, respectively.

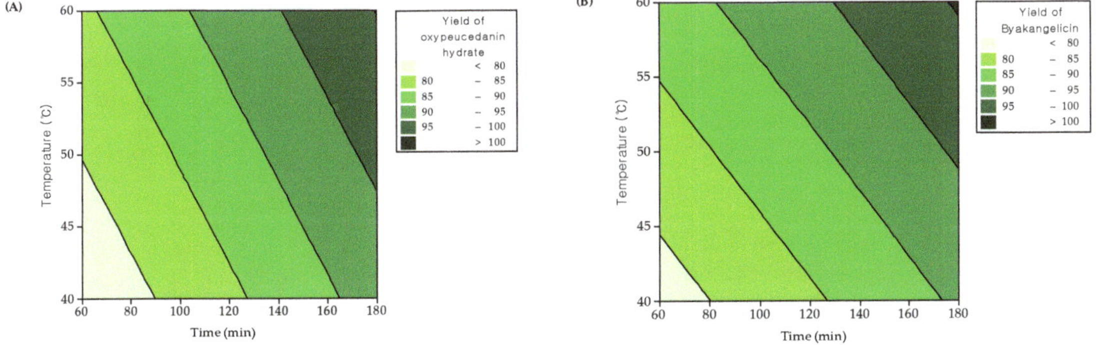

Figure 6. *Cont.*

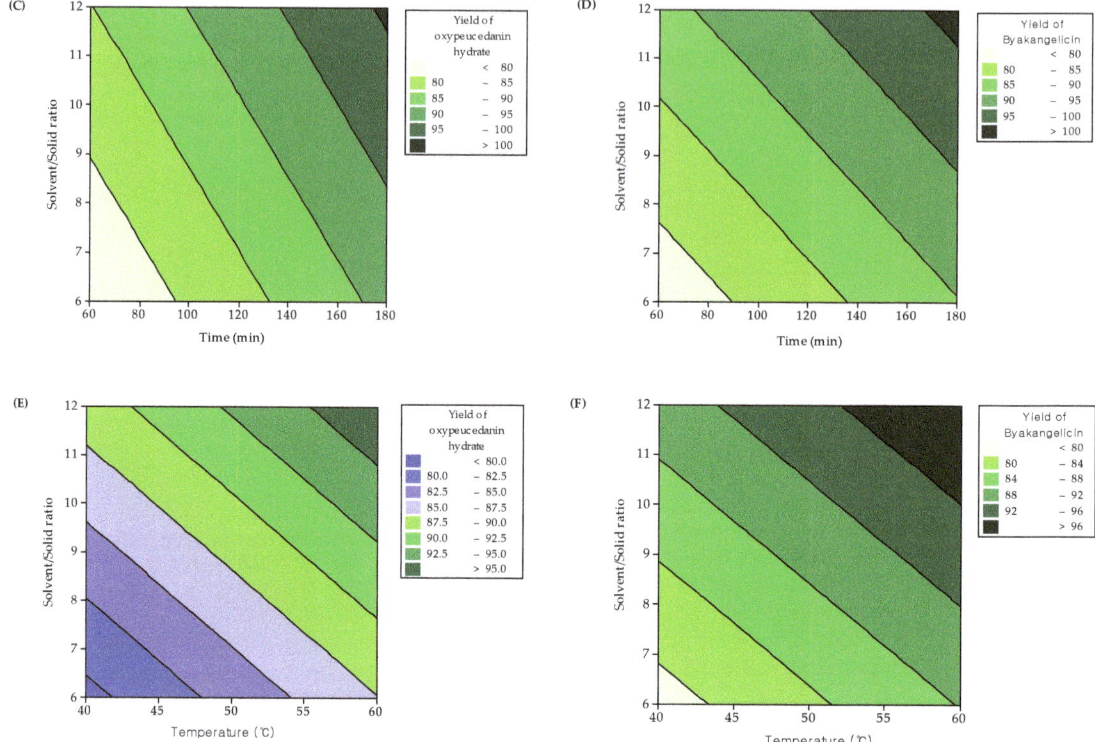

Figure 6. Contour plots of extraction yields of oxypeucedanin hydrate and byakangelicin versus variables. (**A**) effect of temperature and time on extraction yield of oxypeucedanin hydrate, (**B**) effect of temperature and time on extraction yield of byakangelicin, (**C**) effect of solvent/solid ratio and time on extraction yield of oxypeucedanin hydrate, (**D**) effect of solvent/solid ratio and time on extraction yield of byakangelicin, (**E**) effect of solvent/solid ratio and temperature on extraction yield of oxypeucedanin hydrate and (**F**) effect of solvent/solid ratio and temperature on extraction yield of byakangelicin.

2.4. Extraction of Oxypeucedanin Hydrate and Byakangelicin Using Optimal Extraction Conditions

Samples were extracted under optimum extraction conditions proposed by RSM for examining the precision and validity of the method. The proposed optimum conditions were as follows: solvent/solid ratio 8:1, extraction temperature 60 °C and extraction time 180 min. The results predicted by RSM and the experimental results are depicted in Table 4. As observed in Table 4, the experimental values agreed with the predicted values. Accordingly, it is presumed that the RSM model is appropriate to estimate the extraction of oxypeucedanin hydrate and byakangelicin from the roots of *A. dahurica*.

Table 4. Validation of the optimized extraction conditions.

	Predicted Yield of Oxypeucedanin Hydrate (%)	Observed Yield of Oxypeucedanin Hydrate (%)	Predicted Yield of Byakangelicin (%)	Observed Yield of Byakangelicin (%)
Mean	97.99	98.06	98.01	99.52
Standard Deviation		2.37		0.96
Relative Standard Deviation (%)		2.42		0.96

2.5. Isolation of Oxypeucedanin Hydrate and Byakangelicin from the IL Solution Using Back-Extraction

The back-extraction process was conducted for the isolation and enrichment of oxypeucedanin hydrate and byakangelicin from the [Bmim]Tf$_2$N extraction solution. Preliminary investigation was conducted using deionized water (DIW) and hexane as back-extraction solvents because of their capability to form a two-phase system with the [Bmim]Tf$_2$N. From the preliminary experiment, it was found out that DIW exhibited promising isolation performance compared to hexane, as DIW showed high selectivity in recovering oxypeucedanin hydrate and byakangelicin. However, using hexane as a back-extraction solvent led to unsatisfactory results as several components present in the [Bmim]Tf$_2$N solution were simultaneously back-extracted in the hexane phase. This was presumably due to the fact that most components of the [Bmim]Tf$_2$N solution migrated to the hexane phase as they were more highly soluble in hexane than in [Bmim]Tf$_2$N.

Because DIW demonstrates promising outcomes, various aqueous solutions (DIW, 0.01 N NaOH, and 0.01 N HCl) were evaluated for improving the rate of recovery and for the enrichment of target components in the final product. The initial amounts of oxypeucedanin hydrate and byakangelicin in the [Bmim]Tf$_2$N solution were 609.90 µg and 729.81 ug, respectively. The volume of [Bmim]Tf$_2$N solution was 4 mL and the volume of the back-extraction solvent was set as 30 mL (1:7.5 IL/back-extraction solvent). The equilibration time was fixed for 4 h to allow complete phase-equilibration. From the results (Table 5), when 0.01 NaOH was used as a back-extraction solvent, no target compound could be recovered. However, when DIW and 0.01 N HCl were employed, oxypeucedanin hydrate and byakangelicin could be recovered from the [Bmim]Tf$_2$N extraction solution. Nevertheless, the most outstanding performance was acquired when 0.01 N HCl was utilized (Table 5).

Table 5. Recovery of oxypeucedanin hydrate and byakangelicin.

Back-Extraction Solvent	Recovered Amount of Oxypeucedanin Hydrate (µg)	Recovered Amount of Byakangelicin (µg)	Yield of Oxypeucedanin Hydrate (%)	Yield of Byakangelicin (%)
DIW	332.15	300.50	54.46	41.18
0.01 N HCl	517.79	570.98	84.90	78.24
0.01 NaOH	0	0	0	0

In addition, various volumes (10, 20, 30, 40 and 50 mL) of 0.01 N HCl back-extraction solvent were investigated to improve the rate of recovery of oxypeucedanin hydrate and byakangelicin. The volume of [Bmim]Tf$_2$N solution was maintained as 4 mL. Namely, the ratios of IL to back-extraction solvent were as follows: 1:2.5, 1:5, 1:7.5, 1:10, and 1:12.5. The initial amounts of oxypeucedanin hydrate and byakangelicin in the [Bmim]Tf$_2$N solution, the volume of [Bmim]Tf$_2$N solution and the equilibration time were as above. As the results, 40 mL (1:10 IL/back-extraction solvent) was the adequate volume of 0.01 N HCl back-extraction solvent as it exhibited satisfactory recovery yields of both oxypeucedanin hydrate and byakangelicin: 91.99% (561.06 µg) and 89.17% (650.76 µg), respetively (Table 6). This result can be explained by the fact that, when using 0.01 N HCl as back-extraction medium, ion exchange occurred between Cl and Tf$_2$N anions because the interaction energy between the Bmim cation and Cl anion is relatively greater than between the Bmim cation and Tf$_2$N [50]. However, due to the presence of a low concentration of Cl anion in the formed system, it is assumed that both [Bmim]Tf$_2$N and [Bmim]Cl coexisted in the IL phase. The change in the IL solution composition is believed to have influenced the decrease in oxypeucedanin hydrate and byakangelicin solubilities in the IL phase. Consequently, this change has favorized their migration to the aqueous phase. Nevertheless, oxypeucedanin hydrate and byakangelicin possessed inferior solubilities in the back-extraction solution. Therefore, as observed in Table 6, their recovery amounts were enhanced with the increase in the amount of the back-extraction solvent.

Table 6. Recovery of oxypeucedanin hydrate and byakangelicin using distinct volumes of 0.01 N HCl.

Volume of 0.01 N HCl (mL)	Recovered Amount of Oxypeucedanin Hydrate (µg)	Recovered Amount of Byakangelicin (µg)	Yield of Oxypeucedanin Hydrate (%)	Yield of Byakangelicin (%)
10	342.22	257.22	56.11	35.24
20	441.64	538.66	72.41	73.81
30	529.42	580.29	86.81	79.51
40	561.06	650.762	91.99	89.17
50	541.69	660.012	89.47	90.44

Several laboratory (researching) experiments are carried out at small-scale using small volume sizes of materials. However, the moving of promising experiments from small-scale to large-scale is challenging and sometimes unsuccessful due to some limitations encountered. Therefore, in this study, the back-extraction experiment was carried out at a larger scale to confirm whether the results agreed with data obtained at small-scale. The contents (purities) of oxypeucedanin hydrate and byakangelicin in the final product were investigated as well. The initial amounts of oxypeucedanin hydrate and byakangelicin in the [Bmim]Tf$_2$N solution were 5930.02 and 7370.23 µg, respectively. The volume of the back-extraction solvent was 400 mL (1:10 IL/back-extraction solvent), and the equilibration time was fixed for 4 h. As illustrated in Table 7, both oxypeucedanin hydrate and byakangelicin were obtained in satisfactory yield (approximately 90%), and these results agreed with data obtained from small-scale experiments (Table 6). Moreover, the total purity of both components exceeded 80% in the final product, namely, 36.99% oxypeucedanin hydrate and 45.12% byakangelicin.

Table 7. Contents of oxypeucedanin hydrate and byakangelicin in the product.

	Recovered Amount (µg)	Yield (%)	Content (%)
Total extract	14,456.80	-	-
Oxypeucedanin hydrate	5347.09	90.17	36.99
Byakangelicin	6523.39	88.51	45.12

From this experiment, the enrichment of both oxypeucedanin hydrate and byakangelicin, two minor coumarins of the roots of *A. dahurica*, was achieved in a short period and few steps. This is assumed to considerably reduce the time and effort needed for the separation of these two components. HPLC chromatograms of oxypeucedanin hydrate and byakangelicin standards, [Bmim]Tf$_2$N extraction solution and product are illustrated in Figure 7.

2.6. Comparison of the Extraction and Separation Method of Oxypeucedanin Hydrate and Byakangelicin

The roots of *A. dahurica* were extracted with DIW and 50 and 95% ethanol to compare the ability of these frequently employed extraction solvents with that of IL. These solvents were specifically extracting oxypeucedanin hydrate and byakangelicin. Experiments were performed at the same extraction conditions (solvent/solid ratio 8:1, extraction temperature 60 °C and extraction time 180 min) proposed by RSM when IL was employed as the extraction solvent. The results are illustrated in Table 8. As observed, at the same extraction conditions, the IL [Bmim]Tf$_2$N exhibited the highest extraction ability for oxypeucedanin hydrate (98.06%) and byakangelicin (99.52%) compared to other solvents (Table 8).

Table 8. Comparison of the extraction ability of different solvents.

Extraction Solvent	Amount of Oxypeucedanin Hydrate (µg)	Amount of Byakangelicin (µg)	Extraction Yield of Oxypeucedanin Hydrate (%)	Extraction Yield of Byakangelicin (%)
DIW	462.42	625.01	37.90	41.39
50% ethanol	885.39	782.64	72.57	51.83
95% ethanol	946.18	1288.83	77.56	85.35
[Bmim]Tf$_2$N	1194.89	1498.55	98.06	99.52

In addition, efficient, facile and rapid isolation and enrichment techniques of oxypeucedanin hydrate and byakangelicin from *A. dahurica* using the above-mentioned solvents have yet been unreported in the literature. This is assumed to be because of the laborious and exhausting separation techniques of active components from plant raw materials. More importantly, this requires expensive equipment and/or repetitive chromatography. Moreover, the process is demanding, especially when it comes to the separation of minor compounds such as oxypeucedanin hydrate and byakangelicin. However, a facile and rapid method for the separation (enrichment) of minor coumarins, oxypeucedanin hydrate and byakangelicin, from *A. dahurica* using IL as the extraction solvent was successfully achieved. The process was completed in few steps and within record time compared to traditional techniques that require repetitive chromatography. The novel developed approach is believed to considerably reduce the steps and time required for the isolation of oxypeucedanin hydrate and byakangelicin from the roots of *A. dahurica*. Therefore, the developed approach is assumed to overcome the drawbacks of traditional methods by preventing time wasting and by lessening labor cost.

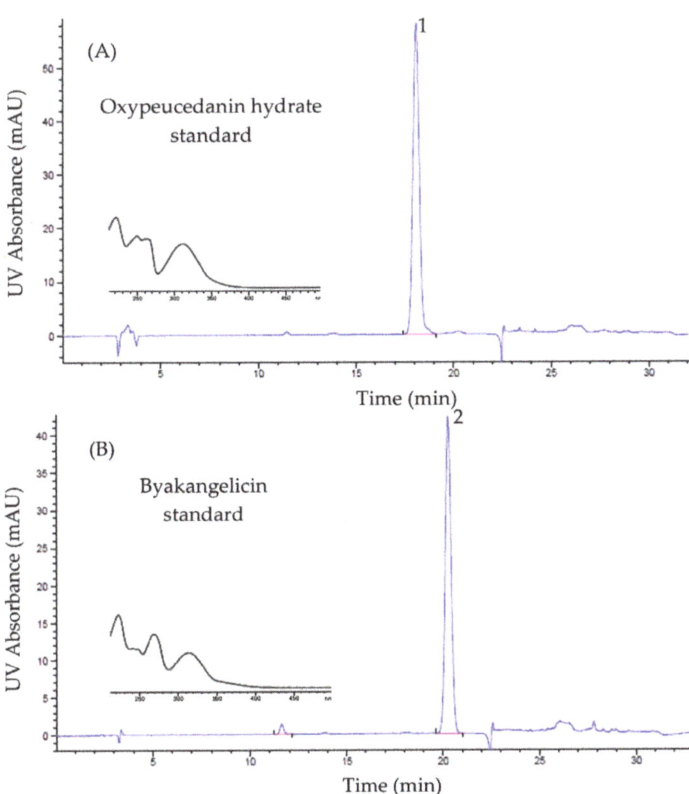

Figure 7. *Cont.*

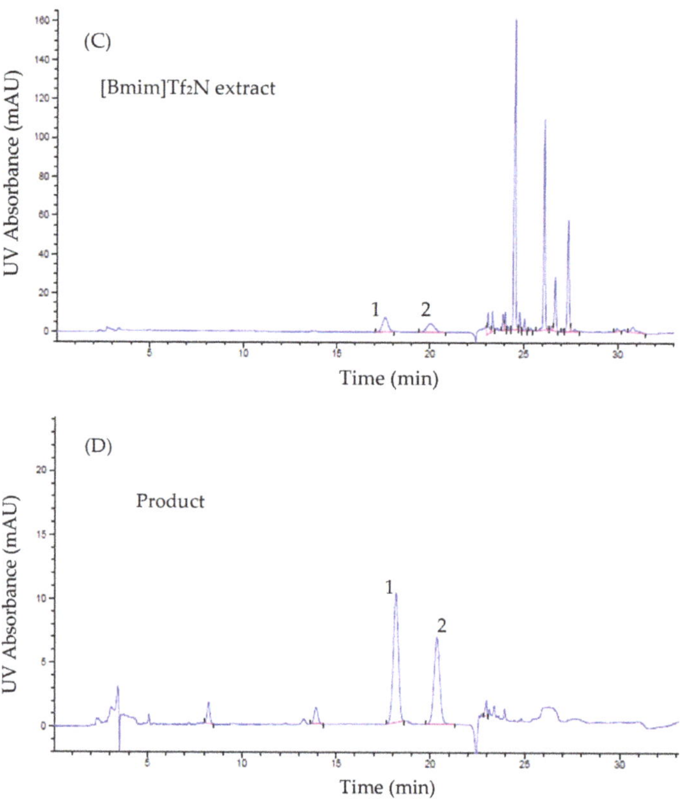

Figure 7. HPLC chromatograms of (**A**) oxypeucedanin hydrate standard, (**B**) byakangelicin standard, (**C**) [Bmim]Tf$_2$N extraction solution and (**D**) product. (254 nm).

3. Materials and Methods

3.1. Reagents and Materials

A. dahurica was purchased from an oriental herbal medicine market in Seoul, Korea, in 2019. Authentication of the plant material was conducted by one of the authors, Professor Kiwon Jung (College of Pharmacy, CHA University, Pocheon-si, Korea). Moreover, voucher specimen (HPC-AR01) was archived in the Herbarium of the College of Pharmacy, CHA University. Oxypeucedanin hydrate and byakangelicin standards (HPLC purity > 95%) were directly isolated from *A. dahurica*. From A-star Co., Ltd. (Ulsan, Korea), 1-Butyl-3-methylimidazolium bis(trifluoromethylsulfonyl) imide [Bmim]Tf$_2$N and 1-Butyl-3-methylimidazolium hexafluorophosphate [Bmim]PF$_6$ ILs were purchased. HPLC grade deionized water (DIW) and acetonitrile and analytical grade ethanol (purity > 98%), hexane and hydrochloric acid were all purchased from Daejung Chemicals (Siheung-si, Korea).

3.2. Methods

3.2.1. Extraction Procedure and Quantitative Evaluation of Oxypeucedanin Hydrate and Byakangelicin in the IL Extraction Solutions

Samples of one gram (1.0 g) of grounded *A. dahurica* roots were mixed with known volumes of two hydrophobic ILs, [Bmim]Tf$_2$N and [Bmim]PF$_6$, and soaked for 30 min at constant rotational speed (500 rpm). The samples were then extracted on a hot plate at 40 °C for 120 min. The purpose was to evaluate the extraction ability of two ILs on the extraction of oxypeucedanin hydrate and byakangelicin, thus selecting the promising IL for further experiments. After selecting the relevant IL, experiments were carried out on the selected IL. Then, the solid/solvent ratio, extraction temperature and time were cautiously investigated, and RSM employing Box–Behnken design was performed to optimize the extraction yields of target compounds. Minitab 16 was the software. After each extraction, samples were vacuum-filtered through 0.45 μm membrane filters and 50 μL of each filtrate was diluted with acetonitrile to make 1 mL solution. Prepared sample solutions were filtered onto 0.45 μm nylon membranes and then analyzed by HPLC to determine the extraction yield. All experiments were conducted in duplicate and average values were reported. The following equation was employed to determine the extraction efficiency:

$$\text{Extraction yield } (\%) = \frac{M_1(W)}{M(W)} \times 100 \tag{3}$$

where M_1 is the content of target compound in the extract, M is the content of the target compound in the plant material and W is the code for weight.

3.2.2. High-Performance Liquid Chromatography (HPLC) Analysis

HPLC measurements were performed on an Agilent 1260 Infinity II LC system (CA, USA) equipped with a G7111A quaternary gradient pump, a G7129C autosampler and a G7115A UV-detector. Compounds were separated in an Aegispak C18-L (5 μm, 4.6 mm × 250 mm, YoungJin Biochrom, Seongnam-si, Korea) analytical column at a flow rate of 1.0 mL/min and a column temperature of 30 °C. The injection volume was 10 μL and the UV detection wavelength was set at 254 nm. The mobile phases consisted of 0.1% phosphoric acid-water (A) and acetonitrile (B). The time program of the gradient elution of acetonitrile was as following: 25–30% (0–18 min), 30–70% (18–20 min), 70% (20–28 min), then followed by a return to the initial conditions 70–25% (28–29 min) and kept 6 min (29–35 min) for the column equilibrium.

The chromatographic plots for oxypeucedanin hydrate and byakangelicin constructed on three consecutive runs showed good linearities (correlation coefficient, $R^2 = 0.999$) within the investigated range (2.5 ug–60 μg·mL^{-1}). Y = 33.18x − 29.51 was the regression equation for oxypeucedanin hydrate while Y = 18.79x − 5.76 was the regression equation for determining byakangelicin. The method also showed good accuracies (mean recoveries: 97.85% and 99.28%) and precisions (RSD%: 3.63 and 2.67) for both oxypeucedanin hydrate and byakangelicin, respectively. As follows, the above HPLC method was confirmed adequate for the quantitative determination of oxypeucedanin hydrate and byakangelicin in further analyses.

3.2.3. Quantitative Evaluation of Oxypeucedanin Hydrate and Byakangelicin

The purpose of this experiment was to determine the total amount of oxypeucedanin hydrate and byakangelicin in the roots of *A. dahurica*. Herein, three different samples of one gram each were extracted with 50 mL of 95% ethanol at room temperature. The extraction process was repeated until oxypeucedanin hydrate and byakangelicin were undetected anymore in the extraction solution. Afterwards, extraction solutions of each sample were separately collected, concentrated under reduced pressure, and further analyzed by HPLC. The average amounts of oxypeucedanin hydrate and byakangelicin in the extract were 1218.50 μg (0.12% in the raw material) and 1505.75 μg (0.15% in the raw material), respectively. The above contents were considered as reference amounts of oxypeucedanin hydrate and byakangelicin in the roots of *A. dahurica* and were used for comparative studies.

3.2.4. Isolation of Oxypeucedanin Hydrate and Byakangelicin from the IL Extraction Solution

The isolation (recovery) of two minor coumarins, oxypeucedanin hydrate and byakangelicin, from the IL extraction solution was performed using the back-extraction method. DIW and hexane were selected for this purpose because of their unique characteristic to form a two-phase system with the employed IL. The ability of each solvent for recovering oxypeucedanin hydrate and byakangelicin from the IL extraction solution was investigated. The separation process was conducted in a separation funnel by adding a known amount of back-extraction solvent and IL extraction solution to the funnel. The funnel was vigorously shaken and then the system was maintained to equilibrate at room temperature. After equilibration, the upper phase was separated from the lower phase. Subsequently, the upper phase was concentrated using rotary evaporator and then dried in vacuum overnight. Moreover, obtained extracts were analyzed by HPLC to determine the rate of recovery of target components. Further, the procedure was optimized to improve the rate of recovery and the contents (purity) of oxypeucedanin hydrate and byakangelicin in the final product. The following equations were utilized to measure the rate of recovery and the contents of both components in the product:

$$\text{Recovery } (\%) = \frac{M_2(W)}{M_1(W)} \times 100 \tag{4}$$

$$\text{Content } (\%) = \frac{M_2(W)}{M_4(W)} \times 100 \tag{5}$$

where M_1, M, and W are the same as in Equation (3). M_2 is the content of target compound in the product and M_4 represents the total amount of the product.

3.2.5. Structural Identification

Solution-state ^{1}H-NMR and ^{13}C-NMR analyses were performed, respectively, in CDCl$_3$ on 800 MHz and 200 MHz NMR spectrometers (Advance, Bruker, Billerica, MA, USA). Precisely, these analyses confirmed the chemical structures of oxypeucedanin hydrate and byakangelicin reference standards isolated from the *A. dahurica*. Obtained data were compared with those reported in the literature [51].

Oxypeucedanin hydrate ^{1}H-NMR and ^{13}C-NMR spectra were identical with those reported in the literature [51]. ^{1}H-NMR (CDCl3, 800 MHz) δ 8.14 (d, *J* = 9.7 Hz, 1H), 7.58 (d, *J* = 2.3 Hz, 1H), 7.13 (s, 1H), 6.97 (dd, *J* = 2.3, 0.8 Hz, 1H), 6.25 (d, *J* = 9.7 Hz, 1H), 4.52 (dd, mboxemphJ = 9.7, 2.9 Hz, 1H), 4.42 (dd, *J* = 9.7, 7.9 Hz, 1H), 3.89 (dd, *J* = 7.8, 2.8 Hz, 1H) 2.94 (bs, 1H), 2.24 (bs, 1H), 1.34 (s, 3H), 1.29 (s, 3H); ^{13}C-NMR (CDCl3, 200 MHz) δ 161.1, 158.1, 152.5, 148.5, 145.3, 139.0, 114.2, 113.0, 107.3, 104.7, 94.8, 76.5, 74.4, 71.6, 26.7, 25.1.

Byakangelicin ^{1}H-NMR and ^{13}C-NMR spectra were identical with those reported in the literature [51]. ^{1}H-NMR (CDCl3, 800 MHz) δ 8.10 (d, *J* = 9.8 Hz, 1H), 7.62 (d, *J* = 2.3 Hz, 1H), 7.00 (d, *J* = 2.3 Hz, 1H), 6.27 (d, *J* = 9.8 Hz, 1H), 4.58 (dd, *J* = 10.2, 2.6 Hz, 1H), 4.25 (dd, *J* = 10.2, 7.9 Hz, 1H), 4.17 (s, 3H), 3.81 (dd, *J* = 7.9, 2.6 Hz, 1H), 1.30 (s, 3H), 1.26 (s, 3H); ^{13}C-NMR (CDCl3, 200 MHz) δ 160.1, 150.2, 145.2, 144.9, 144.0, 139.4, 126.8, 114.5, 112.9, 107.5, 105.3, 76.1, 75.9, 71.5, 60.7, 26.7, 25.0.

4. Conclusions

In this study, a methodology for the enrichment of oxypeucedanin hydrate and byakangelicin from *A. dahurica* has been successfully developed. Herein, IL was the extraction solvent and the back-extraction method was used for both components from the IL solution. Several conditions were studied and optimized. Under optimum separation conditions, the developed approach demonstrated satisfactory extraction efficiency, recovery and enrichment of both components. The extraction and recovery yields were, respectively, 98.06% and 91.99% for oxypeucedanin hydrate, and 99.52% and 89.17% for byakangelicin. The contents of oxypeucedanin hydrate and byakangelicin were 36.99% and 45.12%, respectively. Namely, the total content of both components reached up to 80% in the product. Generally, the separation of minor compounds from plant raw materials is a tedious and time-consuming task which entails repetitive chromatographic experiments. However, employing the proposed method, the enrichment of minor coumarins, oxypeucedanin hydrate and byakangelicin, from *A. dahurica* was achieved in few steps. The novel developed approach considerably reduces the steps and time for isolating oxypeucedanin hydrate and byakangelicin from the roots of *A. dahurica*. Therefore, this approach alleviates the drawbacks of traditional methods. This study also proved the potentiality of IL as an effective solvent for the separation of bioactive components from plant materials. The facile and high accessibility to minor active components is assumed to boost research related to these compounds. Consequently, this inevitably increases their potential application as lead compounds and/or drug candidates in the pharmaceutical industry.

Author Contributions: Conceptualization and methodology, A.N.K., G.H. and K.J.; Analysis and validation, A.N.K., G.H., H.S.K. and K.J.; data curation A.N.K., Y.-G.S. and K.J.; writing—original draft preparation, A.N.K. and K.J.; writing—review and editing, A.N.K., H.S.K., Y.-G.S. and K.J.; supervision, K.J. All authors have read and agreed to the published version of the manuscript.

Funding: This work was supported by the Basic Science Research Program through the National Research Foundation of Korea (NRF) funded by the Ministry of Education (NRF-2020R1F1A1048888), and the Gyeonggi GRRC program (GRRC-CHA2017-A01).

Conflicts of Interest: The authors declare that they have no conflict of interest.

Sample Availability: Samples of the compounds are available from the authors.

References

1. Bai, Y.; Li, D.; Zhou, T.; Qin, N.; Li, Z.; Yu, Z.; Hua, H. Coumarins from the roots of *Angelica dahurica* with antioxidant and antiproliferative activities. *J. Funct. Foods* **2016**, *20*, 453–462. [CrossRef]
2. Deng, G.G.; Wei, W.; Yang, X.W.; Zhang, Y.B.; Xu, W.; Gong, N.B.; Lü, Y.; Wang, F.F. New coumarins from the roots of *Angelica dahurica* var. formosana cv. Chuanbaizhi and their inhibition on NO production in LPS-activated RAW264.7 cells. *Fitoterapia* **2015**, *101*, 194–200. [CrossRef]
3. Li, B.; Zhang, X.; Wang, J.; Zhang, L.; Gao, B.; Shi, S.; Wang, X.; Li, J.; Tu, P. Simultaneous Characterisation of Fifty Coumarins from the Roots of *Angelica dahurica* by Off-line Two-dimensional High-performance Liquid Chromatography Coupled with Electrospray Ionisation Tandem Mass Spectrometry. *Phytochem. Anal.* **2014**, *25*, 229–240. [CrossRef]
4. Liao, Z.-G.; Liang, X.-L.; Zhu, J.-Y.; Zhao, G.-W.; Yang, M.; Wang, G.-F.; Jiang, Q.-Y.; Chen, X.-L. Correlation between synergistic action of Radix *Angelica dahurica* extracts on analgesic effects of Corydalis alkaloid and plasma concentration of dl-THP. *J. Ethnopharmacol.* **2010**, *129*, 115–120. [CrossRef]
5. Yang, W.T.; Ke, C.Y.; Wu, W.T.; Tseng, Y.H.; Lee, R.P. Antimicrobial and anti-inflammatory potential of *Angelica dahurica* and *Rheum officinale* extract accelerates wound healing in *Staphylococcus aureus*-infected wounds. *Sci. Rep.* **2020**, *10*, 5596. [CrossRef]
6. Lee, M.Y.; Seo, C.S.; Lee, J.A.; Lee, N.H.; Kim, J.H.; Ha, H.; Zhen, M.-S.; Son, J.-K.; Shin, H.-K. Anti-asthmatic effects of *Angelica dahurica* against ovalbumin-induced airway inflammation via upregulation of heme oxygenase-1. *Food Chem. Toxicol.* **2011**, *49*, 829–837. [CrossRef]
7. Lee, K.; Shin, M.S.; Ham, I.; Choi, H.-Y. Investigation of the mechanisms of *Angelica dahurica* root extract-induced vasorelaxation in isolated rat aortic rings. *BMC Complement. Altern. Med.* **2015**, *15*, 395. [CrossRef]
8. Pervin, M.; Hasnat, M.D.; Debnath, T.; Park, S.R.; Kim, D.H.; Lim, B.O. Antioxidant, anti-inflammatory and antiproliferative activity of *Angelica dahurica* root extracts. *J. Food Biochem.* **2014**, *38*, 281–292. [CrossRef]
9. Marumoto, S.; Miyazawa, M. β-Secretase inhibitory effects of furanocoumarins from the root of *Angelica dahurica*. *Phytother. Res.* **2010**, *24*, 510–513. [CrossRef]
10. Liang, W.-H.; Chang, T.-W.; Charng, Y.-C. Influence of harvest stage on the pharmacological effect of *Angelica dahurica*. *Bot. Stud.* **2018**, *59*, 14. [CrossRef]
11. Chen, L.; Jian, Y.; Wei, N.; Yuan, M.; Zhuang, X.M.; Li, H. Separation and simultaneous quantification of nine furanocoumarins from *Radix Angelicae dahuricae* using liquid chromatography with tandem mass spectrometry for bioavailability determination in rats. *J. Sep. Sci.* **2015**, *38*, 4216–4224. [CrossRef] [PubMed]
12. Fan, G.; Deng, R.; Zhou, L.; Meng, X.; Kuang, T.; Lai, X.; Zhang, J.; Zhang, Y. Development of a rapid resolution liquid chromatographic method combined with chemometrics for quality control of *Angelica dahurica* radix. *Phytochem. Anal.* **2012**, *23*, 299–307. [CrossRef] [PubMed]
13. Li, F.; Song, Y.; Wu, J.; Chen, X.; Hu, S.; Zhao, H.; Bai, X. Hollow fibre cell fishing and hollow fibre liquid phase microextraction research on the anticancer coumarins of Radix *Angelicae dahuricae* in vitro and in vivo. *J. Liq. Chromatogr. Relat. Technol.* **2019**, *42*, 79–88. [CrossRef]
14. Yang, L.; Li, Q.; Feng, Y.; Qiu, D. Simultaneous Determination of Three Coumarins in *Angelica dahurica* by 1H-qNMR Method: A Fast and Validated Method for Crude Drug Quality Control. *J. Anal. Methods Chem.* **2020**, *2020*, 8987560. [CrossRef]
15. Wang, J.; Peng, L.; Shi, M.; Li, C.; Zhang, Y.; Kang, W. Spectrum Effect Relationship and Component Knock-Out in *Angelica Dahurica* Radix by High Performance Liquid Chromatography-Q Exactive Hybrid Quadrupole-Orbitrap Mass Spectrometer. *Molecules* **2017**, *22*, 1231. [CrossRef]

16. Amponsah, I.K.; Fleischer, T.C.; Dickson, R.A.; Annan, K.; Thoss, V. Evaluation of anti-inflammatory and antioxidant activity of Furanocoumarins and Sterolin from the stem bark of *Ficus exasperata* Vahl (Moraceae). *J. Sci. Innov. Res.* **2013**, *2*, 880–887.

17. Kang, Y.Y.; Song, J.; Kim, J.Y.; Jung, H.; Yeo, W.-S.; Lim, Y.; Mok, H. Byakangelicin as a modulator for improved distribution and bioactivity of natural compounds and synthetic drugs in the brain. *Phytomedicine* **2019**, *62*, 152963. [CrossRef]

18. Li, X.; Shao, S.; Li, H.; Bi, Z.; Zhang, S.; Wei, Y.; Bai, J.; Zhang, R.; Ma, X.; Ma, B.; et al. Byakangelicin protects against carbon tetrachloride–induced liver injury and fibrosis in mice. *J. Cell. Mol. Med.* **2020**, *24*, 8623–8635. [CrossRef]

19. Singh, S.K.; Savoy, A.W. Ionic liquids synthesis and applications: An overview. *J. Mol. Liq.* **2020**, *297*, 112038. [CrossRef]

20. Welton, T. Ionic liquids: A brief history. *Biophys. Rev.* **2018**, *10*, 691–706. [CrossRef]

21. Ahrenberg, M.; Beck, M.; Neise, C.; Keßler, O.; Kragl, U.; Verevkin, S.P.; Schick, C. Vapor Pressure of Ionic Liquids at Low Temperatures from AC-Chip-Calorimetry. *Phys. Chem. Chem. Phys.* **2016**, *18*, 21381–21390. [CrossRef]

22. Wu, H.B.; Zhang, B.; Liu, S.H.; Chen, C.C. Flammability estimation of 1-hexyl-3-methylimidazolium bis(trifluoromethylsulfonyl)imide. *J. Loss Prev. Process Ind.* **2020**, *66*, 104196. [CrossRef]

23. Villanueva, M.; Coronas, A.; García, J.; Salgado, J. Thermal Stability of Ionic Liquids for Their Application as New Absorbents. *Ind. Eng. Chem. Res.* **2013**, *52*, 15718–15727. [CrossRef]

24. Becherini, S.; Mezzetta, A.; Chiappe, C.; Guazzelli, L. Levulinate amidinium protic ionic liquids (PILs) as suitable media for the dissolution and levulination of cellulose. *New J. Chem.* **2019**, *43*, 4554–4561. [CrossRef]

25. Karmakar, A.; Mukundan, R.; Yang, P.; Batista, E.R. Solubility model of metal complex in ionic liquids from first principle calculations. *RSC Adv.* **2019**, *9*, 18506–18526. [CrossRef]

26. Keaveney, S.T.; Haines, R.S.; Harper, J.B. Ionic liquid solvents: The importance of microscopic interactions in predicting organic reaction outcomes. *Pure Appl. Chem.* **2017**, *8*, 745–757. [CrossRef]

27. Claus, J.; Sommer, F.O.; Kragl, U. Ionic liquids in biotechnology and beyond. *Solid State Ion.* **2018**, *314*, 119–128. [CrossRef]

28. Martins, V.L.; Torresi, R.M. Ionic liquids in electrochemical energy storage. *Curr. Opin. Electrochem.* **2018**, *9*, 26–32. [CrossRef]

29. Trujillo-Rodríguez, M.J.; Nan, H.; Varona, M.; Emaus, M.N.; Souza, I.D.; Anderson, J.L. Advances of Ionic Liquids in Analytical Chemistry. *Anal. Chem.* **2019**, *91*, 505–531. [CrossRef]

30. Tampucci, S.; Guazzelli, L.; Burgalassi, S.; Carpi, S.; Chetoni, P.; Mezzetta, A.; Nieri, P.; Polini, B.; Pomelli, C.S.; Terreni, E.; et al. pH-Responsive Nanostructures Based on Surface Active Fatty Acid-Protic Ionic Liquids for Imiquimod Delivery in Skin Cancer Topical Therapy. *Pharmaceutics* **2020**, *12*, 1078. [CrossRef] [PubMed]

31. Santos, M.M.; Alves, C.; Silva, J.; Florindo, C.; Costa, A.; Petrovski, Ž.; Marrucho, I.M.; Pedrosa, R.; Branco, L.C. Antimicrobial Activities of Highly Bioavailable Organic Salts and Ionic Liquids from Fluoroquinolones. *Pharmaceutics* **2020**, *12*, 694. [CrossRef]

32. Mushtaq, M.; Akram, S. Ionic Liquid for the Extraction of Plant Phenolics. In *Nanotechnology-Based Industrial Applications of Ionic Liquids*; Nanotechnology in the Life Sciences; Inamuddin, A.A., Ed.; Spring: Cham, Switzerland, 2020; pp. 81–97.

33. Bogdanov, M.G.; Svinyarov, I. Ionic liquid-supported solid–liquid extraction of bioactive alkaloids. II. Kinetics, modeling and mechanism of glaucine extraction from *Glaucium flavum* Cr. (Papaveraceae). *Sep. Purif. Technol.* **2013**, *103*, 279–288. [CrossRef]

34. Ma, W.; Lu, Y.; Hu, R.; Chen, J.; Zhang, Z.; Pan, Y. Application of ionic liquids-based microwave-assisted extraction of three alkaloids N-nornuciferine, O-nornuciferine, and nuciferine from lotus leaf. *Talanta* **2010**, *80*, 1292–1297. [CrossRef]

35. Wang, L.; Bai, M.; Qin, Y.; Liu, B.; Wang, Y.; Zhou, Y. Application of Ionic Liquid-Based Ultrasonic-Assisted Extraction of Flavonoids from Bamboo Leaves. *Molecules* **2018**, *23*, 2309. [CrossRef]

36. Tan, Z.; Yi, Y.; Wang, H.; Zhou, W.; Wang, C. Extraction, Preconcentration and Isolation of Flavonoids from *Apocynum venetum* L. Leaves Using Ionic Liquid-Based Ultrasonic-Assisted Extraction Coupled with an Aqueous Biphasic System. *Molecules* **2016**, *21*, 262. [CrossRef]

37. Lin, H.; Zhang, Y.; Han, M.; Yang, L. Aqueous ionic liquid based ultrasonic assisted extraction of eight ginsenosides from ginseng root. *Ultrason. Sonochem.* **2013**, *20*, 680–684. [CrossRef]

38. Kiyonga, A.N.; An, J.H.; Lee, K.Y.; Lim, C.; Suh, Y.G.; Chin, Y.W.; Jung, K. Rapid and Efficient Separation of Decursin and Decursinol Angelate from *Angelica gigas* Nakai using Ionic Liquid, (BMIm)BF4, Combined with Crystallization. *Molecules* **2019**, *24*, 2390. [CrossRef]

39. Martins, M.A.R.; Domańska, U.; Schröder, B.; Coutinho, J.A.P.; Pinho, S.P. Selection of Ionic Liquids to be Used as Separation Agents for Terpenes and Terpenoids. *ACS Sustain. Chem. Eng.* **2015**, *4*, 548–556. [CrossRef]

40. Ji, S.; Wang, Y.; Su, Z.; He, D.; Du, Y.; Guo, M.; Yang, D.; Tang, D. Ionic liquids-ultrasound based efficient extraction of flavonoid glycosides and triterpenoid saponins from licorice. *RSC Adv.* **2018**, *8*, 13989–13996. [CrossRef]

41. Dai, Y.; Spronen, J.; Witkamp, G.J.; Verpoorte, R.; Choi, Y.H. Ionic liquids and deep eutectic solvents in natural products research: Mixtures of solids as extraction solvents. *J. Nat. Prod.* **2013**, *76*, 2162–2173. [CrossRef]

42. Ventura, S.P.M.; e Silva, F.A.; Quental, M.V.; Mondal, D.; Freire, M.G.; Coutinho, J.A.P. Ionic-Liquid-Mediated Extraction and Separation Processes for Bioactive Compounds: Past, Present, and Future Trends. *Chem. Rev.* **2017**, *117*, 6984–7052. [CrossRef]

43. Mazzola, P.G.; Lopes, A.M.; Hasmann, F.A.; Jozala, A.F.; Penna, T.C.V.; Magalhaes, P.O.; Rangel-Yagui, C.O.; Pessoa, A. Liquid–liquid extraction of biomolecules: An overview and update of the main techniques. *J. Chem. Technol. Biotechnol.* **2008**, *83*, 143–157. [CrossRef]

44. Absalan, G.; Akhond, M.; Sheikhian, L. Extraction and high performance liquid chromatographic determination of 3-indole butyric acid in pea plants by using imidazolium-based ionic liquids as extractant. *Talanta* **2008**, *77*, 407–411. [CrossRef]

45. Larriba, M.; Omar, S.; Navarro, P.; Garcia, J.; Rodriguez, F.; Gonzalez-Miquel, M. Recovery of tyrosol from aqueous streams using hydrophobic ionic liquids: A first step towards developing sustainable processes for olive mill wastewater (OMW) management. *RSC Adv.* **2016**, *6*, 18751–18762. [CrossRef]
46. Nazet, A.; Sokolov, S.; Sonnleitner, T.; Makino, T.; Kanakubo, M.; Buchner, R. Densities, Viscosities, and Conductivities of the Imidazolium Ionic Liquids [Emim][Ac], [Emim][FAP], [Bmim][BETI], [Bmim][FSI], [Hmim][TFSI], and [Omim][TFSI]. *J. Chem. Eng. Data* **2015**, *60*, 2400–2411. [CrossRef]
47. Fan, W.; Zhou, Q.; Sun, J.; Zhang, S. Density, Excess Molar Volume, and Viscosity for the Methyl Methacrylate +1-Butyl-3-methylimidazolium Hexafluorophosphate Ionic Liquid Binary System at Atmospheric Pressure. *J. Chem. Eng. Data* **2009**, *54*, 2307–2311. [CrossRef]
48. Yang, L.; Wang, H.; Zu, Y.-G.; Zhao, C.; Zhang, L.; Chen, X.; Zhang, Z. Ultrasound-assisted extraction of the three terpenoid indole alkaloids vindoline, catharanthine and vinblastine from *Catharanthus roseus* using ionic liquid aqueous solutions. *Chem. Eng. J.* **2011**, *172*, 705–712. [CrossRef]
49. Passos, H.; Freire, M.G.; Coutinho, J.A.P. Ionic liquid solutions as extractive solvents for value-added compounds from biomass. *Green Chem.* **2014**, *16*, 4786–4815. [CrossRef]
50. Naert, P.; Rabaey, K.; Stevens, C.V. Ionic liquid ion exchange: Exclusion from strong interactions condemns cations to the most weakly interacting anions and dictates reaction equilibrium. *Green Chem.* **2018**, *20*, 4277. [CrossRef]
51. Ishihara, K.; Fukutake, M.; Asano, T.; Mizuhara, Y.; Wakui, Y.; Yanagisawa, T.; Kamei, H.; Ohmori, S.; Kitada, M. Simultaneous determination of byak-angelicin and oxypeucedanin hydrate in rat plasma by column-switching high-performance liquid chromatography with ultraviolet detection. *J. Chromatogr. B* **2001**, *753*, 309–314. [CrossRef]

Article

α-Terpineol: An Aggregation Pheromone in *Optatus palmaris* (Coleoptera: Curculionidae) (Pascoe, 1889) Enhanced by Its Host-Plant Volatiles

José Manuel Pineda-Ríos [1], Juan Cibrián-Tovar [1,*], Luis Martín Hernández-Fuentes [2], Rosa María López-Romero [3], Lauro Soto-Rojas [1], Jesús Romero-Nápoles [1], Celina Llanderal-Cázares [1] and Luis F. Salomé-Abarca [1,*]

[1] Postgrado en Fitosanidad, Programa de Entomología y Acarología, Colegio de Postgraduados Campus Montecillo, Km 36.5 Carretera, Texcoco 56230, Mexico; pinedarmanuel@gmail.com (J.M.P.-R.); rojo@colpos.mx (L.S.-R.); jnapoles@colpos.mx (J.R.-N.); llcelina@colpos.mx (C.L.-C.)

[2] Instituto Nacional de Investigaciones Forestales, Agrícolas y Pecuarias, Progreso Número 5, Barrio de Santa Catarina, Delegación Coyoacán, Ciudad de México 04010, Mexico; hernandez.luismartin@inifap.gob.mx

[3] Postgrado en Edafología, Colegio de Postgraduados Campus Montecillo, Km 36.5 Carretera, Texcoco 56230, Mexico; rosal@colpos.mx

* Correspondence: jcibrian@colpos.mx (J.C.-T.); luis.salome@colpos.mx (L.F.S.-A.); Tel.: +52-155-383-54600 (J.C.-T.); +52-175-810-86324 (L.F.S.-A.)

Citation: Pineda-Ríos, J.M.; Cibrián-Tovar, J.; Hernández-Fuentes, L.M.; López-Romero, R.M.; Soto-Rojas, L.; Romero-Nápoles, J.; Llanderal-Cázares, C.; Salomé-Abarca, L.F. α-Terpineol: An Aggregation Pheromone in *Optatus palmaris* (Coleoptera: Curculionidae) (Pascoe, 1889) Enhanced by Its Host-Plant Volatiles. *Molecules* **2021**, *26*, 2861. https://doi.org/10.3390/molecules26102861

Academic Editor: Angelo Canale

Received: 31 March 2021
Accepted: 6 May 2021
Published: 12 May 2021

Abstract: The Annonaceae fruits weevil (*Optatus palmaris*) causes high losses to the soursop production in Mexico. Damage occurs when larvae and adults feed on the fruits; however, there is limited research about control strategies against this pest. However, pheromones provide a high potential management scheme for this curculio. Thus, this research characterized the behavior and volatile production of *O. palmaris* in response to their feeding habits. Olfactometry assays established preference by weevils to volatiles produced by feeding males and soursop. The behavior observed suggests the presence of an aggregation pheromone and a kairomone. Subsequently, insect volatiles sampled by solid-phase microextraction and dynamic headspace detected a unique compound on feeding males increased especially when feeding. Feeding-starvation experiments showed an averaged fifteen-fold increase in the concentration of a monoterpenoid on males feeding on soursop, and a decrease of the release of this compound males stop feeding. GC-MS analysis of volatiles identified this compound as α-terpineol. Further olfactometry assays using α-terpineol and soursop, demonstrated that this combination is double attractive to Annonaceae weevils than only soursop volatiles. The results showed a complementation effect between α-terpineol and soursop volatiles. Thus, α-terpineol is the aggregation pheromone of *O. palmaris*, and its concentration is enhanced by host-plant volatiles.

Keywords: terpenoid; signaling; chemical cue; kairomone; potentiation

1. Introduction

Most of the members of the Curculionidae family, except for Platypodinae and Scolytinae, are called weevils. This term comes from their characteristic long snouts and capitate antennae with small clubs [1]. When not in use, the antennae are stored in grooved cavities along the snout. As mentioned, these insects possess a long rostrum (snout) with mouthparts at the end of it. [1,2]. Their distinctive snout allows feeding on various plant organs like roots, stems, leaves, flowers, and fruits [1]. Because of the damage caused to diverse plant species, several weevils are considered economically important agronomic pests [1,3–5]. In addition to damage from their feeding behavior, female weevils could oviposit in holes left after feasting on the fruits; thus, new-offspring will feed on the fruit mesocarp and seeds, causing their detachment [6].

In Mexico, several curculios such as *Scyphophorus acupunctatus*, *Rhynchophorus palmarum*, *Anthonomus eugenii*, and *Anthonomus grandis* are considered principal crop pests, and they have been extensively studied; these insects attack agave, coconut palm, chili pepper, and cotton plants, respectively [7–10]. Weevil pests also attack soursop (*Annona muricata* L.), another important Mexican agronomic product. Mexico is the biggest soursop producer worldwide [11, 12]; however, the Annonaceae fruits weevil (*Optatus palmaris* Pascoe) causes significant losses in this crop. This weevil feeds on young leaf buds and flowers, causing detachment from the tree [6]. Annonaceae fruits weevils also feed on fruits, preferring ripe fruits if available, causing external damage to up to ca. 40% of the fruit surface. Additionally, larvae instar feed from the inside of the fruit, destroying the mesocarp and the seeds of the soursop [6]. The weevil attack can also cause detachment of small fruits from the tree [6,13,14].

The life cycle of the Annonaceae fruits weevil last around 215 days; five days as egg, 73 days as larva, 25 days as pupa, and 112 days as an adult. In Mexico, their activity period is reported to be between 8:30 h and 23:00 h for adult specimens [15]. However, another report mentioned that the activity period of this weevil is restricted between 10:00 h and 18:00 h in the field [16]. The presence of adults in the field starts with the beginning of the rainy season (August–November). Once in the field, the adults feed and copulate on ripening soursop fruits, especially close to their harvest point. It is common to find up to 30 adults per fruit or even more depending in the size of the fruit. On the other hand, when they are present in leaf buds and flowers, usually only one specimen is found in those organs. When not copulating, the females mainly dedicate to feed and oviposit. Conversely, the males go to the treetop to rest and feed on young leaf buds and flowers before copulating and feeding on soursop fruits again [15]. Other Annonaceae fruits weevils' alternative host-plants are reported to be cherimoya (*Annona cherimola* Mill), ilama (*Annona macroprophyllata* Donn. Sm.), and custard apple (*Annona reticulata* L.) [17].

Due to its recent detection and fast spread, research on the Annonaceae fruits weevil control is limited to some insecticides rotating use [18]. However, this strategy is not enough to fully control the weevil population level and, thus, the damage caused by them. Typically, effective monitoring and control of weevils include the use of insect traps baited with their aggregation pheromones [19]. The use of semiochemicals could potentially handle or even fully control Annonaceae fruits weevil infestations. However, research about pheromones or kairomone does not exist for this insect species.

Therefore, the feeding behavior of males and females of the anonacea weevil was characterized. The characterization consisted of behavioral assays carried out by olfactometry. Subsequently, chemical analyzes were carried out to identify candidate volatiles functioning as chemical cues. We expected results to provide insight on volatiles mediation of the feeding behavior of the Annonaceae fruits weevil adults. Moreover, the obtained data may serve as the basis for establishing volatile-based management of this pest.

2. Results

The Annonaceae fruits weevil has a long-life cycle, and an artificial diet for its rearing is lacking. Therefore, this research employed wild insects. On the first experimental stage, olfactory two-choice preference experiments were carried out in males and females. Before starting any odor testing, the air in both olfactometer arms was used to check no specific selection by male or female weevils existed. In general, only 6–13% (1 to 2 weevils out of 15) chose an olfactometer arm; the rest showed no response to air stimulus. Similar results were observed for the carrier solvent (hexane) evaporation tests. These selection trials indicated no position, light, or remnant volatile or solvent effects over the behavior of the tested weevils. After the blank test, different odor sources vs. air were used as preference tests separately in male and female weevils. Treatments included pieces of soursop, only non-feeding males, only non-feeding females, males feeding on soursop (feeding males), and females feeding on soursop (feeding females). Results showed that feeding males produced the most attractive aroma for male weevils with 94.99% selection ($F_{4,15} = 11.77$, $p < 0.001$). On the other hand, the preference of males towards soursop and feeding females showed non-significant statistical differences ($F_{4,15} = 11.77$, $p < 0.001$) among treatments (Figure 1). Conversely, females' preference towards feeding males, soursop, and feeding-female volatiles showed non-significant statistical differences among them. That is, female weevils did not display a specific behavior for each tested volatile source. Finally, both males and females showed less preference by volatile emitted by non-feeding females ($F_{4,15} = 11.77$, $p < 0.001$; $F_{4,15} = 26.44$, $p < 0.0001$, respectively).

Based on this information, in a second experimental stage, both males' and females' preference to soursop volatiles was challenged against volatiles from feeding males and females. Moreover, both sexes' preference to volatiles simultaneously emitted by feeding males and feeding females from two different emission arms was tested. Finally, the preference for non-feeding males and non-feeding females was evaluated too. The preference of aroma of feeding males was highly significant (G = 22.64, $p < 0.0001$ by males and G = 48.50, $p < 0.0001$ by females) by both sexes (Figure 1). Females displayed marked preference to the treatments containing male volatiles, especially to those of feeding males. Even so, males did not display significantly differentiated preference when both feeding females and feeding males are simultaneously tested (G = 0.99, $p = 0.31$) (Figure 1). Thus, the higher preference for feeding males and soursop aromas observed in the first and second olfactometry assays demonstrated the presence of an aggregation pheromone and a kairomone which could parallel mediate the feeding behavior of the Annonaceae fruits weevil. All of the non-responding insect values in each experimental set did not show significant differences when analyzed by an ANOVA test. This might mean that the no response behavior of some weevils are related to some intrinsic biological factors of each insect rather than the strength of the response evoked by every odor source (Figure 1).

Based on these results, volatiles emitted by soursop, only males, only females, and feeding females were used as background chromatograms to scrutinize volatiles specifically produced when males feed on soursop, which could potentially function as aggregation pheromone. The volatile fractions were trapped by solid phase microextraction (SPME) and dynamic headspace (DH). Comparing the GC-MS profiles of all of the samples showed the presence of a chromatographic peak exclusively found in males at 10.34 min (Figure 2). Remarkably, this peak area increased from five to fifty times in some individual cases and resulted in a 15-fold average increment of the content of this compound when males fed on soursop (Figure 2). The chromatographic peak was identified as α-terpineol [2-(4-methylcyclohex-3-en-1-yl)propan-2-ol] by matching its spectrum (95%) with that of the NIST library.

Compound identification was also supported by the interpretation of mass losses when looking at the mass spectrum of such metabolite. The main m/z values were 139, 136, 121, 93, 81, 67, and 59. Based on these values, a total loss of the parent and molecular ions was determined. Furthermore, the m/z 139 corresponded to M-15, which is explained by the loss of a methyl group from the molecule. The m/z 136 corresponded to M-18 explained by a water molecule loss typical in alcohols, in this case, α-terpineol. This fact was also supported by m/z 121 (M-33), which indicated the loss of a methyl group and a water molecule typical in alcohols with a methyl group in its chain aliphatic side. The m/z 93 indicated the formation of a $C_7H_9^+$ ion which is a typical mass for terpenoids [20,21]. The m/z 81 was explained by the loss of $C_4OH_{10}^+$, potentially followed by a Mclafferty rearrangement. The m/z 69 value was also explained by the loss of all radicals bounded to the cyclohexene ring of α-terpineol, followed by its rupture through a retro Diels–Alder reaction. The base peak corresponded to m/z 59, representing a $C_3H_7O^+$ fragment, which is also a typical mass for alcohols [21] (Figure 3a). Finally, the chromatographic peak's retention time and mass spectrum were compared with those of a standard compound (Figure 3b).

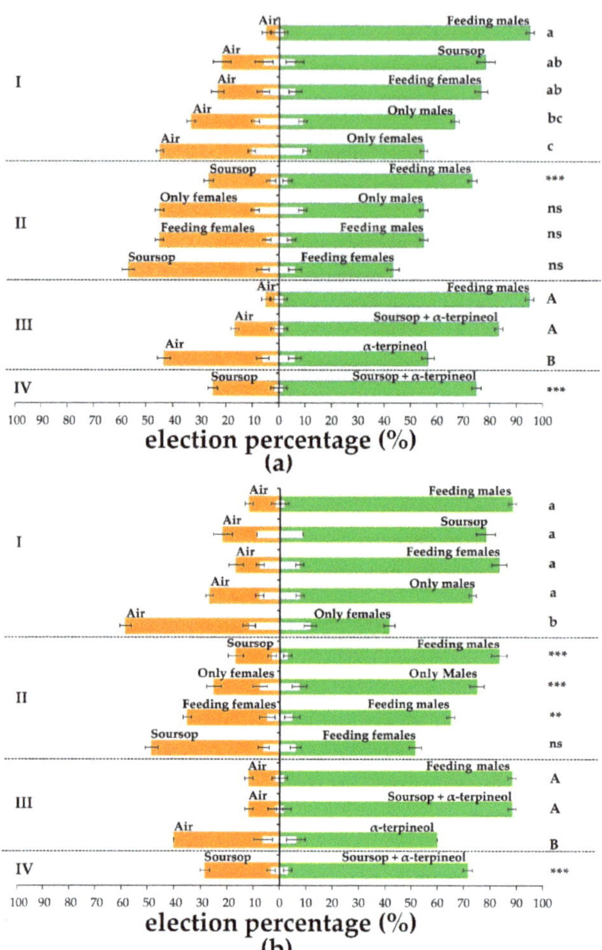

Figure 1. Olfactory preference of males and females of *Optatus palmaris* to different odors as single or double stimulus sources. (**a-I**), male preferences to single stimulus source vs. air. (**a-II**), male preference when two odor source are challenged against each other. (**a-III**), male preference to one stimulus source involving α-terpineol against air. (**a-IV**), male preference to soursop challenged against soursop supplemented with α-terpineol. Data represent the selection percentage of males to an odor stimulus (*n* = 4). Horizontal black bars indicate standard error values. Treatments with the same letters are not significantly different in a (**a-I**) Tukey test (*p* < 0.05), (**a-II**) G-test (G < 0.05), *** < 0.0001, (**a-III**) Tukey test (*p* < 0.05), and (**a-IV**) G-test (G < 0.05), *** < 0.0001. (**b-I**), female preferences to single stimulus source vs. air. (**b-II**), female preference when two odor source are challenged against each other. (**b-III**), female preference to one stimulus source involving α-terpineol against air. (**b-IV**), female preference to soursop challenged against soursop supplemented with α-terpineol. Data represent the selection percentage of males to an odor stimulus (*n* = 4). Horizontal black bars indicate standard error values. Treatments with the same letters are not significantly different in a (**b-I**) Kruskal–Wallis/Bonferroni test (*p* < 0.05), (**b-II**) G-test (G < 0.05), ** < 0.001, *** < 0.0001, (**b-III**) Kruskal-Wallis/Bonferroni test (*p* < 0.05), and (**b-IV**) G-test (G < 0.05), *** < 0.0001. The statistical analysis was performed separately for males' and females' data. White bars represent average values (*n* = 4) of non-responding insects registered in each preference test. Horizontal black bars indicate standard error values. The non-responding insect values were compared for the first three sections (I, II, and III) with an ANOVA test. There were no statistical differences among non-responding insect values in any set of experiments analyzed (*p* > 0.05).

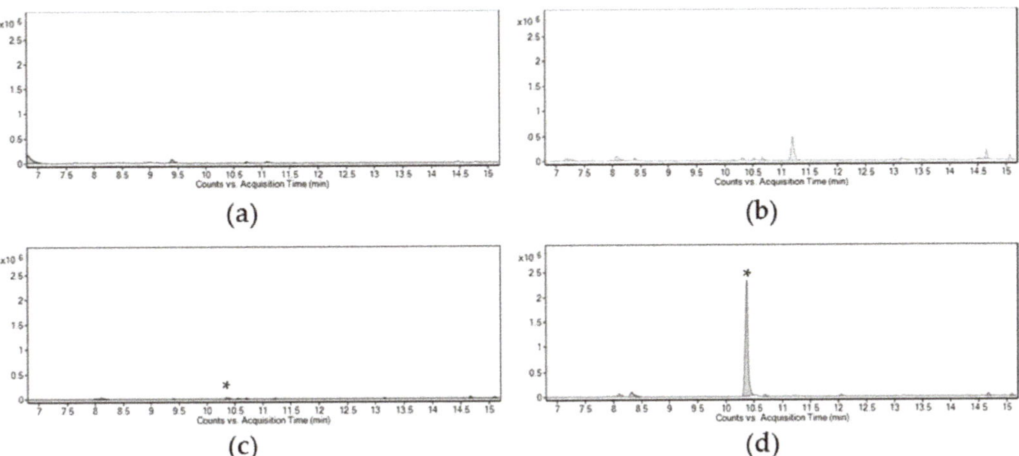

Figure 2. Gas chromatography-mass spectrometry profiles of volatile blends produced by different odor sources. (**a**), soursop, (**b**), insect females feeding on soursop, (**c**), non-feeding males, and (**d**), males feeding on soursop. * indicates the presence of an exclusive peak at 10. 34 min found in males and increased when males feed on soursop.

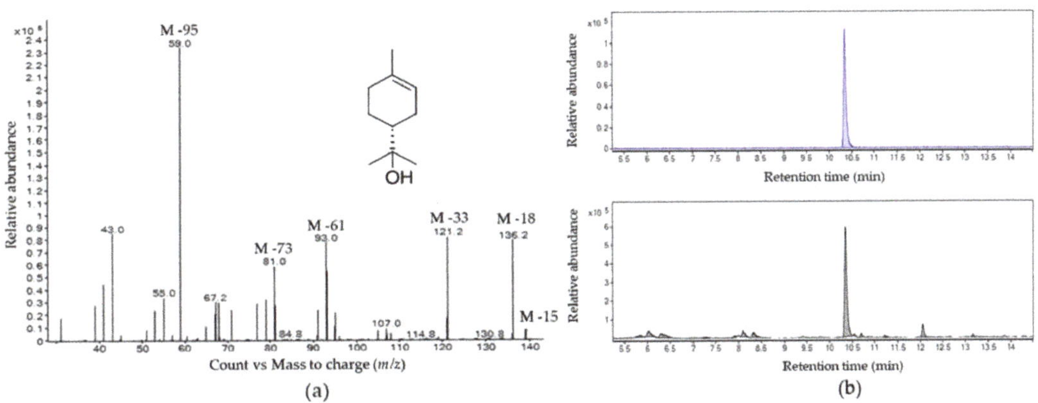

Figure 3. Gas-chromatography-mass spectrometry identification of α-terpineol (**a**), mass spectrum obtained from a chromatographic peak of the volatile blend produced by males feeding on soursop. The matching factor with the library spectrum (NIST 2014) of this metabolite was 95%. (**b**), Retention time comparison between chromatographic peaks of a standard compound (upper chromatogram) and a volatile blend produced by males feeding on soursop (lower chromatogram).

To further correlate the release of α-terpineol with the feeding behavior of the Annonaceae fruits male weevils, its concentration in the headspace was quantified. Thus, matrix recovering effects were determined in order to achieve proper quantitative data. Briefly, some paper disks were loaded with a known concentration of the volatile, then trapped, extracted and quantified. Moreover, the same process was performed with Tenax® cartridges directly loaded with the same compound. The average α-terpineol content of the filter paper extracts was 14.89 ± 2.61 ng/μL, and 14.29 ± 3.18 ng/μL for the cartridges directly loaded with the same amount of α-terpineol. There were no significant differences between these values ($t > 0.05$). Thus, the matrix recovering effect experiments showed that all of the α-terpineol originally loaded in the disks filter papers (18 μg) released to the headspace and moved into the cartridge. However, the concentration of the filter papers extracts (14.89 ± 2.61 ng/μL) and its total extract volume (350 μL), indicate that only 5.21 μg of α-terpineol were recovered from the cartridge under the used elution conditions. That is, 71% of the total amount of this metabolite is kept in the dynamic headspace cartridge. Thus, every initial quantification value was compensated with 71% of its mass.

Subsequently, a set of reversal feeding experiments (feeding-starvation) were performed. Briefly, the production of α-terpineol was determined in males initially feeding on soursop. After 48 h, the food was removed and the α-terpineol production was quantified again. The same experiment was simultaneously performed with males without food and fed with soursop after 48 h. These experiments allowed to characterize the release of α-terpineol in response to the feeding habits of the Annonaceae fruits male weevils. Interestingly, as previously observed, the release of α-terpineol was higher in the initial feeding-males. On the other hand, the non-feeding insects increased this volatile production when fed and, conversely, the initial feeding males reduced their α-terpineol release when food was removed. Regardless of the total amount of this compound produced by non-feeding males fed after a starvation period, the release of this volatile increased 87 ± 3.87% in all the experiments (Figure 4). The same decrease ratio was observed when food was removed from initial feeding males (Figure 4).

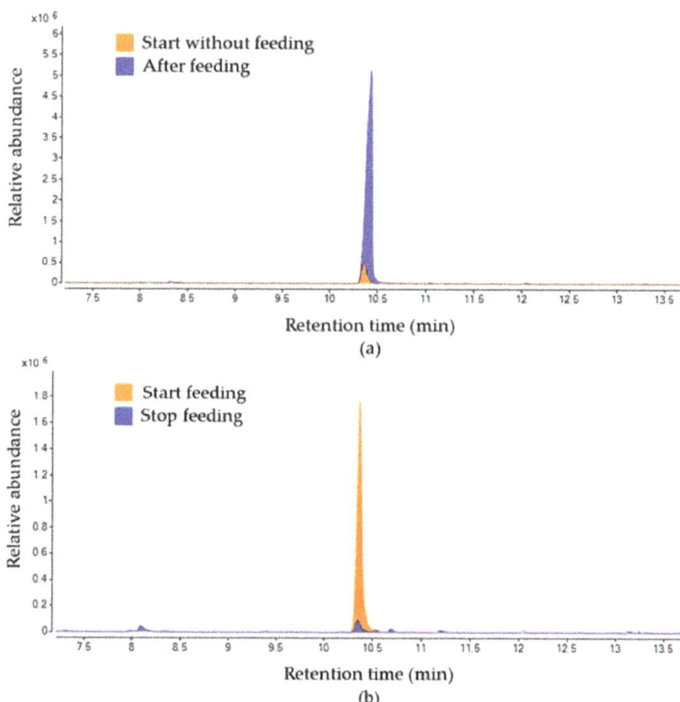

Figure 4. Representative chromatograms of the feeding and starvation effects over the α-terpineol release by *Optatus palmaris*. (**a**), Increase of the production of α-terpineol after *O. palmaris* males feed on soursop. (**b**), Decrease of the production of α-terpineol after *O. palmaris* males stopped feeding on soursop.

Considering the total amount of males (40) used in the bioassays to get the averaged α-terpineol content in the headspace, a release rate of 1.83 ± 0.36 ng per 40 males per second was determined. When considering the production per one equivalent male, a release ratio of 0.05 ± 0.01 ng per equivalent male per second was determined. Thus, solutions ranging between 0.5 and 10 ng/μL and 100 ng/μL were tested in two-choice olfactometry experiments. The range 0.5–2 ng/µL showed no clear selection pattern neither by male nor female weevils. The range 3–5 ng/µL showed attraction to both male and female weevils. Above these concentrations, a repellent effect took place for both sexes, especially at 100 ng/µL. Because of this, all following preferences experiments with this metabolite used solutions at 4 ng/µL. Intriguingly this volatile compound showed only 56.66% and 60% of attraction to males and females, respectively. Therefore, even if all evidence showed this volatile as a metabolite related to the feeding behavior of the Annonaceae fruits weevil, its not-very-high attractiveness indicates rather the presence of other possible volatile compounds potentiating its attraction effects. This volatile compound(s) could be produced either by the insect or by its host-plant. However, the chromatographic scrutiny of the analyzed samples did not show any other potential compound working as a secondary component of the aggregation pheromone of the Annonaceae fruits weevil. Thus, the whole volatile bouquet of soursop was supplemented with 4 ng/µL of α-terpineol, which simulated the effect of feeding males on this fruit. The combination of the soursop volatiles and this metabolite resulted in the increased attraction of males (83.33%) and females (88.33%) to this odor source. Moreover, these values are comparable to those obtained from feeding males showing non-significant statistical differences between them ($p > 0.05$) (Figure 1). Furthermore, when the aroma of soursop and soursop supplemented with α-terpineol were challenged against each other, the latter was at least two times more attractive to both male (74.99%) and female weevils (71.66%) (Figure 1).

The GC-MS analysis of the volatiles fraction trapped by dynamic headspace and solid phase microextraction from soursop fruits, detected more than 200 chromatographic peaks. Unfortunately, just few of them showed a roughly good identification match with the NIST library (>75%). A total of 16 compounds were identified including mono- and sesquiterpenoids, aldehydes, fatty acids, a ketone, and an alcohol. Nonetheless, only three compounds, D-limonene, β-caryophyllene, and nonanal were annotated by comparison with standard compounds (Supplementary Table S1).

3. Discussion

In nature, chemical messengers constitute the language of intra- and inter-kingdom communication [22]. Several of these chemical cues possess a volatile character, producing physiological or behavioral responses in the receptor organism [23]. Volatiles that determine intraspecific interactions are labeled pheromones. Depending on the behavioral outcome, they can be categorized as sexual, aggregation, alarm, attack, or epideictic pheromones, among others [24]. Currently, these volatile compounds, when used in combination with traps, efficiently and effectively detect, monitor, and control insect pests [25].

In an era where resistance to conventional control strategies and chemicals by microbes and insects increases continuously, alternatives become essential to ensure food safety. There are several successful cases of control in this context, especially in lepidopterans and curculios insects, with semiochemicals [9,26]. In this research, volatile compounds mediating the feeding behavior of the Annonaceae fruits weevil (*Optatus palmaris*) were scrutinized for their potential use in managing this insect pest. The first experimental stage showed high preference by male and female weevils to the aroma of soursop and males feeding on this fruit. Our research showed weevils use soursop volatiles to find food based on the attraction to soursop aroma; this behavior resembles behavior caused by kairomone odor [27]. However, host-plant volatile compounds are not the only cue used for food or mate location, as observed in some weevil species. For example, male-produced aggregation pheromones cause attraction and grouping of males and females of the same species to the food source for feeding and mating [19]. The same behavior was recorded in the olfactometry assays as feeding males stimulated higher aroma preference, particularly by male weevils, compared to any other odor source (Figure 1). Lower preference for feeding females further corroborated the results. The weevil attraction to this treatment could result from soursop volatiles only, as showed by the comparison of attraction values of feeding females and soursop tested against air (Figure 1). In the case of females, even if non-significant differences were observed between preferences by feeding males or only soursop volatiles, attraction to feeding males showed a trend to be more preferred. However, the observed values also indicated females possess high host volatiles usage to locate food. In this regard, it has been reported that female weevils locate their host-plant using various volatile cues, including kairomone and aggregation pheromones [28,29].

The observed insect behavior suggested the convenience of volatile chemical analysis. In several weevil species, aggregation pheromones are molecules containing ten carbons or less [7–9]. Because of this, solid-phase microextraction (SPME) polydimethylsiloxane (PDMS) and a carboxen/divinylbenzene in PDMS fibers were chosen to capture volatiles in the 40–275 g/mol range. Dynamic headspace (DH) experiments complemented the SPME data. The chromatographic analyses of all volatile fractions provided one distinctive chromatographic peak, highly amplified only in feeding males. That is, a low-concentration metabolite produced by groups of non-feeding males averaged a 15-fold increase when males fed on soursop. This metabolite was identified as α-terpineol, a common plant monoterpenoid.

It has been reported that some insects can hijack host-plant metabolites for direct or indirect defense or communication molecules [30]. However, GC-MS analyses did not detect α-terpineol (limit of detection = 2.10 ng/µL), in the volatile makeup of soursop. The production of this metabolite by the soursop fruits after damage caused by feeding of the weevils was discarded as the chromatographic profiles of the females feeding on soursop also did not show α-terpineol. Therefore, de novo synthesis by the Annonaceae fruits weevil must occur. Compared to the thousands of terpenoids detected in plants, these metabolites are found only in nine orders of insects [31]. Thus, the total amount of insect-produced terpenoids represent less than 1% of all terpenes found in nature [32]. Coleoptera to which the anonacea weevil belongs is one of these nine orders capable of synthesizing *de novo* terpenoids [31]. Specific examples of insects producing terpenoids as aggregation pheromones components in the Curculionidae family are *Anthonomus eugenii*, *Anthonomus musculus*, *Anthonomus rubi*, and *Anthonomus grandis* [33–35]; these insects attack chili pepper, cranberry, strawberry flowers, and cotton bolls, respectively. Their aggregation pheromones contain terpenoids such as geranic acid, geraniol, grandlure II, III, IV, and lavandulol [33–35]. Furthermore, *Curculio caryae* uses some of these compounds at varying ratios, as aggregation pheromone [36]. To our knowledge, α-terpineol is present only in the cerambycid beetle *Megacyllene caryae* as one of the seven components of its aggregation pheromone [37,38]. Hence, this is the first report α-terpineol is potentially produced as a unique aggregation pheromone component in the Curculionidae family.

Additionally, a set of reverse feeding experiments showed an increased-decreased release relationship of α-terpineol with the feeding behavior of the Annonaceae fruits weevil. These results further supported this metabolite acts as an aggregation pheromone [19,39,40] (Figure 4). Moreover, during insect collections, observations in the field also confirmed that soursop serves as a copulation and feeding site for this weevil. This in-field behavior corresponds to the typical behavioral pattern of insects under an aggregation pheromone effect [19].

However, α-terpineol showed no strong attraction effect to males or females when individually tested against air. The attraction values of this volatile were around 60% for both sexes. Moreover, the tested solution acted as a repellent when the concentration increased above 10 ng/µL. The repellent effect was powerful at 100 ng/µL: all specimens, independently of the sex, selected the air arm of the Y-olfactometer. This behavior does not indicate a preference for air, but rather, it suggests an escape from the concentrated airstream with α-terpineol. The concentration-related anti-aggregation effect of α-terpineol may work as follows: when there is not enough space for feeding and mating because of high insect density in a fruit or tree, a high α-terpineol concentration from the cluster of weevil males signals other conspecifics the lack of space availability [19]. Anti-aggregation effects mitigate intraspecific competition for food and mating [41]. Thus, using a single compound, which role depends on its concentration, represents an advantageous, energy-saving strategy.

The observed, relatively low attraction effect of α-terpineol (60%) indicated that the aggregation effect signaled by feeding males depended on other volatile compounds too. As previously mentioned, GC-MS analyses discarded another compound's possibility complementing the attraction effect of the aggregation pheromone. The combination of α-terpineol with soursop volatiles explored a second possibility. The blend attracted around 40% more than only α-terpineol when challenged against air. Moreover, in a dual choice assay, soursop volatiles supplemented with α-terpineol challenged against only soursop volatiles were twice as attractive to both male and female weevils.

Previous research reported that host-plant volatiles enhance some insect pheromones' activity [42]. In this regard, interaction of volatile compounds from host-plant and insects and its repercussion in the resulting attractiveness have been considerably studied in Coleoptera [31]. For example, boll weevil (*A. grandis*) captures in traps increased significantly when its aggregation pheromone was combined with *trans*-2-hexen-1-ol, *cis*-3-hexen-1-ol, or 1-hexanol; compounds emitted by cotton plants [43]. The same effect happened in traps baited with the plum curculio aggregation pheromone and benzaldehyde, a fruit volatile [44,45]. Similarly, host-plant volatiles synergized the response of the Asian palm weevil (*Rhynchophorus ferrugineus*), the South American palm weevil (*Rhynchophorus palmarum*), and the agave weevil (*Scyphophorus acupunctatus*) [46,47]. These effects are not limited to agronomic crop pests but also extended to forest insects. The captures of traps baited with the aggregation pheromone of *Dendroctonus ponderosae* increased highly (5- to 13-fold) when combined with pine volatiles [48].

Hence, we confirmed α-terpineol as a volatile compound linked to the feeding behavior of males of the Annonaceae fruits weevil. Furthermore, this volatile is used as an aggregation pheromone, which host-plant volatiles enhance its effect. Further field trials corroboration might be required, but these results highlight the potential of using semiochemicals associated with the feeding behavior of the Annonaceae fruits weevil for its management and control.

However, even if the GC-MS analysis of the volatile fraction of soursop fruits detected more than 200 chromatographic peaks, just few of them showed a good identification match with the NIST library. Only three compounds, D-limonene, β-caryophyllene, and nonanal were annotated by comparison with standard compounds (Supplementary Table S1). Furthermore, there was no clear criteria of selection for specific soursop volatile selection for further combinatorial experiments. Moreover, even if identified, D-limonene, and β-caryophyllene are general compounds found in several plant species [49,50], which make them not really soursop specific compounds. These findings suggest the need for developing a specific approach for the characterization of the Annonaceae fruits weevil's kairomone. This, specially to determine specific compounds and ratios which enhance the aggregation pheromone attractiveness.

4. Materials and Methods

4.1. Insect Collection and Species Corroboration

The insects were collected at the municipality of Compostela, Nayarit, México (21°6′17.337″, −105°9′49.917″; 21°6′2.808″, −105°10′40.872″; 21°6′3.8874″, −105°9′48.492″; 21°6′13.572″, −105°9′52.2″). *Optatus palmaris* males and females were manually collected during August 2019 and September 2020. The insects were collected and transported on hermetic plastic bottles (1 L). The bottles were pierced to allow gas exchange for insects' oxygenation. Subsequently, the specimens were sexed according to their morphological features [51]. The weevils were separated by sex and kept separated in entomological cages. They were fed with soursop pieces, and their rearing conditions were set to 25 ± 2 °C, relative humidity between 60 and 70%, and a photoperiod of 12:12 (L:D). The species corroboration was made using morphological characters previously reported [51]. The identified specimens were deposited at the Insect Collection of the Colegio de Postgraduados (CEAM).

4.2. Volatile Capture

Volatile compounds were trapped by solid-phase microextraction (SPME) and dynamic headspace (DH). In both techniques, borosilicate flasks were employed. Material cleaning was performed by subsequent washes with 2% Extran®, distilled water, acetone, and left to dry in a fume hood. Afterward, the material was heated at 300 °C for 3 h.

4.2.1. Solid-Phase Microextraction

Groups of 40 insects, only males or only females, were placed in 250 mL SPME flasks. Polydimethylsiloxane (PDMS/100 μm) and Divinylbenzene/Carboxen/PDMS (50/30 μm) fibers were used to trap volatiles. Both fibers were 1 cm long. Before volatile collection, the SPME fibers were cleaned in the injection port of a gas chromatograph (GC) (HP-6890, Midland, ON, Canada) at 250 °C. Subsequently, the SPME fiber was introduced through the flask septa, which contained a metallic mesh envelope to avoid contact between insects and the fiber. The distance between the SPME fiber and the insects was 2 cm. The volatile collection time was 1 h. Subsequently, the fiber was removed and desorbed in the GC injection port at 250 °C for 2 min for chromatographic analysis. The volatile samples were taken from groups of only males, only females, males feeding on 3 g of soursop, females feeding on 3 g of soursop, and only soursop. Blanks contained volatiles trapped from empty SPME flasks. Ten replicates were performed for each group.

4.2.2. Dynamic Headspace

Groups of 40 insects, only males, only females, males feeding on 8 g of soursop, females feeding on 8 g of soursop, and only soursop were placed in 650 mL Drechsel gas washing bottles (PIREX®, glendla, AZ, USA, EE.UU). Airflow was provided by an air pump (ELITE 802®, Colchester, VT, USA) attached to volatile-free PVC tubing (Nalgene®, Rochester, NY, USA, 180 PVC, 3/16″ ID). The airstream was filtered with a 50 mg Tenax® Baltimore, MD, USA, (60/80) cartridge. The airstream was humidified with a soft bubbling of distilled water. The airflow entering the system was 0.330 L/min controlled with a flowmeter (GILMONT®, London, UK) and calibrated with a manual glass flowmeter (Hewlett-Packard). The volatiles carried out by the airstream were finally captured in a cartridge containing 50 mg of Tenax® (60/80) as adsorbent and 20 mg of glass wool at each end of the cartridge. All of the cartridges were previously washed with 5 mL of hexane and left to dry in a fume hood, to be then heated at 300 °C for 3 h. The volatile collections were carried out for 48 h at the same temperature, humidity, and light conditions as the rearing conditions. Once the volatile collection time was reached, cartridges were eluted with 350 μL of HPLC-grade hexane. Then, 50 μL of hexane were added to increase the recovered volume. Blanks consisted of the elution of cartridges connected to empty bottles. Ten replicates were performed for each group.

4.3. Gas Chromatography Coupled to Mass Spectrometry (GC-MS)

The volatile extracts were analyzed with an HP-6890 gas chromatograph coupled to an HP-5973 single quadrupole mass spectrometry (MS) detector. The GC-MS system was equipped with an HP-5MS column (30 m × 0.250 mm ID, and 0.25 μm stationary phase thickness, J&W Science, Folsom, CA, USA) for sample separation. Helium (99.999% purity) was used as carrier gas at a flow rate of 1 mL/min for all analyses. The oven was programmed to start at 60 °C for 1 min, then increased 8 °C/min up to 90 °C followed by a 1 min hold. Subsequently, the temperature increased again at 5 °C/min up to 190 °C and was held by 1 min. Finally, the temperature increased 10 °C/min up to 250 °C. The injection port was set to 250 °C on splitless mode. The injection volume was 1 μL. For liquid injection and SPME injection, 2- and 0.75-mm ID liners were used, respectively. The ion source and quadrupole temperature of the mass detector were 230 °C and 150 °C, respectively. The transfer line temperature was 280 °C. Ionization energy in EI mode was 70 eV, and the mass data was acquired on SCAN mode (50–550 *m/z*). Chromatographic peak identification was made by comparing their ion spectra with the NIST library (version 2014) and comparing retention time and spectrum with that of a standard compound.

A calibration curve was built to quantify the α-terpineol content in the samples. The concentrations ranged from 2.6 to 1984 ng/μg. Three replicates were made for each point of the calibration curve, and the average value of each was used to construct a final calibration curve for sample quantitation. Linear regression was calculated from the areas of the injected standard compound solutions; the slope equation was obtained, and the area of the samples was used to determine their concentration. The concentration was expressed on ng of α-terpineol per equivalent male per second.

4.3.1. Matrix Recovering Effects

To evaluate possible matrix effects, dynamic headspace experiments were set. First, filter papers (4 cm diameter) were loaded with 18 μg of α-terpineol. Subsequently, the paper disks were placed in 650 mL Drechsel gas washing bottles (PIREX®, EE.UU). The terpineol in the paper disks were trapped in the same manner as for previously described biological samples (n = 4). A second experiment was carried out by loading 600 μg of α-terpineol contained in 50 μL of hexane in a cartridge with 50 mg of Tenax® (60/80) as adsorbent and 20 mg of glass wool at each end. The solvent was left to dry for 3 min and the compound was eluted as previously described for other biological samples (n = 4). Extracts from both experiments were quantified and compared against each other to check if all the α-terpineol was moved from the headspace to the Tenax® cartridge. That is, all the α-terpineol was considered to be moved from the headspace to the cartridge if no significant concentration differences between both extracts were determined. To check matrix effects from the Tenax® cartridge, the total amount of α-terpineol in the whole eluted extract was compared to that of the initial amount loaded in the paper disk of this metabolite (n = 4). The analysis showed that an average of 71% (n = 4) of the total amount of α-terpineol was retained in the cartridge under the elution conditions used in the experiments. This matrix effect was corrected during the calculation of the α-terpineol content by adding this percentage (71%) in the resulting concentration of the samples.

4.3.2. Limit of Detection (LOD) and Limit of Quantification (LOQ)

To determine the LOD and LOQ values for α-terpineol, a 15 points calibration curve ranging from 0.04 to 627.52 ng/μg was constructed. However, only the first 12 points which showed linearity (0.04−79.04 ng/μg). The R^2 coefficient for this linear regression was 0.994, and the slope equation was y = 72391x + 1077.9. The typical standard error xy (Sty) and slope (m) values were calculated for the resulting data set with the "SLOPE" and "STEYX" functions in Microsoft Excel® 2016. The values were used in the following formulas: LOD = 3.3 (Sty/m) and LOQ = 10 (Sty/m).

For this study, the LOD value was equal to 2.10 ng/μL and the LOQ value was equal to 6.36 ng/μL.

4.4. Bioassays

4.4.1. Reversal Feeding Experiments

A reversal feeding experiment was designed to correlate the release of α-terpineol to the behavior of feeding males. Two sets of male weevils were prepared, males feeding on 8 g soursop pieces (initial feeding males) and males without food (no-feeding males). Each insect group contained 40 males, and each group became a replicate. Fifteen replicates were performed for initial feeding and no-feeding male weevils. The volatile collections were carried out as previously described for dynamic headspace. After the volatile collection time was reached, the insects were removed from the flask and transferred to clean ones. Then, food was not provided to the initial feeding males. On the other hand, 8 g of soursop was provided to the previous no-feeding males. The volatile captures started again under the same conditions. The samples were then analyzed, quantified by GC-MS, and compared to each other.

4.4.2. Behavioral Response of Males and Female Weevils to Volatile Compounds

The behavior of males and females Annonaceae fruits weevil was evaluated through their response to different volatile stimuli in two-choice olfactometry bioassays. Choice combinations are presented in Table 1. The olfactometer was a Y-maze model: its main arm was 12 cm long with 2.5 cm of internal diameter (ID); the two choice arms were 10 cm long and 2.5 cm of ID separated at a 45° angle. The bioassay room was covered with black matte paper to avoid light reflections on the walls. Due to the biological behavior of the anonacea weevils, the olfactometer was elevated 6 cm from the table using a white styrofoam bar, just under the end of the choice arms. This elevation also caused a 4.5 cm elevation between the working bench and the arms' union point of the olfactometer (Supplementary Figure S1). The odor sources were placed in 650 mL Drechsel gas washing bottles. The airstream flow used to carry the odor source volatiles from the release point to the corresponding election arm was set the same as for dynamic headspace. When male or female weevils were used as odor source or part of the odor source, they were added as groups of 20 insects. When α-terpineol was tested alone or combined with soursop, it was released from a 4 cm diameter filter paper (Whatman N° 2). The filter paper was previously loaded with 100 μL of a 4 ng/μL α-terpineol hexane solution. The hexane was left to dry for 30 s. The circle paper with α-terpineol was exchanged every 3 tested weevils. In all cases, 2 g of soursop were used in these experiments. Bottles containing the odor sources were also covered with black matte paper to avoid visual and light interference in the bioassays. The light was provided with a white light bar placed in the center of the olfactometry room's ceiling. The olfactometry system was placed right under the light source. To avoid light reflections on the olfactometer glass, a 1 cm thick white styrofoam plate was placed 10 cm above the olfactometer. That is, the styrofoam plate was located between the light source and the olfactometer, thus, getting indirect illumination.

Table 1. Odor source combinations used in olfactometry assays. For all bioassays, when the odor source required it, 2 g of soursop were used. All male and female groups contained 20 specimens. For combinations including α-terpineol, 100 μL of a 4 ng/μL solution were added.

One Odor Source Trials	
Air	Soursop
Air	Males
Air	Females
Air	α-terpineol
Air	soursop + α-terpineol
Air	soursop + males
Air	soursop + females
Two Odor Source Trials	
Soursop	soursop + males
Soursop	soursop + females
Soursop	soursop + α-terpineol
soursop + females	soursop + males
Females	Males

Groups of 15 insects were tested to record the odor preference of the Annonaceae fruits weevil. Each group of 15 male or female weevils was considered one replicate. There were 4 replicates for males and females tested against different odor elections ($15 \times 4 \times 2 = 120$ insects tested per treatment) (Table 1). A single insect at time was released 2 cm inside the main arm of the Y-maze olfactometer. Subsequently, the behavior and choice of the insect were observed and annotated no longer than 10 min. An odor source choice was scored when the insect walked more than half of one of the choice arms and stayed there longer than 1 min. The data was reported as the percentage of weevils selecting a specific odor source. The data represented the choice average of four replicates ± standard error.

All tested insects were subjected to a fasting period of 24 h before being used in a two-choice olfactometry assay. The arms of the olfactometer were rotated every five insects to avoid position effects on arm selection. Additionally, the olfactometer was replaced every two replicates to avoid residual odor interference. Before starting odor tests, the air in both arms was used to check there was no specific selection caused by light reflections or remnant volatiles in the system. Four groups of 20 insects (males or females) were selected and prepared to test one whole treatment or comparison. The already used insects were fed and left to recover during 48 h. The recovered insects were mixed with other non-used insects and randomly select again to form 4 new testing groups. The selection percentage values were calculated only from responding insects (15 insects). Nonetheless, the non-responding insects' percentage was also recorded and reported. The replicates for one treatment. To test remaining hexane in the filter paper effects, clean 4 cm diameter filter paper (Whatman N° 2) were wet with 100 µL of pure hexane. The solvent was left to dry for 30 s and selection between air vs. dried filter paper was assayed. Moreover, the dried filter papers were placed in both arms to determine the presence of selection patterns by male and female weevils.

4.5. Statistical Analysis

All data was subjected to normality and variance homogeneity tests. For olfactometry assays, one odor source vs. air, a one-way ANOVA was used, followed by a Tukey test (α = 0.05). A G-test analyzed olfactometry assays challenging two odor sources at once. Reversal feeding experiment data (measurements over the same specimens) was treated as pairwise samples analyzed by a U-test. For matrix recovering effects, the concentration effects of the extracts were tested by a 2 tails t-test ($t < 0.05$). On the other hand, a Kruskal–Wallis test was performed for data without normal distribution or homogeneous variance. The post hoc analysis was performed with Bonferroni correction (non-response olfactometry data). Statistical differences among non-responding weevils were determined with an ANOVA test. All analyzes were performed in R software [52].

Supplementary Materials: The following are available online. Table S1: GC-MS identification of soursop volatiles trapped by dynamic headspace and solid phase microextraction. Figure S1: Olfactometry system diagram.

Author Contributions: Conceptualization, J.M.P.-R., J.C.-T. and L.F.S.-A.; Data curation, L.S.-R. and L.F.S.-A.; Formal analysis, J.M.P.-R., L.S.-R. and L.F.S.-A.; Investigation, J.M.P.-R., L.M.H.-F., R.M.L.-R., J.R.-N. and C.L.-C.; Supervision, J.C.-T. and L.F.S.-A.; Writing—original draft, J.M.P.-R.; Writing—review and editing, J.C.-T., L.M.H.-F., R.M.L.-R., L.S.-R., J.R.-N., C.L.-C. and L.F.S.-A. All authors have read and agreed to the published version of the manuscript.

Funding: This research received no external funding.

Institutional Review Board Statement: Not applicable.

Informed Consent Statement: Not applicable.

Data Availability Statement: Not applicable.

Acknowledgments: The first author thanks to the Mexican Scientific council (CONACyT) for his PhD scholarship (N. 731909).

Conflicts of Interest: The authors declare no conflict of interest.

Sample Availability: Not applicable.

References

1. Morrone, J.J. Biodiversity of Curculionoidea (Coleoptera) in Mexico. *Rev. Mex. Biodivers.* **2014**, *85*, 312–324. [CrossRef]
2. Hespenheide, H.A. Beetles. In *Encyclopedia of Biodiversity*, 1st ed.; Levin, S.A., Ed.; Academic Press: Princeton, NJ, USA, 2001; Volume 1, pp. 351–358.
3. Servín, V.R.; García, H.J.L.; Tejas, R.A.; Martínez, C.J.L.; Toapanta, M.A. Susceptibility of pepper weevil (*Anthonomus eugenii* Cano) (Coleoptera: Curculionidae) to seven insecticides in rural areas of Baja California Sur, Mexico. *Acta Zool. Mex.* **2008**, *24*, 45–54.

4. Oehlschlager, A.C.; Chinchilla, C.; Castillo, G.; González, L. Control of red ring disease by mass trapping of *Rhynchophorus palmarum* (Coleoptera: Curculionidae). *FLA Entomol.* **2002**, *85*, 507–513. [CrossRef]

5. Ocan, D.; Mukasa, H.H.; Rubaihayo, P.R.; Tinzaara, W.; Blomme, G. Effects of banana weevil damage on plant growth and yield of East African Musa genotypes. *J. Appl. Biol. Sci.* **2008**, *9*, 407–415.

6. Hernández, F.L.M.; Castañeda, V.A.; Urías, L.M.A. Weevil borers in tropical fruit crops: Importance, biology and management. In *Insect Physiology and Ecology*, 2nd ed.; Shields, V.D.C., Ed.; IntechOpen: Rijeka, Croatia, 2017; Volume 3, pp. 154–196.

7. Showler, A.T. Boll weevil (Coleoptera: Curculionidae) damage to cotton bolls under standard and proactive spraying. *J. Econ. Entomol.* **2006**, *99*, 1251–1257. [CrossRef] [PubMed]

8. Esparza, D.G.; Olguín, A.; Carta, L.K.; Skantar, A.M.; Villanueva, R.T. Detection of *Rhynchonphorus palmarum* (Coleoptera:Curculionidae) and identification of associated nematodes in South Texas. *FLA Entomol.* **2013**, *96*, 1513–1521. [CrossRef]

9. Azuara, D.A.; Terán, V.A.P.; Soto, S.A.; Aguilar, P.N.; Martínez, B.L. Evaluación del tipo de trampa, atrayente alimenticio y feromona de agregación en el trampeo del picudo del agave *Scyphophorus acupunctatus* Gyllenhal en Tamaulipas, México. *Entomotropica* **2014**, *29*, 1–8.

10. Bautista, H.C.F.; Cibrián, T.J.; Velázquez, G.J.C.; Rodríguez, G.M.P. Evaluación en campo de atrayentes para la captura de *Anthonomus eugenii* Cano (Coleoptera: Curculionidae). *Rev. Chil. Entomol.* **2020**, *46*, 211–219.

11. Coelho, L.M.A.; Alves, R.E. Soursop (*Annona muricata* L.). In *Postharvest Biology and Technology of Tropical and Subtropical Fruits. Mangosteen to White Sapote*; Yahia, M.E., Ed.; Woodhead Publishing Limited: Cambridge, UK, 2011; Volume 4, pp. 363–391.

12. Jiménez, Z.J.O.; Balois, M.R.; Alia, T.I.; Juárez, L.P.; Sumaya, M.M.T.; Bello, L.J.E. Characterization of soursop fruit (*Annona muricata* L.) in Tepic, Nayarit, Mexico. *Rev. Mex. Cienc. Agríc.* **2016**, *7*, 1261–1270.

13. Corrales, M.G.J. Observations on the biology and behavioral food of *Optatus* sp. (Coleoptera: Curculionidae) in *Annona cherimola* Mill. In *Resúmenes de 2do. Congreso Centroamericano y del Caribe y 3ero. Costarricense de Entomología*; Editorial San José ASENCO: San José, Costa Rica, 1995; p. 6.

14. Hernández-Fuentes, L.M.; Gómez, J.R.; Andrés, A.J. Importance, pest insect and diseases fungus of the cultivation of the soursoup. In *Libro Técnico Núm. 1*, 1st ed.; Instituto Nacional de Investigaciones Forestales, Agrícolas y Pecuarias, Campo Experimental Santiago Ixcuintla: Nayarit, México, 2013; Volume 1, pp. 85–87.

15. Maldonado, E.; Hernández, L.M.; Luna, G.; Gómez, J.R.; Flores, R.J.; Orozco-Santos, M. Bioecology of *Optatus palmaris* Pascoe (Coleoptera:Curculionidae) in *Annona muricata* L. *Southw. Entomol.* **2014**, *39*, 773–782. [CrossRef]

16. Castañeda-Vildózola, A.; Nava-Díaz, C.; Hernández-Fuentes, L.M.; Valdez-Carrasco, J.; Colunga-Treviño, B. New host record and geographical distribution of *Optatus palmaris* Pascoe (Coleoptera: Curculionidae) in Mexico. *Acta Zool. Mex.* **2009**, *25*, 663–666.

17. Castañeda-Vildózola, A.; Morales-Trujillo, M.C.; Franco-Mora, O.; Valdez-Carrasco, J.; Mejía-Carranza, J. A new record of *Optatus palmaris* Pascoe (Coleoptera: Curculionidae) associated with Annona L. (Annonaceae) in the State of Mexico, Mexico. *Revista Chilena de Entomología* **2020**, *46*, 397–400. [CrossRef]

18. Hernández, F.L.M.; Nolasco, G.Y.; Orozco, S.M.; Montalvo, G.E. Toxicidad de insecticidas contra (*Optatus palmaris* Pascoe) en guanábana. *Rev. Mex. Cienc. Agríc.* **2021**, *12*, 49–60.

19. Tewari, S.; Leskey, T.C.; Nielsen, A.L.; Piñero, J.C.; Rodriguez, S.C.R. Integrated pest management: Current concepts and ecological perspective. In *Integrated Pest Management: Current Concepts and Ecological Perspective*, 1st ed.; Abrol, D.P., Ed.; Academic Press: San Diego, CA, USA, 2014; pp. 141–168.

20. Mormann, M.; Salpin, J.Y.; Kuck, D. Gas-phase titration of C7H9+ ion mixtures by FT-ICR mass spectrometry: Semiquantitative determination of ion populations generated by CI-induced protonation of C7H8 isomers and by EI-induced fragmentation of some monoterpenes. *Int. J. Mass Spect.* **2006**, *249–250*, 340–352. [CrossRef]

21. Nicolescu, T.O. Interpretation of mass spectra. In *Mass Spectrometry*; Aliofkhazraei, M., Ed.; Intech Open: London, UK, 2017; pp. 24–78.

22. Beck, J.J.; Torto, B.; Vannette, R.L. Eavesdropping on plant-insect-microbe chemical communications in agricultural ecology: A virtual issue on semiochemicals. *J. Agric. Food Chem.* **2017**, *65*, 5101–5103. [CrossRef] [PubMed]

23. Beck, J.J.; Vannette, R.L. Harnessing insect-microbe chemical communications to control insect pests of agricultural systems. *J. Agric. Food Chem.* **2017**, *65*, 23–28. [CrossRef] [PubMed]

24. Nandagopal, V.; Prakash, A.; Rao, J. Know the pheromones: Basics and its application. *J. Biopestic.* **2008**, *1*, 210–215.

25. Sharma, A.; Sandhi, R.K.; Reddy, G.V.P. A Review of interactions between insect biological control agents and semiochemicals. *Insects* **2019**, *10*, 1–16. [CrossRef] [PubMed]

26. Haenniger, S.; Goergen, G.; Akinbuluma, M.D.; Kunert, M.; Heckel, D.G.; Unbehend, M. Sexual communication of *Spodoptera frugiperda* from West Africa: Adaptation of an invasive species and implications for pest management. *Sci. Rep.* **2020**, *10*, 2892. [CrossRef] [PubMed]

27. Bruce, T.J.A.; Wadhams, L.J.; Woodcock, C.M. Insect host location: A volatile situation. *Trends Plant Sci.* **2005**, *10*, 269–274. [CrossRef]

28. Wang, H.M.; Bai, P.H.; Zhang, J.; Zhang, X.M.; Hui, Q.; Zheng, H.X.; Zhang, X.H. Attraction of bruchid beetles *Callosobruchus chinensis* (L.) (Coleoptera: Bruchidae) to host plant volatiles. *J. Integr. Agric.* **2020**, *19*, 3035–3044. [CrossRef]

29. Webster, B.; Cardé, R. Use of habitat odour by host-seeking insects. *Biol. Rev.* **2016**, *92*, 1241–1249. [CrossRef]

30. Erb, M.; Robert, C.A.M. Sequestration of plant secondary metabolites by insect herbivores: Molecular mechanisms and ecological consequences. *Curr. Opin. Insect. Sci.* **2016**, *14*, 8–11. [CrossRef]

31. Beran, F.; Köllner, T.G.; Gershenzon, J.; Tholl, D. Chemical convergence between plants and insects: Biosynthetic origins and functions of common secondary metabolites. *New Phytol.* **2019**, *223*, 52–67. [CrossRef] [PubMed]

32. Zhao, R.; Lu, L.H.; Shi, Q.X.; Chen, J.; He, Y.R. Volatile terpenes and terpenoids from workers and queens of *Monomorium chinense* (Hymenoptera: Formicidae). *Molecules* **2018**, *23*, 2838. [CrossRef]

33. Eller, F.J.; Bartelt, R.J.; Shasha, B.S.; Schuster, D.J.; Riley, D.J.; Stansly, P.A.; Mueller, T.F.; Shuler, K.D.; Johnson, B.; Davis, J.H.; et al. Aggregation pheromone for the pepper weevil, *Anthonomus eugenii* Cano (Coleoptera: Curculionidae): Identification and field activity. *J. Chem. Ecol.* **1994**, *20*, 1537–1555. [CrossRef] [PubMed]

34. Cross, J.V.; Hall, D.R.; Innocenzi, P.J.; Hesketh, H.; Jay, C.N.; Burgess, C.M. Exploiting the aggregation pheromone of strawberry blossom weevil *Anthonomus rubi* (Coleoptera: Curculionidae): Part 2. Pest monitoring and control. *Crop Prot.* **2006**, *25*, 155–166. [CrossRef]

35. Hardee, D.D.; McKibben, G.H.; Rummel, D.R.; Huddleston, P.M.; Coppege, J.R. Response of boll weevils to component ratios and doses of the Grandlure. *Environ. Entomol.* **1974**, *3*, 135–138. [CrossRef]

36. Hedin, P.A.; Dollar, D.A.; Collins, J.K.; Dubois, J.G.; Mulder, P.G.; Hedger, G.H.; Smith, M.W.; Eikenbary, R.D. Identification of male pecan weevil pheromone. *J. Chem. Ecol.* **1997**, *23*, 965–977. [CrossRef]

37. Lacey, E.S.; Moreira, J.A.; Millar, J.G.; Hanks, L.M. A Male-produced aggregation pheromone blend consisting of alkanediols, terpenoids, and an aromatic alcohol from the Cerambycid Beetle *Megacyllene caryae*. *J. Chem. Ecol.* **2008**, *34*, 408–417. [CrossRef]

38. Mitchell, R.F.; Ray, A.M.; Hanks, L.M.; Millar, J.G. The common natural products (s)-α-terpineol and (e)-2-hexenol are important pheromone components of *Megacyllene antennata* (Coleoptera: Cerambycidae). *Environ. Entomol.* **2018**, *47*, 1547–1552. [CrossRef] [PubMed]

39. Tinzaara, W.; Dicke, M.; Van, H.A.; Gold, C.S. Use of infochemicals in pest management with special reference to the banana weevil, *Cosmopolites sordidus* (Germar) (Coleoptera: Curculionidae). *Int. J. Trop. Insect Sci.* **2002**, *22*, 241–261. [CrossRef]

40. Müller, M.; Buchbauer, G. Essential oil components as pheromones. A review. *Flavour Fragr. J.* **2011**, *26*, 357–377. [CrossRef]

41. Stelinski, L.L.; Oakleaf, R.; Rodríguez, S.C. Oviposition deterring pheromone deposite don blueberry fruit by the parasitic wasp, *Diachasma alloeum*. *Behavior* **2007**, *144*, 429–445.

42. Landolt, P.J.; Phillips, T.W. Host plant influences on sex pheromone behavior of phytophagous insects. *Annu. Rev. Entomol.* **1997**, *42*, 371–391. [CrossRef]

43. Dickens, J.C. Green leaf volatiles enhance aggregation pheromone of boll weevil, *Anthonomus grandis*. *Entomol. Exp. Appl.* **1989**, *52*, 191–203. [CrossRef]

44. Piñero, J.C.; Prokopy, R.J. Field evaluation of plant odor and pheromonal combinations for attracting plum curculios. *J. Chem. Ecol.* **2003**, *29*, 2735–2748. [CrossRef]

45. Piñero, J.C.; Wright, S.E.; Prokopy, R.J. Response of plum curculio (Coleoptera: Curculionidae) to odor-baited traps near woods. *J. Econ. Entomol.* **2001**, *94*, 1386–1397. [CrossRef]

46. Hallett, R.H.; Oehlschlager, A.C.; Borden, J.H. Pheromone trapping protocols for the Asian palm weevil, *Rhynchophorus ferrugineus* (Coleoptera: Curculionidae). *Int. J. Pest Manag.* **1999**, *45*, 231–237. [CrossRef]

47. Cruz, F.J.J.; Figueroa, C.P.; Alcántara, J.J.A.; López, M.V.; Silva, G.F. Vegetal synergists for trapping the adult of *Scyphophorus acupunctatus* Gyllenhal, in pheromone baited traps, in *Agave angustifolia* Haw, in Morelos, México. *Acta Zool. Mex.* **2019**, *35*, 1–9. [CrossRef]

48. Borden, J.H.; Pureswaran, D.S.; Lafontaine, J.P. Synergistic blends of monoterpenes for aggregation pheromones of the mountain pine beetle (Coleoptera: Curculionidae). *J. Econ. Entomol.* **2008**, *101*, 1266–1275. [CrossRef]

49. Erasto, P.; Viljoen, A.M. Limonene—A Review: Biosynthetic, Ecological and Pharmacological Relevance. *Nat. Prod. Commun.* **2008**, *3*, 1193–1202. [CrossRef]

50. Hartsel, J.A.; Eades, J.; Hickory, B.; Makriyannis, A. *Cannabis sativa* and Hemp. In *Nutraceuticals: Efficacy, Safety and Toxicity*, 1st ed.; Gupta, R., Ed.; Academic Press: Princeton, NJ, USA, 2016; Volume 1, pp. 735–754.

51. Champion, G.C. *Biologia Centrali-Americana. Insecta. Coleoptera, Rhynchophora Part 5*; Curculionidae. Curculioninae: Porter, London, 1907; Volume 4, p. 513.

52. R Core Team. *R: A Language and Environment for Statistical Computing*; R Foundation for Statistical Computing: Vienna, Austria, 2020.

Review

A Systematic Approach to *Agastache mexicana* Research: Biology, Agronomy, Phytochemistry, and Bioactivity

Mariana Palma-Tenango [1], Rosa E. Sánchez-Fernández [2] and Marcos Soto-Hernández [3,*]

[1] Facultad de Ciencias, Universidad Nacional Autónoma de México, Ciudad Universitaria, Coyoacán, 04510 Ciudad de México, Mexico; marianapt@ciencias.unam.mx

[2] Laboratorio Nacional de Investigación y Servicios Agroalimentarios y Forestales (LANISAF), Edificio Efraím Hernández Xolocotzi Nivel 1 y 2, Universidad Autónoma Chapingo, 56230 Texcoco, Mexico; resf2012@gmail.com

[3] Posgrado en Botánica, Colegio de Postgraduados-Campus Montecillo, km 36.5, Carretera México-Texcoco, 56230 Texcoco, Mexico

* Correspondence: msoto@colpos.mx; Tel.: +52-595952-0200 (ext. 1361)

Abstract: Mexico is the center of origin of the species popularly known as toronjil or lemon balm (*Agastache mexicana* Linton & Epling). Two subspecies have been identified and are commonly called purple or red (*Agastache mexicana* Linton & Epling subspecies. mexicana) and white (*Agastache mexicana* subspecies xolocotziana Bye, E.L. Linares & Ramamoorthy). Plants from these subspecies differ in the size and form of inflorescence and leaves. They also possess differences in their chemical compositions, including volatile compounds. Traditional Mexican medicine employs both subspecies. *A. mexicana* exhibits a broad range of pharmacological properties, such as anti-inflammatory, anxiolytic, and antioxidant. A systematic vision of these plant's properties is discussed in this review, exposing its significant potential as a source of valuable bioactive compounds. Furthermore, this review provides an understanding of the elements that make up the species' holistic system to benefit from lemon balm sustainably.

Keywords: toronjil; Mexican agastache; aromatic plants

Citation: Palma-Tenango, M.; Sánchez-Fernández, R.E.; Soto-Hernández, M. A Systematic Approach to *Agastache mexicana* Research: Biology, Agronomy, Phytochemistry, and Bioactivity. *Molecules* **2021**, *26*, 3751. https://doi.org/10.3390/molecules26123751

Academic Editors: Young Hae Choi, Young Pyo Jang, Yuntao Dai and Luis Francisco Salomé-Abarca

Received: 25 April 2021
Accepted: 7 June 2021
Published: 20 June 2021

Publisher's Note: MDPI stays neutral with regard to jurisdictional claims in published maps and institutional affiliations.

1. Introduction

Lamiaceae is the eighth most diverse plant family in Mexico and 5.5% of the species worldwide are found in this country. Thus, Mexico may be one of the most important diversification centers [1]. This family contains a wide range of aromatic plants possessing agronomical, pharmacological, and commercial potential. Mexican *Agastache* belongs to this family and its use and commercialization for traditional Mexican medicine make it the most important member of the *Agastache* genus in Mexico [2,3]. The species *Agastache mexicana* divides into two subspecies, based on anatomical characteristics [2] and chemical composition [4]: red lemon balm, *Agastache mexicana* Linton & Epling subspecies mexicana, and white toronjil, *Agastache mexicana* subspecies xolocotziana Bye, E.L. Linares & Ramamoorthy [5].

The species is distributed in the states of Guanajuato, Mexico, Michoacán, Puebla, Querétaro, Hidalgo, Veracruz, Chihuahua, Morelos, and Tlaxcala, as well as in Mexico City (Figure 1). The species concentrates in the volcanic axis of Central Mexico [2,6].

This article presents a review of the research on *Agastache mexicana* from the perspective of a holistic system to understand its components and interactions. A literature search was sourced from meta-analyses of available *Agastache* data. Data were sourced from bibliographic engines like Web of Science®, Scopus®, ScienceDirect®, and Google Scholar® using "*Agastache mexicana*" as a keyword.

265

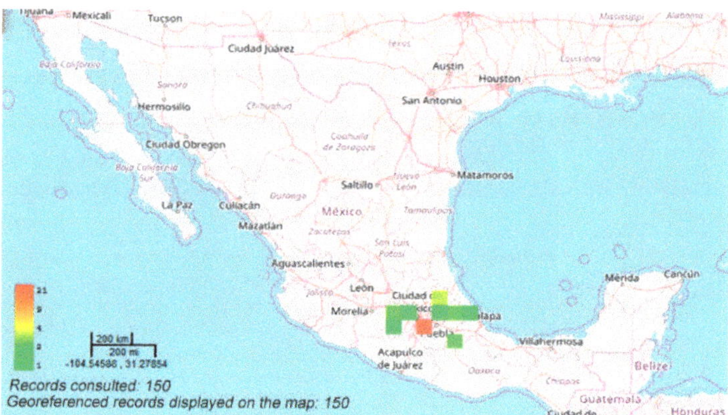

Figure 1. Distribution of *Agastache mexicana* in Mexico [7].

2. A Holistic Approach to *Agastache mexicana* Usage

Systems biology analysis allows the understanding of different biological elements and their interactions with non-biological elements, such as the environment or human impacts (for example, the analysis of various traditional medicine systems like traditional Chinese medicine) [8]. Systems biology, in tandem with reverse pharmacology, may allow discovering new active biological compounds [9].

Life science studies relying on systems biology and holistic approaches shy away from reductionist views and incorporate biological effects and their interaction with the environment [8,9]. A biological system contains numerous components interacting in a vast variety of combinations. Once the components and interactions of a system are known, a system's behavior may be understood [10].

We used systems biology principles for a holistic analysis of different components within the lemon balm plant system and its environment (Figure 2). This system's insights are derived from a general vision that includes the system components' relationships and interactions. This approach may provide new collaborative information, fresh insights, and research prospects for the species.

2.1. Biology

A. mexicana is a native vascular plant of Mexico [11]. It is a perennial herb. Plants of both subspecies have a typical Lamiaceae morphology: opposite, petiolate leaves, a four-angled stem, and numerous trichomes [12]. In the two subspecies, the stem is quadrangular in cross section; in *A. mexicana* ssp. mexicana, the basal and middle part of the stem is purple. The cuticle is smooth, but in the angles, it is observed to be crenulated, with a thickness of 4–8 μm. The three types of epidermal appendages described for the leaf are observed, but uniseriate non-glandular trichomes are abundant in *A. mexicana* ssp. xolocotziana, while they are more scattered and not visible to the naked eye in *A. mexicana* ssp. mexicana [2]. The plant height reaches between 50 and 150 cm [2] and the base chromosome number is 9. The plants have slender and spreading rhizomes [5].

Various plants from the *Agastache* genus are used for bee forage and honey production [13,14]. Toronjil is a honey plant; its flowers produce nectar for bee collection [15].

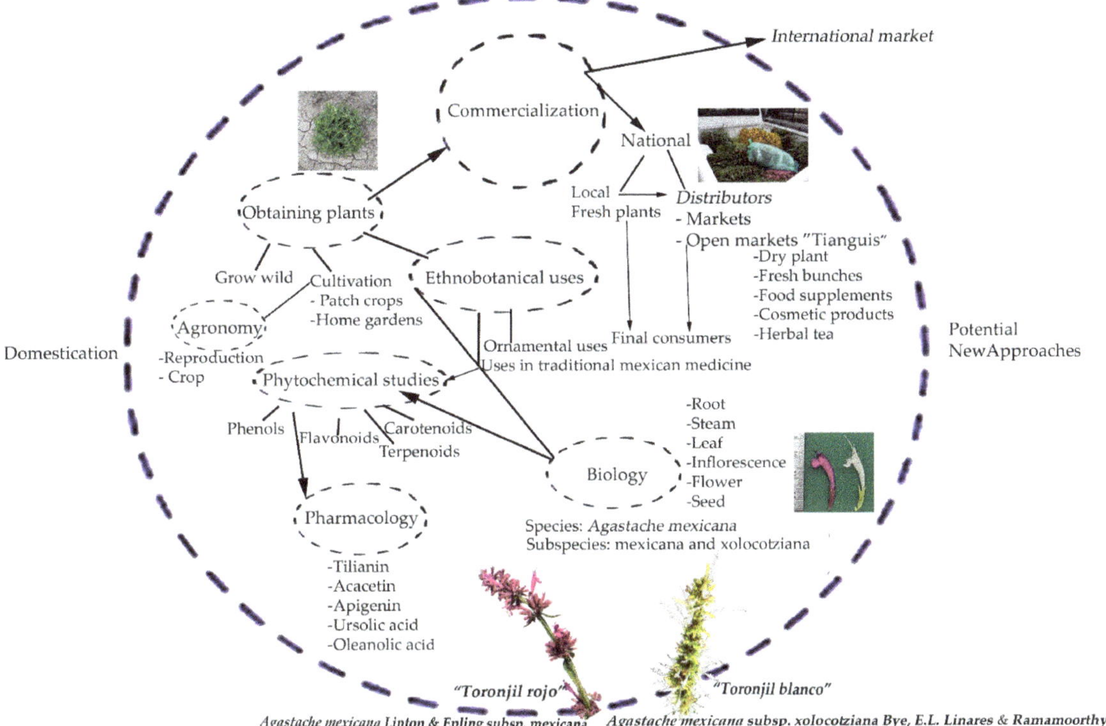

Figure 2. Systems biology approaches of the study of *Agastache mexicana*.

2.1.1. *A. mexicana* ssp. mexicana

The stem of *A. mexicana* ssp. mexicana is erect, branched, and four-angled. The basal and middle part of the stem is purple. The form of the leaves is ovate-lanceolate, measuring 4.4 to 6.3 cm long and 2.1 to 2.5 cm wide [2]. The petiole is 1 cm long [2], and the corolla is purplish red to red [2,5]. The seeds measure approximately 4 to 5 mm (Figure 3).

2.1.2. *A. mexicana* ssp. xolocotziana

The stem of *A. mexicana* ssp. xolocotziana is erect, branched, and four-angled. The form of the leaves is ovate-lanceolate, measuring 4.6 to 6.2 cm long and 1.7 to 3.0 cm wide [2]. The petiole is 1 cm long [2]. Inflorescences end in ramifications of interrupted whorls of cymes with numerous flowers. The calyx is 1.0 to 1.3 cm long and the corolla is white and approximately 2.4 cm long. Stamens are didynamous and exserted. Its anther is approximately 1 mm long. The style is 2.8 cm long, and its tip is bifid with the upper arm slightly longer. Ovules are only 0.5 mm tall [5]. The seeds measure approximately 3 to 4 mm (Figure 4).

Figure 3. Plant biology of *Agastache mexicana* ssp. mexicana.

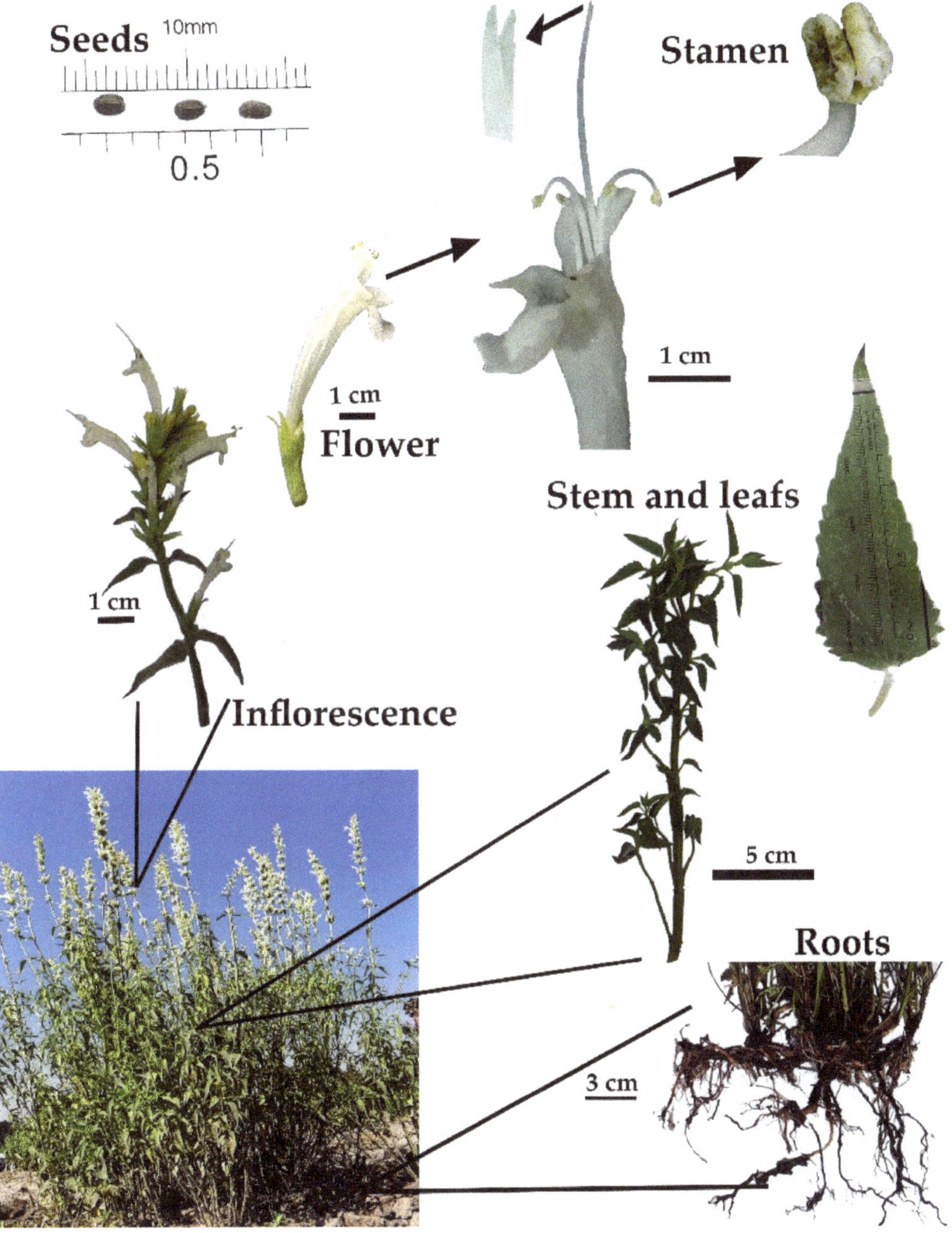

Figure 4. Plant biology of *Agastache mexicana* ssp. xolocotziana.

2.2. Ethnobotanical Uses

The genus *Agastache* includes ornamental plants and aromatic plants that contain essential oils [12]. For example, both *A. mexicana* subspecies have therapeutic and ornamental uses [2]. Knowledge about lemon balm healing properties is cited in sources dating back to pre-Hispanic culture, such as in the De la Cruz Badiano Codex [16]. In the Nahuatl tongue, *A. mexicana* is known as tlalahuehuetl [6]. Traditional Mexican medicine labels the plant as "hot". It is prescribed to cure fright, stomach pain, excessive bile, cough, vomit, chills, and anxiety [2].

The holistic method to study plants with medicinal properties examines the interactions and relationships among the environment's biological and cultural components. Rural and urban populations use this plant for in-home treatments in the form of herbal teas (infusions and decoctions) [17]. In Mexico, *A. mexicana* is identified for its medicinal properties against anxiety and as a sleep-promoting plant [3,16]. The subspecies have specific uses: *A. mexicana* ssp. mexicana is preferred for wound healing, as an antispasmodic agent, and against stomach pain, while *A. mexicana* ssp. xolocotziana is employed to treat heart disease.

Many modern drugs originated from ethnopharmacology and knowledge of traditional medicine [18]. Results from research on the medicinal effects of *A. mexicana* ssp. mexicana and ssp. xolocotziana support their use in traditional medicine as an anxiolytic, tranquilizer, and sedative, as well as a remedy to alleviate "nervousness" [3,19].

2.3. Agronomy

Many medicinal and aromatic plants are industrially sown, but most are still obtained by harvesting wild populations. The need for renewable sources and protection of plant biodiversity creates an opportunity for farmers to grow these crops [20]. In Mexico, over-harvesting of medicinal plants is counteracted by collecting seeds, cuttings, or roots to propagate the plant. Most of these collected samples are planted in small home gardens to be later sown on cultivated fields [21].

A. mexicana is a candidate species for structured cultivation as a source for active principles, extracts, essential oils, and pharmaceutical products [22]. Propagation is mainly asexual [23], through vegetative propagation, and depends on its rhizomes' division, as seed viability is low; *A. mexicana* ssp. xolocotziana seems to have even lower viability. A further complication arises as seed formation is hindered since harvesting occurs during flowering [5]; inflorescences are the main commercialized product. However, red lemon balm exists in wild populations, unlike white lemon balm. A hybridization process between the Mexican subspecies and *Agastache palmeri* possibly originated *A. mexicana* ssp. xolocotziana [23].

A. mexicana blooms from June to November [22]. Subspecies show phenotypic differences in leaf shape, flower color, and flavor [2]. Farmers from Santiago Mamalhuazuca (State of Mexico) have empirically gathered knowledge that the xolocotziana subspecies is more susceptible to extreme temperature and humidity. The cultivation of both subspecies begins in April, and the stems and inflorescences are harvested in November. The rhizome promotes stem sprouting, which allows a new harvest in the following February [23]. No technological packages based on crop physiology, detailing handling on its phenological stages, leading to higher biomass yields or providing information on bioactive production per cultivation area, exist for *A. mexicana* cultivation.

2.4. Commercialization

Ethnobotanical studies have impacted toronjil research. Empirical observations have detected that mexican markets sold a different subspecies from the typical *A. mexicana* subspecies mexicana. Identification of *A. mexicana* subspecies xolocotziana occurred through differential characterization of morphological, chemical, and pharmacological features [2,5]. The commercialization of botanical products promotes the cultural exchange of traditional knowledge and the exploitation of natural resources. Studies illustrate the influence popular markets have on the demand for plants with novel applications. Attention should also focus on the dangers of overcollection of wild species in response to increasing demand and supporting natural habitats' conservation [21].

White and red lemon balm are commercially sown and traded in various Mexican regions, including Hidalgo, Mexico, Morelos, Puebla, and Veracruz. Inflorescence bundles or dried plants are distributed through different regional sales channels in the State of Mexico, Southeast Puebla, Morelos, and Mexico City [24].

2.5. Phytochemical and Biological Activity

2.5.1. Phytochemistry

The *Agastache* genus produces various volatile and non-volatile secondary metabolites, mainly phenylpropanoids and terpenoids. *A. mexicana* contains terpenoid compounds like monoterpenes (limonene, pulegone), sesquiterpenes (β-caryophyllene), diterpenes (breviflorine), triterpenes (ursolic, corosolic, maslinic acids); phenolic and phenylpropanoid compounds like flavones (acacetin) and flavonoids (tilianin, hesperitin); carboxylic acids (9-hexadecenoic acid, butanoic acid); and soluble sugars (glucose, sucrose) [12]. Subspecies mexicana and xolocotziana share common compounds, but have different chemical profiles [3,4].

The chemical composition of essential oils is influenced by the subspecies, environmental conditions of the crop, harvest time and type of extraction [4,12]. Table 1 shows the chemical compositions of different essential oils obtained from *A. mexicana* subspecies. However, some chemical studies do not specify the studied subspecies, which makes it difficult to establish a defined chemical profile for each subspecies. Plants introduced to other countries have essential oils with different chemical compositions. For example, the essential oil of *A. mexicana* grown in Scotland (subspecies not specified) is characterized by pulegone as the main compound, followed by menthone and limonene [25]; in contrast, *A. mexicana* ssp. mexicana plants introduced in Belarus produced methyl eugenol and estragole as the main compounds [26]. Extraction methods also influence the variability of the physical and chemical characteristics of the essential oils, but different distillation apparatus does not affect the quality of *A. mexicana* essential oil [27].

Chemical study of aqueous and organic extracts from aerial plant parts and whole plants led to the isolation of monoterpenes, diterpenes, triterpenes, flavones, and flavonoids. Table 2 shows the chemical compositions of non-polar and polar extracts from *A. mexicana* subspecies. The compounds in both subspecies are tilianin, acacetin, ursolic acid, salvigenine, 5-hydroxy-7,4′dimethoxyflavone, (2-acetyl)-7-O-glucosyl acacetin, diosmetin 7-O-β-D-(6″-O-malonyl)-glucoside, acacetin 7-O-β-glucoside, acacetin 7-O-β-D-(6″-O-malonyl)-glucoside, acacetin-7-O-β-glucoside-D-(2″-acetyl-6″malonyl), diosmetin, gardenin A, 5,6,7,8,3- pentahydroxy-4-methoxy flavone, 8-hydroxy-salvigenin, α-terpineol, and pulegone (Table 2). The concentration of each compound varies in both subspecies [3], tilianin and acacetin are more abundant in the subspecies xolocotziana [28].

Table 1. Composition and bioactivity of essential oils from aerial parts of *Agastache mexicana*.

Taxa	Chemical Composition (Main Compound)	Biological Activity
Agastache mexicana	Pulegone (75.3%), followed by menthone (13.9%) and limonene (3.1%) [25,29] 17 compounds: menthone (42.4%), isomenthone (18.8%) and pulegone (7.3%) [30]. 44 compounds: pulegone (47.77–49.48%), limonene (15.45–15.93%), *cis*-menthone (12.25–12.89%), and *trans*-menthone (2.89–3.17) [27]. Estragole (86.78%), limonene (11.24%) and linalool (1.98%) [4]. Estragole (80.28%), D-limonene (17.56%) and linalyl anthranilate (17.56%) [31].	
Agastache mexicana Linton & Epling ssp. mexicana	19 volatile compounds [1]: geranyl acetate (37.5%), followed by geranial (17%) and geraniol (16%) [22]. 16 compounds: geranyl acetate (61.4%) followed by geranial (11%) and geraniol (8.3%) [22]. 21 compounds: estragole (15.1%) and methyl eugenol (20.8%) [26]. Cultivated: estragole, geraniol, linalool, menthone and pulegone. Encouraged: geraniol and pulegone [23].	Tracheal relaxation in guinea pig model. EC_{50} of 18.25 μg mL^{-1} with contractions induced by carbachol and 13.30 μg mL^{-1} with contractions induced by histamine [31].
Agastache mexicana ssp. xolocotziana Bye, E.L. Linares & Ramamoorthy	Bornyl acetate [5]. Pulegone (80.07%), limonene (9.49%), menthone (7.91%) and isopulegone (2.53%) [4]. Methyl eugenol (36.41%) estragole (27.92%), linalool (10.66%), menthone (10.29%), pulegone (6.46%) and limonene (5.70%) [32]. Estragole, geraniol, linalool, menthone and pulegone [23].	Antifungal activity (MIC): *Aspergillus amylovorus* (0.3 μg mL^{-1}), *A. flavus* (0.3 μg mL^{-1}), *A. nomius* (30 μg mL^{-1}), *A. ostianus* (30 μg mL^{-1}), *Eurotium halophilicum* (30 μg mL^{-1}), *Eupenicillum hirayamae* NRRL 3587 (30 μg m^{L-1}), *E. hyrayamae* NRRL 3588 (0.3 μg mL^{-1}), *E. hyrayamae* NRRL 3589 (30 μg mL^{-1}), *E. hyrayamae* NRRL 3591 (0.3 μg mL^{-1}), *Penicillium cinnamopurpureum* (0.3 μg mL^{-1}), *P. viridicatum* var. ii (30 μg mL^{-1}) [32].

[1] Volatile compounds present in the head space before extraction of the essential oil. EC_{50}: mean effective concentration. MIC: minimum inhibitory concentration.

Table 2. Chemical composition of aqueous and organic extracts of *Agastache mexicana*.

Taxa	Flavonoids	Flavones	Terpenes	Organic Acids	Esters	Alcohols, Aldehydes, and Ketones	Hydrocarbons
Agastache mexicana	Tilianin [33], hesperitin, quercetin [19].		Limonene, linalool, menthone, α-terpineol, pulegone, eugenol [34].				
Agastache mexicana Linton & Epling spp. mexicana	Tilianin [28,35–37], gardenin A, 5-hydroxy-7,4′ dimethoxy flavone [3].	Acacetin [4,28,38], 7-O-glucosyl acacetin, (2-acetyl)-7-O-glucosyl acacetin [4], diosmetin 7-O-β-D-(6″-O-malonyl)-glucoside, acacetin 7-O-β-glucoside, acacetin 7-O-β-D-(6″-O-malonyl)-glucoside, acacetin-7-O-β-glucoside-D-(2″-acetyl-6″ malonyl), diosmetin, 5,6,7,8,3-pentahydroxy-4-methoxy flavone [3], luteolin 7-O-β-D-glucoside, luteolin 7-O-β-D-(6-O-malonyl)-glucoside [3].	Ursolic acid [4,38], oleanolic acid [38], salvigenine, 8-hydroxy-salvigenin [3], estragole, oleanoic acid [4].	Malic acid [3], hexadecanoic acid, 9-hexadecenoic acid [4].	Butanoic acid-hexane-dioctyl, hexanedioc-dioctyl ester, 6-octen-1-ol-3,7-dimethyl propionate [4].	3-methoxy-cinnamaldehyde, 2,6-dimethoxy-4-(2-propenyl)-phenol [4]	9-Eicosyne [4]
Agastache mexicana spp. xolocotziana Bye, E.L. Linares & Ramamoorthy	Tilianin [28], pratol [5], gardenin A, pilosin [3].	Acacetin [3,4,28,39], 5-hydroxy-7,4′ dimethoxy flavone, (2-acetyl)-7-O-glucosyl acacetin [4], acacetin 7-O-β-glucoside, acacetin 7-O-β-D-(6″-O-malonyl)-glucoside, acacetin-7-O-β-glucoside-D-(2″-acetyl-6″-malonyl), diosmetin 7-O-β-D-(6″-O-malonyl)-glucoside, diosmetin, 5,6,7,8,3-pentahydroxy-4-methoxy flavone; diosmetin 7-β-O-glucoside, 8-hydroxy-flavone [4], chrysene [5].	Salvigenine, corosolic acid, maslinic acid [4], ursolic acid [4,39], β-amirin, 8-hydroxy-salvigenin [3], breviflorine [5], nerol, pulegone, camphor, *p*-menth-6-ene-2,8-diol, α-terpineol, isopiperitenone, geraniol, α-terpineol-methyl ether, *p*-menthane-1,8-diol, neryl acetate, thymol acetate, piperitone, *p*-menth-2-ene-1,8-diol, isoeugenol, diosphenol, β-terpinyl acetate, ocimenol, 2,8-dihydroxy-*p*-menth-3-en-5-one, *p*-menth-1-en-7,8-diol, linalool 3,7-oxide, oleic acid [4].	Butanoic acid [4].	Hexadecanoic acid methyl ether, ethyl palmitate [4].	2-hydroxy-6-methoxyacetophenone, 2-pentadecanone [4].	9-octadecyne, 3,3,6-trimethyl 1,5-heptadiene [4].

2.5.2. Biological Activity

The biological activity attributed to *A. mexicana* differs between subspecies because each one has a different chemical profile [4]. It also differs with respect to essential oils or extracts, as well as secondary metabolites present in them. Terpenes and flavonoids, such as ursolic acid, oleanolic acid, acacetin, apigenin, and tilianin, are the most active [33]. In some studies, where the biological activity of *A. mexicana* is determined, the studied subspecies is not specified. Tables 1 and 3 show the biological activities of the essential oils, extracts, and compounds for each subspecies.

Antihypertensive, Vaso-Relaxant, Spasmolytic, and Spasmogenic Properties

Pharmacological studies correlate with the ethnomedicinal uses of *A. mexicana*. In 1982, infusions from both subspecies underwent pharmacological studies. Results showed contrasting effects for each subspecies. It was found that *A. mexicana* ssp. xolocotziana contracted the aorta, bladder, intestinal and uterine muscles, and the heart in experiments with frogs [5]. Studies performed in 2009 and 2010 showed antioxidant and vasoactive activities for *A. mexicana* ssp. mexicana extracts (Table 3) [19]. Additionally, the flavonoid tilianin, extracted from the plant, had antihypertensive and vasorelaxant effects on in vitro experiments, as observed on rat aortic rings and in vivo experiments in spontaneously hypertensive rats (SHR) [35]. A later study validated a liquid chromatographic method to detect tilianin in aqueous and organic (methanolic and hydroalcoholic) extracts of *A. mexicana* ssp. mexicana and correlated the biological activity with tilianin content and extraction conditions. The methanolic extracts had higher concentrations of tilianin and were the more vasorelaxant on thoracic aorta rat rings compared to carbachol, while the methanol extracts from dried biomass at 100, 90, and 50 °C were potent vasorelaxants [40]. Tilianin did not have toxic effects in sub-acute and acute oral administration in mice [36]. The vasorelaxant activity of the dichloromethane soluble extract from *A. mexicana* ssp. mexicana and its components (ursolic acid, oleanolic acid, and acacetin) also showed therapeutic effects: ursolic acid and acacetin had antihypertensive activity [38] (Table 3).

Furthermore, spasmogenic and spasmolytic activities differ between the two subspecies [28]. *A. mexicana* ssp. mexicana extracts were spasmogenic in guinea pig ileum, while *A. mexicana* ssp. xolocotziana extracts had a spasmolytic effect. Additionally, subs. xolocotziana contains a higher amount of acacetin and tilianin. Thus, only *A. mexicana* ssp. xolocotziana should be used to treat gastrointestinal afflictions [28] (Table 3). These results disagree with those reported in 1982 [5], but confirm each toronjil subspecies' contrasting pharmacological effects.

Studies of the effect of *A. mexicana* ssp. xolocotziana hexanic, dichloromethanic and methanolic extracts on tracheal rat rings found a relaxant-like activity [41], while the essential oil of *A. mexicana* ssp. mexicana caused relaxation of guinea pig tracheal tissue. The essential oil contains primarily estragole and D-limonene, which act as relaxants and anti-asthmatic compounds [31] (Tables 1 and 3). These studies support the potential therapeutic use of *A. mexicana* for asthma treatment.

Plant-tissue cultures of *A. mexicana* ssp. mexicana further confirmed the observed in vivo antihypertensive and vasorelaxant effects in SHR [33,35]. Tilianin isolated from methanolic extracts obtained from in vitro plantlets and calli confirmed the conservation of its vasorelaxant effects. The in vitro methanolic extracts contained a higher concentration of tilianin and produced a stronger vasorelaxant effect on aorta rat rings than extracts from wild plants [33] (Table 3).

Analgesic and Anti-inflammatory Properties and Effects in the Central Nervous System

The first pharmacological study on the effects of water-soluble *A. mexicana* extract on the central nervous system showed an anxiogenic-like effect in behavioral experiments at the doses tested in male rats [42]. Chemical and pharmacological studies performed in 2014 to identify the effects aqueous extracts from both subspecies have on the central nervous system found similar chemical profiles but different compound abundances.

Low doses of the extracts produced an anxiolytic effect, but higher doses sedated mice. Flavonoid derivatives may be responsible for the observed pharmacological effect [3]. Additionally, organic extracts of *A. mexicana* ssp. xolocotziana contain acacetin and ursolic acid and produce anxiolytic, spasmolytic, and antinociceptive effects in in vitro and in vivo experiments in mice [39]. In vivo experiments in different pain models in rodents confirmed the antinociceptive effect of organic *A. mexicana* ssp. xolocotziana extracts and ursolic acid [24,39]. Further behavioral experiments in mice determined the anxiolytic effects of *A. mexicana* ssp. mexicana methanolic extract and tilianin; they confirmed lemon balm contains tilianin, an anxiolytic compound, plus acacetin and ursolic acid [37] (Table 3). Taken together, results from pharmacological studies validate the traditional use of toronjil (lemon balm) to relieve gastrointestinal disorders, stomach pain, asthma, anxiety, insomnia, and hypertension.

Antioxidant and Nutraceutical Properties

Traditional Mexican medicine promotes lemon balm as an herbal product. However, herbal products lack strict quality control to guarantee their chemical composition or authenticity for manufacture. Thus, the consumer may not experience the expected therapeutic effect. However, various herbal products containing *A. mexicana* found significant antioxidant activity [43]. Additional reports detailed similar antioxidant activity of *A. mexicana* [22,34] (Table 3).

A. mexicana var. "Sangria" (ssp. mexicana) inflorescences are edible and have nutraceutical potential as they contain sugars and secondary metabolites. Compared to other *Agastache* species and the Lamiaceae family members, lemon balm inflorescences have higher polyphenols and flavonoid content and higher antioxidant properties [22].

Antifungal and Phytotoxic Properties

Aside from its use as a medicinal plant, *A. mexicana* produces bioactive compounds with antifungal activity (Table 1). The essential oil contains monoterpenes and phenylpropanoids, such as estragole and methyl eugenol, which act as antifungal compounds. Notably, the essential oil did not show toxicity against human macrophages and brine shrimp. Research has shown the potential for its use as a non-toxic botanical fungicidal and as an alternative to synthetic fungicides [32]. A recent study tested the effect of adding *A. mexicana* essential oil to wheat grains as a food preservative for flour and dough. The quality of the dough and cookies prepared with the treated flour did not decrease and the growth of fungal pathogens of *Aspergillus*, *Eurotium*, *Eupenicillum* and *Penicillium* species were reduced (see Table 1). Additionally, by day 49, 79.2% of the added amount persisted. These properties indicate the essential oil as a candidate non-toxic food preservative [44].

In addition, the phytotoxic potential of organic extracts obtained with hexane, acetone and ethanol was explored (Table 3). The acetone extract of *A. mexicana* (subspecies not specified) leaves was the most active, with an IC_{50} of 71 µg/mL on the radical growth of *Amaranthus hypochondriacus* L. [45].

Table 3. Biological activity of extracts and compounds isolated from *Agastache mexicana*.

Taxa	Antioxidant	Antimicrobial	Phytotoxic	Central Nervous System	Antihypertensive and Vasorelaxant	Spasmolytic and Antinociceptive
Agastache mexicana	Reduction percentage: **Methanol extract:** DPPH ~93%, ABTS ~99%, and TBARS ~94%. **Eugenol:** DPPH ~94%, ABTS ~98%, and TBARS ~98% [34]. **Aqueous extract:** DPPH (IC_{50} 502.3 µg mL^{-1}) and TEAC (926.9 µmol Trolox g extract^{-1}) [19]. DPPH assay of herbal products containing *A. mexicana*: **Hydroalcoholic extracts** reduction percentage: A, 80.3%; B, 81.4%; C, 80.9%; D, 83.1% [43].	**Aqueous extract** for the synthesis of silver nanoparticles with activity against *Escherichia coli* [46].	Phytotoxic activity at 1000 µg mL^{-1} (% of growth inhibition): **hexane extract** (60.5%) **acetone extract** (85.7%) and **ethanolic extract** (35.5%) on *Amaranthus hypochondriacus* L. **Acetone extract** (48.7%) on *Echinochloa crus-galli* (L.) P Beauv. [45].	**Aqueous extract:** Anxiogenic-like effect in male Wistar rats at doses of 3–12 mg kg^{-1} in elevated plus-maze, forced swimming, and open field tests [42].	Vasorelaxant effect on rat aortic rings: **methanolic extract** of wild plants (Emax = 31.96%, EC_{50} = 113.72 µg mL^{-1}), in vitro plantlets (Emax = 37.0%, EC_{50} = 82.64 µg mL^{-1}) and callus (Emax = 59.64%, EC_{50} = 105.43 µg mL^{-1}) [33]. **Aqueous extract:** EC_{50} 233.7 µg mL^{-1} and Emax 24.9% [19].	
Agastache mexicana Linton & Epling ssp. mexicana	DPPH assay of **hydroalcoholic extract:** IC_{50} 1.4 mg mL^{-1} [22].			Anxiolytic effect in mice: **Methanol extract** and **Tilianin** at dosage of 30 mg kg^{-1} (ip.) or 300 mg kg^{-1} (po.) [37]. **Aqueous extract:** activity at low doses (0.1–10.0 mg kg^{-1}). Reduced motor coordination and sedative-like actions at high doses (100–200 mg kg^{-1}). Toxicity: LD_{50} > 5000 mg kg^{-1} [3].	Vasorelaxant effect in rat aortic rings: **Dichloromethane extract** Emax 76.27%, IC_{50} 189.06 µg mL^{-1} [35]. **Methanolic extract:** Emax 82.3% and EC_{50} 291.25 µg mL^{-1} [40]. **Acacetin:** Emax 63.4% and EC_{50} 210.84 µM. **Ursolic acid:** Emax 86% and EC_{50} 39.56 µM and in vivo antihypertensive action on SHR [38]. **Tilianin** induced NO overproduction in rat aorta: 1.49–0.86 µM of nitrites g^{-1} of tissue and vasorelaxant effect at 0.002–933 µM, Emax 84.7% and EC_{50} 104.4 µg mL^{-1}. Antihypertensive action on SHR at 50 mg kg^{-1} [35,40]. LD_{50} of 6624 mg kg^{-1} in mice and antihypertensive effect (ED_{50} 53.51 mg kg^{-1}) in SHR [36].	**Methanolic extract:** spasmogenic effect on guinea pig ileum. Maximal contractile response with 316 µg mL^{-1} (60%) [28].

Table 3. *Cont.*

Taxa	Antioxidant	Antimicrobial	Phytotoxic	Central Nervous System	Antihypertensive and Vasorelaxant	Spasmolytic and Antinociceptive
Agastache mexicana ssp. xolocotziana Bye, E.L. Linares & Ramamoorthy				Anxiolytic effect in mice: **Acacetin** at dosage of 100–300 mg kg^{-1} in mice [39]. **Aqueous extract** activity at low doses (0.1–10.0 mg kg^{-1}). Reduced motor coordination and sedative-like actions at high doses (100–200 mg kg^{-1}). Toxicity: LD$_{50}$ of 3807 mg kg^{-1} [3].	Relaxant effect on rat tracheal rings. **Hexane extract:** Emax 100.16% and EC$_{50}$ 219 µg mL^{-1}. **Dichloromethane extract:** Emax 97.78% and EC$_{50}$ 320.8 µg mL^{-1}. **Methanol extract:** Emax 75.54% and EC$_{50}$ 644.44 µg mL^{-1} [41].	Spasmolytic effect on guinea pig ileum: **Methanolic extract** maximal relaxant effects: 100 µg mL^{-1} (72.6%)–316.2 µg mL^{-1} (68.6%) [28]. **Acacetin** IC$_{50}$ of 1.1 µM and antinociceptive activity in mice (ED$_{50}$ 2 mg kg^{-1}). **Ursolic acid:** spasmolytic response and antinociceptive effect: ED$_{50}$ 3 mg kg^{-1} [39] and 2 mg kg^{-1} in mice [24]; ED$_{50}$ 44 mg kg^{-1} in rats [24]. Writhing test in mice: maximum latency at 300 mg kg^{-1} and antinociceptive response of extracts: **hexane** 73% (ED$_{50}$: 56.68 mg kg^{-1}), **ethyl acetate** 90% (ED$_{50}$: 31.81 mg kg^{-1}), and **methanol** 48% (ED$_{50}$: 253.25 mg kg^{-1}). Anti-inflammatory activity on the rat paw and formalin tests. Plantar test: antinociceptive responses of **hexane extract** from 30 to 300 mg kg^{-1} [39].

* DPPH: 1,1-Diphenyl-2-picrylhydrazyl; ABTS: 2,2′-azino-bis (3-ethylbenzothiazoline-6-sulfonic acid), TBARS: thiobarbituric acid reactive substance, TEAC: Trolox equivalent antioxidant capacity, NO: nitric oxide, LD$_{50}$: median lethal dose, IC$_{50}$: mean inhibitory concentration, EC$_{50}$: mean effective concentration, Emax: maximum relaxant effect, ED$_{50}$: median effective dose, SHR: spontaneously hypertensive rats, ip: intraperitoneal injection, po: oral administration.

3. Potential and Perspectives

The holistic approach to studying the *A. mexicana* species focuses on biology, ethnobotany, chemical composition, and biological activity. The species has potential pharmacological uses as a source of bioactive compounds, such as tilianin, acacetin, apigenin, ursolic acid, and oleanolic acid [33] in areas such as drug development, disease modeling, and other biological explorations [47].

Agro-industrial applications and the production of essential oils require greater knowledge and understanding of endemic and native species of cultivated aromatic plants, such as *Agastache mexicana*. It is also essential to develop appropriate technologies for industrial applications and products. The information provided in this review supports the cultivation of lemon balm to take advantage of the plant, extracts, and essential oils. Red lemon balm has high essential oil yields, averaging 2.26%, regardless of the type of distillation device [27], while white lemon balm yields about 1.2%. This essential oil has proven antifungal activity against eleven strains isolated from wheat grains during storage [32]. Extracts from the leaves of *A. mexicana* contain reducing compounds like phenols and flavonoids and have been successfully used to provide a reducing medium for the synthesis of nanoparticles as well as their stability [46].

Progress has been made in the botanical and anatomical differentiation of the two identified subspecies. Evidence from biotechnological studies show that *A. mexicana* plant tissue cultures have great potential as a source of tilianin and other bioactive compounds [33], but information is scarce in terms of a technological package of cultivation and standardization of its components. There is phenotypic variability between subspecies and populations concerning wild or cultivated plants [23]. These results suggest that there may be genetic variability and the potential for genetic improvement of *A. mexicana* to increase plant biomass, improve resistance to climatic factors, resistance to pests and diseases. Furthermore, this variability could allow for the development of populations with specific chemotypes. For this reason, a holistic approach to the study of the species could help visualize a broader panorama that allows the sustainable use of lemon balm.

Author Contributions: Conceptualization, M.P.-T.; writing-original draft preparation, M.P.-T., R.E.S.-F. and M.S.-H.; writing-review and editing, M.P.-T. and M.S.-H. All authors have read and agreed to the published version of the manuscript.

Funding: This research received no external funding.

Data Availability Statement: Publicly available datasets were analyzed in this study (Figure 1). This data can be found here: https://datosabiertos.unam.mx/biodiversidad/ (accessed on 20 April 2021).

Acknowledgments: The authors would like to thank Manuel Jimenez Vasquez for his support in producing Figures 3 and 4.

Conflicts of Interest: The authors declare no conflict of interest.

References

1. Martínez-Gordillo, M.; Fragoso-Martínez, I.; García-Peña, M.D.R.; Montiel, O. Géneros de Lamiaceae de México, diversidad y endemismo. *Rev. Mex. Biodivers.* **2013**, *84*, 30–86. [CrossRef]
2. Santillán, M.A.; López, M.E.; Aguilar, S.; Aguilar, A. Estudio etnobotánico, arquitectura foliar y anatomía vegetativa de *Agastache mexicana* ssp. mexicana y A. mexicana ssp. xolocotziana. *Rev. Mex. Biodivers.* **2008**, *79*, 513–524.
3. Estrada-Reyes, R.; López-Rubalcava, C.; Ferreyra-Cruz, O.A.; Dorantes-Barrón, A.M.; Heinze, G.; Moreno Aguilar, J.; Martínez-Vázquez, M. Central nervous system effects and chemical composition of two subspecies of *Agastache mexicana*; An ethnomedicine of Mexico. *J. Ethnopharmacol.* **2014**, *153*, 98–110. [CrossRef]
4. Estrada-Reyes, R.; Aguirre Hernández, E.; García-Argáez, A.; Soto Hernández, M.; Linares, E.; Bye, R.; Heinze, G.; Martínez-Vázquez, M. Comparative chemical composition of *Agastache mexicana* subsp. mexicana and A. mexicana subsp. xolocotziana. *Biochem. Syst. Ecol.* **2004**, *32*, 685–694. [CrossRef]
5. Bye, R.; Linares, E.; Ramamoorthy, T.P.; García, F.; Collera, O.; Palomino, G.; Corona, V. *Agastache mexicana* Subs. xolocotziana (Lamiaceae). A new taxon from mexican medicinal plants. *Phytologia* **1987**, *62*, 156–163.
6. Martínez-Gordillo, M.; Bedolla-García, B.; Cornejotenorio, G.; Fragoso-Martínez, I.; García-Peña, M.D.R.; González-Gallegos, J.G.; Lara-Cabrera, S.I.; Zamudio, S. Lamiaceae de México. *Bot. Sci.* **2017**, *95*, 780–806. [CrossRef]

7. Universitarios, D.G. de R. Dirección General de Repositorios Universitarios, Universidad Nacional Autónoma de México. Portal de Datos Abiertos UNAM, Colecciones Universitarias. Available online: https://datosabiertos.unam.mx/ (accessed on 21 March 2021).

8. Verpoorte, R.; Crommelin, D.; Danhof, M.; Gilissen, L.J.W.J.; Schuitmaker, H.; van der Greef, J.; Witkamp, R.F. Commentary: "A systems view on the future of medicine: Inspiration from Chinese medicine"? *J. Ethnopharmacol.* **2009**, *121*, 479–481. [CrossRef] [PubMed]

9. Fierascu, R.C.; Fierascu, I.; Ortan, A.; Georgiev, M.I.; Sieniawska, E. Innovative approaches for recovery of phytoconstituents from medicinal/aromatic plants and biotechnological production. *Molecules* **2020**, *25*, 309. [CrossRef]

10. Mushtaq, M.Y.; Verpoorte, R.; Kim, H.K. Zebrafish as a model for systems biology. *Biotechnol. Genet. Eng. Rev.* **2013**, *29*, 187–205. [CrossRef]

11. Villaseñor, J.L. Checklist of the native vascular plants of Mexico. *Rev. Mex. Biodivers.* **2016**, *87*, 559–902. [CrossRef]

12. Zielińska, S.; Matkowski, A. Phytochemistry and bioactivity of aromatic and medicinal plants from the genus *Agastache* (Lamiaceae). *Phytochem. Rev.* **2014**, *13*, 391–416. [CrossRef]

13. Ayres, G.S.; Widrlechner, M.P. The Genus Agastache as Bee Forage: A Historical Perspective. *Am. Bee J.* **1994**, *134*, 341–348.

14. Anand, S.; Deighton, M.; Livanos, G.; Morrison, P.D.; Pang, E.C.K.; Mantri, N. Antimicrobial activity of *Agastache* honey and characterization of its bioactive compounds in comparison with important commercial honeys. *Front. Microbiol.* **2019**, *10*, 1–16. [CrossRef]

15. Flora Melífera de la Ciudad de México. In *Fortalecimiento de la Producción Apícola en Suelo de Conservación de la Ciudad de MéxicoN*; Food and Agriculture Organization of the United Nations: Rome, Italy, 2020. [CrossRef]

16. Gutiérrez, S.L.G.; Chilpa, R.R.; Jaime, H.B. Medicinal plants for the treatment of "nervios", anxiety, and depression in Mexican Traditional Medicine. *Rev. Bras. Farmacogn.* **2014**, *24*, 591–608. [CrossRef]

17. Bye, R.A. Medicinal plants of the sierra madre: Comparative study of tarahumara and Mexican market plants. *Econ. Bot.* **1986**, *40*, 103–124. [CrossRef]

18. Patwardhan, B.; Vaidya, A.; Chorghade, M.; Joshi, S. Reverse Pharmacology and Systems Approaches for Drug Discovery and Development. *Curr. Bioact. Compd.* **2008**, *4*, 201–212. [CrossRef]

19. Ibarra-Alvarado, C.; Rojas, A.; Mendoza, S.; Bah, M.; Gutiérrez, D.M.; Hernández-Sandoval, L.; Martínez, M. Vasoactive and antioxidant activities of plants used in Mexican traditional medicine for the treatment of cardiovascular diseases. *Pharm. Biol.* **2010**, *48*, 732–739. [CrossRef] [PubMed]

20. Lubbe, A.; Verpoorte, R. Cultivation of medicinal and aromatic plants for specialty industrial materials. *Ind. Crops Prod.* **2011**, *34*, 785–801. [CrossRef]

21. Linares, E.; Bye, R. Traditional Markets in Mesoamerica: A Mosaic of History and Traditions. *Ethnobot. Mex.* **2016**, 151–177. [CrossRef]

22. Najar, B.; Marchioni, I.; Ruffoni, B.; Copetta, A.; Pistelli, L.; Pistelli, L. Volatilomic analysis of four edible flowers from agastache genus. *Molecules* **2019**, *24*, 4480. [CrossRef]

23. Carrillo-Galván, G.; Bye, R.; Eguiarte, L.E.; Cristians, S.; Pérez-López, P.; Vergara-Silva, F.; Luna-Cavazos, M. Domestication of aromatic medicinal plants in Mexico: Agastache (Lamiaceae)- A n ethnobotanical, morpho-physiological, and phytochemical analysis. *J. Ethnobiol. Ethnomed.* **2020**, *16*, 1–16. [CrossRef]

24. Verano, J.; González-trujano, M.E.; Déciga-campos, M.; Ventura-martínez, R.; Pellicer, F. Pharmacology, Biochemistry and Behavior Ursolic acid from *Agastache mexicana* aerial parts produces antinociceptive activity involving TRPV1 receptors, cGMP and a serotonergic synergism. *Pharmacol. Biochem. Behav.* **2013**, *110*, 255–264. [CrossRef] [PubMed]

25. Svoboda, K.P.; Gough, J.; Hampson, J.; Galambosi, B. Analysis of the essential oils of some Agastache species grown in Scotland from various seed sources. *Flavour Fragr. J.* **1995**, *10*, 139–145. [CrossRef]

26. Kovalenko, N.A.; Supichenko, G.N.; Leontiev, V.N.; Shutova, A.G. Composition of essential oil of plants some species of the genus Agastache L. introduced in Belarus. *Proc. Natl. Acad. Sci. Belarus Biol. Ser.* **2019**, *64*, 147–155. [CrossRef]

27. Jadczak, P.; Bojko, K.; Wesołowska, A. Chemical composition of essential oils isolated from Mexican giant hyssop [*Agastache mexicana* (Kunth.) Link. & Epling.] via hydrodistillation in Deryng and Clevenger apparatuses. *Ann. Hortic.* **2017**, *27*, 11–17. [CrossRef]

28. Ventura-Martínez, R.; Rodríguez, R.; González-Trujano, M.E.; Ángeles-López, G.E.; Déciga-Campos, M.; Gómez, C. Spasmogenic and spasmolytic activities of *Agastache mexicana* ssp. mexicana and A. mexicana ssp. xolocotziana methanolic extracts on the guinea pig ileum. *J. Ethnopharmacol.* **2017**, *196*, 58–65. [CrossRef] [PubMed]

29. Manjarrez, A.; Mendoza, A. The volatile oils of Agastache mexicana (Benth) Epling and Cunila lythrifolia Benth. *Perfum. Essent. Oil. Rec.* **1966**, *57*, 561–562.

30. Myadelets, M.A.; Vorobyeva, T.A.; Domrachev, D.V. Composition of the Essential Oils of Some Species Belonging to Genus Agastache Clayton ex Gronov (Lamiaceae) Cultivated under the Conditions of the Middle Ural. *Chem. Sustain. Dev. 21* **2013**, *21*, 397–401.

31. Navarrete, A.; Ávila-Rosas, N.; Majín-León, M.; Balderas-López, J.L.; Alfaro-Romero, A.; Tavares-Carvalho, J.C. Mechanism of action of relaxant effect of agastache mexicana ssp. Mexicana essential oil in guinea-pig trachea smooth muscle. *Pharm. Biol.* **2017**, *55*, 96–100. [CrossRef]

32. Juárez, Z.N.; Hernández, L.R.; Bach, H.; Sánchez-arreola, E.; Bach, H. Antifungal activity of essential oils extracted from *Agastache mexicana* ssp. xolocotziana and Porophyllum linaria against post-harvest pathogens. *Ind. Crop. Prod.* **2015**, *74*, 178–182. [CrossRef]

33. Carmona-Castro, G.; Estrada-Soto, S.; Arellano-García, J.; Arias-Duran, L.; Valencia-Díaz, S.; Perea-Arango, I. High accumulation of tilianin in in-vitro cultures of *Agastache mexicana* and its potential vasorelaxant action. *Mol. Biol. Rep.* **2019**, *46*, 1107–1115. [CrossRef] [PubMed]

34. Esquivel-Gutiérrez, E.R.; Coria-Orozco, E.; Torres-Martínez, R.; Hernández-García, A.; Ríos-Chávez, P.; Manzo-Ávalos, S.; Saavedra-Molina, A.; Salgado-Garciglia, R. Antioxidant effects of Agastache mexicana extracts: An in vitro approach. *FASEB J.* **2017**, *31*, lb117. [CrossRef]

35. Hernández-Abreu, O.; Castillo-España, P.; León-Rivera, I.; Ibarra-Barajas, M.; Villalobos-Molina, R.; González-Christen, J.; Vergara-Galicia, J.; Estrada-Soto, S. Antihypertensive and vasorelaxant effects of tilianin isolated from Agastache mexicana are mediated by NO/cGMP pathway and potassium channel opening. *Biochem. Pharmacol.* **2009**, *78*, 54–61. [CrossRef] [PubMed]

36. Hernández-Abreu, O.; Torres-Piedra, M.; García-Jiménez, S.; Ibarra-Barajas, M.; Villalobos-Molina, R.; Montes, S.; Rembao, D.; Estrada-Soto, S. Dose-dependent antihypertensive determination and toxicological studies of tilianin isolated from *Agastache mexicana*. *J. Ethnopharmacol.* **2013**, *146*, 187–191. [CrossRef]

37. González-Trujano, M.E.; Ponce-mu, H.; Hidalgo-figueroa, S.; Navarrete-Vaquez, G.; Estrada-Soto, S. Depressant effects of Agastache mexicana methanol extract and one of major metabolites tilianin. *Asian Pac. J. Trop. Med.* **2015**, 185–190. [CrossRef]

38. Flores-Flores, A.; Hernández-Abreu, O.; Rios, M.Y.; León-Rivera, I.; Aguilar-Guadarrama, B.; Castillo-España, P.; Perea-Arango, I.; Estrada-Soto, S. Vasorelaxant mode of action of dichloromethane-soluble extract from Agastache mexicana and its main bioactive compounds. *Pharm. Biol.* **2016**, *54*, 2807–2813. [CrossRef] [PubMed]

39. González-Trujano, M.E.; Ventura-Martínez, R.; Chávez, M.; Díaz-Reval, I.; Pellicer, F. Spasmolytic and antinociceptive activities of ursolic acid and acacetin identified in Agastache mexicana. *Planta Med.* **2012**, *78*, 793–799. [CrossRef] [PubMed]

40. Hernández-Abreu, O.; Durán-Gómez, L.; Best-Brown, R.; Villalobos-Molina, R.; Rivera-Leyva, J.; Estrada-Soto, S. Validated liquid chromatographic method and analysis of content of tilianin on several extracts obtained from Agastache mexicana and its correlation with vasorelaxant effect. *J. Ethnopharmacol.* **2011**, *138*, 487–491. [CrossRef] [PubMed]

41. Sánchez-Recillas, A.; Mantecón-Reyes, P.; Castillo-España, P.; Villalobos-Molina, R.; Ibarra-Barajas, M.; Estrada-Soto, S. Tracheal relaxation of five medicinal plants used in Mexico for the treatment of several diseases. *Asian Pac. J. Trop. Med.* **2014**, *7*, 179–183. [CrossRef]

42. Molina-Hernández, M.; Téllez-Alcántara, P.; Martínez, E. *Agastache mexicana* may produce anxiogenic-like actions in the male rat. *Phytomedicine* **2000**, *7*, 199–203. [CrossRef]

43. Salazar-Aranda, R.; de la Torre-Rodríguez, Y.C.; Alanís-Garza, B.A.; Pérez-López, L.A.; Waksman-de-Torres, N. Evaluación de la actividad biológica de productos herbolarios comerciales Ricardo. *Med. Univ.* **2009**, *11*, 156–164.

44. Juárez, Z.N.; Bach, H.; Bárcenas-Pozos, M.E.; Hernández, L.R. Impact of the Persistence of Three Essential Oils with Antifungal Activities on Stored Wheat Grains, Flour, and Baked Products. *Foods* **2021**, *10*, 213. [CrossRef] [PubMed]

45. Santiago, R.; Rojas, I.; Arvizu, G.; Muñoz, D.; Pérez, D.; Sucilla, M. Caracterización del potencial fitotóxico de *Agastache mexicana* (kunth.) Lint et Epling. *Investig. Univ. Multidiscip.* **2005**, *4*, 14–20.

46. López, J.L.; Baltazar, C.; Torres, M.; Ruíz-Baltazar, A.; Esparza, R.; Rosas, G. Biosynthesis of Silver Nanoparticles Using Extracts of Mexican Medicinal Plants. In *Characterization of Metals and Alloys*; Campos Pérez, R., Cuevas Contreras, A., Muñoz Esparza, R., Eds.; Springer International Publishing: Cham, Switzerland, 2017; p. 255, ISBN 9783319316949.

47. Calvo-Irabien, L.M. Native Mexican aromatic flora and essential oils: Current research status, gaps in knowledge and agro-industrial potential. *Ind. Crops Prod.* **2018**, *111*, 807–822. [CrossRef]

Review

Alkaloids of the Genus *Datura*: Review of a Rich Resource for Natural Product Discovery

Maris A. Cinelli * and A. Daniel Jones *

Department of Biochemistry and Molecular Biology, Michigan State University, East Lansing, MI 48824, USA
* Correspondence: cinellim@msu.edu or secondandone@gmail.com (M.A.C.); jonesar4@msu.edu (A.D.J.);
  Tel.: +1-906-360-8177 (M.A.C.); +1-517-432-7126 (A.D.J.)

Abstract: The genus *Datura* (Solanaceae) contains nine species of medicinal plants that have held both curative utility and cultural significance throughout history. This genus' particular bioactivity results from the enormous diversity of alkaloids it contains, making it a valuable study organism for many disciplines. Although *Datura* contains mostly tropane alkaloids (such as hyoscyamine and scopolamine), indole, *beta*-carboline, and pyrrolidine alkaloids have also been identified. The tools available to explore specialized metabolism in plants have undergone remarkable advances over the past couple of decades and provide renewed opportunities for discoveries of new compounds and the genetic basis for their biosynthesis. This review provides a comprehensive overview of studies on the alkaloids of *Datura* that focuses on three questions: How do we find and identify alkaloids? Where do alkaloids come from? What factors affect their presence and abundance? We also address pitfalls and relevant questions applicable to natural products and metabolomics researchers. With both careful perspectives and new advances in instrumentation, the pace of alkaloid discovery—from not just *Datura*—has the potential to accelerate dramatically in the near future.

Keywords: alkaloid; Solanaceae; tropane; indole; pyrrolidine; *Datura*

Citation: Cinelli, M.A.; Jones, A.D. Alkaloids of the Genus *Datura*: Review of a Rich Resource for Natural Product Discovery. *Molecules* **2021**, *26*, 2629. https://doi.org/10.3390/molecules26092629

Academic Editors: Young Hae Choi, Young Pyo Jang, Yuntao Dai and Luis Francisco Salomé-Abarca

Received: 27 March 2021
Accepted: 25 April 2021
Published: 30 April 2021

Publisher's Note: MDPI stays neutral with regard to jurisdictional claims in published maps and institutional affiliations.

1. Introduction

Perhaps no plants on Earth have been more famous—and infamous—throughout history than those in the genus *Datura*. Naturalized throughout the temperate regions of the world, *Datura* plants, like their other relatives in the family *Solanaceae*, are a rich source of bioactive phytochemicals. Although phenolics, steroids, acylsugars, amides, and other compounds from these plants have been isolated and characterized, it is the *alkaloids* found in *Datura* that have cemented these plants' role in the remedies, religions, history, and lore of different cultures in all corners of the world.

As some of the tropanes, pyrrolidines, indoles, and other alkaloids present in the plant yield both medicinal use and insidious toxicity, *Datura* has also been the subject of fairly intensive study. For over a century, a complex, multidisciplinary effort with contributions from drug discovery and pharmacology, plant breeding and bioengineering, evolutionary biology, analytical chemistry, and ethnobotany has sought to elucidate the peculiar and powerful properties of these plants and the substances within that cause them. This comprehensive review covers the alkaloids of *Datura*, the tools used to detect and identify them, the factors affecting their content and composition, and challenges facing the plant natural products field as a whole. This information could aid anyone interested in discovering novel alkaloids or studying plant breeding or engineering systems for alkaloid production.

2. History and Taxonomy

2.1. Medicine and Culture

The historical significance of *Datura* is both medicinal and cultural. *Datura* plant parts, extracts, and preparations have been used as medicines by humans for millennia. Although comprehensive analysis of the medicinal uses of *Datura* is beyond the scope of this article, the reviews of Maheshwari [1], Benitez [2], and Batool [3] offer additional insights on this topic. *Datura* is described in ancient Chinese, traditional African, and Ayurvedic medicine. These plants were also used medicinally by indigenous Americans, and their properties were also described in Medieval European texts. *Datura* was a remedy for many conditions including pain, bruises, wound infections, swellings and boils, rheumatism, and toothaches, and *Datura* cigarettes were smoked to alleviate asthma and breathing problems [1,2]. Extracts and compounds (both alkaloids and other components) derived from *Datura* have also been investigated for, among others, anticancer, antibacterial, antiviral, and antifungal activities [1,3].

The genus' most famous alkaloidal constituents are the tropane alkaloids atropine (racemic **1**, or hyoscyamine, (*S*)-**1**, see below for clarification) and scopolamine (**2**, or hyoscine), which are shown in Figure 1. Tropane alkaloids (bicyclic [3.2.1] alkaloids) are known across many plant genera beyond *Datura*, and it is worth noting that the primary commercial sources of atropine and scopolamine are usually the genus *Duboisia* (and sometimes *Brugmansia*), and not *Datura* [4]. Alkaloid **2** is a muscarinic antagonist used to treat nausea, vomiting, and motion sickness, whereas atropine (**1**) is a similar anticholinergic agent, and is used in the treatment of certain poisonings and heart conditions, and to dilate the pupils in ophthalmology. Both of these drugs are listed on the World Health Organization's Model List of Essential Medicines [5], and synthetic or semisynthetic derivatives of these drugs are also used.

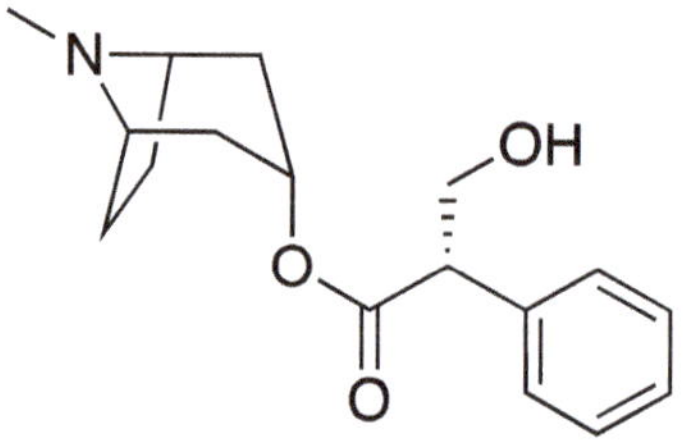

Figure 1. Hyoscyamine [(*S*)-1, the *S*-enantiomer of atropine, (**1**) and scopolamine (**2**), the most well-known alkaloids of the genus *Datura*.

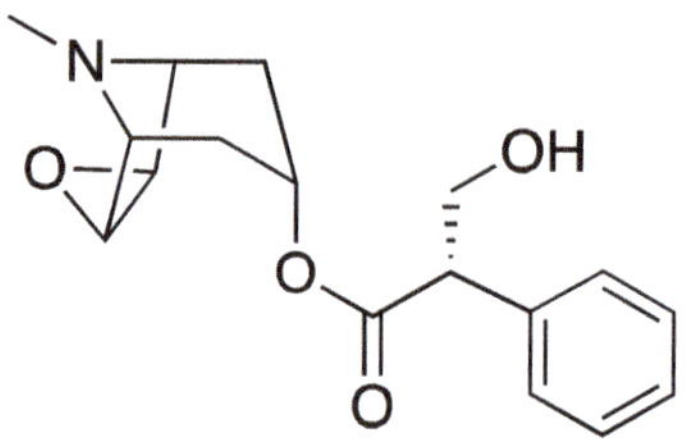

As *Datura* contains high levels of these potent alkaloids (and many others whose bioactivity is ill-defined), all parts of the plants are considered toxic, and can easily poison both humans and livestock animals if accidentally or deliberately ingested [6,7]. *Datura* poisoning in humans causes dilated pupils, increased heart rate, blood pressure, and body temperature, dry mucous membranes, confusion, depression, and incoherence [8,9]. Severe symptoms can cause heart arrhythmias, coma [8], respiratory depression, and death. In one 26-year period in the United States, *Datura* species were responsible for 20% of fatal plant poisonings in humans, despite their consumption being uncommon [7]. *Datura* poisoning also causes characteristic delirium and hallucinations. Aggression, memory and recognition loss, picking at imaginary objects, delusions of being attacked by animals, public indecency, and other erosions of social inhibition have been reported [10–14]. Many names for *Datura* in different languages, such as *devil's trumpet, malpitte* (Afrikaans, "crazy seeds", and *torna-loca* (Spanish, "maddening plant") reflect this particular property, and *Datura* has sometimes been used recreationally to induce these hallucinations. Due to their toxicity and ability to render subjects delirious and unaware, *Datura* species have also been used as a deliberate poison, especially to facilitate robberies or sexual assault [2,4,9,15].

Datura also has religious and cultural significance. As this is often the case in the same cultures where it is used medicinally, remedy, magic, tradition, and religion are sometimes difficult to separate. Indigenous American peoples brewed *Datura* into tea or chewed parts of the plant, utilizing its hallucinogenic and euphorigenic properties in initiation, divination, or luck-bringing ceremonies, or to bring skill or strength prior to hunting [16,17]. In archeological sites in the Southwestern United States and Central America, ceramic forms resembling the spiky fruit of *D. inoxia* (or a similar species) were found [17], and in California, chewed masses of *D. wrightii* were found alongside cave paintings believed to represent the flower of this plant [16]. Plants matching the description of *D. metel* appear in Hindu iconography depicting the god Shiva [18], while in Haiti, *D. stramonium* is associated with catatonia and zombies related to vodou [19]. In Europe, where *Datura* was believed to have been introduced by Roma immigrants, the plants, along with other toxic nightshades, became tied to witchcraft, the devil, and other dark or occult notions [12,15].

2.2. Description and Taxonomy

Datura species are herbaceous annuals or in some places, perennials. The plants are characterized by trumpet-shaped flowers, often malodorous foliage, and spiny fruits (excluding the smooth-fruited *D. ceratocaula*). The representative morphologies of *D. stramonium* and *D. metel* are shown in Figure 2. *Datura* is classified as a "nightshade" in the family Solanaceae, sub-family *Solanoideae*, and tribe Datureae. Datureae is comprised of four accepted sections, three of which belong to the genus *Datura*—*Dutra*, *Stramonium*, and *Ceratocaulis* [20]. The genus was originally known as Stramonium, but Linnaeus, in 1737, renamed it *Datura*, a likely Latinization of the Sanskrit *dhattura* [21], a word used to refer to the plant (probably *D. metel*) in classical texts.

Figure 2. Flower of *D. metel* (**left**) and fruit of *D. stramonium* var. *stramonium* (**right**).

There are generally considered to be nine species of Datura, although anywhere from 9–15 are reported in the literature, which can lead to confusion. *Datura stramonium* (jimsonweed, section *Stramonium*) is naturalized throughout the world. This species has four varieties (vars. *stramonium*, *tatula*, *godronii*, and *inermis*) that differ in their flower and fruit color and morphology [22]. Clustered in the same section (*Stramonium*) as Jimsonweed is *Datura ferox* (fierce thornapple). *Datura quercifolia* (from the Latin "oak-leaved") is a shrub-like, American *Datura* species that is not extensively studied, much like *Datura discolor*. Closely related to *D. discolor* is *D. inoxia* (often spelled *innoxia*, Latin for "innocent" or "innocuous"), which is native to and found throughout the Americas. *Datura wrightii*, ("Wright's *Datura*" or "Sacred *Datura*"), is often polymorphic, but it is distinguishable from *D. inoxia* [23]. Another plant related to *D. inoxia* and *D. wrightii* (and also in section Dutra) is *D. metel*, a cultivated ornamental with many varieties, including *alba*, *fastuosa*, *rubra*, *metel*, and *muricata* [24]. *Datura leichhardtii* (section *Dutra*) has two subspecies (*leichhardtii* and ssp. *pruinosa*) and is characterized by its small, yellowish-white flowers. *Datura ceratocaula* (from the Latin "horn-stalked") is an aquatic, hollow-stemmed *Datura* species native to Mexico and South America and is the lone member of section *Ceratocaulis*.

Datura taxonomy is notoriously confusing. Several other described species were later reclassified. For example, the name *D. meteloides* is ambiguous; it is usually classified as *D. wrightii*, but occasionally as *D. inoxia* [6,25,26]. *D. lanosa* was reclassified as either *D. wrightii* or *D. inoxia*, whereas *D. velutinosa* and *D. guyaquilensis* are generally accepted to be *D. inoxia* [6,25]. *D. kymatocarpa* and *D. reburra* are usually described as conspecific with *D. discolor*, although phenetic comparisons suggest that they are actually all disparate species [27]. Additionally, different varieties of a single species are sometimes classified as separate species (e.g., *D. fastuosa* for *D. metel* var. *fastuosa* or *D. tatula* for *D. stramonium* var. *tatula*). Additionally, members of the related arboreous South American genus *Brugmansia* (the fourth section of the tribe *Datureae*) were previously classified in the genus *Datura*. but these two genera have been accepted as morphologically distinct since the 1970s [28].

3. Alkaloid Isolation and Purification, and Analytical Techniques for Detection, Quantification, and Identification

Before carrying out further discussion of the structures or origin of *Datura* alkaloids, it is helpful to review both the methods used to isolate and/or purify them from plants, and the techniques and instrumentation used to detect or quantify alkaloids and gain structural insights about the molecules themselves. The natural products field is beholden to analytical chemistry, and the pace of alkaloid discovery has accelerated dramatically since the 1950s with the development of modern instrumentation and advent of new methodologies in chromatography and spectroscopy. The following section applies only to *Datura* and its alkaloids; excellent reviews for more general alkaloid analyses, or analysis of those from other genera, have been written by Christen [29,30] and Petruczynik [31].

3.1. Extraction and Purification of Alkaloids

Extractions of alkaloids from *Datura* plant tissues have been performed on fresh, flash-frozen, or dried *Datura* tissue. Tissue is often macerated or powdered (sometimes mechanically, such as the ball-milling technique used by Moreno-Pedraza [32]), and then treated with an appropriate solvent to extract the alkaloids. Most solvents described in the literature are relatively non-reactive (i.e., will not degrade the components or significantly change the chemical composition of the sample). Examples include methanol [33] (though extractions with 100% methanol may lead to artifactual reactions, particularly with aldehydes [34]), methanol/water or ethanol/water mixtures [35,36], 5% concentrated ammonium hydroxide in ethanol [37], or ethyl acetate (following alkalization) [38]. Some reports describe the addition of acid to the solvent, as alkaloids are more soluble in water or polar solvents when protonated: Witte [39] used 2 M hydrochloric acid, while Doncheva and colleagues [40] used 3% sulfuric acid, although strong acidic and basic conditions present a risk that reactive groups may decompose during extraction.

Samples are sometimes heated or agitated to facilitate extraction. Temerdashev [41] compared different extraction methods for extracting *D. metel*, including those with heating and shaking, and found that a 1:1 mixture of 0.1 M hydrochloric acid/70% ethanol in water, heated to 60 °C in a water bath, yielded maximum recovery of alkaloids hyoscyamine and scopolamine. Djilani et al. [42] also compared extraction methods, and found that classical room-temperature solvent extraction, Soxhlet (reflux-temperature continuous solid-liquid) extraction, and solvents containing surfactants did not dramatically differ in their efficiency of extracting alkaloids from *D. stramonium*. This should not come as a great surprise, since the known *Datura* alkaloids range from polar to semi-polar compounds soluble in a wide range of solvents.

Following removal of insoluble materials by filtration or centrifugation, extracts are analyzed (such as in GC-MS or LC-MS metabolic profiling studies) without further processing, but additional measures can be taken to remove potentially interfering non-alkaloid substances. If an acidic extraction medium is used, it may be extracted with a nonpolar solvent (such as hexanes) to remove nonbasic lipids [43]. Since many *Datura* alkaloids are usually organic-soluble in uncharged free base forms, polar aqueous extracts can be basified (with, for example, sodium hydroxide) and extracted with a semipolar solvent such as chloroform [43], dichloromethane [44], or ethyl acetate [45]. More common is the application of the basic extract to solid-phase extraction columns, where the alkaloids are then eluted with dichloromethane or a similar solvent, as described by Berkov and colleagues (see below). In a recent report, Ciechomska [46] and colleagues developed a highly replicable, microwave-assisted extraction followed by solid-phase extraction to isolate atropine and scopolamine from *D. metel*, a technique which the authors report to consume less sample and less time than solid-liquid or Soxhlet extraction. It is worth noting that all published extraction methods should be approached critically. While certain conditions may work well for targeted compounds (such as **1** and **2**), other alkaloids may degrade under acidic conditions, whereas esters may be hydrolyzed under basic conditions, particularly at elevated temperatures. More labile alkaloids could potentially be missed in analysis because of harsh extraction conditions.

More specific purifications of individual alkaloids have been carried out using various methods, which include classical (usually normal phase) column chromatography, preparative or semi-preparative HPLC, preparative TLC, recrystallization, and other specialized stepwise protocols. Liu and colleagues [47] purified four indole alkaloids (**3–6**, Figure 12) from *D. metel* using silica gel column chromatography (eluting with chloroform/methanol), followed by preparative C_{18} HPLC and Sephadex column chromatography, to yield the indole alkaloids in 96.4–97.8% purity, although the yield was very low—only milligram quantities of the alkaloids were isolated from 7 kg of *D. metel* seeds. Welegergs et al. [48] purified a trisubstituted tropane alkaloid (β-**7**) from *D. stramonium* seeds using a combination of silica gel chromatography and preparative TLC; Siddiqui and co-workers similarly used preparative TLC (chloroform/ether) to isolate the nonpolar benzoyltropane datumetine (**8**) from *D. metel*, which they then recrystallized as a high-purity hydrochloride salt [45]. In a unique example of column chromatography, Beresford and Woolley first used a column of Kieselguhr (diatomaceous earth) buffered with phosphate (pH 6.8) to fractionate *D. ceratocaula* aerial part extract into three "bases" when eluting with light petrol followed by ether; alkaloid **8** (SI Table S1) was isolated from one fraction [49]. A method for isolation of high purity **2** (>99%) from "Hindu Datura" (presumably *D. metel*) flowers was recently reported by Fu [50]. Following several liquid-liquid extraction, basification, and back-extraction steps, the crude alkaloid residue was applied to a bed of D151 resin (a macroporous, cation-exchange resin), and then desorbed using an acidic aqueous solution (pH 1). Basification, desalting, and treatment with hydrobromic acid afforded crystalline scopolamine hydrobromide.

3.2. Crude Detection, Thin-Layer Chromatographic (TLC), Colorimetric, Densitometric, and Optical Methods

Early detections of *Datura* alkaloids employed colorimetric or spectrophotometric means, some of which are still employed today. Fuller and Gibson used the Vitali-Morin reaction on extracts of *D. stramonum* var. *tatula* [51]. In this reaction, the crude alkaloid extract was treated with nitric acid and then acetone in the presence of potassium hydroxide—the reaction of acetone with nitrated aromatic groups in alkaloids results in a red color, which can be quantified spectrophotometrically at 540 nm. This procedure is limited to aromatic alkaloids and, therefore, under-quantifies total alkaloid content. Alkaloids have also been titrated with sulfuric acid and bromocresol green [52], or with sulfuric acid and then back-titrated with sodium hydroxide and methyl red [53]. The Mayer reagent (mercuric chloride and potassium iodide) forms a white precipitate upon reaction with alkaloids; it has even been used for preparative scale alkaloid isolation from *D. stramonium* [42]. The most common reagent still used for alkaloid detection is the Dragendorff reagent (potassium bismuth iodide). This has broad reactivity for *Datura* alkaloids, including tropanes, indoles, and pyrrolidines, but it does not detect calystegines [54].

Alkaloids in crude mixtures can be separated and detected by TLC. Most frequently, silica-backed glass TLC plates are used. Visualization of alkaloids may use short-wave UV light, iodine [45], iodoplatinate or Dragendorff reagent, or densitometry and digital processing [55,56]. Advanced TLC methods have been published: Sharma et al. [55] employed high-performance TLC with an automatic TLC sampler and densitometric measurements following Dragendorff staining. This technique allowed the authors to determine alkaloid "fingerprints" for morphotypes of *D. metel* and quantify atropine and scopolamine, which separated very well using this method. Malinowska et al. [56] found that ETLC, or planar electrochromatography, was effective at resolving the alkaloids of *D. inoxia* [57]. Although TLC can be performed very quickly, it often has poor sensitivity: stain colors can fade, excessive background staining often occurs, and many components are not UV-active (or stain-reactive), making exact quantitative measurements very difficult. Duez et al. [58] however, described how staining TLC plates with *p*-dimethylaminobenzaldehyde (following development with a mixture of 1,1,1-trichloroethane and diethylamine) and using scanning densitometry resulted in alkaloid separation and quantification power comparable to HPLC, although it is worth noting that this report was published before the advent of many modern HPLC instruments.

Although it is uncommon, infrared (IR) spectroscopy has been applied to *Datura* alkaloids. Alkaloids (*S*)-**1** and **2** have characteristic IR bands at 1732 and 1170 cm^{-1} (indicative of esters, with the former being the carbonyl C=O stretch), 635 and 696 cm^{-1} (out-of-plane monosubstituted aromatic bending), and 858 cm^{-1} (scopolamine epoxide). By examining these bands in the ATR-FTIR (attenuated total reflectance Fourier transform IR) spectra of leaf powder of several nightshade genera (*Atropa*, *Datura*, *Duboisia*, *Hyoscyamus*, and *Solanum*), Naumann and colleagues were able to group similar spectra and chemotaxonomically organize these genera, as well as develop calibration curves for quantifying **1** and **2** in nightshades [59]. Ultraviolet (UV)-visible spectroscopy, by contrast, is commonly employed, and almost always in tandem with HPLC, although the UV–VIS spectra of individual compounds are sometimes collected as part of characterization [47]. Monitoring is usually performed at a wavelength of 210 nm, as most UV-active functionalities, including carbonyl groups, absorb in this region [60]. Wavelengths of 254 or 259 nm are also used, although these are more specific for aromatic alkaloids [35,58]. Two drawbacks of monitoring via UV–VIS spectroscopy are (a) that some alkaloids do not have functional groups that absorb UV light in a useful wavelength range, so they cannot be detected by this method; and (b) other compounds containing UV-active functional groups (e.g., phenolics, unsaturated fatty acids, amino acids, steroids, and terpenes) can interfere, especially if chromatographic resolution is poor.

As most alkaloids are chiral compounds, polarimetry (the rotation of plane-polarized light) is sometimes used in elucidation of absolute configuration as part of characterization. One elegant example is found in the work of Beresford and Woolley [49] who isolated alkaloid **9** (above and below) from *D. ceratocaula*. This alkaloid contains two chiral moieties (the tropane and the methylbutyric acid), which were first separated by hydrolysis of the ester bond. The dihydroxytropane was then esterified using tigloyl chloride, and polarimetry confirmed an identical specific rotation to that reported in the literature for (−)-3α,6β-ditigloyloxytropane, confirming the absolute stereochemistry of the tropane. Similarly, the acyl portion was confirmed as (+)-2-methylbutyric acid by polarimetry and comparison to the literature.

3.3. Gas Chromatography-Mass Spectrometry (GC-MS)

Mass spectrometry is extremely useful and powerful for alkaloid profiling, as many alkaloids, especially tropanes and pyrrolidines, lack useful UV or fluorescent chromophores but form characteristic fragment ions whose masses are diagnostic of their substitution patterns. Putative structures can be assigned from these fragment ions and with the aid of standards [on the basis of both their fragmentation and retention time (LC) or retention index (GC)]. GC-MS has been the most widely employed technique for studying *Datura* alkaloids. It has been used for assessing the impact of experimental variables on alkaloid content, examining the metabolites of transformed root cultures, discovering novel compounds, studying alkaloid biosynthesis or metabolism, metabolite profiling, and for chemotaxonomy [22,40,61–64].

Usually plant extracts or compounds are analyzed as-is, although pre-derivatization has been performed [38,41,65,66] by treatment with *N,O*-bis(trimethylsilyl)trifluoroacetamide (BSTFA) or *N*-methyl-*N*-trimethylsilyltrifluoroacetamide (MSTFA). These reagents convert free alcohols to less polar trimethylsilyl (TMS) ether derivatives, including those in alkaloids like (*S*)-**1** and **2**. Silylation improves thermal stability, improves chromatography, and reduces analyte polarity.

Columns used for GC alkaloid analysis typically contain nonpolar polysiloxane stationary phases such as SPB1 [37], SE-54 [67], HP-5, HP-1, HP-19091S-433, OV-1, or RTX-5 columns [22,38,41,65,66]. Mass spectrometry is almost always performed in GC/MS analyses using 70 eV Electron Ionization (EI) and forms true molecular ions (M^+) from loss of an electron, though these often fragment extensively. An advantage of GC-MS is that fragment ion masses of alkaloids are often diagnostic for structural features. Some characteristic fragment ions for tropane, nortropane, epoxytropane, and pyrrolidine alkaloids from *Datura* are given in Table 1 [68,69]. For example, mass spectra of compounds containing a *N*-methylpyrrolidine ring contain a prominent fragment (often the base peak) at *m/z* 84, whereas EI mass spectra of tropane alkaloids often contain fragment ions at *m/z* 124, 113, 112, 96, 95, 94, 83, and 82 that reflect decorations on the tropane ring system. Positions of substitutions on tropane rings and the identities of side chains can often be deduced by analyzing both the abundances of certain fragment ions as well neutral mass losses from the M^+ ion (Table 1). For example, disubstituted tropanes containing a free hydroxyl group at position 3 frequently exhibit a base peak at *m/z* 113 [40,70] but other disubstitution patterns yield a base peak in the spectrum at *m/z* 94. Compounds containing a tigloyl group show a characteristic neutral loss of 99 Daltons (Da), and those containing tropic acid dehydrate and lose formaldehyde via McLafferty rearrangement [37]. In addition, many alkaloid standards are commercially available, allowing for comparisons of retention times, retention indices, and reference EI mass spectra.

Table 1. Comparison of tropane and pyrrolidine alkaloid fragment ions obtained by GC-(EI) and LC-(ESI)-MS.

Class	Representative Structure, Where R = Acyl or Hydroxyl	Diagnostic GC-MS (EI) Fragmentation (m/z, M^+)	Diagnostic LC-MS (ESI) Fragmentation (m/z)	References
Monosubstituted tropane		141, 140, 124 (usually base peak), 96, 94 (*N*-methylpyridinium cation), 83, 82 (base peak when no ester at C-3), 42	142 (loss of anhydro-acid), 124 (loss of neutral acid), 93, 91 (loss of methylamine, dehydrogenation), 77	[37,68]
Disubstituted tropane	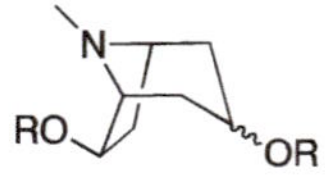	156, 140, 138, 122, 113 (prominent if 3-hydroxyl), 96 (3-hydroxyl), 95, 94 (usually base peak if not 3-hydroxyl), 55, 42	158 (sometimes, dihydroxytropane), 140 (loss of anhydro-acid), 122 (loss of neutral acid), 91 (loss of methylamine)	[39,40,68–70]
Trisubstituted tropane (non-epoxy)	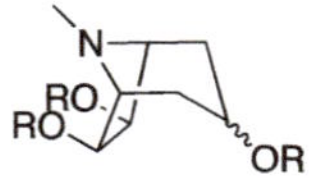	154, 138, 113 (if 3-hydroxyl) 94 (often base peak)	156, 138, 120, sometimes 94, 93 or 91	[37,39,40,64,68,69]
Epoxytropane		154, 94 (often base peak), 55, 42	156, 138, generally little fragmentation beyond that.	[37,69]
3-Monosubstituted Nortropane	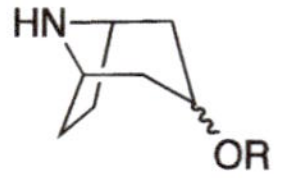	156, 138, 110 (base peak), 80, 68	128 (loss of anhydro-acid), 110 (loss of neutral acid), 93, 91	[40,61,68]
3-substituted 6,7-Epoxynortropane		122 (base peak), 124, 94, 80	142, 124 (norscopine)	[61,68,71,72]

Table 1. *Cont.*

Class	Representative Structure, Where R = Acyl or Hydroxyl	Diagnostic GC-MS (EI) Fragmentation (m/z, M$^+$)	Diagnostic LC-MS (ESI) Fragmentation (m/z)	References
3-substituted 6,7-Dehydrotropane		94 (base peak)	140, (loss of anhydro acid) 122 (loss of neutral acid), 93, 91	[64]
Tigloyl esters		Loss of 99	Loss of 100	[37]
Tropic acid derivative [(S)-1 or substituted on tropic acid]	R$_1$ = tropane R$_2$ = OH or ester	271 (dehydration or elimination of substituent), loss of 30 (formaldehyde via McLafferty rearrangement), 133, 121, 103 (derived from apotropic acid)	272 (dehydration or elimination of substituent), 103, loss of 166.064 or 148.044)	[37,68,69,73]
Phenylacetate esters	R$_1$ = tropane or substituted tropane	91 (derived from benzyl group)	Loss of 136.05 (neutral acid)	[37,61]
Pyrrolidine	R = other groups	Often low-intensity M$^+$; 85, 84 (usually base peak), 83, 82, 70, 55, 32	84 (*N*-methylpyrrolinium ion), loss of 83.	[68]

An excellent chemotaxonomic study performed by Doncheva and co-workers in 2006 investigated the alkaloid composition of twenty plant accessions belonging to all four sections of the tribe *Datureae*, and identified various alkaloids in roots and leaves [40]. Sixty-six tropane alkaloids were detected, with considerable variation among the minor alkaloids across all sections and species. *Brugmansia* could be easily distinguished from *Datura*, as the former contained high contents of 3,6-disubstituted tropanes and pseudotropine derivatives. Section *Brugmansia*'s alkaloid profile was similar to *D. metel* var. *fastuosa* (section *Dutra*) suggesting a relationship between those two sections, whereas section *Ceratocaulis'* profile was distinct (few alkaloids, especially epoxytropanes, in the roots) and on the basis of this analysis, should be considered separate from the rest of the sections.

Although commonly used, GC-MS has many limitations for metabolite discovery. First, nonvolatile alkaloids such as underivatized polar compounds, or those of high molecular weight, may not elute as intact molecules and may not be reliably detected [68]. Second, because GC-MS usually employs high temperatures in both the columns and injection ports, there is potential for thermal decomposition of some compounds. Glycosides, *N*-oxides, or other sensitive functionalities can be "burned off" in injection ports, and dehydrations, eliminations, and intramolecular reactions occur frequently for such compounds. Some compounds frequently identified in *Datura* samples (see below) may be formed during GC-MS, and whether they are purely artifactual or actually occur *in planta* [39,61,64] is the subject of ongoing dispute. Therefore, quantitative analyses using GC-MS may overrepresent certain alkaloids while underrepresenting others, and using another technique (such as LC-MS) for validation may be beneficial. Another caveat of mass spectrometry (including GC-MS) is that fragment ion masses provide virtually no information about absolute configuration (i.e., enantiomers) unless a chiral GC (or LC) column or chiral derivatizing reagent is used.

3.4. Liquid Chromatography-Mass Spectrometry (LC-MS) and LC-Tandem Mass Spectrometry (LC-MS/MS)

High-performance liquid chromatography-MS (HPLC-MS) and ultrahigh-performance liquid chromatography (UHPLC-MS), are increasingly used as alternatives to GC-MS. LC-MS has been used to quantify alkaloids in *Datura* nectar [74], track the effect of water volume [32], geography and natural selection [75], herbivory, and plant age [76] on alkaloid content and composition, for investigating how animals metabolize *Datura* components [77], for historical investigation [16], and for general metabolic profiling [36,78,79]. LC-MS methods employ a solid stationary phase (column) and a high-pressure (HPLC; <6000 psi, UHPLC; to ~15,000 psi) delivery of a liquid mobile phase, which the analytes are dissolved in as they move through the column. An advantage of LC-MS is that it can be performed at lower (ca. 35–50 °C) and ambient temperatures, and avoids thermal degradation. A downside of LC-MS is that LC separations are less efficient than capillary GC separations, though a wider range of experimental parameters are available including a variety of mobile phase options, and exploration of the range of conditions needed to resolve analytes can be time-consuming. As in GC, enantiomers cannot be identified or differentiated unless chiral stationary phases are used (vide infra).

Reverse-phase C_{18} (octadecyl-capped silica)-based columns are frequently used for analysis and separation of *Datura* alkaloids. A pentafluorophenyl phase column has also been used to analyze tropane alkaloids [10] following the recognition that this column type is more retentive toward positively-charged analytes [80,81], and can be used in both reversed-phase and hydrophilic interaction chromatography (HILIC) modes. UHPLC columns [36,77] have also been used for profiling of alkaloids; UHPLC columns have smaller particle sizes to provide more efficient separations, often with shorter analysis times. Common polar mobile phases for HPLC-MS and UHPLC-MS include water/methanol or water/acetonitrile, with formic acid (0.1–3%), acetic acid, or ammonium formate [82–84] added. These additives serve to keep basic alkaloids protonated, which reduces chromatographic streaking and tailing but also reduces retention on reversed phase separations.

Profiling of *Datura* alkaloids using LC separations coupled to mass spectrometry (LC-MS) has been performed on a variety of mass spectrometry platforms including triple-quadrupole [10], ion trap [65], hybrid triple-quad/linear ion trap [36], time-of-flight (ToF) [75], quadrupole time-of-flight (QToF) hybrid, or orbitrap instruments [32,36,77,85]. Ionization usually has involved electrospray ionization (ESI), although APCI (atmospheric pressure chemical ionization) has also been used [16,84].

By ESI, most alkaloids are highly ionizable and give $[M+H]^+$, or sometimes $[M+Na]^+$, $[M+K]^+$, or $[M+H+CH_3CN]^+$ pseudomolecular ions [76,83] with minimal formation of fragment ions. Alkaloid discovery is aided when LC-MS uses instruments with high mass resolution that provide accurate mass measurements that can distinguish isobaric compounds (i.e., those with common nominal masses but different elemental formulas), whereas experiments that employ more energetic ion activation methods (such as MS^E) yield abundant fragments [86].

Tandem mass spectrometry (MS/MS), where a given ion is first selected (for example, by the quadrupole filter of a QToF instrument, and then fragmented using collision-induced dissociation) is an incredibly powerful technique for gaining structural insights into alkaloids and aiding in annotation by associating fragment ions with specific precursor ions.

Much like the EI fragments obtained from GC-MS, ESI fragments (like those obtained by LC-MS) are also often diagnostic, but the ion fragmentation processes differ because EI-generated molecular ions are radicals, whereas ESI-generated ions are not. Some relevant positive-mode ESI fragment ion masses (from MS/MS) are also given in Table 1. For example, as observed in EI fragmentation, *N*-methylpyrrolidine-containing alkaloids frequently yield a *m/z* 84 fragment (*N*-methylpyrrolinium ion) or show neutral losses of 83 Da, whereas tigloyl esters are often eliminated as discrete units of 100 Da. Acyltropane substitution patterns are also easily differentiated by the fragments obtained when the acyl groups are lost as either neutral acids or ketenes/anhydro groups. An example of the fragment ions obtained from high-resolution ESI-MS/MS of hyoscyamine (detected in the chromatogram of *D. metel*) is shown in Figure 3.

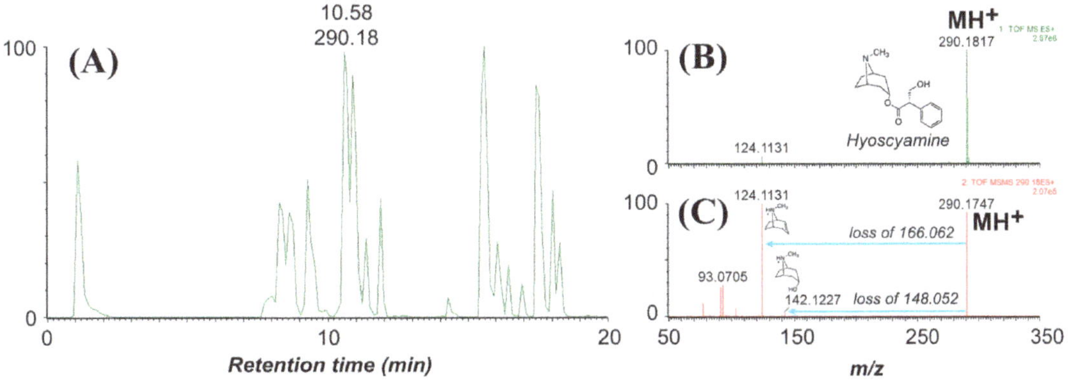

Figure 3. HPLC/ESI-MS base peak intensity chromatogram of a methanol-water extract of three-week-old *D. metel* root (left) using a C_{18} LC column on a Waters Xevo G2-XS QToF high-resolution mass spectrometer. The analysis was performed using data-dependent MS/MS in positive-ion mode. (**A**) Survey scan base peak intensity chromatogram. (**B**) The survey scan spectrum for hyoscyamine (retention time 10.58 min). (**C**) MS/MS product ion spectrum for *m/z* 290 for hyoscyamine shows characteristic monosubstituted tropane alkaloid fragment ions at *m/z* 142 and 124.

Examination of the peak (*m/z* 290) in the chromatogram corresponding to hyoscyamine shows characteristic monosubstituted tropane fragments at *m/z* 124 and 142 [10,69]. The *m/z* 124 fragment results from neutral loss of tropic acid (166.063 Da theoretical mass), and loss of the anhydro-tropic acid fragment (148.052 Da) yields *m/z* 142; additional odd-mass fragments at *m/z* 93 and 91 result from loss of the nitrogen (as methylamine). The low mass defects (portion of mass that follows the decimal place) of the neutral losses derived from tropic acid suggest that these fragments are relatively hydrogen poor (e.g., aromatic). Although less predictable, fragments obtained from acylium ions (arising from elimination and dehydration of acyl groups) are sometimes visible and can provide information, along with the mass defects, about the nature of the acids esterified to the tropane core.

LC-MS has been used in remarkably diverse ways to characterize *Datura* metabolites. In one 2020 study, Hui and colleagues [36] used UHPLC-QToF MS/MS (positive-ion mode) to profile roots, leaves, stems, flowers, seeds and peels (possibly pericarp) of *D. metel*, and identified 65 alkaloids, which they classified as indole, amide, or tropane. This group also developed a UHPLC-Q-TRAP MS/MS technique to quantify 22 selected *Datura* metabolites (mostly hydroxycinnamic acid amides, although some *beta*-carbolines/indoles and tropanes were included). This method, employing a five-minute LC run, utilized multiple-reaction-monitoring (MRM) data acquisition. It was observed that seeds contained the highest amounts of the selected alkaloids; a dependence on geography was also observed.

An elegant example of HPLC-MS with a chiral column was described in a recent manuscript by Marín-Sáez [87]. In this method, a CHIRALPAK-AY3 column (silica support with amylose tris-5-chloro-2-methylphenylcarbamate, eluting with ethanol/0.1% diethanolamine), was used to separate the enantiomers of atropine. ESI-MS/MS was used to monitor the alkaloids, using the MRM transition *m/z* 290>124 characteristic of atropine/hyoscyamine. Excellent resolution of the two enantiomers was observed, and it was determined that *D. stramonium* seeds grown in Spain contained (–)-hyoscyamine (*S*)-**1**: (+)-hyoscyamine in a 90:1 ratio.

3.5. Other Mass Spectrometry Methods

In addition to GC-MS and LC-MS, newer methods have been developed with the goal of eliminating processing, extraction and chromatographic steps entirely. One such method is called direct analysis in real-time mass spectrometry (or DART-MS), where a whole tissue sample is inserted between the ion source and the mass spectrometer inlet. Molecules in the tissue are then protonated by water clusters (arising from carrier helium and atmospheric water), and, as in ESI, molecules more basic than water readily ionize. This method allows for rapid development of "heat maps" or "fingerprints" that show the spatial distribution of metabolites in whole tissues, which can rapidly identify plant species. Lesiak and Musah applied this method to *D. stramonium*, *D. inoxia*, and *D. ferox* seeds, which were sliced transversely and analyzed using soft-ionization positive-mode DART-MS [88]. Spectra of *Datura* seeds show characteristic tropane fragment ions (e.g., *m/z* 174, 158, 142, 124), and the different seed ions clustered on the basis of species by principal component analysis.

In another study by the same group [89], a method called Laser Ablation DART Imaging (LADI-MS) was developed. In this technique, a *Datura* seed was embedded in silicon putty on a laser imaging platform. A UV laser pulse then scanned over the sample, producing an analyte-rich ablation plume, which was carried through a tube using helium to the DART ion source-inlet space (as described above). As the composition of the ablation plume may differ spatially across the sample, the distribution of any given analyte can be effectively "imaged". For example, in *D. leichhardtii* seeds, a different spatial distribution was observed for arginine (a biosynthetic precursor to tropanes), tropine (α-**10**), and hyoscyamine ((*S*)-**1**)/littorine (**11**, SI Table S1), which suggests the presence of sophisticated cellular transport machinery and/or specialized localization of alkaloid biosynthetic enzymes. Moreno-Pedraza elaborated on this technique by developing a UV laser desorption method that uses a low-temperature, 3D-printed plasma probe for sample ionization (known as laser-desorption-low temperature plasma mass spectrometry), which removes the need for sample matrices or organic solvents entirely. This technique was employed to image the distribution of tropane alkaloids and their fragment ions in cross sections of whole *D. stramonium* fruits and seeds [90].

Another method used for direct imaging is Desorption ESI (or DESI), where droplets of solvent are sprayed directly at an intact tissue sample or section, and the impact of the droplets produces gas-phase ions that are then analyzed. Cooks and colleagues [91] analyzed *D. stramonium* roots and seeds using this technique. With 1:1 methanol:water as the spray solvent, they were able to detect fifteen of the nineteen reported alkaloids in *D. stramonium* roots and confirm their identities by their fragment ion masses.

3.6. Nuclear Magnetic Resonance (NMR) Spectroscopy

NMR is frequently used for structure elucidation, and along with X-ray crystallography, is the most powerful tool for demonstrating molecular connectivity. Nonetheless, as the structures of many alkaloids (especially tropanes and pyrrolidines) are putative, proposed on the basis of mass spectral fragmentation (or comparison of retention time to standards), NMR is generally only used on isolated and purified alkaloids (see below).

Typically, the first NMR experiment performed during structural elucidation is proton (^{1}H-NMR) spectroscopy, which yields signals that reflect environments of the hydrogens (protons) in the molecule. ^{1}H-NMR (or proton NMR) yields distinctive signals for the tropane nucleus [40], and analysis of the splitting patterns has been frequently used to determine the configuration at position 3 (e.g., α-**10**, or tropine-derived, or β-**10**, or pseudotropine, derived; see Figure 5). Additional information about groups esterified to the tropane core can be gained from examining the olefinic (5–7 ppm) and aromatic (6–8 ppm) regions. An example of a ^{1}H-NMR spectrum of a tropane alkaloid (**2**) is shown in Figure 4. Carbon-13 (^{13}C-NMR) spectra are also commonly obtained, although these experiments have lower sensitivity in part because NMR-active ^{13}C nuclei are less abundant than protons. ^{13}C spectra are also collected as part of DEPT (distortionless enhancement via polarization transfer) experiments. These experiments (DEPT-45, DEPT-90, and DEPT-135) distinguish ^{13}C signals on the basis of the number of attached protons. In cases where the structure is more complex (as in many tropane alkaloids), two-dimensional NMR experiments that provide more explicit information about the connectivity of atoms, such as total correlated spectroscopy (TOCSY, H-H), correlation spectroscopy (COSY, H-H), heteronuclear single quantum coherence (HSQC), heteronuclear multiple-bond correlation (HMBC, H-C, or H-N), or various through-space experiments dependent on the nuclear Overhauser effect (such as NOESY or ROESY). Although the last experiments are not commonly included in the *Datura* alkaloid literature, they have been employed for structural determination of other tropane alkaloids [30].

The largest advantage of NMR spectroscopy is that, in addition to the myriad structural information it provides, it is generally nondestructive, and samples can be recovered after analysis. Nonetheless, it can be very time-consuming, especially when performing ^{13}C-based or 2D-experiments on low field strength instruments, and its sensitivity is much lower than mass spectrometry. To perform structural determination of an alkaloid by NMR, generally 1–5 mg of a compound is required, which often makes isolation (and purification) of enough compound from a natural source a major barrier to NMR experiments. Additionally, as most alkaloids are chiral molecules, a downside of NMR, like mass spectrometry, is the inability of the method to establish absolute stereochemical configuration (e.g., enantiomeric identity or ratio)—that is usually established by chiral HPLC and comparison to a standard, circular dichroism or similar methods, X-ray crystallography, or chemical derivatization (although often not used for *Datura* alkaloids) [47,49,92]. However, in the presence of so-termed chiral shift reagents, enantiomers can be differentiated and quantified by NMR. Lanfranchi et al. [44] added one such reagent, Yb(hfc)$_3$, to a solution of atropine trifluoroacetate. This reagent causes deshielding of certain carbons, and the splitting of certain ^{13}C signals into discrete peaks corresponding to each enantiomer, from which enantiomeric ratio can then be determined. This method was optimized with different ratios of atropine enantiomers and then applied to crude *D. stramonium* seed extract, which revealed that (*S*)-(–)-hyoscyamine (*S*)-**1** was the dominant enantiomer by a factor of 20:1 (in agreement with the chiral HPLC results, *vide supra*) [87].

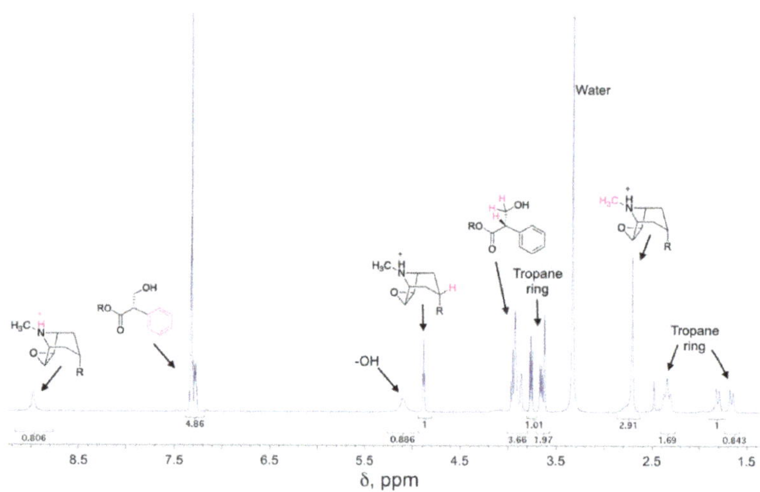

Figure 4. ^{1}H-NMR spectrum (500 MHz, in DMSO-d_6) of scopolamine (**2**). Indicated peaks correspond to the structural features highlighted in magenta.

NMR has also been used in studies of the metabolism and biosynthesis of alkaloids in *Datura* plants and root cultures derived from *Datura*. Robins et al. [63] fed *D. stramonium* root cultures with NMR-active ^{13}C-labeled sodium acetate, hygrine (**12**, Figure 11), and several pyrrolidine esters, all of which could be hypothetical precursors of tropane alkaloids. By monitoring incorporation of the ^{13}C label into tropane alkaloids by NMR, it was determined that hygrine and *N*-methyl-2-pyrrolidinyl acetate ester were not direct precursors of tropane alkaloids (hygrine was only found incorporated into cuscohygrine, **13**, Figure 11), but 1-methyl-2-pyrrolidinyl-3-oxobutanoate (see below) and sodium acetate likely were. ^{15}N-NMR spectroscopy was also used to study the de-differentiation of *D. stramonium* root cultures fed with ^{15}N-labeled nitrate or ammonium salts. When these cultures were treated with various hormones, the cells formed a suspension culture and lost most of their ability to synthesize tropane alkaloids. Instead, ^{1}H-^{15}N-HMBC spectra (see below) of extracts showed that these cells accumulated *gamma*-aminobutyric acid and *N*-acetylputrescine [62,93] suggesting that the nitrogen balance of these cells had shifted away from alkaloid biosynthesis.

The literature contains several examples of applying these experiments to the structural determination of alkaloids. Siddiqui et al. [45] isolated a *para*-methoxybenzoic acid tropane ester from *D. metel*, which was appropriately named datumetine (**8**). ^{1}H-NMR revealed a triplet consistent with the methine of a 3α-tropyl ester and a characteristic downfield doublet-of-doublets suggestive of a *para*-substituted benzoate, which was confirmed by ^{13}C-NMR. Additional information about the structure (such as the presence of methines indicative of a tropane bridgehead (2- and 5-positions) and the *para*-methoxyl group was revealed by DEPT. Similarly, Welegergs and colleagues isolated a trisubstituted tropane alkaloid from Ethiopian *D. stramonium* seeds (β-**7**), and they used a similar combination of ^{1}H, ^{13}C, and DEPT-135 NMR experiments to assign the structure of the isolated alkaloid as **7** (SI Table S3) [48]. Liu and co-workers utilized ^{1}H-^{1}H COSY and ^{1}H-^{13}C-HMBC, and ^{1}H-^{13}C-HSQC, in addition to ^{1}H and ^{13}C-NMR, to establish connectivity of the atoms in the daturametelindole alkaloids (**3–6**) isolated, obviously, from *D. metel*), several of which possess a unique spirocyclic oxindole core [47].

4. Alkaloids of the Genus *Datura*—Tropane Alkaloids

4.1. History—And Some Troublesome Aspects of Identifying and Reporting Tropane Alkaloids

Although the medicinal and psychotropic effects of *Datura* species have been known throughout history, the active alkaloid constituents were a mystery for centuries. Hyoscyamine [(*S*)-**1**)] and scopolamine (**2**) were not isolated from any plant until 1833 and 1881, respectively [6], and are named after *Hyoscyamus* and *Scopolia*—the genera from which they were first isolated. The principal alkaloid isolated from *D. stramonium* was originally called "daturine". Daturine was later determined to be identical to two other alkaloids, "duboisine" (presumably isolated from the genus *Duboisia*) and hyoscyamine (which was established as being "chemically identical" to atropine, **1**). The observation that *Datura* actually contained (*S*)-**1** was made in 1880 [94], and appears to have been to be common knowledge by 1901 [95] but references to (*S*)-**1**'s possible presence in *Datura* existed as early as 1850! Although research on *Datura* alkaloids has been conducted for over 170 years, the majority of new alkaloids were identified from 1970 to the present.

There are several caveats that must be acknowledged when perusing the *Datura* alkaloid literature and examining the annotations and identifications of tropane alkaloids and the species that they are isolated from. First, several *Datura* species very closely resemble each other and may be polymorphic (i.e., *D. metel*, *D. wrightii*, and *D. inoxia*), which is complicated further by ambiguous or obsolete names in the literature (e.g., *D. meteloides*). Some of these similar species (such as *D. inoxia* and *D. wrightii*) also have significant overlap between their native or naturalized geographic ranges [23,25]. Without a definitive botanical or genetic identification of a wild plant accession, it may be difficult to tell exactly which species of *Datura* is being studied.

Second, not all *Datura* species have been studied equally with regards to alkaloids. Some species, like *D. stramonium*, have been profiled extensively using multiple analytical methods, while others, like *D. discolor* and *D. quercifolia*, are relatively uninvestigated. Although a specific alkaloid may be listed as absent from a certain species in the literature (or in SI Tables S1–S5), this may not mean that the alkaloid is truly absent; it may instead mean that it has not been discovered in that species to date, or the analytical methods used simply have not detected it.

Third, the tropane literature is rife with structural ambiguity. Tropane, tropine (α-**10**), and pseudotropine (β-**10**) are achiral *meso*-compounds that are bisected by an internal plane of symmetry (Figure 5) and many 3-monoacylated compounds are also *meso* compounds (unless they contain chiral acyl groups, like hyoscyamine (*S*)-**1**). Additional functionalization at position 6 or 7, however, can desymmetrize the molecule [92]. Therefore, 3,6- and 3,7-disubstituted tropanes (and some asymmetrically substituted 3,6,7-trisubtituted tropanes) form enantiomers. As substitutions on positions 6 and 7 (as far as those have been isolated) are always *exo* (β), asymmetrically substituted tropanes are only diastereomers if position 3 differs in its α/β configuration—see Figure 5. As mentioned above, enantiomers cannot be differentiated on the basis of achiral LC-MS, GC-MS, or standard NMR techniques used for structural elucidation. Since the majority of *Datura* alkaloids have been annotated based on GC-MS data, there is considerable uncertainty regarding absolute configurations. Stereochemistry often seems to be assigned arbitrarily, or by comparison to retention indices of standards whose structures themselves are not definitively assigned. In the following sections (and the tables and figures referenced herein), the structures of tropane alkaloids are described as reported by the original authors, although the stereochemical assignments may not have been definitively established.

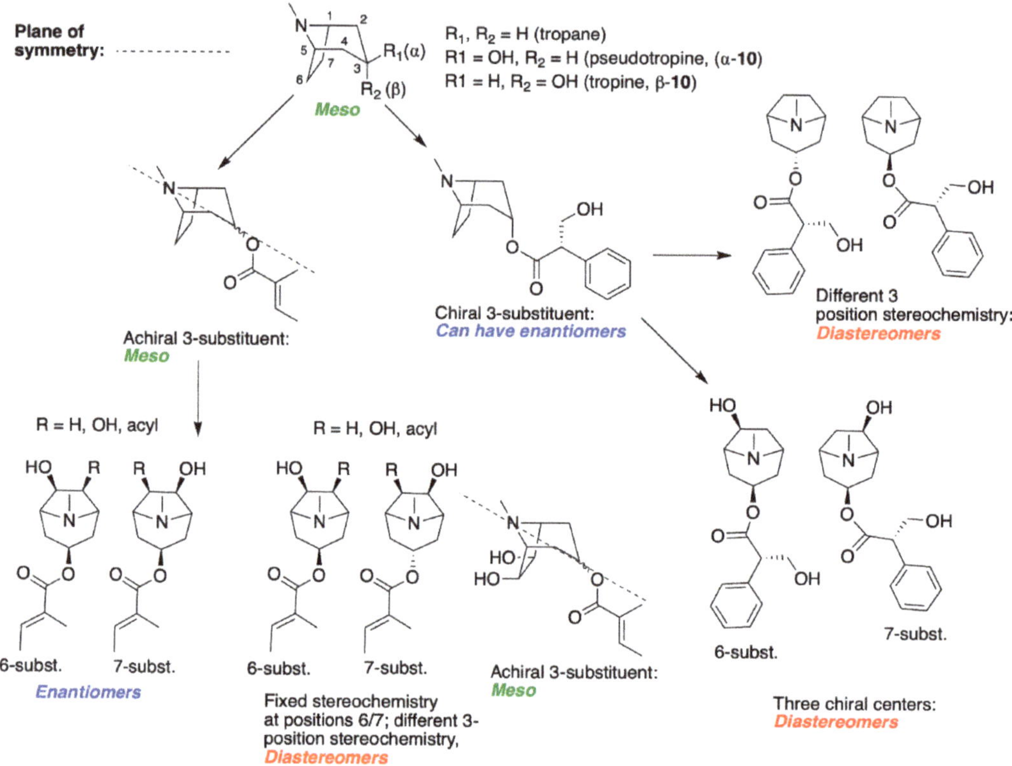

Figure 5. Chirality in tropane alkaloids and relationships between meso compounds, enantiomers, and diastereomers for substituted tropanes.

For the purposes of brevity, classes of tropanes (organized into subgroups roughly following the conventions of Lounasmaa and Tamminen [96]) that contain many different compounds (monosubstituted, disubstituted, trisubstituted and epoxytropanes, and nortropanes) are organized into SI Tables S1–S5 (in the Supporting Information). It may be helpful for the reader to refer to these tables frequently while going through the following sections, as alkaloids are referenced in the text using numerical descriptors, some of which are only found in these tables. The structures of other alkaloids of note are depicted in Figures 7–13 (*below*).

4.2. Biosynthesis of the Tropane Core

Datura species start producing tropane alkaloids as early as six days after germination [72], or when the germinated roots reach approximately 3 mm long [97]. An excellent, in-depth review of the biosynthesis of tropane alkaloids was provided by Kohnen-Johannsen and Kayser [12], and an overall scheme is shown in Figure 6. The roots are believed to be the principal site of tropane alkaloid biosynthesis, and the precursors of the tropane (and pyrrolidine) alkaloids are ultimately the amino acids arginine, glutamine, and ornithine.

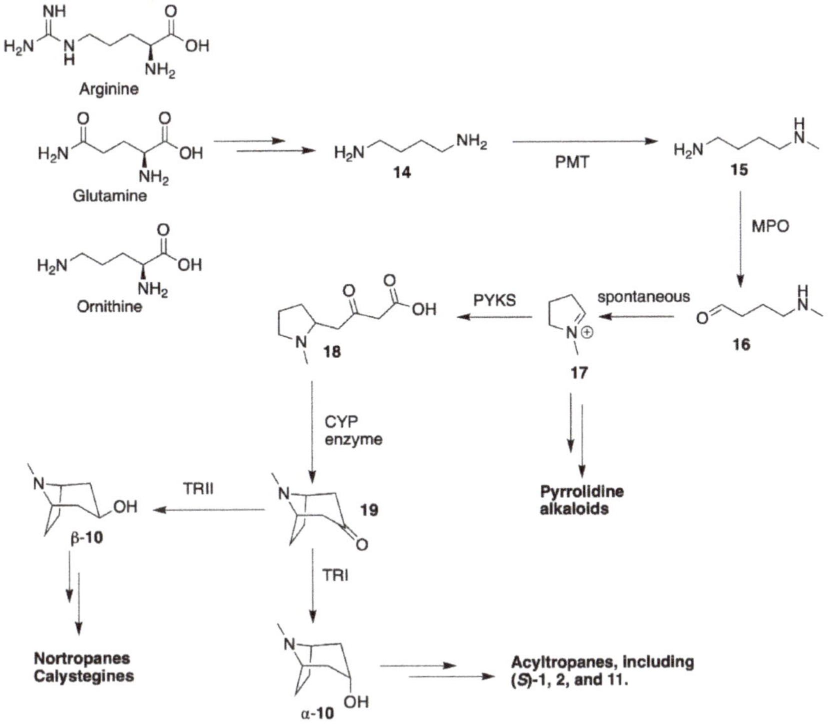

Figure 6. Biosynthesis of tropane and pyrrolidine alkaloids in *Datura*.

These amino acids can be converted to putrescine (**14**, Figure 6) through the action of several enzymes. The enzyme putrescine methyltransferase (PMT) [62,98] *N*-monomethylates putrescine to yield **15**, which is then oxidatively deaminated by methylputrescine oxidase (MPO). The resultant unstable amino aldehyde (**16**) spontaneously cyclizes into *N*-methylpyrrolinium ion (**17**). This pyrrolinium species is electrophilic and can react with a variety of nucleophiles. Pyrrolidine, ecgonine, and tropinone-derived alkaloids are all derived from this intermediate. To form tropinone-derived alkaloids (which represent the majority of the tropane alkaloids present in *Datura* species), this intermediate condenses with two molecules of malonyl-Coenzyme A (CoA) to yield the pyrrolidine oxobutanoic acid **18**. The enzyme that catalyzes this condensation in *Atropa belladonna* roots was recently revealed to be a polyketide synthase (*AbPYKS*) [99], and similar enzyme activity probably serves the same function in *Datura*. This intermediate was identified as a tropane alkaloid precursor in *D. stramonium* root cultures, as ^{13}C from a labeled ethyl ester precursor of **18** that was fed to the cultures was found incorporated into (S)-**1** and other tropanes [63]. Similar results were observed in feeding experiments using hydroponically-grown *D. inoxia* [100].

The intermediate **18** is then cyclized, with concomitant oxidation and decarboxylation, to afford tropinone (**19**). In *A. belladonna*, the enzyme that catalyzes this cyclization is a cytochrome P450 (called *AbCYP82M3*) [99]; no analogous enzyme has been isolated or purified from the genus *Datura* to date. Tropinone (**19**) has several fates: it can be converted to tropine (α-**10**, the 3α-anomer) by tropinone reductase I (TRI), or pseudotropine (β-**10**, 3β-anomer) by tropinone reductase II (TRII); both of these enzymes have been isolated from *D. stramonium* root cultures [101]. Tropine (α-**10**) gives rise to many acyltropines, including hyoscyamine (S)-**1**, littorine (**11**), datumetine (**8**), and more functionalized compounds such as scopolamine (**2**), meteloidine (α-**20**, SI Table S3), anisodamine (**21**, Figure 7), and anisodine (**22**) (SI Table S4). Pseudotropine also yields aliphatic esters such as tigloidine (β-**23**), and is the precursor of the polyhydroxylated calystegines. More information about the biosynthesis of more specific alkaloids is provided under individual subheadings below.

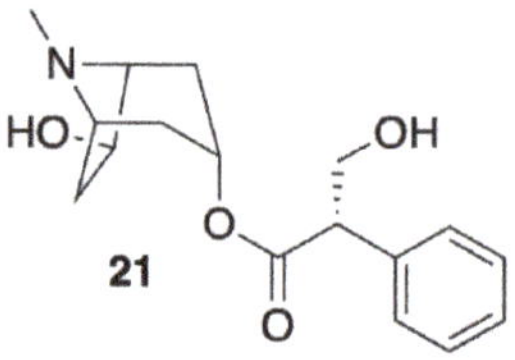

Figure 7. 3,7-Disubstituted tropanes found in *Datura*.

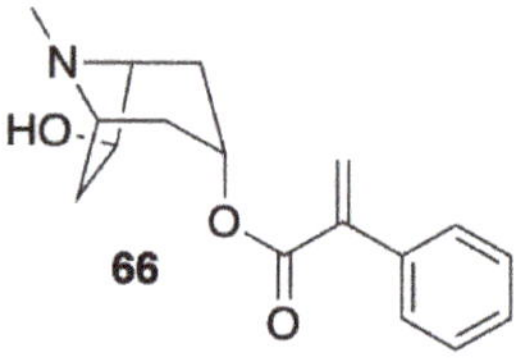

4.3. Monosubstituted Tropanes

All reported monosubstituted tropanes, for the exception of 7-hydroxytropane (See SI Table S1) are substituted in the 3-position, and include **19**, α-**10**, β-**10**, and esters derived from the latter two compounds (**1, 8, 11, 23–39**). Numerous monoacylated tropanes have been identified in *Datura* species (mostly by GC-MS), from both intact plant parts and transformed root cultures. Most of these alkaloids have α stereochemistry at the 3-position, and are derived from α-**10**, but a significant number of β-**10**-derived alkaloids, such as tigloidine (β-**23**), have also been detected. Unfortunately, in many cases, the stereochemistry at the 3 position has not been reported; in Table S1 these instances are cited under "α or unspecified".

Robins [102] reported that different *Datura* acyltransferases act depending on whether the substrate is α-**10** or β-**10**. Individual acyltransferases also may have promiscuous substrate specificity. For example, tigloyl pseudotropine acyltransferase isolated from *D. stramonium* root cultures has the highest activity with tigloyl-CoA, but can also use acetyl, propionyl, isobutyroyl, and senecioyl-CoAs (among other aliphatic substrates) to acylate β-**10**, suggesting that the abundance of a given acyltropane may in part arise from competition between different substrates in the plant's acyl-CoA (or other acyl donor) pool [102,103]. The aliphatic CoAs utilized by these enzymes are derived from short-chain (C_2-C_5) carboxylic acids. Generally, aliphatic acyl groups containing more than five carbons have not been observed in *Datura* alkaloids. Most short-chain carboxylic acids are derived from amino acids. 2-Methylbutanoyl and tigloyl groups (both very common in *Datura* tropanes) are synthesized from isoleucine [49,104], whereas isobutyroyl and 3-methylbutanoyl (isovaleroyl) groups are derived from valine and leucine, respectively [105,106].

Most of *Datura*'s aromatic monoacyl tropanes (SI Table S1) contain phenylacetyl (**30**), 2-phenylpropionyl (**34**), phenyllactoyl (littorine, **11**), or tropic acid moieties [(*S*)-**1** and its derivatives]. One interesting exception is the hydroxylated phenylacetate derivative homatropine (**31**), which is a synthetic drug. Temerdashev [41] detected this alkaloid in *D. metel* extracts using LC-MS, but did not elaborate on how it was identified as homatropine. Another exception is datumetine (**8**), a 4-methoxybenzoyl ester isolated from *D. metel* [45] and characterized by NMR. This alkaloid is unique because benzoyl and substituted benzoyl esters are not found in *Datura*, let alone in most of the Solanaceae; these esters are more common in the Erythroxylaceae (e.g., cocaine) and Convolvulaceae [96].

All phenyllactoyl and tropic acid moieties in *Datura* alkaloids are ultimately derived from phenylalanine, via reduction of phenylpyruvate. In *A. belladonna* roots, an aromatic amino acid aminotransferase (termed *Ab-ArAT*) is found co-expressed with tropane biosynthesis genes; it is anticipated that a similar "ArAT enzyme" produces phenylpyruvate in *Datura* [107]. Post-phenylpyruvate reduction, phenyllactate is then esterified to tropine to form littorine (**11**) by littorine synthase. This enzyme is a specific acyltransferase that uses phenyllactoylglucose as the acyldonor [108]. Hyoscyamine ((*S*)-**1**) is present in all *Datura* species (and root cultures derived from these species), and is frequently one of the most abundant alkaloids, although its abundance depends on the species, organ and ontogenetic stage examined. Hyoscyamine (*S*)-**1** is sometimes interchangeably referred to as "atropine" in the literature, although this is a misnomer, as "atropine" (**1**) refers to pharmaceutical preparations of racemic hyoscyamine, whereas the tropic acid moiety is reported to be almost exclusively (*S*)-(−)- in *Datura stramonium* [44,87].

The tropic acid moiety is derived from the phenyllactate of **11**, isomerized by a P450 enzyme called littorine mutase. The kinetics of (*S*)-**1** formation (i.e., isomerization of the phenyllactate of **11**) in *Datura* (specifically, in *D. inoxia* root cultures) are reported to be slow (around half the rate of the synthesis of **11** itself), suggesting that this may be the rate-limiting step in the synthesis of (*S*)-**1** [109]. The mechanism of this isomerization is still disputed, but computational studies suggest that a P450-mediated concerted cationic rearrangement to yield hyoscyamine aldehyde is a low-energy process [110]. Indeed, masses consistent with hyoscyamine aldehyde (which is then presumably reduced to **11**) have been observed by LADI-MS in *D. leichhardtii* seeds [89].

The tropic acid moiety itself is also further elaborated in *Datura* (SI Table S1); it is hydroxylated at the α-position (alkaloid **35**) in *D. stramonium*, *D. wrightii*, *D. ceratocaula*, and *D. inoxia*, methylated (**36**) in *D. inoxia* and *D. stramonium*, and acetylated (**37**) in most species. One unique hyoscyamine derivative is the *O*-formyl ester of hyoscyamine (α-**38**) observed in *D. ceratocaula*; GC-MS revealed a M$^+$ ion of *m/z* 317, which showed a characteristic loss of HCOO• (45 Da) to yield protonated apoatropine (*m/z* 272) [73]. This formyl ester (to date) is unreported in other species. The dehydration product of hyoscyamine (or elimination product of other hyoscyamine derivatives), called apohyoscyamine or apoatropine (**39**, SI Table S1)) has also been commonly observed in GC-MS profiles of *Datura* species. There is debate as to whether this alkaloid legitimately exists in nature, or is an artifact produced by thermal dehydration of (*S*)-**1** (or a derivative) in heated injectors during GC-MS. For example, Dupraz states that **39** was not detected by HPLC analysis of extracts of *D. quercifolia* (root cultures and intact plants), but **39** was detected when the same extracts were analyzed by GC-MS, suggesting it is artifactual [67]. By contrast, in Temerdashev's 2012 study, **39** was observed in LC-MS traces of *D. metel* extract, suggesting that either it is formed in planta or during extraction (rather than during analysis) [41].

4.4. Disubstituted Tropanes

Many *Datura* tropane alkaloids are hydroxylated (or acylated) or epoxidized on the tropane core. Most documented tropane hydroxylation (and epoxidation) occurs in positions 6 and 7, exclusively in the *exo* (β) position, and is catalyzed by the enzyme Hyoscyamine 6β-Hydroxylase (H6H). This enzyme, a nonheme, iron-containing 2-oxoglutarate-dependent dioxygenase, has been cloned from *D. inoxia*, and recombinant H6H from *D. metel* has been expressed in *E. coli* [111,112]. In both cases, the isolated H6H enzyme could also convert (*S*)-**1** to scopolamine (**2**), showing that it can construct epoxytropanes as well. The *D. metel* enzyme also could produce both 6-hydroxyhyoscyamine (**40**, SI Table S2) in addition to scopolamine, but the epoxidase activity (discussed in more detail below) was much lower than the hydroxylase activity.

Table S2 summarizes known *Datura* disubstituted tropane alkaloids (**40–65**). Like their monosubstituted counterparts, disubstituted tropanes are found in all *Datura* species (and in root cultures derived from those species). In addition to 3,6-dihydroxytropane (**41**), these alkaloids contain either one ester and one hydroxyl group or two esters. Interestingly, other than 3,6-diacetoxy (**42**) and -ditigloyloxy (α/β-**43**) tropanes, duplicate ester groups (e.g., diisobutryloxy, diisovaleroyl, or dipropionyloxy) have not been reported, suggestive of enzymatic recognition. Most disubstituted tropanes are aliphatic, although hydroxylated and acylated derivatives of phenylacetoxytropane, apohyoscyamine, and hyoscyamine (such as **40**, **55–57**, **62–65**) have also been found in several species.

Although almost all disubstituted *Datura* tropanes are 3,6-disubstituted, several alkaloids exhibit the rare 3,7-disubstitution pattern. 7-Hydroxyhyoscyamine (**21**, Figure 7), also known as anisodamine, is distinguishable from 6-hydroxyhyoscyamine (**40**) by its GC retention index on an achiral column, as it is a diastereomer of **40**, rather than an enantiomer. This alkaloid has also been detected in *D. inoxia* [113,114], *D. ceratocaula* [73], and *D. stramonium* [72,115]. 7-Hydroxyapoatropine (**66**) [72], was also observed in *D. stramonium*. Two unique disubstituted tropane alkaloids are 3-formyl-6-tigloyloxy tropane (**54**, from *D. stramonium*), which is the only reported alkaloid of this type containing a formate ester, and 3-hydroxy-6-(2-methylbutanoyloxy)-tropane (α-**9**, SI Table S2), which was isolated from *D. ceratocaula*. Beresford and Wooley [49] determined the stereochemistry of the 2-methylbutanoate side chain to be (+)-(*S*). The stereochemistry of 2-methylbutanoate side chains has not been determined or reported in other instances, although it is plausible that they are also (+)-(*S*), if they are derived from *L*-isoleucine as is proposed (vide supra).

4.5. Trisubstituted and Epoxytropanes

More elaborate trisubstituted and epoxidized tropane alkaloids have also been found in *Datura*, suggesting that the enzymes that perform hydroxylation, epoxidation, and acylation have evolved to recognize a variety of substrates. 3,6,7-Trisubstituted tropanes (**7, 20, 67–77**) are in summarized in Table S3; 3-substituted 6,7-epoxytropanes are likewise in Table S4. All trisubstituted tropanes detected in or isolated from *Datura* species contain either two acyl groups and one hydroxyl group (as in the very abundant and common 3,6-ditigloyloxy-7-hydroxytropane, α-**72**), or two hydroxyl groups and one acyl group, such as 3-tigloyloxy-6,7-hydroxytropane (α-**21**) which was first isolated from *D. meteloides* and thus given the common name meteloidine.

As with the disubstituted tropanes, substituents at positions 6 and 7 have been exclusively assigned β-stereochemistry, whereas position 3 can be α or β, and the only substituents found to repeat in the same compound are tigloyl (as in α/β-**72**) and hydroxyl (as in **20, 67**, and **75**, SI Table S3). Interestingly, 3,6,7-triacylated tropanes have never been reported in the genus *Datura* [4,96]. In fact, they are not reported to exist in the *Solanaceae* at all; the only representative of this class thus far has been found in *Erythroxylum zambesiacum* (Erythroxylaceae) [116]. Additionally, only two unique trisubstituted tropanes from *Datura* are derived from (*S*)-**1**. Vitale and co-workers [37] detected (by GC/MS) an alkaloid they proposed to be 7-hydroxy-6-propenyloxy-3-tropoyloxytropane (α-**77**, SI Table S3) in the seeds of *D. ferox*. The structure was proposed on the basis of a) loss of formaldehyde by McLafferty rearrangement (consistent with tropic acid derivatives), and b) loss of 72 Da corresponding to the propenyloxy group. Welegergs et al. [48] isolated 3-(3′-methoxytropoyloxy)-6-tigloyloxy-7-hydroxytropane (β-**7**) and assigned its structure by NMR. The 3-position stereochemistry for this alkaloid is drawn as β, although the report does not specify how this assignment was confirmed, nor is the stereochemistry of the tropic acid side chain explicitly specified.

The most abundant epoxytropane in the genus *Datura* is scopolamine (**2**), which is found in all species and hairy root cultures derived therefrom. Most epoxytropanes (SI Table S4) are aromatic, derived from tropic (**2, 22, 81**), apotropic (**82**), 3-phenylpropanoic (**83**), or phenylacetic acid esters (**84**), although 3-tigloyl-, 3-hydroxy-(scopine), and 3-acetyl-6,7-epoxytropanes (**80, 78**, and **79**), respectively) have also been observed in *D. stramonium* and/or *D. inoxia*. As described above, epoxidation, like hydroxylation, is catalyzed by H6H. Epoxidation can occur by dehydrogenation of hydroxytropanes, and in vitro experiments have shown that 6,7-dehydrotropanes are also recognized as epoxidation substrates [12]. Ushimaru and colleagues [117] synthesized a series of tropane analogues as mechanistic probe substrates for H6H. The use of these probes revealed that tropane epoxidation does *not* occur via a dihydroxytropane intermediate, but by radical abstraction of a β-hydrogen one carbon removed from the hydroxyl group. The substrate configuration required for epoxidation is rather strict—the configuration between the β-hydrogen and hydroxyl group must be *syn*-periplanar for epoxidation to occur. If there is no hydroxyl group (i.e., monosubstituted tropane), hydroxylation will occur instead, and if the hydroxyl and β-hydrogen cannot adopt a *syn*-periplanar configuration, dihydroxylation predominates over epoxidation [117].

4.6. Dehydrotropanes, Nortropanes, and Calystegines

A rarer class of tropanes are those with a double bond in the tropane nucleus (dehydrotropanes). The first dehydrotropane discovered in *Datura* was 2,3-dehydrotropane (**85**, Figure 8), which was detected by GC-MS in diploid and tetraploid hairy root cultures derived from *D. stramonium* [118]. The remaining known alkaloids of this class (**86–90**) are 6,7-dehydrotropanes, which were found in *D. stramonium* (**87**), *D. inoxia* (**88**), or both (**85**, **86**, **89**, **90**) by El Bazaoui and co-workers [64,113] Virtually nothing is known (or even proposed) about the biosynthetic origin these alkaloids, although, like apotropoyl derivatives, it is considered that they could be thermal artifacts formed by dehydration or elimination of hydroxyl or ester groups during GC-MS. In LC-MS studies from our own laboratory, however, we have detected dehydrotropanes in neutral water/methanol extracts of both *D. stramonium* and *D. metel* parts, suggesting that some of these alkaloids are formed *in planta*; De-la-Cruz's (2020) LC-MS observations also support this [75].

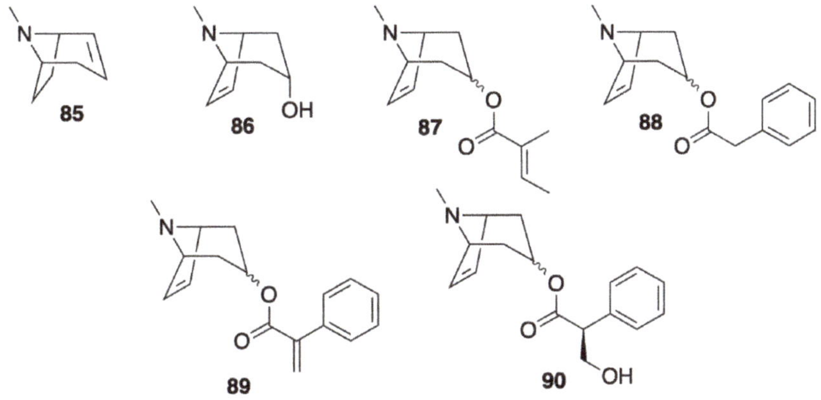

Figure 8. Dehydrotropanes identified in *Datura* species.

Nortropanes (Table S5) are tropane alkaloids lacking the methyl group on the tertiary amine. The enzyme responsible for tropane *N*-demethylation has not been isolated from *Datura* or characterized biochemically, although it may act in an oxidative fashion similar to the enzyme that demethylates nicotine [119]. All nortropanes (**91–104**) detected in *Datura*, for the exception of 3-tigloyloxynortropane (**92**) and 3-acetoxynortropane (**91**), are aromatic, and many are epoxy or apotropyl derivatives. It is worth noting that although various other thermal reactions are possible in the GC-MS, thermal tropane demethylation is not considered likely, so observations of presence or absence of nortropanes can be considered a true metabolic difference and not artifactual. In a 2006 chemotaxonomic study by Doncheva et al. [40] the distribution of nortropanes was heavily dependent on the part of the plant studied. Nortropanes were found only in the leaves of *Datura* species; they were not detected in the roots of multiple varieties or accessions of *D. stramonium*, *D. ferox*, *D. inoxia*, *D. wrightii*, *D. leichhardtii*, *D. metel*, or *D. ceratocaula*.

Calystegines are a special class of norpseudotropanes [119] sometimes referred to as "sugar mimic" alkaloids because of their extensive hydroxylation and ability to inhibit various glycosidases [54]. Calystegines are challenging to detect because they have high aqueous solubility that impairs liquid-liquid extraction into organic solvents, do not run well on GC without prior derivatization, lack useful UV chromophores, and do not stain with Dragendorff reagents. Although the genes required to produce calystegine precursors (pseudotropine, β-**10**) are present in *Datura*, reports of these alkaloids are rare for the genus. Nash [54] reported calystegine B2 (**105**, Figure 9) in *D. wrightii* leaves, whereas Drager and colleagues [120] reported the presence of calystegines in *D. stramonium* leaves and roots, although their contents were too low (<0.5 μg/g tissue) to quantify reliably, and they were not detected in cultured roots of *D. stramonium* [121].

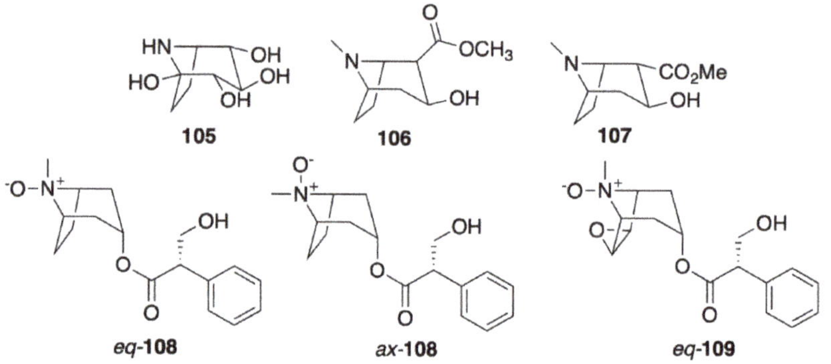

Figure 9. Calystegine, ecgonine alkaloids, and tropane *N*-oxides found in *Datura* species. Note that the "pseudo-" nomenclature for ecgonine alkaloids refers to the stereochemistry of the 2-carboxylate, *not* the 3-hydroxyl.

4.7. Ecgonine Derivatives, Tropane N-Oxides, and Cyclic and Dimeric Tropane Alkaloids

The ecgonine alkaloids (2-carboxy-3-hydroxytropanes) are the class of tropane derivatives that includes cocaine and its derivatives and precursors. These alkaloids, which share similar pyrrolidine-based biosynthetic origin to tropinone-derived alkaloids [12] are more common in the Erythroxylaceae than the Solanaceae [4]. There are several exceptions: methylecgonine (**106**, Figure 9) was observed by Berkov in the roots and stems of *D. cerato-caula*, with the stems having a slightly higher content as assessed by GC-MS (% of total ion chromatogram) [76]. Similarly, **106** was reported in low concentrations in *D. stramonium* stems and roots, and *D. inoxia* roots, stems, leaves, and flowers—accessions of both species were harvested from Northwestern Morocco [64,113]. Additionally, **106** was found in both diploid and tetraploid hairy root cultures derived from *D. stramonium* [119]. Methylpseu-doecgonine (**107**) was also discovered in *D. stramonium* and it was identified as such by comparison of the natural alkaloid with all four synthesized isomers of methylecgonine by GC-MS [122]. This alkaloid was also found in *D. stramonium*, *D. stramonium* var. *tatula*, and *D. inoxia* harvested from Algeria [115].

Although unusual, the tropane nitrogen can become oxidized, yielding a tropane *N*-oxide (such as hyoscyamine *N*-oxide). Both an axial (*ax*-**108**) and an equatorial (*eq*-**108**) isomer of this alkaloid exist (Figure 9) and are separable; both have been isolated from *D. stramonium* root, stem, leaf, flower, and pericarp; the equatorial isomer of scopolamine *N*-oxide (*eq*-**109**) was isolated from the same plant and parts [123]. Tropane *N*-oxides are problematic to detect, isolate, and characterize because of their poor solubility in common extraction solvents, such as ether, and their propensity to thermally deoxygenate and demethylate during GC-MS, which makes this a poor detection method [68]. As such, their structures were assigned by NMR. Interestingly, Vitale [37] reported that no tropane *N*-oxides were detected in *D. ferox* seeds; there is little comment in the literature regarding the presence of these derivatives in other *Datura* species.

The cyclic tropane (+)-scopoline (**110**, Figure 10), also known as oscine, was first isolated from the genus *Datura* by Heusner in 1954 [124]. Since then, it has been observed in *D. stramonium* [68,72,74], *D. inoxia* [113,114], and *D. wrightii* [40]. A second cyclic tropane, cyclotropine (**111**), was observed in *D. stramonium* and *D. inoxia* [113,114]. Ionkova and Jenett-Siems propose that these compounds may be artificially produced during the isolation procedure (arising from degradation of epoxytropanes followed by intramolec-ular nucleophilic attack of the 3-hydroxyl on the epoxide), or during GC-MS, although both of these alkaloids were also observed by LC-MS during De-la-Cruz's profiling of *D. stramonium* leaves [61,75,122].

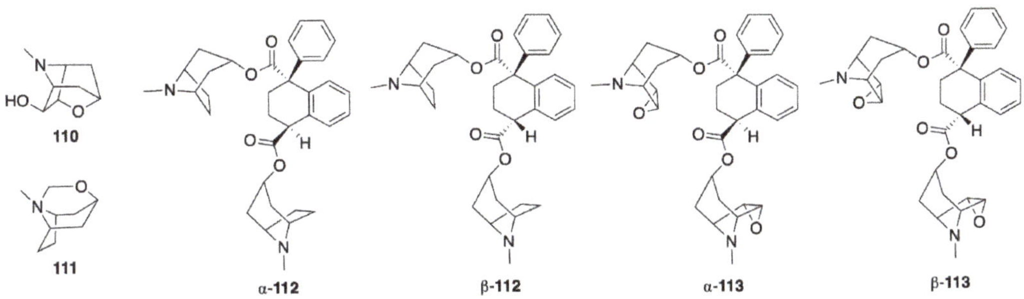

Figure 10. Cyclic and dimeric tropane alkaloids from *Datura*.

D. stramonium contains the dimeric alkaloid belladonnine (α/β-**112**, Figure 10), which consists of two tropine (α-**10**) moieties esterified to a tetrahydronaphthalene diacid known as isatropic acid. Aripova and Yunusov propose that isotropic acid may be formed by photodimerization of atropic acid (dehydrated tropic acid, the hydrolysis product of apohyoscyamine (**39**, SI Table S1)) [125]. Prolonged heating of hyoscyamine can also produce this dimer. An analogous ester of scopine (**78**, SI Table S4) and isatropic acid, called scopadonnine (α/β-**113**) was isolated from *D. inoxia* seeds. This alkaloid, like **112**, is naturally isolated as a mixture of α-(*anti*) and β-(*syn*) isomers, and the structure of β-scopadonnine (α/β-**113**) was unambiguously determined by *N*-methylation with methyl iodide, recrystallization of the di-quaternary salt, and X-ray crystallography [126].

5. Alkaloids of the Genus *Datura*—Non-Tropane Alkaloids

Thus far, this review has classified compounds using the classical definition of alkaloids—natural products containing a basic (and often heterocyclic) nitrogen, or those compounds biogenetically related to basic alkaloids. In addition to tropanes, there are many other types of nitrogenous compounds in *Datura* species, such as hydroxycinnamic amides, amino acids, diketopiperazines, and their derivatives; those are not considered alkaloids for the purposes of this discussion, although other studies have used the term "alkaloid" more broadly [36,77,127].

5.1. Pyrrolidines

Pyrrolidine alkaloids (Figure 11) are products of the reaction of electrophilic *N*-methylpyrrolinium ion (**17**, Figure 6) with various nucleophiles. The most common pyrrolidine alkaloid in the genus *Datura* is hygrine (**12**), which results from the condensation of **17** and acetoacetate (derived from two molecules of acetyl-CoA) [19]. Although originally believed to be such, this alkaloid is not a tropane precursor, but rather a by-product of the tropane biosynthetic pathway [63]. Two enantiomers of **12** are possible, although the literature does not explicitly elaborate on which are found in *Datura* (it is drawn as racemic). *D. inoxia* and *D. metel* var. *fastuosa* roots, but not leaves, are reported to contain small amounts of hygrine [40]. This alkaloid was also detected in the seeds of *D. ferox* from Argentina [37], *D. stramonium* from Egypt and Mexico [22,75], in *D. leichhardtii* seeds [87], and in root cultures derived from *D. stramonium*, *D. inoxia*, *D. metel* var *fastuosa*, *D. quercifolia*, *D. ferox*, and *D. wrightii*) [61,128].

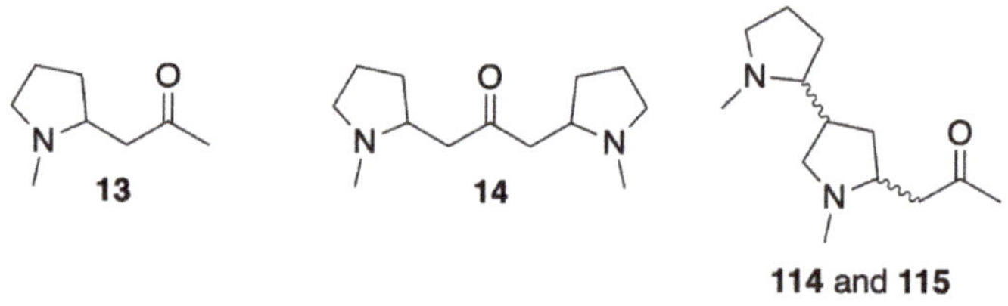

Figure 11. *N*-methylpyrrolinium ion (**17**)-derived pyrrolidine alkaloids.

Cuscohygrine (**13**) results from condensation of oxobutanoate ester **18** (Figure 6) with **17**, followed by decarboxylation, [99] although labeling studies suggest it can also be formed from **12** [63]. Although it can exist as two diastereomers (two pairs of enantiomers), no references are made to the specific isomer(s) found in *Datura*, although both isomers of **12** and **18** convert to **13** [63,129]. This alkaloid has been found in *D. inoxia* [39,114], *D. stramonium* [68,75] *D. metel* var. *fastuosa*, *D. leichhardtii*, and *D. ferox* [40]. It is found in only low amounts in *D. ceratocaula* [73], but it is reported to be the most abundant alkaloid in *D. discolor* roots (around 20% of the total alkaloid content) [130]. Cuscohygrine (**13**) is also found in root cultures derived from *D. stramonium, D. ferox, D. wrightii, and D. inoxia*, and is 10% of the total alkaloid content in the last species [118,128].

Two other pyrrolidine alkaloids, called *N*-methylpyrrolidinylhygrines A and B (**114** and **115**) were detected by GC-MS in 1987 in *D. inoxia* roots [47]. On the basis of the fragment ions in mass spectra, the structures were only putative, but it was proposed that there are two isomers. Two isomers were also found in *D. inoxia* root cultures [61] and later in *D. ceratocaula* [73]. It is almost certain that *Datura* species contain more undiscovered pyrrolidine alkaloids. In Witte's GC study [39] of *D. inoxia* roots, nitrogen-specific detection suggested that one alkaloid (m/z 414, M$^+$) contained more than one nitrogen, and yielded EI-MS fragments of m/z 124 and 84, suggesting that this alkaloid contained both pyrrolidine and tropane moieties. This is unusual, as tropane alkaloids containing additional nitrogen atoms (excluding dimeric tropanes) have not been reported in the Solanaceae. "Alkaloid 414" has also been recognized in the transformed roots of *Brugmansia* and intact plants and root cultures of *Hyoscyamus* species but no structure was proposed [128,131].

5.2. Indoles and Beta-Carbolines

The second major non-tropane class of alkaloids found in the genus *Datura* are those based on indole or oxindole cores (Figure 12). In 2020, four racemic indole alkaloids (daturametelindoles **3–6**) were isolated from an ethanolic extract of *D. metel* seeds by Liu et al. [47] who used 2D-NMR, chiral LC, polarimetry, and circular dichroism spectra to resolve structures and stereochemistry. Interestingly, the *n*-butyl ether of compound **5**, to the outsider, appears to be a potential artifact of isolation, as *n*-butanol was used during fractionation, but a 2019 study of rodent metabolism of *D. metel* metabolites detected **5** in an aqueous extract of the seeds (where no *n*-butanol was used), suggesting that the *n*-butyl group occurs naturally in the plant [77].

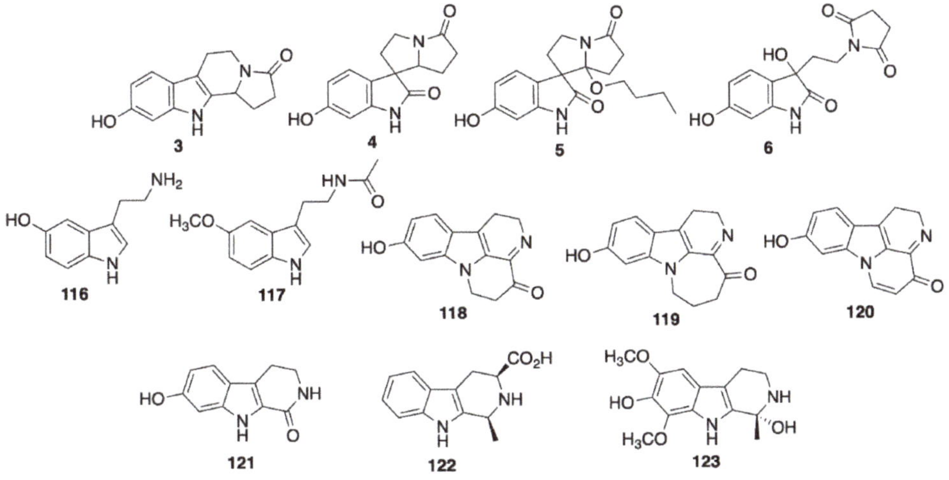

Figure 12. Indole and *beta*-carboline alkaloids found in *D. stramonium* and *D. metel*.

While daturametelindoles B-D are oxindoles, daturametelindole A (**3**) appears to be hydroxytryptamine-derived. In that context, it is interesting that Murch and colleagues (2009) found the hydroxytryptamine serotonin (**116**) along with melatonin (**117**) in the flowers and buds of *D. metel* var. *fastuosa*. Interestingly, the compounds were most abundant in immature flower buds, and levels decreased as the buds matured and flowered. Cold stress also significantly increased melatonin concentration in the buds, suggesting that it may be part of a protective or stress response [132].

Beta-carbolines (similar in structure to the *Peganum harmala* alkaloids) have also been observed in *Datura*. In 1981, Maier and co-workers sliced *D. stramonium* (all varieties) seeds and noted that the cut surface had a distinctive blue-green fluorescence under UV light. This fluorescence was attributed to two compounds later isolated by a combination of column and preparative thin-layer chromatography—the *beta*-carbolines fluorodaturatin (**118**) and homofluorodaturatin (**119**) [133]. A 1,2-dehydro analogue (**120**) was subsequently reported in 1984 [134]. Similar *beta*-carboline structures are reported in *D. metel*. In Hui's metabolomics study of this plant, alkaloids **121** and **122** were noted, but they appear to be in low abundance in the reported chromatograms [36]. Okwu and Igara also isolated what they proposed to be the antibacterial *beta*-carboline alkaloid **123** from *D. metel* leaves from a Nigerian plant accession, although the authors did not use any additional NMR techniques (such as COSY, TOCSY, or HMBC) to confirm connectivity, so their structure should be treated as putative [135].

5.3. Miscellaneous Alkaloids

Kariñho-Betancourt [76] reports that the toxic glycoalkaloid solanine (**124**, Figure 13) is present in the leaves of most *Datura* species, albeit as a minor alkaloid whose concentration changes very little with age. Solanine is very common in the Solanaceae and is also found in tomatoes, potatoes, and eggplant, among other species. In Witte's [39] GC study, tyramine (**125**) was found in *D. inoxia* leaves and flowers. Similarly, a joint Japanese-Bangladeshi group [136] also identified tyramine as in *D. stramonium* leaves while performing bioassay-guided fractionation in search for compounds that would induce apoptosis in cancer cells. The presence of free tyramine in these plants is unsurprising, as it is commonly incorporated into hydroxycinammic acid amides and similar compounds. The same group [136] identified the pyridine trigonelline (**126**) in the same screen; although, like tyramine, this compound is found across many genera and plant families and is not unique to or characteristic of *Datura*. Hui et al. [36] reported that the tetrahydroisoquinoline derivative chenoalbicin (**127**) is present in many parts of *D. metel* but is the highest in the seeds. This compound was identified by matches with its exact mass, retention time, and two MRM transitions from *m/z* 607>353 and *m/z* 607>277 matching a standard.

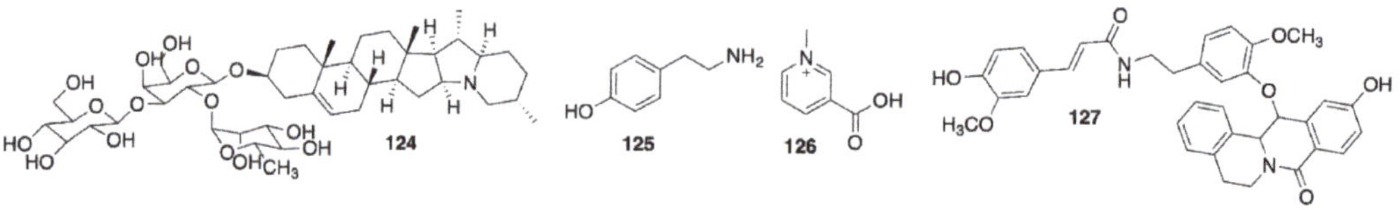

Figure 13. Miscellaneous alkaloids identified in *Datura*.

6. Factors Affecting Alkaloid Content and Composition

As described in the previous sections, the most obvious factor influencing alkaloid composition within the genus *Datura* seems to be genotype and tissue type [34,35,40,74, 128,137,138]. Nonetheless, a multitude of other variables can affect alkaloid identity and diversity, total alkaloid content, and ratios between different alkaloid levels in *Datura*; those variables are briefly summarized in the following subsections.

6.1. Location within the Plant

Although all parts of all *Datura* plants contain alkaloids, the roots are considered the primary site of alkaloid biosynthesis, and alkaloids are then trafficked throughout the plant. This was largely confirmed by the elegant grafting studies performed by James [53], where *D. inoxia* and *A. belladonna* roots and aerial parts (species which contain very different alkaloid profiles) were grafted onto each other. The resulting aerial tissues contained the alkaloid composition and ratio of the species used as the rootstock.

As roots are the ultimate origin of most alkaloids, their diversity and concentrations are usually highest in these parts. For example, Evans and Somanabandhu [130] found that the alkaloid content in the roots of *D. discolor* was 0.31% of the dry weight, as compared to only 0.17% in the aerial parts. Doncheva [40] also reported that the roots of various *Datura* species exhibited a greater "variety of alkaloids" than the leaves, although some leaf-specific alkaloids (such as nortropanes) were also found. Similarly, the aerial parts of *D. inoxia* contain mostly (*S*)-**1** and **2**, whereas the alkaloid content of the roots was much more diverse and included various pyrrolidine alkaloids in addition to tropanes (*vide supra*) [39].

In contrast to other studies, Miraldi et al. [65] described that roots of adult *D. stramonium* originating in Italy generally do not contain appreciable amounts of alkaloids, but also propose that alkaloids may be synthesized in the roots of younger plants and then elaborated in the aerial parts after transport, a hypothesis which was also supported by Berkov [72] and others.

There is considerable variation in alkaloid content between the different aerial parts of individual plants as well: (*S*)-**1** and **2** contents were found to vary significantly between the aerial parts of adult *D. stramonium* plants, with the highest concentration of these alkaloids being found in the seeds, and the lowest in the fruit pericarp and stem [65]. A similarly high concentration of scopolamine was observed in *D. metel* seeds of four varieties, but flowers and leaves of the same plants contained lower amounts [24]. Witte [39] describes similar differences between the aerial parts of *D. inoxia*. Although tyramine (**125**, Figure 13) was present in both flowers and leaves, it was one of the three major constituents of the flowers, whereas the major components of the stems are (*S*)-**1**, **2**, and α-**20** (meteloidine, SI Table S3).

6.2. Plant Age

One of the earliest studies that quantified alkaloid content with respect to age was that of Fuller and Gibson [51] who spectrophotometically measured (*S*)-**1** concentration in *D. stramonium* var. *tatula* plants at different weeks of growth. They found that (*S*)-**1** concentration increased in both roots and leaves until around 10 weeks of age, after which it appeared to plateau. There are many contemporary (and conflicting reports) concerning the relationship between plant age and alkaloid content. Kariñho-Betancourt [76] reported that tropane alkaloid concentration is generally higher during the reproductive stages of all species of *Datura* plants studied (as assessed by LC-MS), whereas Miraldi [65] stated that young *D. stramonium* stems and leaves contained considerably higher amounts of (*S*)-**1** and **2** than the analogous parts of adult plants. A 1989 study [35], on the other hand, revealed that the levels of these alkaloids actually fluctuate in different plant parts over time, suggesting that alkaloid levels reflect a complex interplay between synthesis, transport, and degradation. Jakabová et al. [139] surveyed (*S*)-**1** and **2** contents in *D. metel*, *D. stramonium*, and *D. inoxia*, and compared those plants harvested in Hungary in either the summer and autumn, and the autumn samples (older plants) contained considerably lower alkaloid content than the summer plant samples. It is possible that in this case, seasonal changes also influence alkaloid content, in addition to ontogenetic stage (vide infra).

The exact composition and type of alkaloids found changes with plant age as well: in the roots of yellowed, senile *D. stramonium* plants harvested in autumn, only five alkaloids were detected by GC-MS, suggesting very slow tropane alkaloid synthesis (or shunting of alkaloids into the fruits), whereas forty-two different tropane alkaloids were detected in the roots while the fruits were maturing [72]. This same study also found that in the aerial parts of *D. stramonium*, the percentage of 6,7-epoxytropanes (of all total alkaloids) decreased with age, whereas the percentage of monosubstituted tropane alkaloids increased correspondingly, which suggests possible degradation/recycling or altered alkaloid trafficking as the plant ages.

6.3. Ploidy of Plants

The alkaloid content can also be influenced by the chromosome copy number (ploidy) of the plant. Berkov and Philipov [140] collected natural diploid *D. stramonium* in Bulgaria and artificially created autotetraploid plants by treatment of seedlings with colchicine. While the overall alkaloid profile was not significantly different, tetraploid plants had a higher concentration of alkaloids in roots and leaves, as well as a higher scopolamine/hyoscyamine ratio, suggesting that tetraploid plants may produce more scopolamine. In hairy root cultures established from both diploid and tetraploid *D. stramonium* plants (via *Agrobacterium* transformation), alkaloid content was measured on the 3rd, 9th, and 15th day of growth. At 15 days of age (the stationary phase of the culture), the diploid hairy root cultures contained more than twice as many different alkaloids as the tetraploid cultures—mostly minor acyltropane derivatives [118]. Similarly, hygrine (**12**) and cuscohygrine (**13**, Figure 11), and methylecgonine (**106**, Figure 10) were found in the diploid, but not the tetraploid cultures. In contrast, the diploid cultures contained less hyoscyamine than the tetraploid cultures at the stationary phase, possibly as a result of its derivatization into other alkaloids. These results suggest that the ploidy of *Datura* may be an important variable to consider for metabolic engineering.

6.4. Light and Water Amount

Plant growth conditions can also significantly affect alkaloid production. *D. metel* plants exposed to long days and bright light in Cosson's 1969 study accumulated more **2** than plants kept in dim light [141]. It was later suggested that flower formation (which happens during lower light periods and can be inhibited by bright light) somehow negatively affects the enzymatic epoxidation of (*S*)-**1** [33]. Demeyer and Dejaegere [142] also observed that high-energy light caused a redistribution of alkaloids (from the leaves and stems to the fruit and seeds) in *D. stramonium* var. *tatula*. A very recent publication, however, reported that while both *D. stramonium* and *D. inoxia* plants kept in low light had altered morphology (lower mass and larger leaf areas), the overall concentrations of **1** and **2** in the leaves did not change among any of the three light conditions tested [35].

An extensive study by Moreno-Pedraza [32] showed that alkaloid content in *D. stramonium* correlated with soil moisture. Trays of growing plants were irrigated with 500, 1000,

1500, and 2000 mL of water every eight days, and alkaloids were then quantified in different tissues by LC-MS. The relationship between alkaloid content and water amount was highly variable. For example, the average littorine (**11**), scopolamine (**2**), and acetylhyoscyamine (**37**) content decreased in all organs when the water volume increased beyond 1500 mL. Hyoscyamine amounts, by contrast, showed a parabolic dependence, being lowest at very low (500 mL) and very high (2000 mL) water conditions. In other cases, the correlation seems to be organ-specific: one isomer of hydroxyhyoscyamine (possibly anisodamine, **21**, Figure 7) peaked in amount at 1500 mL water in leaves and roots, but peaked at 2000 mL water in stems.

6.5. Chemical Additives

Alkaloid amounts can also be altered by addition of salts, hormones, growth retardants, elicitors, or other exogenous chemicals. A 1985 study [143] reported that salt stress (addition of 0.1538 M NaCl during growth) enhanced **2** accumulation in the young leaves of *D. inoxia*, although the positive correlation between growth and overall alkaloid content did not change. A later study treated *D. stramonium* var. *tatula* with various mineral solutions. These solutions contained with one dominant anion (nitrate, sulfate, or $H_2PO_4^-$) and one dominant cation (K^+, Ca^{2+}, or Mg^{2+}). Plants treated with nitrate-containing ion pairs had increased hyoscyamine yield (likely as a result of increased nitrogen availability), whereas **2** amounts tended to correlate more with growth stage than the presence or absence of additives [33].

In another study published by the same group, Demeyer et al. [144] treated growing *D. stramonium* with the hormones cytokinin (DMAA) and auxin (IAA). The content of (*S*)-**1** was significantly higher in the leaves of DMAA-treated plants, possibly because of increased metabolism or reduced senescence. In IAA-treated plants, the (*S*)-**1** and **2** contents were lower than the control plants in the whole plant until 20 weeks. In the control plants, alkaloid contents decreased after 20 weeks of age, whereas in the IAA-treated plant, they increased. The authors hypothesized that IAA promotes the pentose phosphate pathway and the glyoxylate cycle, which results in excess glutamate and energy, which are converted into ornithine and, thus, shunted into the putrescine (**14**)/*N*-methylpyrrolinium ion (**17**) pathway (Figure 6).

Gupta and Madan [145] sprayed the aerial parts of *D. metel* var. *fastuosa* with two growth retardants (CCC and B-995), and found that at 20,000 ppm, both agents increased alkaloid content in the plants' leaves. This may be because those agents behave as gibberellin antagonists; adding a gibberellin (GA3) to hairy root cultures derived from *D. quercifolia* resulted in a concentration-dependent decrease of (*S*)-**1** production [74]. Interestingly, a 2018 study revealed that addition of acetylsalicylic acid and salicylic acid to hairy root cultures obtained from *D. stramonium*, *D. stramonium* var. *tatula*, and *D. inoxia* significantly increased (*S*)-**1** production, peaking at an additive concentration of ~0.1 mM) [115]. The use of these elicitors to increase alkaloid production has been reported in different species, and they are proposed to act at the transcriptional level. In a 2009 study, it was found that the addition of *Agrobacterium rhizogenes* to the nutrient solution of hydroponically grown *D. inoxia* resulted in increased accumulation of (*S*)-**1**, **2**, and enhanced tropane hydroxylation and esterification [146].

6.6. Geography, Altitude, Climate, and Season

Geography appears to significantly influence alkaloid composition. This appears intuitive at first glance, as the different climates, altitudes, insect populations, and local microenvironments of different areas could impart different selection pressures on plants. Alexander's review [6] summarizes the alkaloid contents of *Datura* plants of the same species isolated from different geographical locations, many of which display striking variance in alkaloid composition. For example, Berkov and colleagues [22,114] noticed that both *D. stramonium* and *D. inoxia* (roots, leaves, and seeds) had different patterns of alkaloids when they were grown in Egypt (versus grown in Bulgaria). In *D. inoxia*, scopine

(**78**, SI Table S4) was the predominant alkaloid found in Egyptian plant leaves, but (*S*)-**1** was instead predominant in Bulgarian plant leaves. The authors also reported that they did not find ten of the reported alkaloids, including **12** (Figure 11) found in Witte's *D. inoxia* study, which used plants grown in a greenhouse in Germany [39]. *D. ferox* [147] also showed geographical variation: both **2** and total alkaloid content varied depending on which province of Argentina the samples were collected from. *D. metel* flowers collected in Shandong, Changde, and Guangdong (China) differed in their **21** (Figure 7) **2**, and (*S*)-**1** contents [60], and similar differences were also observed for specimens of *D. metel* collected in different locations in India [52]. Interestingly, the same authors also discovered that plants grown at high altitude locations (Darjeeling and Pachmari) contained considerably higher alkaloid percentages than those grown at lower altitudes (Poona (Pune) and Bombay (Mumbai)).

Similarly, Karnick and Saxena also found a seasonal periodicity to total alkaloid content in *D. metel* grown in India [52]. Cold or rainy weather favored greater total alkaloid percentages in the roots and stems when compared to hot weather. By contrast, rainy weather favored higher alkaloid content in leaves, and alkaloid contents decrease in roots and leaves during the hot seasons as they accumulate in fruits and seeds, although in this case, it is difficult to distinguish which effects result from seasonal changes and which are derived from plant age.

6.7. Insect Herbivory

As alkaloids are predicted to be defense compounds, several studies on the relationship between insect herbivory and alkaloid production in *Datura* have been conducted. In Shonle and Bergelson's [148] study of *D. stramonium* populations planted in Illinois (US), plants preyed upon by herbivores displayed different selection pressures on certain alkaloids. Hyoscyamine (*S*)-**1** came under positive selection (meaning an increase in fitness for those plants), whereas **2** came under negative selection, suggesting that it was costly for the plant to make and conferred some fitness disadvantage, which lead the authors to believe that it may have actually stimulated certain insects to feed. In a study of Mexican *D. stramonium* populations [149], the reverse was observed: herbivores selected for a reduction in (*S*)-**1** concentration, suggesting it was ineffective at deterring herbivory, whereas **2** appeared effective against generalist but not specialized herbivores (as its concentration was increased in plants fed upon by these insects). The authors claim that this lends support to the "evolutionary arms-race" hypothesis, where *Datura* species develop more progressively functionalized alkaloids (such as trisubstituted or epoxytropanes) to circumvent herbivore resistance to simpler compounds (such as (*S*)-**1**). Interestingly, a 2018 study [150] reported that the growth of two species of caterpillars—one that co-evolved with *D. wrightii* and one that didn't—was unaffected by **2**. This observation suggests that either (a) **2** may serve another purpose altogether—perhaps deterring herbivory or seed predation by mammals, who are affected very strongly by **2**, or (b) it is other *Datura* alkaloids that have anti-insect activity.

Taken together, these studies on the many variables affecting alkaloids provide two valuable conclusions. First, it may be possible to select for the production of certain alkaloids (or increased yields of them) by changing conditions under which plants are grown. Second, because *Datura* alkaloid composition is so sensitive to environmental changes, any study that seeks to assess genome-level (or other biochemical) changes must use plant groups grown and kept under identical and controlled conditions.

7. Conclusions and Future Challenges: Stereochemistry, Dereplication, and Observer Bias

As is the case in almost all natural products research, the isolation and identification of metabolites is still the major bottleneck in studying *Datura*. To understand how alkaloids are synthesized, what their ultimate roles in the plant may be, how they relate to substances found in similar plants, or what bioactivities they may have, more of their structures need to be known, and known *unambiguously* beyond the putative annotations that dominate the literature.

First, *Datura* literature is full of proposed uncharacterized alkaloids, described over the years in terms such as "alkaloid A", "alkaloid 325", "Alkaloid 414", and under catch-all headings such as "uncharacterized bases" [22,39,151,152]. What are these alkaloids, and what other yet-undiscovered compounds accompany them? With the full array of tools available to the modern alkaloid hunter, such as preparative LC columns available in a variety of phases, high-resolution mass spectrometry and MS/MS, and advanced and sensitive NMR techniques, these questions are becoming easier to answer. Special attention, however, must be paid to the absolute configuration of compounds, especially tropane alkaloids, where a multitude of different isomers can be possible and confusion regarding the exact stereochemistry of standards has mounted over the years. Techniques such as chemical derivatization, polarimetry, and NMR chiral shift reagents have all been employed successfully, and it is also worth noting that modern and sophisticated optical techniques, such as vibrational circular dichroism (VCD—circular dichroism applied to the infrared and near-infrared light regions) have been applied to tropane alkaloids, just not to those specifically isolated from or unique to *Datura*. For example, Reina et al. [153] used density functional theory calculations to predict probable conformations and VCD spectra of certain disubstituted tropanes and tropane mixtures from *Schizanthus*, and then were able to assign absolute configuration by comparison of experimental spectra to both the calculated and to literature spectra for similar alkaloids. In another study, the same group [154] also added chemical derivatization to this procedure: Compound **40** (SI Table S2) was first converted to **42** by base hydrolysis followed by treatment with acetic anhydride, and then the absolute configuration of **42** was determined as above. The authors found that the VCD spectra of **40** and **42** had several common bands that are proposed to be directly related to the absolute configuration of the tropane nucleus. Humam et al. [155] also utilized a similar "compute-and-compare" procedure with electronic circular dichroism (which uses light in the UV–VIS range) instead of VCD. These avenues should definitely be explored more for determination of absolute configuration of *Datura*-specific alkaloids.

Another prominent challenge in natural products discovery, that runs hand-in-hand with identification, is that of dereplication [156]. How do researchers avoid spending time, money, and resources isolating (or worse yet, chasing the bioactivity of) a compound that turns out to be already known? This is especially a problem for tropane (and other) alkaloids, where isomerism complicates things immensely. For example, a given GC/MS or LC/MS chromatogram may yield multiple peaks that share common molecular and fragment-ion masses. Standards may not be readily available to sort out which isomer is which, synthesizing them all may be cost-prohibitive or labor-intensive, and it may be near-impossible to obtain a large enough amount of a compound for NMR. Even MS/MS, as useful as it is, is not 100% replicable, as spectra can vary between ion source and individual instrument platforms, and can also be affected by ion source conditions, solvents, salt, or matrix effects, so spectral database comparisons have limits in distinguishing compounds [156]. Special attention must therefore be paid to utilizing and improving those dereplication techniques which are most rapid, high-throughput, and invariant.

An additional complication in both identification and dereplication is that even samples that appear "pure" by HPLC may be a mixture of inseparable isomers or other compounds, which is often an unpleasant surprise late in the isolation workflow. One recently developed technique that could aid in situations such as these is ion mobility spectrometry (IMS), which is now integrated into several mass spectrometry platforms. Often coupled to LC, this mass spectrometry technique is a form of post-ionization processing where ions move differently in a "drift tube" based on their shape, charge and size (called a collision cross-section)—effectively gas phase electrophoresis performed in tandem with mass spectrometry in a time-scale consistent with chromatographic separations and mass spectrometry data acquisition. This processing adds an additional degree of separation, and the data is often plotted with the LC chromatogram, yielding a two-dimensional plot of *retention time* versus *drift time*. IMS can aid in separating isomers and revealing new compounds that may not be detected by conventional LC or LC-MS methods [157].

Additionally, IMS provides an additional "data point" in any compound's given molecular signature, which is valuable in dereplication—two compounds may have identical masses, fragmentation patterns, and retention times, yet differ in their IMS drift times, and are therefore likely different compounds [158]. This method was applied to *beta*-carboline alkaloids from *Peganum harmala*, where more alkaloids were differentiated and identified using the 2D-combined LC-IMS method than by LC or IMS alone [157].

Another emerging issue in natural products research is over-reliance on databases. We live in the Digital Age, and databases such as PubChem, ChemSpider, Metlin, and KEGG are easily used, valuable repositories of information, and software and websites (such as MetFrag or Progenesis QI) are increasingly used to interface with and match empirical mass spectral (and other) data to those in databases. If these results are not examined critically (and within the context of biochemical plausibility) misannotations and misidentifications can (and do) result. Several metabolite profiling studies using database matching [78,79] have presented questionable identifications of components of ipecac, Parkinson's disease drugs, veterinary antibiotics, bacterial-derived organoboron compounds, streptomycin derivatives, and antidepressant metabolites in *Datura* samples, while failing to identify (*S*)-**1**, **2**, or other commonly reported *Datura* alkaloids. An additional example of output provided by computational analysis open to discussion is found in the study of Moreno-Pedraza [32], where the program MetFamily was used to group *Datura* metabolites into putative classes on the basis of their MS/MS fragmentation, which the authors divide into four classes of "atropine-like" compounds, among others. While the study itself appears scientifically sound with many valid conclusions, some of the species classified by Met-Family as "atropine-like", such as $C_{28}H_{24}O_5$, $C_{28}H_{38}O_5$ (Withanolide B), and $C_{25}H_{33}FO_4$, are, obviously, *not* tropane alkaloids if these formulae are correct (lacking nitrogen), even if something about their fragmentation may be tropane-like in nature. Further development and refinements of similar computational approaches are encouraged with appropriate testing and validation.

These examples also speak to the difficulty in overcoming observer bias in metabolomics. When do we trust what our software and databases tell us, and when must we rely on intuition? Since many metabolite studies focus explicitly on one class of compounds, the interpretation of our data (especially mass spectral data) is often clouded by human judgment, operating within the framework of our often-limited knowledge across an entire class of compounds. It is often proposed that if a compound exhibits tropane-like mass spectra, it must be a tropane. While assumptions like these may be a good starting point, how many alkaloids have been missed over the years simply because something about their behavior (unusual masses, retention times, mass defects, or fragmentation) does not fit our hyper-focused definition of how an alkaloid should behave? Additionally, if we do not recognize and report a certain compound because it does not meet some threshold for annotation as that compound, is it actually not present, or are we simply not looking for it? These considerations should be taken to encourage deposition of raw analytical data into shared repositories.

Finally, the question of bioactivity must be addressed. Thirty-four percent of new drugs approved between 1981 and 2010 [159] have a direct natural product origin, and *Datura* has been used medicinally for millennia. Nonetheless, many modern studies on the medicinal properties of *Datura* still employ whole-plant part extracts, and comparatively little is known about the active alkaloidal constituents beyond atropine and scopolamine—two alkaloids out of potentially hundreds across the genus, where there is a plethora of interesting bioactivity waiting to be discovered. In addition to their anticholinergic and psychoactive activity, some tropane alkaloids have shown antiviral, antifungal, and antimicrobial properties, while others can reverse multidrug resistance in cancer cells [48,160,161], while the pyrrolidine, indole, and *beta*-carboline alkaloids are also largely unexplored. More screening of isolated *Datura* alkaloids against diverse targets should be performed, and as their structures are elucidated, it will become easier to perform structure-activity relationship studies and design novel semi-synthetic or synthetic alkaloid mimics.

With new advances in analytical instrumentation (and careful study by many researchers) more than one-hundred alkaloids in multiple classes have been described in the genus *Datura*, and successful efforts have been made to determine precise conditions that affect their presence and amounts. Nonetheless, great untapped potential and promise remains in these plants, and the future of alkaloid research remains bright. It is, however, only when we use both critical thought and the full arrays of tools and techniques available to us that we might take our place in the long history of this fascinating and storied genus. After all, who knows what medicines, mysticism, and magic still lie in wait?

Supplementary Materials: The following are available online: Table S1: Monosubstituted tropane alkaloids described in *Datura*, Table S2: Disubstituted tropane alkaloids from *Datura*, Table S3: 3,6,7-Trisubstituted Tropane alkaloids from *Datura*, Table S4: 3-Substituted 6,7-epoxytropane alkaloids from *Datura*, Table S5: Nortropanes detected in *Datura*.

Author Contributions: Conceptualization: M.A.C. and A.D.J.; methodology: M.A.C. and A.D.J.; experimental analysis: M.A.C.; writing—original draft preparation: M.A.C.; writing—review and editing: M.A.C. and A.D.J.; All authors have read and agreed to the published version of the manuscript.

Funding: This research was funded by a grant from the National Science Foundation, grant number IOS-1546617, (Robert L. Last, PI). A.D.J. acknowledges support from Michigan AgBioResearch through the USDA National Institute of Food and Agriculture, Hatch project number MICL02474.

Institutional Review Board Statement: Not applicable.

Informed Consent Statement: Not applicable.

Data Availability Statement: Not applicable.

Acknowledgments: The authors thank Cornelius Barry and members of the Barry laboratory (Radin Sadre, Hannah Parks, and Josh Grabar) for assistance with *Datura* cultivation and for generously providing a sample of scopolamine for the NMR spectroscopy in Figure 4, and Profesor Robert Last for valuable guidance. We thank Daniel Holmes and Li Xie of the Max T. Rogers NMR facility for assistance with NMR spectroscopy.

Conflicts of Interest: The authors declare no conflict of interest.

Abbreviations

Aromatic amino acid aminotransferase (ArAT), atmospheric pressure chemical ionization (APCI), coenzyme A (CoA), correlation spectroscopy (COSY), Dalton (Da), desorption ESI (DESI), dimethylaminopurine (cytokinin) (DMAA), direct analysis in real-time (DART), distortionless enhancement via polarization transfer (DEPT), electron ionization (EI), electrospray ionization (ESI), planar electrochromatography or thin-layer electrochromatography (ETLC), gas chromatography-mass spectrometry (GC-MS), heteronuclear multiple bond correlation (HMBC), heteronuclear single quantum coherence (HSQC), high-performance liquid chromatography (HPLC), high-performance TLC (HPTLC), hyoscyamine 6β-hydroxylase (H6H), indole-3-acetic acid (auxin) (IAA), infrared (IR), ion-mobility spectrometry (IMS), laser ablation dart imaging (LADI), liquid chromatography-mass spectrometry (LC-MS), mass spectrometry (MS), methylputrescine oxidase (MPO), multiple reaction monitoring (MRM), tandem mass spectrometry (MS/MS), nuclear magnetic resonance (NMR), nuclear Overhauser effect spectroscopy (NOESY), putrescine *N*-methyltransferase (PMT), quadrupole time-of-flight mass spectrometry (QToF), rotating frame Overhauser enhancement spectroscopy (ROESY), selective reaction monitoring (SRM), time-of-flight mass spectrometry (TOF), thin-layer chromatography (TLC), total correlation spectroscopy (TOCSY), tropinone reductase (TR), ultrahigh-performance liquid chromatography (UHPLC), ultraviolet (UV), vibrational circular dichroism (VCD).

References

1. Maheshwari, N.; Khan, A.; Chopade, B.A. Rediscovering the medicinal properties of *Datura* sp.: A review. *J. Med. Plants Res.* **2013**, *7*, 2885–2897.
2. Benítez, G.; March-Salas, M.; Villa-Kamel, A.; Cháves-Jiménez, U.; Hernández, J.; Montes-Osuna, N.; Moreno-Chocano, J.; Cariñanos, P. The genus *Datura* L. (Solanaceae) in Mexico and Spain—Ethnobotanical perspective at the interface of medical and illicit uses. *J. Ethnopharmacol.* **2018**, *218*, 133–151. [CrossRef]
3. Batool, A.; Batool, Z.; Qureshi, R.; Raja, N.I. Phytochemicals, pharmacological properties and biotechnological aspects of a highly medicinal plant: *Datura stramonium*. *J. Plant Sci.* **2020**, *8*, 29–40.
4. Griffin, W.J.; Lin, G.D. Chemotaxonomy and geographical distribution of tropane alkaloids. *Phytochemistry* **2000**, *53*, 623–637. [CrossRef]
5. WHO. *World Health Organization Model List of Essential Medicines, 21st List, 2019*; World Health Organization: Geneva, Switzerland, 2019; License: CC BY-NC-SA 3.0 IGO.
6. Alexander, J.; Benford, D.; Cockburn, A.; Cravedi, J.-P.; Dogliotti, E.; Di Domenico, A.; Férnandez-Cruz, M.L.; Fürst, P.; Fink-Gremmels, J.; Galli, C.L.; et al. Scientific opinion of the panel on contaminants in the food chain on a request from the European Commission on tropane alkaloids (from *Datura* sp.) as undesirable substances in animal feed. *EFSA J.* **2008**, *691*, 1–55.
7. Kerchner, A.; Farkas, A. Worldwide poisoning potential of *Brugmansia* and *Datura*. *Forensic Toxicol.* **2020**, *38*, 30–41. [CrossRef]
8. Diker, D.; Markovitz, D.; Rothman, M.; Sendovski, U. Coma as a presenting sign of *Datura stramonium* seed tea poisoning. *Eur. J. Intern. Med.* **2007**, *18*, 336–338. [CrossRef]
9. Kanchan, T.; Atreya, A. Datura: The Roadside Poison. *Wilderness Environ. Med.* **2016**, *27*, 441–443. [CrossRef] [PubMed]
10. Ricard, F.; Abe, E.; Duverneuil-Mayer, C.; Charlier, C.; de la Grandmaison, G.; Alvarez, J.C. Measurement of atropine and scopolamine in hair by LC-MS/MS after *Datura stramonium* chronic exposure. *Forensic Sci. Int.* **2012**, *223*, 256–260. [CrossRef]
11. Hanna, J.P.; Schmidley, J.W.; Braselton, W.E. *Datura* Delirium. *Clin. Neuropharmacol.* **1992**, *15*, 109–113. [CrossRef] [PubMed]
12. Kohnen-Johannsen, K.L.; Kayser, O. Tropane alkaloids: Chemistry, pharmacology, biosynthesis, and production. *Molecules* **2019**, *24*, 796. [CrossRef] [PubMed]
13. Marc, B.; Martis, A.; Moreau, C.; Arlie, G.; Kintz, P.; Leclerc, J. Acute *Datura stramonium* poisoning in an emergency department. *Presse Med.* **2007**, *36*, 1399–1403. [CrossRef] [PubMed]
14. Beverley, R.; Elkins, W.M. *The History and Present State of Virginia, in Four Parts*; R. Parker: London, UK, 1722; p. 24.
15. Müller, J.L. Love potions and the ointment of witches: Historical aspects of the nightshade alkaloids. *Clin. Toxicol.* **1998**, *36*, 617–627. [CrossRef]
16. Robinson, D.W.; Brown, K.; McMenemy, M.; Dennany, L.; Baker, M.J.; Allan, P.; Cartwright, C.; Bernard, J.; Sturt, F.; Kotoula, E.; et al. *Datura* quids at Pinwheel Cave, California, provide unambiguous confirmation of the ingestion of hallucinogens at a rock art site. *Proc. Natl. Acad. Sci. USA* **2020**, *49*, 31026–31037. [CrossRef]
17. Litzinger, W.J. Ceramic evidence for prehistoric *Datura* use in North America. *J. Ethnopharmacol.* **1981**, *4*, 57–74. [CrossRef]
18. Geeta, R.; Gharaibeh, W. Historical evidence for a pre-Columbian presence of *Datura* in the Old World and implications for a first millennium transfer from the New World. *J. Biosci.* **2007**, *32*, 1227–1244. [CrossRef]
19. Seigler, D.S. Pyrrolidine, tropane, piperidine, and polyketide Alkaloids. In *Plant Secondary Metabolism*; Springer Science+Business Media: New York, NY, USA, 1998; pp. 531–545.
20. Mace, E.S.; Gebhardt, C.G.; Lester, R.N. AFLP analysis of genetic relationships in the tribe Datureae (Solanaceae). *Theor. Appl. Genet.* **1999**, *99*, 634–641. [CrossRef]
21. Gerlach, G.H. *Datura Innoxia*—A potential commercial source of scopolamine. *Econ. Bot.* **1948**, *2*, 436–454. [CrossRef]
22. Berkov, S.; Zayed, R.; Doncheva, T. Alkaloid patterns in some varieties of *Datura stramonium*. *Fitoterapia* **2006**, *77*, 179–182. [CrossRef] [PubMed]
23. Verloove, F. Datura wrightii (Solanaceae), a neglected xenophyte, new to Spain. *Bouteloua* **2008**, *4*, 37–40.
24. Hiraoka, N.; Tashimo, K.; Kinoshita, C.; Hiro'oka, M. Genotypes and alkaloid contents of *Datura metel* varieties. *Biol. Pharm. Bull.* **1995**, *19*, 1086–1089. [CrossRef] [PubMed]
25. Hammer, K.; Romeike, A.; Tittel, C. Vorarbeiten zur monographichen Darstellung von Wildpflanzensortimenten: Datura L., sections *Dutra* Bernh., *Ceratocaulis* Bernh. et *Datura*. *Kulturpflanze* **1983**, *31*, 13–75. [CrossRef]
26. Haegi, L. Taxonomic account of *Datura L.* (Solanaceae) in Australia with a note on *Brugmansia* Pers. *Aust. J. Bot.* **1976**, *24*, 415–435. [CrossRef]
27. Cavazos, M.L.; Jiao, M.; Bye, R. Phenetic analysis of *Datura* section *Dutra* (Solanaceae) in Mexico. *J. Linn. Soc. Bot.* **2000**, *133*, 493–507. [CrossRef]
28. Lockwood, T.E. Generic recognition of *Brugmansia*. *Bot. Mus. Leaf. Harv. Univ.* **1973**, *23*, 273–284.
29. Christen, P.; Bieri, S.; Berkov, S. Methods of analysis: Tropane alkaloids from plant origin. In *Natural Products*; Ramawat, K.G., Mérillon, J.M., Eds.; Springer: Berlin/Heidelberg, Germany, 2013; pp. 1009–1048.
30. Christen, P.; Cretton, S.; Humam, M.; Bieri Muñoz, O.; Joseph-Nathan, P. Chemistry and biological activity of alkaloids from the genus *Schizanthus*. *Phytochem. Rev.* **2020**, *19*, 615–641. [CrossRef]
31. Petruczynik, A. Analysis of alkaloids from different chemical groups by different liquid chromatography methods. *Cent. Eur. J. Chem.* **2012**, *10*, 802–835. [CrossRef]

32. Moreno-Pedraza, A.; Gabriel, J.; Treutler, H.; Winkler, R.; Vergara, F. Effects of water availability in the soil on tropane alkaloid production in cultivated *Datura stramonium*. *Metabolites* **2019**, *9*, 131. [CrossRef]

33. Demeyer, K.; Dejaegere, R. Influence of the ion-balance in the growth medium on the yield and alkaloid content of *Datura stramonium* L. *Plant Soil* **1989**, *114*, 289–294. [CrossRef]

34. Maltese, F.; Van der Kooy, F.; Verpoorte, R. Solvent derived artifacts in natural products chemistry. *Nat. Prod. Commun.* **2009**, *4*, 447–454.

35. Hirano, I.; Iida, H.; Ito, Y.; Park, H.-D.; Takahashi, K. Effects of light conditions on growth and defense compound contents of *Datura inoxia* and *D. stramonium*. *J. Plant. Res.* **2019**, *132*, 473–480. [CrossRef]

36. Hui, Y.S.; Yan, L.; Qi, W.; Ping, S.Y.; Wei, G.; Yuan, L.; You, Y.B.; Xue, K.H. Identification and quantification of alkaloid compounds from different parts and production areas of *Datura metel* L. *Heterocycles* **2020**, *100*, 468–584.

37. Vitale, A.A.; Acher, A.; Pomilio, A.B. Alkaloids of *Datura ferox* from Argentina. *J. Ethnopharmacol.* **1995**, *49*, 81–89. [CrossRef]

38. Sramska, P.; Maciejka, A.; Topolewska, A.; Stepnowski, P.; Halinski, L.P. Isolation of atropine and scopolamine from plant material using liquid-liquid extraction and EXtrelut®columns. *J. Chromatogr. B* **2017**, *1043*, 202–208. [CrossRef]

39. Witte, L.; Müller, K.; Arfmann, H.-A. Investigation of the alkaloid pattern of *Datura innoxia* plants by capillary gas-liquid-chromatography-mass-spectrometry. *Planta Med.* **1987**, *53*, 192–197. [CrossRef]

40. Doncheva, T.; Berkov, S.; Philipov, S. Comparative study of the alkaloids in the tribe Datureae and their chemosystematic significance. *Biochem. Syst. Ecol.* **2006**, *34*, 478–488. [CrossRef]

41. Temerdashev, A.Z.; Kolychev, I.A.; Kiseleva, N.V. Chromatographic determination of some tropane alkaloids in *Datura metel*. *J. Anal. Chem.* **2012**, *67*, 960–966. [CrossRef]

42. Djilani, A.; Legseir, B.; Soulimani, R.; Dicko, A.; Younos, C. New extraction technique for alkaloids. *J. Braz. Chem. Soc.* **2006**, *17*, 418–520. [CrossRef]

43. Payne, J.; Hamill, J.D.; Robins, R.J.; Rhodes, M.J.C. Production of hyoscyamine by 'hairy root' cultures of *Datura stramonium*. *Planta Med.* **1987**, *53*, 474–478. [CrossRef] [PubMed]

44. Lanfranchi, D.A.; Tomi, F.; Casanova, J. Enantiomeric differentiation of atropine/hyoscyamine by ^{13}C NMR spectroscopy and its application to *Datura stramonium* extract. *Phytochem. Anal.* **2010**, *21*, 597–601. [CrossRef] [PubMed]

45. Siddiqui, S.; Sultana, N.; Ahmed, S.S.; Haider, S. Isolation and structure of a new alkaloid datumetine from the leaves of *Datura metel*. I. *J. Nat. Prod.* **1986**, *49*, 511–513. [CrossRef]

46. Ciechomska, M.; Wozniakiewicz, M.; Nowak, J.; Swiadek, K.; Bazylewicz, B.; Koscielniak, P. Development of a microwave-assisted extraction of atropine and scopolamine from Solanaceae family plants followed by a QuEChERS cleanup procedure. *J. Liq. Chromatogr. Relat. Technol.* **2016**, *39*, 538–548. [CrossRef]

47. Liu, Y.; Jiang, H.-B.; Liu, Y.; Algradi, A.M.; Naseem, A.; Zhou, Y.-Y.; She, X.; Li, L.; Yang, B.-Y.; Kuang, H.-X. New indole alkaloids from the seeds of *Datura metel*. L. *Fitoterapia* **2020**, *146*, 104726. [CrossRef] [PubMed]

48. Welegergs, G.G.; Hulif, K.; Mulaw, S.; Gebretsadik, H.; Tekluu, B.; Temesgen, A. Isolation, structural elucidation and bioactivity studies of tropane derivatives of alkaloids from seeds extract of Datura stramonium. *Sci. J. Chem.* **2015**, *3*, 78–83. [CrossRef]

49. Beresford, P.J.; Woolley, J.G. 6β-[2-methylbutanoyloxy]tropan-3α-ol, a new alkaloid from *Datura ceratocaula*, structure and biosynthesis. *Phytochemistry* **1974**, *13*, 2511–2513. [CrossRef]

50. Fu, C.; Zhang, W.; Wu, Z.; Chen, P.; Hiu, A.; Zheng, Y.; Li, H.; Xu, K. A novel process for scopolamine separation from *Hindu Datura* extracts by liquid-liquid extraction, microporous resins, and crystallization. *Sep. Sci. Technol.* **2020**, *55*, 922–931. [CrossRef]

51. Fuller, W.C.; Gibson, M.R. The Arginase-Alkaloid Relationship in *Datura tatula* L. *J. Am. Pharm. Assoc.* **1952**, *41*, 263–266. [CrossRef]

52. Karnick, C.R.; Saxena, M.D. On the variability of alkaloid production in *Datura* species. *Planta Med.* **1970**, *18*, 266–269. [CrossRef] [PubMed]

53. James, G.M.; Thewlis, B.H. The separation and identification of solanaceous alkaloids from normal and grafted plants. *New Phytol.* **1952**, *51*, 250–255. [CrossRef]

54. Nash, R.J.; Rothschild, M.; Porter, E.A.; Watson, A.A.; Waigh, R.D.; Waterman, P.G. Calystegines in *Solanum* and *Datura* species and the death's head hawk-moth (*Acherontia atropus*). *Phytochemistry* **1993**, *34*, 1281–1283. [CrossRef]

55. Sharma, V.; Sharma, N.; Singh, B.; Gupta, R.C. Cytomorphological studies and HPTLC fingerprinting in different plant parts of three wild morphotypes of *Datura metel* L. "Thorn Apple" from North India. *Int. J. Green. Pharm.* **2009**, *3*, 40–46. [CrossRef]

56. Malinowska, I.; Studzinski, M.; Niezabitowska, K.; Gadzikowska, M. Comparison of TLC and different micro TLC techniques in analysis of tropane alkaloids and their derivatives mixture from *Datura Inoxia* Mill. extract. *Chromatographia* **2013**, *76*, 1327–1332. [CrossRef] [PubMed]

57. Malinowska, I.; Studzinski, M.; Niezabitowska, K.; Malinowski, H. Planar electrochromatography and thin-layer chromatography of tropane alkaloids from *Datura Inoxia* Mill. extract. in pseudo-reversed-phase systems. *JPCJ Planar. Chromat.* **2016**, *29*, 38–44. [CrossRef]

58. Duez, P.; Chamart, S.; Hanocq, M.; Molle, L.; Vanhaelen, M.; Vanhaelen-Fastré, R. Comparison between thin-layer chromatography-densitometry and high-performance liquid chromatography for the determination of hyoscyamine and hyoscine in leaves, fruit and seeds of *Datura* (*Datura* spp.). *J. Chromatogr.* **1985**, *329*, 415–421. [CrossRef]

59. Naumann, A.; Kurtze, L.; Krähmer, A.; Hagels, H.; Schulz, H. Discrimination of Solanaceae taxa and quantification of scopolamine and hyoscyamine by ATR-FTIR spectroscopy. *Planta Med.* **2014**, *80*, 1315–1320. [CrossRef]

60. Li-Yi, H.; Guan-De, Z.; Yu-Yi, T.; Sagara, K.; Oshima, T.; Yoshida, T. Reversed-phase ion-pair high-performance liquid chromatographic separation and determination of tropane alkaloids in Chinese solanaceous plants. *J. Chromatogr.* **1989**, *481*, 428–433. [CrossRef]

61. Ionkova, I.; Witte, L.; Alfermann, A.W. Spectrum of tropane alkaloids in transformed roots of *Datura innoxia* and *Hyoscyamus x györffyi* cultivated in vitro. *Planta Med.* **1994**, *60*, 382–384. [CrossRef] [PubMed]

62. Ford, Y.-Y.; Ratcliffe, R.G.; Robins, R.J. In vivo NMR analysis of tropane alkaloid metabolism in transformed root and de-differentiated cultures of *Datura stramonium*. *Phytochemistry* **1996**, *43*, 115–120. [CrossRef]

63. Robins, R.J.; Abraham, T.W.; Parr, A.J.; Eagles, H.; Walton, N.J. The biosynthesis of tropane alkaloids in *Datura stramonium*: The identity of the intermediates between *N*-methylpyrrolinium salt and tropinone. *J. Am. Chem. Soc.* **1997**, *119*, 10929–10934. [CrossRef]

64. El Bazaoui, A.; Bellimam, M.A.; Soulaymani, A. Nine new tropane alkaloids from *Datura stramonium* L. identified by GC/MS. *Fitoterapia* **2011**, *82*, 193–197. [CrossRef] [PubMed]

65. Miraldi, E.; Masti, A.; Ferri, S.; Comparini, I.B. Distribution of hyoscyamine and scopolamine in *Datura stramonium*. *Fitoterapia* **2001**, *72*, 644–648. [CrossRef]

66. Patterson, S.; O'Hagan, D. Biosynthetic studies on the tropane alkaloid hyoscyamine in *Datura stramonium*; hyoscyamine is stable to in vivo oxidation and is not derived from littorine via a vicinal interchange process. *Phytochemistry* **2002**, *61*, 323–329. [CrossRef]

67. Dupraz, J.-M.; Christen, P.; Kapetanidis, I. Tropane alkaloids in transformed roots of *Datura quercifolia*. *Planta Med.* **1994**, *60*, 158–162. [CrossRef] [PubMed]

68. Philipov, S.; Berkov, S.; Doncheva, T. GC-MS survey of *Datura stramonium* alkaloids. *Compt. Rend. Acad. Bulg. Sci.* **2007**, *60*, 239–250.

69. Zhou, M.; Ma, X.; Sun, J.; Ding, G.; Ciu, Q.; Miao, Y.; Hou, Y.; Jiang, M.; Bai, G. Active fragments-guided drug discovery and design of selective tropane alkaloids using ultra-high performance liquid chromatography-quadrupole time-of-flight tandem mass spectrometry coupled with virtual calculation and biological evaluation. *Anal. Bioanal. Chem.* **2017**, *409*, 1145–1157. [CrossRef] [PubMed]

70. Christen, P.; Roberts, M.; Phillipson, J.D.; Evans, W.C. Alkaloids of hairy root cultures of a *Datura candida* hybrid. *Plant Cell Rep.* **1990**, *9*, 101–104. [CrossRef] [PubMed]

71. Chen, H.; Chen, Y.; Wang, H.; Du, P.; Han, F.; Zhang, H. Analysis of scopolamine and its eighteen metabolites in rat urine by liquid chromatography-tandem mass spectrometry. *Talanta* **2005**, *67*, 984–991. [CrossRef]

72. Berkov, S.; Doncheva, T.; Philipov, S.; Alexandrov, K. Ontogenetic variation of the tropane alkaloids in *Datura stramonium*. *Biochem. Syst. Ecol.* **2005**, *33*, 1017–1029. [CrossRef]

73. Berkov, S. Alkaloids of *Datura ceratocaula*. *Z. Naturforsch.* **2003**, *58c*, 455–458. [CrossRef]

74. Boros, B.; Farkas, A.; Jakabová, S.; Bacskay, I.; Kilár, F.; Felinger, A. LC-MS quantitative determination of atropine and scopolamine in the floral nectar of *Datura* species. *Chromatographia* **2010**, *71*, S43–S49. [CrossRef]

75. De-la-Cruz, I.M.; Cruz, L.L.; Martínez-García, L.; Valverde, P.L.; Flores-Ortiz, C.M.; Hernández-Portilla, L.B.; Núñez-Farfán, J. Evolutionary response to herbivory: Population differentiation in microsatellite loci, tropane alkaloids and leaf trichome density in *Datura stramonium*. *Arthropod Plant Interact.* **2020**, *14*, 21–30. [CrossRef]

76. Kariñho-Betancourt, E.; Agrawal, A.A.; Halitschke, R.; Núñez-Farfán, J. Phylogenetic correlations among chemical and physical plant defenses change with ontogeny. *New Phytol.* **2015**, *206*, 796–806. [CrossRef] [PubMed]

77. Xia, C.; Liu, Y.; Qi, H.; Niu, L.; Zhu, Y.; Lu, W.; Xu, X.; Su, Y.; Yang, B.; Wang, Q. Characterization of the metabolic fate of *Datura metel* seed extract and its main constituents in rats. *Front. Pharmacol.* **2019**, *10*, 571. [CrossRef]

78. Tapfuma, K.I.; Mekuto, L.; Makatini, M.M.; Mavumengwana, V. The LC-QTOF-MS/MS analysis data of detected metabolites from the crude extract of *Datura stramonium* leaves. *Data Brief* **2019**, *25*, 104094. [CrossRef] [PubMed]

79. Satpute, S.B.; Vanmare, D.J. Chemical profile of *Datura stramonium* L. by using HPLC-MS spectra. *Int. Res. J. Sci. Eng.* **2018**, *6*, 231–235.

80. Bell, D.S. and Jones, A.D. Solute attributes and molecular interactions contributing to "U-shaped" retention on a fluorinated high-performance liquid chromatography stationary phase. *J. Chromatogr. A* **2005**, *1073*, 99–109. [CrossRef] [PubMed]

81. Bell, D.S.; Cramer, H.M.; Jones, A.D. Rational method development strategies on a fluorinated stationary phase: Mobile phase ionic concentration and temperature effects on the separation of ephedrine alkaloids. *J. Chromatogr. A* **2005**, *1095*, 113–118. [CrossRef]

82. Papadoyannis, I.N. Instrumental analytical techniques used for Datura alkaloid analysis. *Instrum. Sci. Technol.* **1994**, *22*, 241–258. [CrossRef]

83. El Bazaoui, A.; Stambouli, H.; Bellimam, M.A.; Soulaymani, A. Determination of tropane alkaloids in seeds of *Datura stramonium* L. by GC/MS and LC/MS. *Ann. Toxicol. Anal.* **2009**, *21*, 183–188.

84. Gerber, R.; Naudé, T.W.; de Kock, S.S. Confirmed *Datura* poisoning in a horse most probably due to *D. ferox* in contaminated tef hay. *Tydskr, S. Afr. Vet. Ver.* **2006**, *77*, 86–89. [CrossRef]

85. Marín-Sáez, J.; Romero-González, R.; Garrido Frenich, A. Multi-analysis determination of tropane alkaloids in cereals and solanaceaes seeds by liquid chromatography coupled to single stage Exactive-Orbitrap. *J. Chromatogr. A* **2017**, *1518*, 46–58. [CrossRef]

86. Plumb, R.S.; Johnson, K.A.; Rainville, P.; Smith, B.W.; Wilson, I.D.; Castro-Perez, J.M.; Nicholson, J.K. UPLC/MS^E; a new approach for generating molecular fragment information for biomarker structure elucidation. *Rapid Commun. Mass Spectrom.* **2006**, *20*, 1989–1994. [CrossRef]
87. Marín-Sáez, J.; Romero-González, R.; Garrido Frenich, A. Enantiomeric determination and evaluation of the racemization process of atropine in Solanaceae seeds and contaminated samples by high-performance liquid chromatography-tandem mass spectrometry. *J. Chromatogr. A.* **2016**, *1474*, 79–84. [CrossRef]
88. Lesiak, A.D.; Musah, R.A. Rapid high-throughput species identification of botanical material using direct analysis in real time high resolution mass spectrometry. *J. Vis. Exp.* **2016**, *116*, e54197. [CrossRef]
89. Fowble, K.L.; Teramoto, K.; Cody, R.B.; Edwards, D.; Guarrera, D.; Musah, R.A. Development of "laser ablation direct analysis in real time imaging" mass spectrometry: Application to spatial distribution mapping of metabolites along the biosynthetic cascade leading to synthesis of atropine and scopolamine in plant tissue. *Anal. Chem.* **2017**, *89*, 3421–3429. [CrossRef] [PubMed]
90. Moreno-Pedraza, A.; Rosas-Román, I.; Garcia-Rojas, N.S.; Guillén-Alonso, H.; Ovando-Vázquez, C.; Díaz-Ramírez, D.; Cuevas-Contreras, J.; Vergara, F.; Marsch-Martínez, N.; Molina-Torres, J.; et al. Elucidating the distribution of plant metabolites from native tissues with laser desorption low-temperature plasma mass spectrometry imaging. *Anal. Chem.* **2019**, *91*, 2734–2743. [CrossRef] [PubMed]
91. Talaty, N.; Takáts, Z.; Cooks, R.G. Rapid in situ detection of alkaloids in plant tissue under ambient conditions using desorption electrospray ionization. *Analyst* **2005**, *130*, 1624–1633. [CrossRef]
92. Humam, M.; Shoul, T.; Jeannerat, D.; Muñoz, O.; Christen, P. Chirality and numbering of substituted tropane alkaloids. *Molecules* **2011**, *16*, 7199–7209. [CrossRef] [PubMed]
93. Fliniaux, O.; Mesnard, F.; Raynaud-Le Grandic, S.; Baltora-Rosset, S.; Bienaimé, C.; Robins, R.J.; Fliniaux, M.-A. Altered nitrogen metabolism associated with de-differentiated suspension cultures derived from root cultures of *Datura stramonium* studied by heteronuclear multiple bond coherence (HMBC) NMR spectroscopy. *J. Exp. Bot.* **2004**, *55*, 1053–1060. [CrossRef]
94. Sharp, G.; Edinburg, M.D. Our present knowledge of the mydriatic group. *Pharm. Era* **1897**, *18*, 587–588.
95. Dunstan, W.R.; Brown, H.V. The alkaloid of *Hyoscyamus muticus* and of *Datura stramonium* grown in Egypt. *J. Chem. Soc. Trans.* **1901**, *79*, 71–74. [CrossRef]
96. Lounasmaa, M.; Tamminen, T. The tropane alkaloids. In *The Alkaloids*; Bossi, A., Ed.; Academic Press: New York, NY, USA, 1993; Volume 44, pp. 1–114.
97. James, W.O. Demonstration of alkaloids in Solanaceous meristems. *Nature* **1946**, *4011*, 377–378. [CrossRef] [PubMed]
98. Walton, N.J.; Peerless, A.C.J.; Robins, R.J.; Rhodes, M.J.C.; Boswell, H.D.; Robins, D.J. Purification and properties of putrescine *N*-methyltransferase from transformed roots of *Datura stramonium* L. *Planta* **1994**, *193*, 9–15. [CrossRef]
99. Bedewitz, M.A.; Jones, A.D.; D'Auria, J.C.; Barry, C.S. Tropinone synthesis via an atypical polyketide synthase and P450-mediated cyclization. *Nat. Commun.* **2018**, *8*, 5281. [CrossRef] [PubMed]
100. Abraham, T.W.; Leete, E. New intermediate in the biosynthesis of the tropane alkaloids in *Datura innoxia*. *J. Am. Chem. Soc.* **1995**, *117*, 8100–8105. [CrossRef]
101. Dräger, B. Tropinone reductase in root cultures of *Hyoscyamus niger*, *Atropa belladonna*, and *Datura stramonium*. *Planta Med.* **1990**, *56*, 499. [CrossRef]
102. Robins, R.J.; Bachman, P.; Robinson, T.; Rhodes, M.J.C.; Yamada, Y. The formation of 3α- and 3β-acetoxytropanes by *Datura stramonium* transformed root cultures involves two acetyl-CoA-dependent acyltransferases. *FEBS Lett.* **1991**, *292*, 293–297. [PubMed]
103. Rabot, S.; Peerless, A.C.J.; Robins, R.J. Tigloyl-CoA: Pseudtropine acyl transferase—an enzyme of tropane alkaloid biosynthesis. *Phytochemistry* **1995**, *39*, 315–322. [CrossRef]
104. McGaw, B.A.; Woolley, J.G. Non-reversibility of the isoleucine-acetate pathway in *Datura*. *Phytochemistry* **1977**, *16*, 1711–1713. [CrossRef]
105. Lucas, K.A.; Filley, J.R.; Erb, J.M.; Graybill, E.R.; Hawes, J.W. Peroxisomal metabolism of propionic acid and isobutyric acid in plants. *J. Biol. Chem.* **2007**, *292*, 24980–24989. [CrossRef]
106. Ning, J.; Moghe, G.D.; Leong, B.; Kim, J.; Ofner, I.; Wang, Z.; Adams, C.; Jones, A.D.; Zamir, D.; Last, R.L. A feedback-insensitive isopropylmalate synthase affects acylsugar composition in cultivated and wild tomato. *Plant Physiol.* **2015**, *169*, 1821–1835. [CrossRef]
107. Bedewitz, M.A.; Góngora-Castillo, E.; Uebler, J.B.; Gonzales-Vigil, E.; Wiegert-Rininger, K.E.; Childs, K.L.; Hamilton, J.P.; Vaillancourt, B.; Yeo, Y.-S.; Chappell, J.; et al. A root-expressed *L*-phenylalanine: 4-hydroxyphenylpyruvate aminotransferase is required for tropane alkaloid biosynthesis in *Atropa belladonna*. *Plant Cell* **2014**, *26*, 3745–3762. [CrossRef] [PubMed]
108. Qiu, F.; Zeng, J.; Wang, J.; Huang, J.-P.; Zhou, W.; Yang, C.; Lan, X.; Chen, M.; Huang, S.-X.; Kai, G.; et al. Functional genomics analysis reveals two novel genes required for littorine biosynthesis. *New Phytol.* **2020**, *225*, 1906–1914. [CrossRef]
109. Lanoue, A.; Boitel-Conti, M.; Portais, J.-C.; Laberche, J.-C.; Barbotin, J.-N.; Christen, P.; Sangwan-Norreel, S. Kinetic study of littorine rearrangement in *Datura innoxia* hairy roots by ^{13}C NMR Spectroscopy. *J. Nat. Prod.* **2002**, *65*, 1131–1135. [CrossRef] [PubMed]
110. Sandala, G.M.; Smith, D.M.; Radom, L. The carbon-skeleton rearrangement in tropane alkaloid biosynthesis. *J. Am. Chem. Soc.* **2008**, *130*, 10684–10690. [CrossRef]

111. Li, Q.; Zhu, T.; Zhang, R.; Bu, Q.; Yin, Q.; Zhang, L.; Chen, W. Molecular cloning and functional analysis of hyoscyamine 6β-hydroxylase (H6H) in the poisonous and medicinal plant *Datura innoxia* mill. *Plant Physiol. Biochem.* **2020**, *153*, 11–19. [CrossRef] [PubMed]
112. Pramod, K.K.; Singh, S.; Jayabaskaran, C. Biochemical and structural characterization of recombinant hyoscyamine 6β-hydroxylase from *Datura metel* L. *Plant Physiol. Biochem.* **2011**, *48*, 966–970. [CrossRef]
113. El Bazaoui, A.; Bellimam, M.A.; Soulaymani, A. Tropane alkaloids of *Datura innoxia* from Morocco. *Z. Naturforsch.* **2012**, *67c*, 8–14. [CrossRef]
114. Berkov, S.; Zayed, R. Comparison of tropane alkaloid spectra between *Datura innoxia* grown in Egypt and Bulgaria. *Z. Naturforsch.* **2004**, *59c*, 184–186. [CrossRef]
115. Harfi, B.; Khelifi, L.; Khelifi-Slaoui, M.; Assaf-Ducrocq, C.; Gontier, E. Tropane alkaloids GC/MS analysis and low dose elicitors' effects on hyoscyamine biosynthetic pathway in hairy roots of Algerian *Datura* species. *Sci. Rep.* **2018**, *8*, 17951. [CrossRef]
116. El-Iman, Y.M.A.; Evans, W.C.; Grout, R.J.; Ramsey, K.P.A. Alkaloids of *Erythroxylum zambesiacum* root-bark. *Phytochemistry* **1987**, *26*, 2385–2389. [CrossRef]
117. Ushimaru, R.; Ruszczycky, M.; Chang, W.-C.; Yan, F.; Liu, Y.-N.; Liu, H.W. Substrate conformation correlates with the outcome of hyoscyamine 6β-hydroxylase catalyzed oxidation reactions. *J. Am. Chem. Soc.* **2018**, *140*, 7433–7436. [CrossRef]
118. Pavlov, A.; Berkov, S.; Weber, J.; Bley, T. Hyoscyamine biosynthesis in *Datura stramonium* hairy root in vitro systems with different ploidy levels. *Appl. Biochem. Biotechnol.* **2009**, *157*, 210–225. [CrossRef] [PubMed]
119. Dräger, B. Chemistry and biology of calystegines. *Nat. Prod. Rep.* **2004**, *21*, 211–223. [CrossRef]
120. Dräger, B.; van Almsick, A.; Mrachatz, G. Distribution of calystegines in several Solanaceae. *Planta Med.* **1995**, *61*, 577–579. [CrossRef]
121. Dräger, B.; Funck, C.; Höhler, A.; Mrachatz, G.; Nahrstedt, A.; Portsteffen, A.; Schaal, A.; Schmidt, R. Calystegines as a new group of tropane alkaloids in Solanaceae. *Plant Cell, Tissue Organ Cult.* **1994**, *38*, 235–240. [CrossRef]
122. Jenett-Siems, K.; Weigl, R.; Böhm, A.; Mann, P.; Tofern-Reblin, B.; Ott, S.C.; Ghomian, A.; Kaloga, M.; Siems, K.; Witte, L.; et al. Chemotaxonomy of the pantropical genus *Merremia* (Convolvulaceae) based on the distribution of tropane alkaloids. *Phytochemistry* **2005**, *66*, 1448–1464. [CrossRef]
123. Phillipson, J.D.; Handa, S.S. N-oxides of hyoscyamine and hyoscine in the Solanaceae. *Phytochemistry* **1975**, *14*, 999–1003. [CrossRef]
124. Heusner, A. Notiz zur Stereochemie von Scopin and Scopolin. *Chem. Ber.* **1954**, *7*, 1063–1067. [CrossRef]
125. Aripova, S.F.; Yunusov, S.S. Dimeric tropane alkaloids. *Chem. Nat. Compd.* **1991**, *27*, 389–395. [CrossRef]
126. Aripova, S.F.; Tashkhodzhaev, B. α-and β-scopadonnines from *Datura inoxia* seeds. *Chem. Nat. Compd.* **2004**, *27*, 464–468. [CrossRef]
127. Yang, B.-Y.; Xia, Y.-G.; Wang, Q.-H.; Dou, D.-Q.; Kuang, H.-X. Two new amide alkaloids from the flower of *Datura metel* L. *Fitoterapia* **2010**, *81*, 1003–1005. [CrossRef] [PubMed]
128. Parr, A.J.; Payne, J.; Eagles, J.; Chapman, B.T.; Robins, R.J.; Rhodes, M.J.C. Variation in tropane alkaloid accumulation within the Solanaceae and strategies for its exploitation. *Phytochemistry* **1990**, *29*, 2545–2550. [CrossRef]
129. McGaw, B.A.; Woolley, J.G. The separate roles of hygrine enantiomers in the biosynthesis of tropane alkaloids. *J. Pharm. Pharmacol.* **1997**, *29*, 16. [CrossRef] [PubMed]
130. Evans, W.C.; Somanabandhu, A. Alkaloids of *Datura discolor*. *Phytochemistry* **1974**, *13*, 304–305. [CrossRef]
131. El-Shazly, A.; Tei, A.; Witte, L.; El-Domiaty, M.; Wink, M. Tropane alkaloids of *Hyoscyamus boveanus*, *H. muticus*, and *H. albus* from Egypt. *Z. Naturforsch.* **1997**, *52c*, 529–539. [CrossRef]
132. Murch, S.J.; Alan, A.R.; Cao, J.; Saxena, P.K. Melatonin and serotonin in flowers and fruits of *Datura metel* L. *J. Pineal Res.* **2009**, *47*, 277–283. [CrossRef]
133. Maier, I.; Jurenitsch, J.; Heresch, F.; Haslinger, E.; Schulz, G.; Pöhm, M.; Jentzsch, K. Fluorodaturatin und Homofluorodaturatin—zwei neue β-carbolinderivate in Samen von *Datura stramonium* L. *var. stramonium*. *Monatsh. Chem.* **1981**, *112*, 1425–1439. [CrossRef]
134. Jurenitsch, J.; Piccardi, K.; Pöhm, M. 1,2-Dehydrofluorodaturatine—A further β-carboline derivative from seeds of *Datura stramonium* L. var. *stramonium*. *Sci. Pharm.* **1984**, *52*, 301–309.
135. Okwu, D.E.; Igara, E.C. Isolation, characterization, and antibacterial activity screening of a new β-carboline alkaloid from *Datura metel* Linn. *Der Chemica Sinica* **2011**, *2*, 261–267.
136. Karmakar, U.K.; Toume, K.; Ishikawa, N.; Arai, M.A.; Sadhu, S.K.; Ahmed, F.; Ishibashi, M. Bioassay-guided isolation of compounds from *Datura stramonium* with TRAIL-resistance overcoming activity. *Nat. Prod. Commun.* **2016**, *11*, 185–187. [CrossRef] [PubMed]
137. Nuhu, H.; Ghani, A. Alkaloid content of the leaves of three Nigerian Datura species. *Nig. J. Nat. Prod. And Med.* **2002**, *6*, 15–18.
138. Houmani, Z.; Cosson, L.; Houmani, M. *Datura ferox* L. and *D. quercifolia* Kunth (Solanaceae) in Algeria. *Flora Mediterr.* **1999**, *9*, 57–60.
139. Jakabová, S.; Vincze, L.; Farkas, A.; Kilár, F.; Boros, B.; Felinger, A. Determination of tropane alkaloids atropine and scopolamine by liquid chromatography-mass spectrometry in plant organs of *Datura* species. *J. Chromatogr. A* **2012**, *1232*, 295–301. [CrossRef] [PubMed]

140. Berkov, S.; Philipov, S. Alkaloid production in diploid and autotetraploid plants of *Datura stramonium*. *Pharm. Biol.* **2002**, *40*, 617–621. [CrossRef]

141. Cosson, L. Influence de l'éclairement sur les variations ontogéniques de teneurs en scopolamine et en hyoscyamine des feuilles de *Datura metel*. *Phytochemistry* **1969**, *8*, 2227–2233. [CrossRef]

142. Demeyer, K.; Dejaegere, R. The influence of the Ca^{2+}/K^+ balance and light energy on alkaloid content and partitioning in *Datura stramonium*. *Aust. J. Bot.* **1997**, *45*, 81–101. [CrossRef]

143. Brachet, J.; Cosson, L. Changes in the total alkaloid content of *Datura innoxia* Mill. subjected to salt stress. *J. Exp. Bot.* **1986**, *37*, 650–656. [CrossRef]

144. Demeyer, K.; Van de Velde, H.; Dejaegere, R. Influence of IAA and DMAA on hyoscyamine and scopolamine production in *Datura stramonium*. *Planta Med.* **1989**, *55*, 693. [CrossRef]

145. Gupta, S.; Madan, C.L. Effect of growth retardants on growth and alkaloid formation in *Datura metel* var. *fastuosa*. *Planta Med.* **1975**, *28*, 193–200. [CrossRef]

146. Jousse, C.; Vu, T.D.; Tran, T.L.M.; Al Balkhi, M.H.; Molinié, R.; Boitel-Conti, M.; Pilard, S.; Mathiron, D.; Hehn, A.; Bourgaud, F.; et al. Tropane alkaloid profiling of hydroponic *Datura innoxia* Mill. plants inoculated with *Agrobacterium rhizogenes*. *Phytochem. Anal.* **2010**, *21*, 118–127. [CrossRef]

147. Padula, L.Z.; Bandoni, A.L.; Rondina, R.V.D.; Coussio, J.D. Quantitative determination of total alkaloids and scopolamine in *Datura ferox* growing in Argentina. *Planta. Med.* **1976**, *29*, 357–360. [CrossRef]

148. Shonle, I.; Bergelson, J. Evolutionary ecology of the tropane alkaloids of *Datura stramonium* L. (Solanaceae). *Evolution* **2000**, *54*, 778–788. [CrossRef] [PubMed]

149. Castillo, G.; Cruz, L.L.; Tapia-López, R.; Olmedo-Vicente, E.; Carmona, D.; Anaya-Lang, A.L.; Fornoni, J.; Andraca-Gómez, G.; Valverde, P.L.; Núñez-Farfán, J. Selection mosaic exerted by specialist and generalist herbivores on chemical and physical defense of *Datura stramonium*. *PLoS ONE* **2014**, *9*, e102478. [CrossRef] [PubMed]

150. Wilson, J.K.; Tseng, A.S.; Potter, K.A.; Davidowitz, G.; Hildebrand, J.G. The effects of the alkaloid scopolamine on the performance and behavior of two caterpillar species. *Arthropod Plant Interact.* **2018**, *12*, 21–29. [CrossRef] [PubMed]

151. Philipov, S.; Berkov, S. GC-MS investigation of tropane alkaloids in *Datura Stramonium*. *Z. Naturforsch.* **2002**, *57c*, 559–561. [CrossRef] [PubMed]

152. Evans, W.C.; Ghani, A.; Woolley, V.A. Distribution of littorine and other alkaloids in the roots of *Datura* species. *Phytochemistry* **1972**, *11*, 2527–2529. [CrossRef]

153. Reina, M.; Burgueño-Tapia, E.; Bucio, M.A.; Joseph-Nathan, P. Absolute configuration of tropane alkaloids from *Schizanthus* species by vibrational circular dichroism. *Phytochemistry* **2010**, *71*, 810–815. [CrossRef]

154. Muñoz, M.A.; Muñoz, O.; Joseph-Nathan, P. Absolute configuration determination and conformational Analysis of (−)-(3*S*,6*S*)-3α,6β-diacetoxytropane using vibrational circular dichroism and DFT techniques. *Chirality* **2010**, *22*, 234–241. [CrossRef]

155. Humam, M.; Christen, P.; Muñoz, O.; Hostettmann, K.; Jeannerat, D. Absolute configuration of tropane alkaloids bearing two α,β-unsaturated ester functions using electronic CD spectroscopy: Application to (*R,R*)-*trans*-3-hydroxysenecioyloxy-6-senecioloxytropane. *Chirality* **2008**, *20*, 20–25. [CrossRef]

156. Hubert, J.; Nuzillard, J.-M.; Renault, J.-H. Dereplication strategies in natural product research: How many tools and methodologies behind the same concept? *Phytochem. Rev.* **2017**, *16*, 55–95. [CrossRef]

157. Wang, Z.; Kang, D.; Jia, X.; Zhang, H.; Guo, J.; Liu, C.; Meng, Q.; Liu, W. Analysis of alkaloids from *Peganum harmala* L. sequential extracts by liquid chromatography coupled to ion mobility spectrometry. *J. Chromatogr. B* **2018**, *1096*, 73–79. [CrossRef] [PubMed]

158. Schrimpe-Rutletdge, A.C.; Sherrod, S.D.; McLean, J.A. Improving the discovery of secondary metabolite natural products using ion mobility-mass spectrometry. *Curr. Opin. Chem. Biol.* **2018**, *42*, 160–166. [CrossRef] [PubMed]

159. Harvey, A.L.; Edrada-Ebel, R.; Quinn, R.J. The re-emergence of natural products for drug discovery in the genomics era. *Nat. Rev. Drug Discov.* **2016**, *14*, 111–129. [CrossRef] [PubMed]

160. Chávez, D.; Cui, B.; Chai, H.-B.; García, R.; Mejía, M.; Farnsworth, N.R.; Cordell, G.A.; Pezzuto, J.M.; Kinghorn, A.D. Reversal of multidrug resistance by tropane alkaloids from the stems of *Erythroxylum rotundifolium*. *J. Nat. Prod.* **2002**, *65*, 606–610. [CrossRef] [PubMed]

161. Özçelik, B.; Kartal, M.; Orhan, I. Cytotoxicity, antiviral, and antimicrobial activities of alkaloids, flavonoids, and phenolic acids. *Pharm. Biol.* **2011**, *49*, 396–402. [CrossRef]

MDPI

St. Alban-Anlage 66

4052 Basel

Switzerland

Tel. +41 61 683 77 34

Fax +41 61 302 89 18

www.mdpi.com

Molecules Editorial Office

E-mail: molecules@mdpi.com

www.mdpi.com/journal/molecules